harr

Springer Series on
ATOMIC, OPTICAL, AND PLASMA PHYSICS   35

Springer
Berlin
Heidelberg
New York
Hong Kong
London
Milan
Paris
Tokyo

Physics and Astronomy

ONLINE LIBRARY

http://www.springer.de/phys/

Springer Series on
ATOMIC, OPTICAL, AND PLASMA PHYSICS

The Springer Series on Atomic, Optical, and Plasma Physics covers in a comprehensive manner theory and experiment in the entire field of atoms and molecules and their interaction with electromagnetic radiation. Books in the series provide a rich source of new ideas and techniques with wide applications in fields such as chemistry, materials science, astrophysics, surface science, plasma technology, advanced optics, aeronomy, and engineering. Laser physics is a particular connecting theme that has provided much of the continuing impetus for new developments in the field. The purpose of the series is to cover the gap between standard undergraduate textbooks and the research literature with emphasis on the fundamental ideas, methods, techniques, and results in the field.

Series homepage – http://www.springer.de/phys/books/ssaop/

Vols. 1–26 of the former Springer Series on Atoms and Plasmas are listed at the end of the book

J. Ullrich   V.P. Shevelko   (Eds.)

Many-Particle Quantum Dynamics in Atomic and Molecular Fragmentation

With 179 Figures

Springer

seplae
phys.

Professor Dr. Joachim Ullrich
Max-Planck Institut für Kernphysik
Saupfercheckweg 1, 69117 Heidelberg, Germany
E-mail: joachim.ullrich@mpi-hd.mpg.de

Professor Dr. Viatcheslav Shevelko
P.N. Lebedev Physical Institute
Leninskii prospect 53, 119991 Moscow, Russia
E-mail: shev@sci.lebedev.ru

ISSN 1615-5653

ISBN 3-540-00667-2 Springer-Verlag Berlin Heidelberg New York

Library of Congress Cataloging-in-Publication Data

Many-particle quantum dynamics in atomic and molecular fragmentation / J. Ullrich, V.P. Shevelko (eds.).
p. cm. – (Springer series on atomic, optical, and plasma physics, ISSN 1615-5653 ; 35)
Includes bibliographical references and index.
ISBN 3-540-00667-2 (alk. paper)
1. Collisions (Atomic physics) 2. Atomic and molecular fragmentation. 3. Many-body problem. 4. Quantum
theory. I. Ullrich, J. (Joachim) II. Shevel'ko, V.P. (Viatcheslav Petrovich) III: Series.
QC794.6.C6M34 2003
539.7'57–dc21   2003050419

Springer-Verlag Berlin Heidelberg New York
a member of BertelsmannSpringer Science+Business Media GmbH

http://www.springer.de

© Springer-Verlag Berlin Heidelberg 2003
Printed in Germany

Typesetting: Camera-ready copies by the author
Final procesing: PTP-Berlin Protago-TeX-Production GmbH, Berlin
Cover concept by eStudio Calmar Steinen
Cover design: design & production GmbH, Heidelberg

Printed on acid-free paper   57/3141/YU - 5 4 3 2 1 0

9/12/03
AR.

Preface

This book aims to give a comprehensive view on the present status of a tremendously fast-developing field – the quantum dynamics of fragmenting many-particle Coulomb systems. In striking contrast to the profound theoretical knowledge, achieved from extremely precise experimental results on the *static* atomic and molecular structure, it was only three years ago when the three-body fundamental *dynamical* problem of breaking up the hydrogen atom by electron impact was claimed to be solved in a mathematically consistent way.

Until now, more "complicated", though still fundamental scenarios, addressing the complete fragmentation of the "simplest" many-electron system, the helium atom, under the action of a time-dependent external force, have withstood any consistent theoretical description. Exceptions are the most "trivial" situations where the breakup is induced by the impact of a single real photon or of a virtual photon under a perturbation caused by fast, low-charged particle impact. Similarly, the dissociation of the "simplest" molecular systems like H_2^+ or HD^+, fragmentating in collisions with slow electrons, or the H_3 molecule breaking apart into two or three "pieces" as a result of a single laser-photon excitation, establish a major challenge for state-of-the-art theoretical approaches.

In the recent past, essentially since less than a decade ago, the field was revolutionized by the invention of advanced, innovative experimental imaging and projection techniques such as "Reaction Microscopes" – the "bubble chambers of atomic and molecular physics" – or similar projection techniques for fast-moving ion beams. Now these methods enable the measurement of the vector momenta of several fragments (ions, electrons, molecular ions) with unprecedented large solid angles, often reaching a hundred per cent of 4π at extreme precision: Energy resolutions below $1\,\mathrm{meV}$ are achieved for slow electrons, while ion momenta are recorded at the $\mu\mathrm{eV}$ level and below, corresponding to a temperature of a few mK. Often, even "kinematically complete" measurements are feasible, where the complete many-particle final quantum state is mapped and can be visualized, providing the ultimate benchmark for comparison with theory.

Imaging techniques, on the one hand, essentially rely on advanced methods for precise preparation of the initial quantum state and, thus, are intimately "entangled" and immensely boosted by the present explosion-like advances in the development of cooling techniques: Electron and laser cooling in storage rings, cold molecular ion sources in combination with small linear traps for cold molecular ions, the use of laser cooling to prepare ultra cold atomic targets or even the Bose–Einstein condensates have just emerged. On the other hand, these techniques have profited tremendously by the availability of more sophisticated tools to induce the fragmentation process in a dynamically refined way, to interrogate the breakup time dependence and, maybe in the near future, to even control fragmentation pathways. Among the more sophisticated techniques are intense third-generation synchrotron radiation sources, shorter and shorter, few femto- or even attosecond ("designed") pulses, pulse-trains and pump-probe arrangements from table-top lasers, upcoming femtosecond VUV free-electron lasers (FEL), as well as intense beams of highly charged ions providing subattosecond delta pulses to explore inner atomic or molecular dynamics on an ultrashort timescale.

In parallel, substantial progress has been achieved in the theoretical treatment of fragmenting Coulomb systems, driven by conceptual innovations as well as by the dramatic increase of computational capabilities in recent years. For example, the exterior complex scaling method along with the excessive use of massively parallel supercomputers allowed one to solve the fundamental three-particle Coulomb breakup of atomic hydrogen by slow-electron impact mentioned above. Convergent close-coupling calculations, as well as hyperspherical R-matrix methods, the latter combined with semiclassical outgoing waves, were finally able to reliably predict fully differential fragmentation patterns for photo double ionization of the simplest correlated two-electron system, the helium atom. Within the last few years, these methods have been successfully applied to describe double ionization by charged-particle impact at high velocities and the first successful attempts have been undertaken to implement higher-order contributions at lower velocities. Moreover, other sophisticated methods were developed or applied in the recent past; among them are the S-matrix approaches to describe the interaction of strong laser fields with atoms, numerical grid methods to directly integrate the Schrödinger equation, hidden crossing techniques at low collision energies, time-dependent density functional methods to approach "true" many-electron problems, the Green-function theory for N-particle finite systems, as well as semiclassical and classical Monte Carlo approximations.

While it is certainly impossible to cover all recent achievements in one single book that, therefore, necessarily has to remain incomplete, we have tried, at least within our personal subjective perspective, to report and review the most basic methods, techniques and advances as well as on the most significant areas of research by attracting international experts and young researchers. In this book, after a detailed introduction into basic kinematic concepts of atomic and molecular fragmentation reactions in Part I, some of the prominent experimental techniques, underlying concepts, present challenges and future prospects are discussed in Part II. Various state-of-the-art theoretical approaches are introduced in Part III followed by different applications in Part IV for fragmentation reactions, induced by time-dependent perturbations such as single photons, short-pulse lasers or fast as well as slowly moving electrons and ions.

Care was taken to organize the sequence of chapters and each single contribution in such a way that the nonexpert reader or undergraduate student being familiar with the basic concepts of atomic, molecular and optical physics could be guided in a comprehensive, pedagogical way from the fundamental methods to the most recent techniques and then to the cutting edge of research in the field, experimentally as well as theoretically. At the University of Heidelberg, this book serves as accompanying material for a newly established Graduate School for Atomic, Molecular and Optical Physics.

Heidelberg,
May 2003

Joachim Ullrich
Viatcheslav Shevelko

Contents

XII   Contents

List of Contributors

L. Adoui
CIRIL /CEA/CNRS/ISMRA/
Université de Caen
Rue Claude Bloch BP 5133
F-14070 Caen cedex 5
France
adoui@ganil.fr

W. Becker
Max-Born-Institut
Max-Born-Str. 2a
12489 Berlin
Germany
wbecker@mbi-berlin.de

J. Berakdar
Max-Planck-Institut
für Mikrostrukturphysik
Weinberg 2
06120 Hallee
Germany jber@mpi-halle.de

H. Bräuning
Institut für Kernphysik
Strahlenzentrum
der Justus-Liebig Universität
Leihgesterner Weg 217
D-35392 Giessen
Germany
Harald.Braeuning
@strz.uni-giessen.de

I. Bray
ARC Physics and Energy Studies
School of Mathematical and Physical
Sciences Murdoch University
90 South Street Murdoch
Perth 6150
Western Australia
I.Bray@murdoch.edu.au

A. Cassimi
CIRIL /CEA/CNRS/ISMRA/
Université de Caen
Rue Claude Bloch BP 5133
F-14070 Caen cedex 5
France
cassimi@ganil.fr

J. Chesnel
CIRIL /CEA/CNRS/ISMRA/
Université de Caen
Rue Claude Bloch BP 5133
F-14070 Caen cedex 5
France
jean-yves.chesnel@ismra.fr

A. Dorn
Max-Planck-Institut für Kernphysik
Postfach 103980
D-69029 Heidelberg
Germany
alexander.dorn@mpi-hd.mpg.de

R. Dörner
Institut für Kernphysik
University Frankfurt
August Euler Str. 6
60484 Frankfurt
Germany
doerner@hsb.uni-frankfurt.de

D. Dowek
Laboratoire des
Collisions
Atomiques et Moléculaires
Bat. 351
Université Paris-Sude
91405 ORSAY Cedex
France
dowek@lcam.u-psud.fr

F.H.M. Faisal
Universität Bielefeld
Fakultät für Physik
Universitätsstr. 25
D-33615 Bielefeld
Germany
ffaisal@physik.uni-bielefeld.de

D. Fischer
Max-Planck-Institut für Kernphysik
Postfach 103980
D-69029 Heidelberg
Germany
daniel.fischer@mpi-hd.mpg.de

F. Fremont
CIRIL /CEA/CNRS/ISMRA/
Université de Caen
Rue Claude Bloch BP 5133
F-14070 Caen cedex 5
France
francois.fremont@ismra.fr

B. Gervais
CIRIL/CEA/CNRS/ISMRA/
Université de Caen
Rue Claude Bloch BP 5133
F-14070 Caen cedex 5
France
gervais@ganil.fr

Yu.S. Gordeev
A.F.Ioffe Physical-technical Institute
Laboratory of Atomic Collisions in
Solids 1
94021 St.-Petersburg
Russia mikoush@eos.ioffe.rssi.ru

S.P. Goreslavski
Moscow Engineering Physics
Institute
Kashirskoe Shosse 31
115409 Moscow
Russia
sgoreslavski@mtu-net.ru

H. Helm
Universität Freiburg
Fakultät für Physik
Hermann-Herder-Str. 3
D-79104 Freiburg
Germany
Hanspeter.Helm
@physik.uni-freiburg.de

D. Hennecart
CIRIL /CEA/CNRS/ISMRA/
Université de Caen
Rue Claude Bloch BP 5133
F-14070 Caen cedex 5
France
dominique.hennecart@ismra.fr

O. Jagutzki
Institut für Kernphysik
University Frankfurt
August Euler Str. 6
60484 Frankfurt
Germany
jagutzki@hsb.uni-frankfurt.de

A.K. Kazansky
Fock Institute of Physics
The University of St.-Petersburg
St.-Petersburg 198504
Russia
Andrey.Kazansky@pobox.spbu.ru

T. Kirchner
Institut für Theoretische Physik TU
Clausthal Leibnizstrasse 10
D-38678 Clausthal-Zellerfeld
Germany
tkirchner@pt.tu-clausthal.de

A. Knapp
Institut für Kernphysik
University Frankfurt
August Euler Str. 6
60484 Frankfurt
Germany
knapp@hsb.uni-frankfurt.de

A. Kheifets
Institute of Advanced Studies
Research School of Physical Sciences
The Australian National University
Canberra ACT 0200
Australia
A.Kheifets@anu.edu.au

H. Kollmus
GSI Planckstrasse 1
D-64291 Darmstadt
Germany
H.Kollmus@gsi.de

R. Kopold
Max-Born-Institut
Max-Born-Str. 2a
12489 Berlin
richard.kopold@onlinehome.de

G. Laurent
CIRIL /CEA/CNRS/ISMRA/
Université de Caen
Rue Claude Bloch BP 5133
F-14070 Caen cedex 5
France
glaurent@ganil.fr

H.-J. Lüdde
Institut für Theoretische Physik
der Johann Wolfgang Goethe
Universität
Robert-Mayer Strasse 8
60054 Frankfurt am Main
Germany luedde
@th.physik.uni-frankfurt.de

J.H. Macek
University of Tennessee
Physics Department
Knoxville Tennessee 37996
USA
jmacek@utk.edu

L. Malegat
Laboratoire d'Interaction du
Rayonnement
X Avec la Matière (UMR 8624 du
CNRS)
Université Paris-Sud, Bat. 350
91405 Orsay
France
laurence.malegat@lixam.u-psud.fr

C.W. McCurdy
Lawrence Berkeley National
Laboratory
Computing Sciences
Berkeley CA 94720 and
University of California
Davis Department of Applied
Sciences Davis CA 95616
USA
cwmccurdy@lbl.gov

V. Mergel
Institut für Kernphysik
University Frankfurt
August Euler Str. 6
60484 Frankfurt
Germany
mergel@hsb.uni-frankfurt.de

C. Miron
LURE Bâtiment 209d
B.P. 34
Centre Universitaire
91898 Orsay Cedex
France and
CEA/DSM/SPAM and
LFP (CNRS URA 2453)
Bâtiment 522 CEN Saclay
91191 Gif sur Yvette Cedex
France
catalin.miron@lure.u-psud.fr

P. Morin
Synchrotron Soleil
L'Orme des Merisiers
Saint-Aubin - BP 48
91192 Gif-sur-Yvette Cedex
France
paul.morin@synchrotron-soleil.fr

R. Moshammer
Max-Planck-Institut für Kernphysik
Postfach 103980
D-69029 Heidelberg
Germany
robert.moshammer@mpi-hd.mpg.de

U. Müller
Universität Freiburg
Fakultät für Physik
Hermann-Herder-Str. 3
D-79104 Freiburg
Germany
ulrich.mueller@uni-freiburg.de

G.N. Ogurtsov
A.F.Ioffe Physical-technical Institute
Laboratory of Atomic Collisions in
Solids
194021 St.-Petersburg
Russia
mikoush@eos.ioffe.rssi.ru

S.Y. Ovchinnikov
310 South College
Physics Department
University of Tennessee
Knoxville TN 37922-1501
USA
serge@charcoal.phys.utk.edu

S.P. Popruzhenko
Moscow Engineering Physics
Institute
Kashirskoe Shosse 31
115409 Moscow
Russia
poprz@theor.mephi.ru

T.N. Rescigno
Lawrence Berkeley National
Laboratory
Computing Sciences Berkeley
CA 94720 and
Lawrence Livermore National
Laboratory
Physics and Advanced Technologies
Livermore CA 94551
USA
tnr@llnl.gov

H. Rottke
Max-Born-Institut
Max-Born-Str. 2a
D-12489 Berlin-Adlershof
Germany
rottke@mbi-berlin.de

L.Ph.H. Schmidt
Institut für Kernphysik der
J.W. Goethe-Universität
August Euler-Str. 6
D-60486 Frankfurt am Main
schmidt@hsb.uni-frankfurt.de

H. Schmidt-Böcking
Institut für Kernphysik
University Frankfurt
August Euler Str. 6
D-60484 Frankfurt
Germany
schmidtb@hsb.uni-frankfurt.de

P. Selles
LIXAM (Laboratoire d'Interaction
du rayonnement X Avec la
Matière)
UMR8624 du CNRS
Université Paris-Sud Bat 350
91405 Orsay Cedex
France
Patricia.Selles@lixam.u-psud.fr

V.P. Shevelko
P.N. Lebedev Physical Institute
Leninskii prospect 53
119991 Moscow
Russia
shev@sci.lebedev.ru

D. Schwalm
Max-Planck-Institut für Kernphysik
Postfach 103980
D-69029 Heidelberg
Germany
dirk.schwalm@mpi-hd.mpg.de

M. Simon
LURE Bâtiment 209d
B.P. 34
Centre Universitaire
91898 Orsay Cedex
France and
CEA/DSM/SPAM and
LFP (CNRS URA 2453)
Bâtiment 522 CEN Saclay
91191 Gif-sur-Yvette Cedex
France
marc.simon@lure.u-psud.fr

P. Sobocinski
CIRIL /CEA/CNRS/ISMRA/
Université de Caen
Rue Claude Bloch BP 5133
F-14070 Caen cedex 5
France
przemek.sobocinski@ismra.fr

L. Spielberger
Institut für Kernphysik
University Frankfurt
August Euler Str. 6
60484 Frankfurt
Germany
spielberger@ikf.uni-frankfurt.de

A. Stelbovics
Murdoch University
South Street
Murdoch
Western Australia 6150
stelbovi@fizzy.murdoch.edu.au

M. Tarisien
CIRIL /CEA/CNRS/ISMRA/
Université de Caen
Rue Claude Bloch BP 5133
F-14070 Caen cedex 5
France
tarisien@ganil.fr

K. Taylor
Queen's University Belfast
University Road
Belfast BT7 1NN
Northern Ireland
United Kingdom
k.taylor@qub.ac.uk

J. Ullrich
Max-Planck-Institut für Kernphysik
Postfach 103980
D-69029 Heidelberg
Germany
Joachim.Ullrich@mpi-hd.mpg.de

T. Weber
Institut für Kernphysik
University Frankfurt
August Euler Str. 6
60484 Frankfurt
Germany
weber@hsb.uni-frankfurt.de

E. Weigold
AMPL Research School of Physical
Sciences & Engineering
Australian National University
Canberra ACT 0200
Australia Erich.Weigold@anu.edu.au

R. Wester
Max-Planck-Institut für Kernphysik
Postfach 103980
D-69029 Heidelberg
Germany
Roland.Wester@mpi-hd.mpg.de

A. Wolf
Max-Planck-Institut für Kernphysik
Postfach 103980
D-69029 Heidelberg
Germany
andreas.wolf@mpi-hd.mpg.de

D. Zajfman
Department of Particle Physics
Weizmann Institute of Science
Rehovot 76100
Israel
fndaniel@wicc.weizmann.ac.il

List of Units and Notations

Fundamental Atomic Constants

c = 299 792 458 $\mathrm{m\,s^{-1}}$ Speed of light in vacuum

$\hbar$ = 1.054 571 596(82) $\times 10^{-34}$ J s   Planck constant divided by 2π

$\hbar c$ = 197.326 960 2(77) MeV fm   Conversion constant

e = 1.602 176 462(63) $\times 10^{-19}$ C   Electron charge magnitude

m_e= 0.510 998 902(21) MeV/c^2 Electron mass

m_p= 938.271 998(38) MeV/c^2 Proton mass

u = 931.494 013(37) MeV/c^2 Unified atomic mass unit (mass of ^{12}C atom)/12

$\alpha = e^2/\hbar c$

    = 1/137.035 999 76(50)   Fine-structure constant

$r_e = e^2/m_e c^2$

    = 2.817 940 285(31) $\times 10^{-15}$ m   Classical electron radius

$a_o = \hbar^2/m_e e^2$

    = 0.529 177 208 3(19) $\times 10^{-10}$ m   Bohr radius

Ry= $m_e e^4/2\hbar^2$

    = 13.605 691 72(53) eV   Rydberg energy

k_B = 8.617 342(15) $\times 10^{-5}$ $\mathrm{eV\,K^{-1}}$ Boltzmann constant

These constants are extracted from the set of constants recommended for international use by the Committee on Data for Science and Technology (CODATA) based on the "CODATA recommended values of the fundamental physical constants: 1998" by P.J. Mohr and B.N. Taylor, Rev. Mod. Phys. **72**, 351 (2000).

NOTATIONS

b Impact parameter

c Speed of light

E_e Electron energy

E_γ Photon energy

I_p Ionization potential

$\boldsymbol{K}_a$ Incident-electron momentum (in electron-impact collisions)

K_b	Fast-scattered electron momentum (in electron-impact collisions)
K_c	Slowly emitted electron momentum (in electron-impact collisions)
m_e	Electron mass
M	Ion mass
M_P	Projectile mass
M_R	Recoil-ion mass
$M_{a,b,c}$	Fragment-ion masses
P_{ei}	Momentum of the ith emitted electron
P_P	Projectile momentum
P_R	Recoil-ion momentum
P_γ	Photon momentum, $P_\gamma = E_\gamma/c$
q	Momentum transfer
Q	Change in internal electron energy (inelasticity)
r_i	Coordinates of the ith electron
SC	Cross section:
DDSC	Double-differential cross section
FDSC	Fully differential cross section
	Five-fold differential cross section
SDSC	Single-differential cross section
TDSC	Triple-differential cross section
U_p	Ponderomotive potential
v_P	Projectile velocity
$v_\perp, P_\perp$	Velocity/momentum projection in the direction transverse to the beam propagation (in charged particle impact)
$v_\parallel, P_\parallel$	Velocity/momentum projection along the beam propagation (in charged particle impact)
α	Fine-structure constant
$\boldsymbol{\alpha}$	Dirac matrix
β	Relativistic factor, $\beta = v_P/c$
γ	Relativistic factor, $\gamma = (1 - \beta^2)^{-1/2}$
ΔE_P	Energy loss of the projectile
ΔP_P	Projectile-momentum change $\Delta P_P = -q$
ϵ	Photon polarization vector
φ_{ei}	Azimuthal emission angel of the ith electron
φ_P	Projectile azimuthal scattering angel
θ_{ei}	Polar emission angel of the ith electron
θ_P	Projectile scattering angel
σ	Cross section
τ	Collision time
ω	Photon frequency
ω_L	Laser frequency
Ω_i	Detection solid angle of the ith electron

1 Kinematics of Atomic and Molecular Fragmentation Reactions

V.P. Shevelko and J. Ullrich

1.1 General Considerations

The kinematics of fragmentation processes induced by single collisions of ions, electrons and photons with atoms and molecules is considered in this chapter, i.e. the rearrangement of colliding particles in a variety of reactions such as electron capture, ionization, Compton scattering, and molecular dissociation. The determination of three-dimensional momentum vectors $\boldsymbol{P}_j = (P_{xj},\ P_{yj},\ P_{zj})$, $j = 1, 2, \ldots, \mathcal{N}$ of all $\mathcal{N}$ fragments after the collision allows one to obtain detailed information on the kinematics, the final charge states and the change of internal electronic energies of the colliding partners. The quantum states of the projectile and the target before the collision are assumed to be known and well prepared.

The kinematics of atomic and molecular many-particle fragmentation reactions is treated on the basis of the equations following from the energy, mass, and momentum conservation laws. Here, more attention is paid to the fragmentation of atoms compared to molecular dissociation because, up to now, atomic reactions have been investigated in more detail, experimentally as well as theoretically. In general, molecules are quantum objects with properties strongly different from those of atoms and ions. Among them, one has to mention the presence of two or more nuclear centers, the internuclear motion, the additional vibrational and rotational states in the energy-level structure, small values of excitation energies and more possible channels for dissociation reactions. All these reasons make a molecular fragmentation much more complicated for investigation compared to atomic breakup processes.

As follows from the kinematics equations for atomic fragmentation, the ionized target atom after the reaction, called the *recoil ion*, is often the most interesting object because its charge gives the multiplicity of the process, i.e., the number of active electrons, and the momentum, transferred to it by the incident projectile, provides unique information on the dynamical mechanisms of the reaction. Therefore, considerable attention is paid to the determination and interpretation of the recoil-ion momentum and its projections. The momenta of emitted electrons become important in investigations of the electronic correlations, which are a topic of increasing interest in atomic physics of multielectron processes.

Atomic units are used, $m_e = e = \hbar = 1$, until otherwise indicated. Here m_e and e denote the electron mass and charge, respectively, and $\hbar$ the Planck constant divided by 2π.

1.1.1 Definitions and Parameters

The most detailed information on many-particle atomic reactions can be obtained from fully differential cross sections, i.e. differential in all observables of the final state. In principle, any cross section integrated over one or more observables usually hides some characteristics of the process, therefore, only fully differential cross sections completely reflect all important physical mechanisms of the many-particle process and provide the most serious test to the many-particle theory. If these cross sections can be measured, one speaks about *kinematically complete* experiments, which means that momenta and, thus, angles and energies of all involved particles are observed in coincidence, except the spin momenta because of the lack of efficient spin-sensitive electron detectors at the present time.

In general, for a kinematically complete experiment with $\mathcal{N}$ particles created in the final channel, it is necessary to measure the $3\mathcal{N}$–4 momentum (x, y, z)-components provided the internal (electronic) energy change is known; the other four components are found from the momentum and energy conservation laws. For example, in fast *heavy-particle* collisions, the coincident detection of the recoil-ion and electron momenta gives, via momentum and energy conservation, the only practical access to the tiny change of the projectile momentum and, thus, to a kinematically complete experiment.

Kinematically complete experiments, monitoring simultaneously a large part of the final-state momentum space, became possible using new methods like the cold target recoil ion momentum spectroscopy (COLTRIMS), where the recoiling target-ion momenta are measured with a high precision, or *Reaction Microscopes* where, in addition, the momenta of several electrons are recorded (see Chap. 2). Both COLTRIMS and Reaction Microscopes are very effective tools for the investigation of many-particle fragmentation reactions, as long as relatively small momentum transfers between colliding partners dominate.

Being a novel momentum-space imaging technique, Reaction Microscopes allow one to measure three-dimensional momentum vectors of the recoiling target ion, other heavy ionic particle fragments and emitted electrons with extremely high resolution and unprecedented completeness. For example, these techniques enable one to measure indirectly projectile scattering angles as small as $\theta_P \approx 10^{-9}$ rad, to separate experimentally electron–electron and electron–nuclear interactions in ion–atom collisions, or to separate the contribution to double photoionization caused by the usual photoabsorption or by Compton scattering (see review articles [1–3]). Because of the unique combination of high resolution and a solid angle of nearly 4π, Reaction Microscopes are ideally suited for *multiple* coincidence studies of all kinds of

Table 1.1. Collisional parameters and detection characteristics of fragmentation experiments performed so far using COLTRIMS and Reaction Microscopes (status in May 2003)

Parameter, range	Value, explanation
Incident particles:	Electrons, protons, antiprotons, highly charged ions, photons, strong laser pulses
Targets:	Atoms, molecules from supersonic jets or magneto-optical traps
Atomic processes:	Single– and multiple–electron capture and loss, ionization, photoionization, strong-field single and multiple ionization
Incident-electron energy:	$500\,\mathrm{eV} \leq E_e \leq 2\,\mathrm{keV}$
Incident-ion energy:	$1\,\mathrm{keV/u} \leq E_\mathrm{P} \leq 1\,\mathrm{GeV/u}$
Incident-photon energy:	$20\,\mathrm{eV} \leq E_\mathrm{ph} \leq 100\,\mathrm{keV}$
Incident-laser intensity:	$10^{13} \leq I \leq 2 \times 10^{15}\,\mathrm{W/cm}^2$
Incident-laser pulsewidth:	$25\,\mathrm{fs} \leq t \leq 150\,\mathrm{fs}$
Projectile scattering angle:	$10^{-9}\,\mathrm{rad} \leq \theta_\mathrm{P} \leq 10^{-4}\,\mathrm{rad}$
Projectile momentum change:	$10^{-9} \leq \Delta P_\mathrm{P}/P_\mathrm{P} \leq 10^{-4}$
Target recoil-ion energy:	$10^{-6}\,\mathrm{eV} \leq E_\mathrm{R} \leq 1\,\mathrm{eV}$
Projectile energy change:	$\Delta E_\mathrm{P}/E_\mathrm{P} \geq 10^{-10}$
Final momenta of all reaction products:	A few a.u.
Resolution of the recoil-ion momentum:	$\delta P_\mathrm{R} \gtrsim 0.06\,\mathrm{a.u.}$
Magneto-optical trap (MOT):	$\delta P_\mathrm{R} \gtrsim 0.03\,\mathrm{a.u.}$
Resolution of the recoil-ion energy:	$\delta E_\mathrm{R} \gtrsim 1.2\,\mu\mathrm{eV}$
Supersonic gas–jet:	Temperature $T_\mathrm{jet} > 50\,\mathrm{mK}$, energy of cooled target atoms $E_\mathrm{T} \approx 1\,\mu\mathrm{eV}$
Magneto-optical trap:	Temperature $T_\mathrm{MOT} \approx \mu\mathrm{K}$, energy of cooled target atoms $E_\mathrm{T} \approx 10^{-10}\,\mathrm{eV}$
Momentum spread of the cold atomic target in a gas–jet:	$\Delta P_\mathrm{jet} \geq 0.05\,\mathrm{a.u.}$
Momentum spread of the cold electrons in a gas–jet:	$\Delta P_\mathrm{jet} \approx 10^{-6}\,\mathrm{a.u.}$
Resolution of electron momentum and energy:	$\delta P_e \geq 0.01\,\mathrm{a.u.}$, $\delta E_e \geq 1.4\,\mathrm{meV}$
Recoil-ion detection solid angle:	$\Delta\Omega/4\pi \geq 98\%$
Electron detection solid angle:	$\Delta\Omega/4\pi \geq 50\%$
Multiplicity for ions:	2 hits
Multiplicity for electrons:	10 hits

atomic collision processes. The main collisional and detection parameters of fragmentation experiments carried out using COLTRIMS and Reaction Microscopes are given in Table 1.1.

1.1.2   Fragmentation Reactions for Atomic Targets

For ionic projectiles, multielectron fragmentation processes arising in collisions of a projectile ion X^{q+} and a neutral target atom A can be written in the general form:

$$X^{q+} + A \rightarrow X^{q'+} + A^{m+} + (q' - q + m)e^- , \tag{1.1}$$

where $X^{q'+}$ denotes the scattered projectile, A^{m+} the recoiling target ion and $q' - q + m$ the number of emitted electrons. Let us introduce three numbers, n_T, n_P, and n_C denoting, respectively, the number of electrons emitted from the target (T) and projectile (P) to continuum states, and the number of electrons (C) captured by the projectile (see Fig. 1.1). The total number N of electrons ejected to the continuum is equal to

$$N = n_P + n_T . \tag{1.2}$$

The following main fragmentation atomic reactions can be distinguished:
1. *pure target ionization*:

$$X^{q+} + A \rightarrow X^{q+} + A^{m+} + me^- , \tag{1.3}$$
$$n_P = 0 , \quad n_T = m , \quad n_C = 0 ;$$

2. *pure projectile ionization*:

$$X^{q+} + A \rightarrow X^{q'+} + A + (q' - q)e^- , \tag{1.4}$$
$$n_P = q' - q , \quad n_T = 0 , \quad n_C = 0 ;$$

3. *electron capture* (always connected with target ionization):

$$X^{q+} + A \rightarrow X^{(q-m)+} + A^{m+} , \tag{1.5}$$
$$n_P = 0 , \quad n_T = 0 , \quad n_C = m .$$

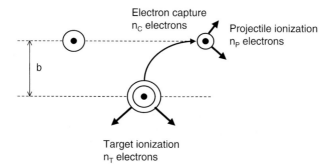

Fig. 1.1. Fragmentation reactions in an ion–atom collision. Interaction of a projectile ion, approaching a target atom at an impact parameter b, can lead to a capture of n_C electrons from the target, ionization of n_T electrons of the target, or ionization of n_P electrons of the projectile

In general, all reactions might occur simultaneously, i.e., the target as well as the projectile might be ionized and, at the same time, electrons might be transferred or excited during the collision. Usually, one only distinguishes in the literature:

4. *projectile plus target ionization or "loss ionization"*:

$$X^{q+} + A \rightarrow X^{q'+} + A^{m+} + (q' - q + m)e^-\,, \tag{1.6}$$

$$n_{\mathrm{P}} = q' - q\,, \quad n_{\mathrm{T}} = m\,, \quad n_{\mathrm{C}} = 0\,;$$

5. *target ionization plus capture or "transfer ionization"*:

$$X^{q+} + A \rightarrow X^{q'+} + A^{m+} + (q - q' + m)e^-\,, \tag{1.7}$$

$$n_{\mathrm{P}} = 0\,, \quad n_{\mathrm{T}} = m + n_C = m + (q - q')\,, \quad n_C = q - q'\,.$$

For photon impact, the following breakup reactions may be distinguished:

6. *photon-impact ionization: photoeffect*:

$$\hbar\omega + A \rightarrow A^{m+} + me^-\,; \tag{1.8}$$

7. *photon-impact ionization: Compton effect*:

$$\hbar\omega_{\mathrm{i}} + A \rightarrow \hbar\omega_{\mathrm{f}} + A^{m+} + me^-\,, \tag{1.9}$$

where indices i and f refer to the frequencies of the incident and scattered photons;

8. *multiphoton ionization*:

$$n\hbar\omega + A \rightarrow A^{m+} + me^-\,. \tag{1.10}$$

Finally, if the projectile is an electron or an antiproton, capture as well as projectile ionization reactions cannot occur, and one obtains:

9. *electron-impact ionization*:

$$e^- + A \rightarrow e^- + A^{m+} + me^-\,. \tag{1.11}$$

10. *antiproton-impact ionization*:

$$\bar{p} + A \rightarrow \bar{p} + A^{m+} + me^-\,. \tag{1.12}$$

For electron impact, 'exchange reactions' should be considered in addition, where the projectile electron stays at the target in the final state and one or more target electrons are emitted (the so-called *electron-exchange effect*). Until now, positron-induced reactions have not yet been investigated in kinematically complete experiments and, thus, are not considered here.

1.2   Particle Kinematics: Fragmentation of Atoms

1.2.1   Momentum and Energy Conservation Equations in the Nonrelativistic Case

For any fragmentation reaction, the initial momentum vector $\boldsymbol{P}$ of the incident particle (ion, electron or photon) and those of the *collision fragments*, i.e., the recoil ion, electrons, scattered projectile, and photons, are related through the momentum and energy conservation laws following from general symmetry considerations in classical mechanics. In the nonrelativistic approach, the velocity v of a particle with the rest mass M, its mechanical momentum P, and its kinetic energy E_{kin} are related by:

$$E_{\mathrm{kin}} = \frac{P^2}{2M} = \frac{Mv^2}{2}.\tag{1.13}$$

For photons having zero rest mass, one has:

$$E_\gamma = P_\gamma/c\tag{1.14}$$

with c being the speed of light.

For two colliding particles, the momentum conservation law can be written as:

$$\boldsymbol{P}_{\mathrm{P}}^{\mathrm{i}} + \boldsymbol{P}_{\mathrm{R}}^{\mathrm{i}} = \boldsymbol{P}_{\mathrm{P}}^{\mathrm{f}} + \boldsymbol{P}_{\mathrm{R}}^{\mathrm{f}} + \sum_{j=1}^{N} \boldsymbol{P}_{ej}^{\mathrm{f}} + \sum_{l=1}^{N'} \boldsymbol{P}_{\gamma l}\,,\tag{1.15}$$

where $\boldsymbol{P}_{\mathrm{P}}^{\mathrm{i}}$ and $\boldsymbol{P}_{\mathrm{P}}^{\mathrm{f}}$ denote the momentum vectors of the projectile before i and after f the collision, $\boldsymbol{P}_{\mathrm{R}}^{\mathrm{i}}$ the momentum vector of the target atom, and $\boldsymbol{P}_{\mathrm{R}}^{\mathrm{f}}$ the momentum vector of the recoil ion, respectively. In the laboratory frame, where before the collision the target is at rest, one has $\boldsymbol{P}_{\mathrm{R}}^{\mathrm{i}} = 0$. $\boldsymbol{P}_{ej}$ and $\boldsymbol{P}_{\gamma l}$ are the momenta of N and N' additional electrons and photons, respectively, emitted into the continuum during the reaction. In what follows, we will assume that the sum over N' in (1.15) is so small that it might be neglected; this assumption is well fulfilled in most realistic situations.

In any collision, the important quantity of the reaction is the *momentum transfer* given by:

$$\boldsymbol{q} = (q_\perp, q_{||}) = \boldsymbol{P}_{\mathrm{P}}^{\mathrm{i}} - \boldsymbol{P}_{\mathrm{R}}^{\mathrm{f}}\,,\tag{1.16}$$

where $q_{||}$ and $q_\perp$ are the *longitudinal* and *transverse* components of the vector $\boldsymbol{q}$, i.e., along or transverse to the projectile propagation, respectively. For the future consideration, we will choose the following directions for the x, y, z axes (Fig. 1.2): the z-axis denotes the projectile-beam direction, the y-axis the target-beam direction, assuming that the target particles come out from a supersonic jet; the x-axis is perpendicular to the (y, z)-plane. Strictly speaking, the momenta $\boldsymbol{P}_{\mathrm{R}}^{\mathrm{i}}$ of atoms in the target beam are nonzero in the laboratory. But, since they are known very precisely, they are usually subtracted from the

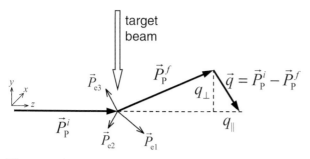

Fig. 1.2. A beam of target atoms propagates along the y-axis, and the projectile beam along the z-axis. The projectile with the initial momentum $\boldsymbol{P}_P^i$ has a momentum $\boldsymbol{P}_P^f$ after a scattering on the target atom. The vector $\boldsymbol{q} = \boldsymbol{P}_P^i - \boldsymbol{P}_P^f$ of the momentum transfer has the longitudinal $q_\parallel$ and transverse $q_\perp$ projections on the z-axis and the (x, y)-plane, respectively. The vectors $\boldsymbol{P}_{ej}$ denote the momenta of ejected electrons after the collision

measured data, i.e., a transformation into the standard "laboratory system" is performed. For simplicity, we will consider below that the target momenta are equal to zero before the collision.

For two colliding particles, the energy conservation law can be written as:

$$E_P^i + E_R^i + E_{\text{bind}}^i = E_P^f + E_R^f + E_{\text{bind}}^f + \sum_{j=1}^{N} E_{ej}^f , \qquad (1.17)$$

or

$$E_P^i + E_R^i = E_P^f + E_R^f + Q + \sum_{j=1}^{N} E_{ej}^f , \qquad (1.18)$$

where E_P^i and E_R^i are the kinetic energies of the projectile and the target before i and after f the collision, respectively, $E_e = P_e^2/2m_e$ is the electron kinetic energy in the continuum and $E_{\text{bind}} < 0$ is the total (negative) energy of all bound electrons before i and after f the collision. We define E_{bind}^f to be the final binding energy of all electrons 'directly' after the collision, i.e., 'before' any de-excitation of excited states via photon or Auger-electron emission has occured. Thus, apart from bremsstrahlung contributions, which are negligibly small at present collision energies, the energies of emitted photons do not have to be accounted for in (1.17) and (1.18).

In (1.18), we introduced the so-called Q-value given by

$$Q = E_{\text{bind}}^f - E_{\text{bind}}^i , \qquad (1.19)$$

which denotes the change in internal energies of the projectile and the target, i.e., the energy difference between final and initial electronic (in general, multielectronic) states of colliding particles before the decay of excited states. The Q-value is a very important quantity called the *inelasticity* of

electron capture

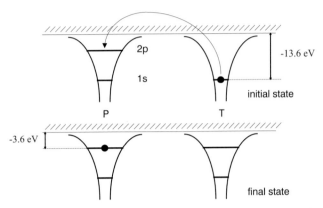

Fig. 1.3. Scheme of the nonresonant electron capture $H^+ + H(1s) \rightarrow H(2p)^* + H^+$ in collisions of protons with hydrogen atoms. In the initial state, the proton captures the 1s-electron from hydrogen in its ground state (*upper picture*) and creates the excited hydrogen atom $H^*(2p)$ in the final state (*lower picture*). This is an endothermic reaction because the Q-value of $+10.2\,\mathrm{eV}$ is taken from the relative motion of proton and the hydrogen atom to excite the electron from the lower 1s state to the upper 2p state

the process. Reactions with positive Q-values, $Q > 0$, are called *endothermic reactions* and are associated with absorption of kinetic energy from the relative motion of the projectile–target system and "transfer" of this energy into electronic excitation of the bound states. The reactions with the negative Q-value, $Q < 0$, are called *exothermic reactions*, where the released kinetic energy is distributed between reaction fragments. For example, the electron capture by an incident proton from the ground-state hydrogen $H(1s)$ into the 2p excited state

$$H^+ + H(1s) \rightarrow H^*(2p) + H^+$$

is an example of the endothermic reaction because the inelasticity is equal to $Q = -3.4\,\mathrm{eV} - (-13.6\,\mathrm{eV}) = +10.2\,\mathrm{eV}$ (see Fig. 1.3).

Capture of the electron into the same 1s-state of the projectile

$$H^+ + H(1s) \rightarrow H(1s) + H^+$$

corresponds to a zero Q-value and constitutes an example of the *resonant electron capture*.

As is seen from (1.15), the initial state is represented by two colliding particles and six momentum (x, y, z)-components, while the final state refers to $\mathcal{N} = N + 2$ fragments and $3\mathcal{N}$ momentum components, where N is the number of free electrons created during the fragmentation reaction. Since the momentum and energy conservation laws are described by four equations

(1.15)–(1.17), only $3\mathcal{N}$–4 of $3\mathcal{N}$ momenta are independent, i.e. the kinematics of the reaction is fully described by $3\mathcal{N}$–4 independent components. If the Q-value is known, only these $3\mathcal{N}$–4 components have to be determined in the experiment in order to perform a "kinematically complete" measurement.

1.2.2 Transverse- and Longitudinal-Momentum Balances

Since for charged-particle impact, the z-axis, oriented along the projectile-propagation direction, is the symmetry axis and, consequently, the x- and y directions contain identical information, cylindrical coordinates are often used. Presenting all momentum vectors as the sum of the longitudinal $\boldsymbol{P}_{||}$ and transverse vectors $\boldsymbol{P}_\perp$

$$\boldsymbol{P} = \boldsymbol{P}_{||} + \boldsymbol{P}_\perp \,, \text{ with } \boldsymbol{P}_{||} = (0,\, 0,\, P_z)\,, \text{ and } \boldsymbol{P}_\perp = (P_x,\, P_y,\, 0)\,, \qquad (1.20)$$

one can write the general set of equations equivalent to (1.15) for the vectors $\boldsymbol{P}_{||}$ and $\boldsymbol{P}_\perp$ in the form:

$$0 = \boldsymbol{P}_{\mathrm{P}\perp}^{\mathrm{f}} + \boldsymbol{P}_{\mathrm{R}\perp}^{\mathrm{f}} + \sum_{j=1}^{N} \boldsymbol{P}_{\mathrm{e}\perp j}^{\mathrm{f}}\,, \qquad (1.21)$$

$$\boldsymbol{P}_{\mathrm{P}||}^{\mathrm{i}} = \boldsymbol{P}_{\mathrm{P}||}^{\mathrm{f}} + \boldsymbol{P}_{\mathrm{R}||}^{\mathrm{f}} + \sum_{j=1}^{N} \boldsymbol{P}_{\mathrm{e}||j}^{\mathrm{f}}\,, \qquad (1.22)$$

where $\boldsymbol{P}_{\mathrm{P}\perp}^{\mathrm{i}} = \boldsymbol{P}_{\mathrm{R}}^{\mathrm{i}} = 0$ in the laboratory frame. Rewriting both equations so that the recoil-ion momentum, which is usually measured in the experiment, is expressed as a function of the other quantities, yields for the momentum in the final state:

$$\boldsymbol{P}_{\mathrm{R}\perp}^{\mathrm{f}} = - \left(\boldsymbol{P}_{\mathrm{P}\perp}^{f} + \sum_{j=1}^{N} \boldsymbol{P}_{\mathrm{e}\perp j}^{\mathrm{f}} \right) = \boldsymbol{q}_\perp - \sum_{j=1}^{N} \boldsymbol{P}_{\mathrm{e}\perp j}^{\mathrm{f}}\,, \qquad (1.23)$$

$$\boldsymbol{P}_{\mathrm{R}||}^{\mathrm{f}} = \boldsymbol{P}_{\mathrm{P}||}^{\mathrm{i}} - \left(\boldsymbol{P}_{\mathrm{P}||}^{\mathrm{f}} + \sum_{j=1}^{N} \boldsymbol{P}_{\mathrm{e}||j}^{\mathrm{f}} \right) = \boldsymbol{q}_{||} - \sum_{j=1}^{N} \boldsymbol{P}_{\mathrm{e}||j}^{\mathrm{f}}\,. \qquad (1.24)$$

In this form, one can immediately characterize two specific situations. For pure electron–capture reactions (no electrons in the continuum, $n_{\mathrm{P}} = n_{\mathrm{T}} = 0$), or for ionization reactions, where the total transverse momentum of the emitted electrons $\left| \sum_{j=1}^{N} \boldsymbol{P}_{\mathrm{e}\perp j}^{\mathrm{f}} \right|$ is much smaller than the heavy particle momenta, one can put $\boldsymbol{P}_{\mathrm{R}\perp}^{f} \simeq -\boldsymbol{P}_{\mathrm{P}\perp}^{\mathrm{f}}$, i.e., the projectile and the target have equal but opposite (in sign) transverse momenta after the collision.

The second situation is realized for collisions at small impact parameters or at small projectile energies where the ("screened") nuclear–nuclear interaction dominates the many-particle momentum balance. At intermediate and high projectile velocities v_P, the total transverse momenta of all products are usually of the same order of magnitude and, therefore, the total momentum $\left| \sum_{j=1}^{N} \boldsymbol{P}_{e\perp j}^{f} \right|$ cannot be neglected (see [1]).

Note that the cylindrical coordinates are not well adapted for fragmentation induced by photon absorption or in strong laser fields, since here the symmetry is broken due to the light polarization in the (x, y)-plane. However, for charged-particle impact, especially if the projectile momentum is large as for heavy-ion projectiles, further advantages emerge from this separation since in many situations, as we will demonstrate in the next section, transverse and longitudinal momentum components are decoupled and contain different physical information on the fragmentation reaction.

1.2.3   Fast Ion–Atom Collisions: Small Momentum, Energy, and Mass Transfers

In the previous sections, exact equations for momenta and energies of particles before and after fragmentation were used that can be applied for any incident projectile. In this section, we concentrate on fast heavy-ion–induced breakup reactions of atoms where substantial approximations can be applied to (1.21) and (1.22) leading to considerable simplifications and providing physical insight into the fragmentation dynamics.

For most reactions occurring in ion–atom collisions, a typical momentum transferred to the target from the projectile is of the order of a few atomic units (a.u.); 1 a.u. of momentum corresponds to 13.606 eV of electron energy. Typical energy transfers range between a few tens up to a few hundreds of electronvolts. Hence, even taking a "light" ionic projectile, a proton, at moderate incident energy of 1 MeV (i.e., a velocity of 6.3 a.u.) its initial energy and momentum ($P_P^i = 6.3 \times 1836$ a.u. $= 11\,567$ a.u.) is at least three orders of magnitude larger than that of an electron moving with the same velocity. Going to collisions with "heavy ions", say Au^{50+}, at intermediate energies of one GeV, the relative energy and momentum change of the projectile, $\Delta E_P/E_P^i$ and $\Delta P_P/P_P^i$, become as small as 10^{-6}. Furthermore, one can see that the final energy of the target ion of typically less than 1 eV, is much smaller compared to the projectile initial energy E_P^i, ranging from keV to GeV (see Table 1.1). Hence, one has in a very good approximation for the projectile energy

$$E_R^f \ll E_P^i = (P_P^i)^2/2M_P^i \approx E_P^f = (P_P^f)^2/2M_P^f \,, \tag{1.25}$$

where M_P^i and M_P^f are the projectile initial and final masses, respectively. Therefore, we can neglect the change of the target energy compared to the initial and final energies of the projectile. However, we note that the change of the projectile energy, as well as of the projectile momentum, is considered to be finite in the following. For small momentum transfers, i.e., for small projectile scattering angles θ_P, one has for the projectile-momentum components:

$$P^f_{P\perp} \ll P^f_{P\|}, \quad P^f_{P\perp}/P^f_{P\|} \approx \theta_P, \tag{1.26}$$

$$E^f_P = \frac{1}{2M^f_P} \left((P^f_{P\|})^2 + (P^f_{P\perp})^2 \right) \approx \frac{{P^f_{P\|}}^2}{2M^f_P}, \tag{1.27}$$

$$P^f_{P\|} \approx \sqrt{2M^f_P E^f_P}. \tag{1.28}$$

Taking (1.26)–(1.28) into account and neglecting E^f_R due to (1.25), one has from (1.18):

$$P^f_{P\|} = \left[2M^f_P \left(E^i_P - Q - \sum_{j=1}^{N} E^f_{ej} \right) \right]^{1/2}, \quad N = n_P + n_T. \tag{1.29}$$

If n_C electrons are captured and n_P electrons of the projectile are ionized in the collision, one can write for the projectile mass in the final state (in atomic units $m_e = 1$):

$$M^f_P = M^i_P + n_C - n_P \tag{1.30}$$

and for the longitudinal component of the projectile momentum

$$P^f_{P\|} \approx \left[(2M^i_P E^i_P) \left(1 + \frac{\Delta M_P}{M^i_P} \right) \left(1 - \frac{\Delta E_P}{E^i_P} \right) \right]^{1/2}$$

$$= \left[(2M^i_P E^i_P) \left(1 + \frac{n_C - n_P}{M^i_P} \right) \left(1 - \frac{Q + \sum_{j=1}^{N} E^f_{ej}}{E^i_P} \right) \right]^{1/2}, \tag{1.31}$$

$$\Delta M_P = M^f_P - M^i_P, \quad \Delta E_P = E^i_P - E^f_P. \tag{1.32}$$

Expanding the last equation over two small parameters, $\Delta M_P/M^i_P \ll 1$ and $\Delta E_P/E^i_P \ll 1$, one has for the longitudinal component of the projectile momentum in the final state:

$$P^f_{P\|} = P^i_{P\|} \left[1 + \frac{1}{2} \left(\frac{n_C - n_P}{M^i_P} - \frac{Q + \sum_{j=1}^{N} E^f_{ej}}{E^i_P} \right) \right]. \tag{1.33}$$

For the change of the projectile momentum along the beam direction, introducing the projectile velocity v_P as

$$v_P = \frac{P^i_{P\|}}{M^i_P}, \tag{1.34}$$

one finally obtains:

$$\Delta P_{P\|} \equiv P^f_{P\|} - P^i_{P\|} = \frac{v_P}{2}(n_C - n_P) - \frac{1}{v_P} \left(Q + \sum_{j=1}^{n_P+n_T} E^f_{ej} \right). \tag{1.35}$$

This equation for the change of the projectile longitudinal momentum is valid in the nonrelativistic approximation under assumptions of small projectile-energy change compared to the incoming projectile kinetic energy, small scattering angles and small mass transfers. The first term in (1.35), called the *mass-transfer* term, describes the projectile momentum change due to electron capture or loss. The last two terms are responsible for delivering the potential and kinetic energy to the target and projectile electrons by the projectile.

1.2.4   Fast Ion–Atom Collisions: Recoil-Ion Momenta

As was noted before, for the majority of fragmentation reactions the momentum transfer to the target is of the order of a few atomic units only, i.e., small. Therefore, the projectile momentum change, compared to its initial momentum $P_P^i = P_{P||}^i = M_P^i \cdot v_P$ is small for heavy projectiles because of their large masses M_P^i, even at very moderate collision velocities of a few atomic units. Thus, its direct determination via measurement of the final energy and scattering angle of the projectile is impossible in most cases. For example, a momentum change of 1 a.u. in a single capture reaction in a collision between the lightest projectile, namely a proton, and a helium target at 1 MeV ($v_P \simeq 6.3$ a.u.) will result in a momentum change in any direction of less than $1/(1836 \times 6.3) \approx 9 \times 10^{-5}$. A maximum scattering angle could be less than 0.1 mrad, i.e., 1 mm deflection on a distance of 10 m which is impossible to measure with sufficient resolution. Even the most sophisticated storage-ring devices are just able to deliver MeV ion beams with an initial momentum spread of the order of $\Delta P_P^i / P_P^i \sim 10^{-5}$. Therefore, the measurement of the momentum change of the *recoil* ion, which is at rest in the laboratory system in the initial state and, thus, is accessible with high accuracy, has become a key technique for the investigation of fast heavy-ion–induced fragmentation reactions.

Using the results of the previous sections, we will now find all three components of the recoil-ion momentum $\boldsymbol{P}_R^f$ with the initial vectors $\boldsymbol{P}_P^i$ and $\boldsymbol{P}_R^i$ given by

$$\boldsymbol{P}_P^i = \left(0,\, 0,\, \sqrt{2M_P^i E_P^i}\right), \quad \boldsymbol{P}_R^i = (0,\, 0,\, 0) . \tag{1.36}$$

Inserting (1.35) into (1.24), we obtain for the longitudinal recoil-ion momentum in the final state:

$$P_{R||}^f = \frac{n_P - n_C}{2} v_P + \frac{Q}{v_P} + \sum_{j=1}^{N=n_P+n_T} \left(\frac{E_{ej}^f}{v_P} - P_{e||j}^f\right) . \tag{1.37}$$

Separating only terms with n_P in the last sum, one can rewrite those as:

$$\sum_{j=1}^{n_P} \left(\frac{E_{ej}^f}{v_P} - P_{e||j}^f\right) \approx \sum_{j=1}^{n_P} \left(\frac{P_{e||j}^{f\,2}}{2v_P} - P_{e||j}^f\right) . \tag{1.38}$$

Really, using the approximation for small projectile scattering angles and small energy transfers to the projectile electrons it is evident that ionized projectile electrons move with about the projectile velocity into the forward direction (the so-called *cusp electrons*). Therefore, for collisions at large velocity v_P, the total energy of these electrons is

$$\frac{1}{2}\sum_{j=1}^{n_P}\left(P_{exj}^{f\;2}+P_{eyj}^{f\;2}+P_{ezj}^{f\;2}\right)\approx\frac{1}{2}\sum_{j=1}^{n_P}P_{e\|j}^{f\;2}\,. \qquad (1.39)$$

Transformation from the laboratory system to the projectile frame (PF) yields for $P_{e\|j}^{f}$ (see the Lorentz transformation (1.56) and (1.57) for the relativistic case):

$$P_{e\|j}^{f}=m_e v_P-(P_{e\|j}^{f})_{PF}\,. \qquad (1.40)$$

Finally, substituting (1.40) into (1.37), one has for the recoil-ion longitudinal momentum:

$$P_{R\|}^{f}=\frac{Q}{v_P}-n_C\frac{v_P}{2}+\sum_{j=1}^{n_T}\left(\frac{E_{ej}^{f}}{v_P}-P_{e\|j}^{f}\right)+\sum_{l=1}^{n_P}\frac{(E_{ej}^{f})_{PF}}{v_P}\,. \qquad (1.41)$$

Sometimes (see [2]), the recoil-ion longitudinal momentum (1.41) is conveniently presented as the sum of three contributions – electron capture (ec), target ionization (ion) and projectile ionization (loss) (see the processes (1.3)–(1.5)):

$$P_{R\|}=P_{R\|}^{ec}+P_{R\|}^{ion}+P_{R\|}^{loss}\,, \qquad (1.42)$$

$$P_{R\|}^{ec}=\frac{Q_{ec}}{v_P}-n_C\frac{v_P}{2}\,, \qquad (1.43)$$

$$P_{R\|}^{ion}=\sum_{j=1}^{n_T}\left(\frac{E_{ej}^{f}-E_{j}^{bind}}{v_P}-P_{e\|j}^{f}\right)\,, \qquad (1.44)$$

$$P_{R\|}^{loss}=\sum_{l=1}^{n_P}\frac{(E_{el}^{f}-E_{l}^{bind})_{PF}}{v_P}\,, \qquad (1.45)$$

where Q_{ec} is a part of the inelasticity responsible for electron capture and $E^{bind}<0$ are the binding energies of the bound electrons in the projectile and the target, respectively. All terms are written in the laboratory frame except the last one in (1.45), which is given in the projectile frame.

Now, we obtain x- and y-components and, thus, the *transverse* component of the recoil-ion momentum. Substitution of (1.36) into the momentum conservation law (1.15) gives:

$$P_{Rx}^{f}=-(P_{Px}^{f}+\sum_{j=1}^{N}P_{exj}^{f})\,, \qquad (1.46)$$

$$P_{Ry}^{f}=-(P_{Py}^{f}+\sum_{j=1}^{N}P_{eyj}^{f})\,. \qquad (1.47)$$

Employing the approximation of high projectile energy, one has:

$$P^f_{Rx} = -(M^f_P \cdot v_P \cdot \mathrm{tag}\theta_P \cdot \cos\varphi_P + \sum_{j=1}^{N} P^f_{exj}), \qquad (1.48)$$

$$P^f_{Ry} = -(M^f_P \cdot v_P \cdot \mathrm{tag}\theta_P \cdot \sin\varphi_P + \sum_{j=1}^{N} P^f_{eyj}), \qquad (1.49)$$

where φ_P is the azimuthal scattering angle of the projectile with respect to the x-axis in the xy-plane. Often, if a scattering plane is not defined, one uses only the transverse projection:

$$P^f_{R\perp} = \sqrt{P^f_{Rx}{}^2 + P^f_{Ry}{}^2} = -P^f_{P\perp} - \sum_{j=1}^{N} P^f_{e\perp j}$$

$$\approx -M_P \cdot v_P \cdot \theta_P - \sum_{j=1}^{N} P^f_{e\perp j}, \qquad (1.50)$$

where P^f_{Rx} and P^f_{Ry} components are given by (1.48) and (1.49).

Obviously, for small final electron momenta compared to the transverse momentum transfer or for pure electron capture reactions with $N = n_P + n_T = 0$, the projectile polar scattering angle θ_P can be directly deduced from the recoil-ion momentum and from the known projectile velocity and mass. Thus, in this case, different information on the reaction can be obtained from the longitudinal and transverse momenta of the recoil ion, respectively, which, moreover, are completely *decoupled* under the approximations made.

Whereas the inelasticity of the reaction, the final electronic bound states, and spectroscopic information can be deduced from the longitudinal recoil-ion momentum, the projectile scattering angle, the dynamics of the reaction, and, for known scattering potential, the impact parameter dependence is connected to the transverse recoil-ion momentum in the final state.

1.2.5 Relativistic Case

In the case of relativistic projectile velocities, the situation is more complicated compared to nonrelativistic collisions where the relation between the projectile *kinetic* energy and its momentum is given by the simple formula (1.13). In the relativistic case, the particle momentum P is related to its *total* energy E_{tot} by (see (1.13)):

$$E^2_{tot} = P^2c^2 + (M_0c^2)^2, \quad P = \gamma M_0 v, \qquad (1.51)$$

$$E_{kin} = E_{tot} - M_0c^2 = (\gamma - 1)M_0c^2, \qquad (1.52)$$

$$\gamma = \frac{1}{\sqrt{1 - \beta^2}}, \quad \beta = v_P/c, \qquad (1.53)$$

where M_0 and M_0c^2 are the *rest mass* and *rest energy*, respectively, v is the particle velocity, c is the speed of light and γ and β are the *relativistic factors*. The rest masses for electron and proton are equal to $511\,\mathrm{keV}$ and $938.3\,\mathrm{MeV}$, respectively.

At small velocities $v_\mathrm{P} \to 0$, the total energy of a particle approaches its rest energy:

$$E_{\mathrm{tot}}(v_\mathrm{P} \to 0) = M_0c^2\left(1 + \frac{1}{2}\beta^2 + \ldots\right) = M_0c^2 + \frac{1}{2}M_0v_\mathrm{P}^2 + \ldots. \qquad (1.54)$$

At the extreme relativistic limit, $v_\mathrm{P} \to c$, one has from (1.51)–(1.53):

$$E_{\mathrm{tot}}(v \to c) = E_{\mathrm{kin}} = Pc. \qquad (1.55)$$

For relativistic projectile velocities v_P, one has to make the Lorentz transformation of the momentum and energy from the laboratory frame to the moving projectile frame (PF)

$$P_{\mathrm{PF}} = \gamma\left(P - v_\mathrm{P}E/c^2\right), \qquad (1.56)$$

$$E_{\mathrm{PF}} = \gamma\left(E - v_\mathrm{P}P\right), \qquad (1.57)$$

and to account for the rest masses of the active n_P and n_C electrons in (1.17) in the form:

$$(E_\mathrm{P}^{\mathrm{i}})_{\mathrm{kin}} = (E_\mathrm{P}^{\mathrm{f}})_{\mathrm{kin}} + Q + \sum_{j=1}^{N} E_{ej}^{\mathrm{f}} + (n_\mathrm{P} - n_\mathrm{C})c^2. \qquad (1.58)$$

Then the basic equations for the change of the projectile momentum (1.35) and the recoil-ion longitudinal momentum (1.41) can be written as [4]:

$$\Delta P_{\mathrm{P}\|} = \frac{v_\mathrm{P}}{1 + \gamma^{-1}}(n_\mathrm{C} - n_\mathrm{P}) - \frac{1}{v_\mathrm{P}}\left(Q + \sum_{j=1}^{n_\mathrm{P}+n_\mathrm{T}} E_{ej}^{\mathrm{f}}\right), \qquad (1.59)$$

$$P_{\mathrm{R}\|}^{\mathrm{f}} = \frac{Q}{v_\mathrm{P}} - n_\mathrm{C}\frac{v_\mathrm{P}}{1 + \gamma^{-1}} + \sum_{j=1}^{n_\mathrm{T}}\left(\frac{E_{ej}^{\mathrm{f}}}{v_\mathrm{P}} - P_{e\|j}^{\mathrm{f}}\right) + \sum_{l=1}^{n_\mathrm{P}}\frac{(E_{ej}^{\mathrm{f}})_{\mathrm{PF}}}{\gamma v_\mathrm{P}}. \qquad (1.60)$$

1.3 Ion–Atom Collisions: Illustrative Examples

1.3.1 Single-Electron Capture

The single-electron capture reaction

$$\mathrm{X}^{q+} + \mathrm{A} \to \mathrm{X}^{(q-1)+} + \mathrm{A}^+$$

is the simplest process since there are only two heavy particles involved with no free electrons in both the initial and final channels. So, in essence, we are dealing with an inelastic two-particle collision.

In this case, we have $\mathcal{N} = 2$ particles in the final state and, if the Q-value is not known, $3\mathcal{N} - 3 = 3$ unknown momentum components: $P^f_{R||}$, P^f_{Rx}, and P^f_{Ry}. For the longitudinal recoil-ion momentum $P^f_{R||}$, one has from (1.41):

$$P^f_{R||} = -\Delta P_{P||} = \frac{Q}{v_P} - \frac{v_P}{2}, \quad Q = E^X_{bind} - E^A_{bind}, \tag{1.61}$$

where E^X_{bind} and E^A_{bind} are the binding energies of the captured electron in the scattered ion $X^{(q-1)+}$ and the target atom A, respectively. For the Q-value one obtains:

$$Q = P^f_{R||} \cdot v_P + \frac{v_P^2}{2}. \tag{1.62}$$

As is seen from (1.62), the longitudinal recoil-ion momentum $P^f_{R||}$ can take only *discrete* values for fixed projectile velocity because the Q-value is quantized. Measuring the longitudinal recoil-ion momentum allows one to determine the Q-value with high accuracy and, therefore, to get information about the final state of the captured electrons. Moreover, if the resolution is high enough, one can even extract spectroscopic information, as has been demonstrated recently [5] in COLTRIMS experiments on single-electron capture in $Ne^{7+} + He$ collisions at velocity $v_P = 0.355$ a.u. (see Fig. 1.4) where a precision in the energy determination of populated states of up to a few meV was obtained.

According to (1.62), the Q-resolution for electron capture is given by

$$\left| \frac{\delta Q}{Q} \right| = \left| \frac{\delta P^f_{R||}}{P^f_{R||}} \right| + 3 \left| \frac{\Delta P^i_P}{P^i_P} \right|, \tag{1.63}$$

where $\delta P^f_{R||}$ denotes the resolution of the recoil-ion momentum and $\Delta P^i_P / P^i_P$ the initial state momentum spread of the projectile ions. In COLTRIMS, using a supersonic jet-target, a momentum resolution up to $\delta P^f_{R||} \approx 0.06$ a.u. can be achieved. With MOTRIMS, a resolution of $\delta P^f_{R||} \approx 0.03$ a.u. has been obtained recently. The quantity $\Delta P^i_P / P^i_P$ is of the order of 10^{-4} for a well–collimated ion beam and may be as small as 10^{-6} in the advanced cooler storage rings (see Table 1.1). Let us estimate the Q-resolution for two capture reactions with different Q-values and at different projectile velocities, for example:

$$H^+ + H(1s) \rightarrow H(2p) + H^+, \quad E_P = 2\,keV/u, \, Q = 10.2\,eV,$$

$$Ne^{10+} + H(1s) \rightarrow Ne^{9+}(3p) + H^+, \quad E_P = 1.6\,MeV/u, \quad Q = -130\,eV.$$

Using (1.62) and (1.63), one has for these reactions (in atomic units): $v_P = 0.28$, $Q = 0.38$, $P^f_{R||} = 1.2$, $\delta Q/Q \approx 0.042$ and $v_P = 8.0$, $Q = -4.8$, $P^f_{R||} = -4.6$, $\delta Q/Q \approx 0.011$, respectively. Therefore, in both cases

$$\left| \frac{\delta Q}{Q} \right| \approx \left| \frac{\delta P^f_{R||}}{P^f_{R||}} \right| \gg \left| \frac{\Delta P^i_P}{P^i_P} \right|, \tag{1.64}$$

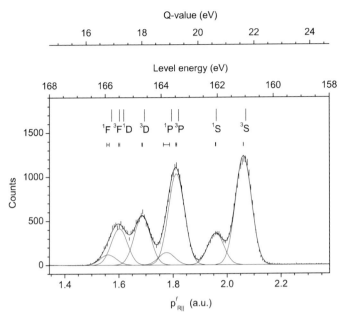

Fig. 1.4. The Q-value spectrum as a function of the longitudinal recoil Ne^{6+} ion momentum $P_{R\parallel}^{f}$ in a single-electron capture $Ne^{7+} + He \rightarrow Ne^{6+}(2s4\ell^{1,3}L) + He^{+}$ measured at projectile velocity $v_P = 0.355$ a.u. using the high-resolution COLTRIMS technique. The Q-resolution was $\delta Q \approx 0.02$ a.u. ≈ 0.54 eV. The scale of energy levels, counted from the ground state of Ne^{6+}, is also shown together with identification of the $2s4\ell^{1,3}L$ singlet and triplet states ($L = S, P, D,$ and F): *upper marks – multiconfiguration Hartree–Fock calculations* [6], *lower marks with error bars – experimental results* [5]

i.e., the longitudinal recoil-ion momentum resolution defines the resolution in the Q-value in most realistic situations. We note, however, that in some cases, e.g., at relatively large projectile velocities v_P, the second term in (1.63) might be of the same order as the first one and, therefore, should be accounted for as well.

As follows from (1.26), (1.48), and (1.49) for pure electron-capture reactions, the transverse momenta of the projectile and the target in the final state are exactly identical

$$P_{P\perp}^{f} = -P_{R\perp}^{f} = M_P \cdot v_P \cdot \theta_P, \tag{1.65}$$

and are uniquely related to the projectile scattering angle θ_P, which makes it possible to determine θ_P with a high accuracy. Again, from this equation one obtains the projectile scattering–angle resolution

$$\delta\theta_P = -\frac{\delta P_{R\perp}^{f}}{P_P^{i}} + \frac{\Delta P_P^{i}}{P_P^{i}} \cdot \theta_P. \tag{1.66}$$

Typically, one has $\delta P_{R\perp}^f \approx 0.1$ a.u. and $P_{R\perp}^f \approx 1$ a.u., so from (1.66) it follows that for small scattering angles considered, $\theta_P < 10^{-2}$ rad, the second term in the sum (1.66) can be neglected, and one obtains

$$\delta\theta_P \approx -\frac{\delta P_{R\perp}^f}{P_P^i} , \tag{1.67}$$

i.e., in pure capture reactions, the resolution of the recoil-ion transverse momentum defines the scattering angle resolution. For example, in 1 GeV/u uranium ion impact with $P_P^i \approx 10^8$ a.u., the scattering angle resolution of $\delta\theta_P$ might be as good as a nanoradian (1 mm deflection on 1000 km distance!).

If some electrons are emitted to the continuum after the capture collision, e.g. like in transfer ionization, then (1.62) and (1.65) are not valid, and the recoil-ion momentum is no longer equivalent to that of the projectile. Still, using Reaction Microscopes, the determination of the Q-value and of the projectile scattering angle is possible if the momenta of the emitted electrons are determined in coincidence.

As is demonstrated in Fig. 1.4, COLTRIMS has developed to be a valuable and competitive spectroscopy technique for the energy determination of excited states, populated during slow ion–atom collisions reaching an accuracy in the line determination of a few meV. Using precooled jets, this accuracy can be improved by a factor of about five. Using a MOT as a target an increase in resolution by a factor of more than 10 (up to 30) might be envisaged in the near future making this technique the ultimate precision spectroscopy tool for highly excited states. Compared to other excitation (beam-foil) and spectroscopy (optical, VUV, X-ray) techniques there are further considerable advantages. The excited states are well defined, there are no unknown spectator electrons due to the known charge state of the ion. Many transitions are measured simultaneously, and one is not restricted by transition (dipole) selection rules. Any excited state that is populated in a capture reaction is accessible with the same spectrometer irrespective of the actual excitation energy. Finally, high-lying doubly and triply excited states will become accessible in combination with effective ion sources for slow highly charged ions such as the electron-beam ion trap (EBIT) (see, e.g., [7]) or the HITRAP–ion trap facility for experiments with highly-charged ions [8].

1.3.2   Target Ionization

In the pure m-electron ionization of the target

$$X^{q+} + A \rightarrow X^{q+} + A^{m+} + me^- ,$$

the longitudinal recoil-ion momentum $P_{R||}^f$ is defined by the general formula followed from (1.41):

$$P_{R||}^f = \frac{Q}{v_P} + \sum_{j=1}^{m}\left(\frac{E_{ej}}{v_P} - P_{e||j}^f\right) = \sum_{j=1}^{m}\left(\frac{E_{ej}^f - E_j}{v_P} - P_{e||j}^f\right) , \tag{1.68}$$

$$Q = -\sum_{j=1}^{m} E_j \,, \tag{1.69}$$

where E_{ej}^f and $P_{e\|j}^f$ are kinetic energy and longitudinal momentum of the j-th electron, respectively, and $E_j < 0$ are the sequential ionization potentials of emitted electrons. If excitation of the remaining target ion also takes place, the sum of all excitation energies has to be considered for the Q-value in addition. *Vice versa*, for known number of emitted electrons and measurement of their longitudinal momentum components, the total excitation energy in collisions with multielectron targets became accessible for the first time [9].

In the general case, all terms in the sum (1.68) can give a substantial contribution. If the projectile velocity is high and the energy of ionized electrons is small, however, as is typical for most impact ionization reactions, then (1.68) reads

$$P_{R\|}^f \approx -\sum_{j=1}^{m} P_{e\|j}^f \,, \tag{1.70}$$

i.e., the longitudinal recoil-ion momentum reflects the total sum of the longitudinal momenta of all ejected electrons irrespective of the distribution of these momenta among particular electrons. It was demonstrated (see [10]) that detailed information on the dynamics as well as sensitive checks of the theory for m-fold electron ionization can be obtained in this way without the need to detect all ejected electrons in coincidence.

Figure 1.5 shows the longitudinal-momentum distributions of recoil ions and emitted electrons for single ionization of He atoms by Se^{28+} at the ion energy of 3.6 MeV/u. The electrons are emitted mainly in the forward direction (positive $P_\|$ values), while recoil ions emerge in the backward direction (negative $P_\|$ values) showing a distinguished forward-backward asymmetry of the ionization fragments. This asymmetry is ascribed to the strong long-range Coulomb potential of the outgoing highly charged Se ions, the so-called *post-collision interaction*, and its direction depends on the sign of the projectile charge (see [12] for details). Evidently, even at this moderate projectile velocity of $v_P = 12$ a.u., (1.70) holds very well, demonstrated here in a kinematically complete measurement where the electron and recoil-ion momenta were measured in coincidence using a Reaction Microscope. The physics seen in these results, namely a strong forward emission characteristics of the electrons, is discussed in detail in Chap. 21.

The longitudinal-momentum $P_{R\|}^f$ distributions of recoil ions created in multiple ionization are shown in Fig. 1.6 for 5.9 MeV/u U^{65+}+Ne collisions emitting up to six electrons. As is seen, recoil ions are ejected with negative longitudinal momenta, i.e., in the backward direction, with a more pronounced feature for high electron multiplicity; thus, electrons are emitted in the forward hemisphere. According to (1.70), the measurement of the longitudinal recoil-ion $P_{R\|}^f$ momenta gives information about the longitudinal

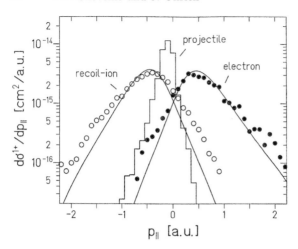

Fig. 1.5. Longitudinal-momentum distributions of recoil He$^+$ ions and emitted electrons for a single ionization of He by 3.6 MeV/u Se^{28+} ion impact. *Symbols* – experimental data obtained using a Reaction Microscope, *solid curves* – classical-trajectory Monte Carlo (CTMC) calculations multiplied by a factor of 1.4; *histogram* in the middle of the figure shows the indirectly measured (1.68) longitudinal-momentum change of the projectiles. Its width is mainly determined by the combined resolution of the electron and recoil-ion longitudinal momenta. From [11]

sum momentum of the emitted electrons. Results shown in Fig. 1.6 represent the first experimental proof that the electron longitudinal sum momentum is directed into the forward hemisphere for multiple ionization by a fast highly charged ion impact.

1.4   Photon–Atom Collisions

1.4.1   Photoeffect

In the case of multielectron ionization by photoabsorption

$$\hbar\omega + A \rightarrow A^{m+} + me^-$$

the momentum and energy conservation laws can be written as [13]:

$$\boldsymbol{P}_\gamma + \boldsymbol{P}_R^i = \boldsymbol{P}_R^f + \sum_{j=1}^{m} \boldsymbol{P}_{ej}^f , \tag{1.71}$$

$$E_\gamma + E_{\text{bind}}^i = E_R + E_{\text{bind}}^f + \sum_{j=1}^{m} E_{ej} , \tag{1.72}$$

where, as before, $E_{\text{bind}}^{i,f} < 0$ is the sum of all binding energies of the electrons before and after the collision, m is the number of emitted electrons, E_R is

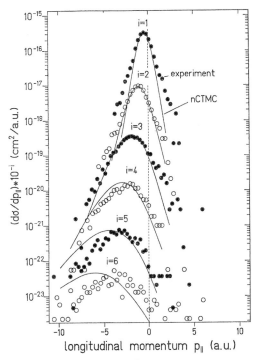

Fig. 1.6. Distributions of the recoil-ion longitudinal momenta $P_{R\parallel}^{f}$ in i-fold ionization of Ne colliding with 5.9 MeV/u U^{65+} ions. *Solid* and *open circles* – experimental data obtained with a high-resolution RIMS, *solid curves* – CTMC calculations; *dashed straight line* in the middle of the figure corresponds to the zero recoil-ion momentum (see Fig. 1.5). From [10]

the kinetic energy of the recoil ion, and $E_\gamma = P_\gamma c$ is the photon energy. As before, the direction of the photon propagation is along the z-axis, so one has for the photon and target momentum (x, y, z)-components:

$$\boldsymbol{P}_\gamma = (0,\, 0,\, E_\gamma/c)\,, \quad \boldsymbol{P}_R^i = (0,\, 0,\, 0)\,. \tag{1.73}$$

For single ionization of He by absorption of one photon

$$\hbar\omega + \mathrm{He} \rightarrow \mathrm{He}^+ + e^-\,, \tag{1.74}$$

the conservation laws are written as:

$$0 = P_{Rx}^{f} + P_{ex}^{f}\,, \tag{1.75}$$

$$0 = P_{Ry}^{f} + P_{ey}^{f}\,, \tag{1.76}$$

$$E_\gamma/c = P_{Rz}^{f} + P_{ez}^{f}\,, \tag{1.77}$$

$$E_\gamma + E_{bind}^{i} = E_R + E_e + E_{bind}^{f}\,. \tag{1.78}$$

Taking into account that in the nonrelativistic approximation, the particle kinetic energy is related to its momentum by (1.13), one has:

$$E_\gamma + E_{\text{bind}}^{\text{i}} - E_{\text{bind}}^{\text{f}}$$
$$= \frac{(P_{\text{R}x}^{\text{f}})^2 + (P_{\text{R}y}^{\text{f}})^2 + (P_{\text{R}z}^{\text{f}})^2}{2M_{\text{R}}} + \frac{(P_{\text{e}x}^{\text{f}})^2 + (P_{\text{e}y}^{\text{f}})^2 + (P_{\text{e}z}^{\text{f}})^2}{2m_{\text{e}}} , \qquad (1.79)$$

where M_{R} denotes the mass of the He$^+$ ion. Using (1.75)–(1.78), one obtains:

$$(P_{\text{R}x}^{\text{f}})^2 + (P_{\text{R}y}^{\text{f}})^2 + \left(P_{\text{R}z}^{\text{f}} - \frac{M_{\text{R}}}{m_{\text{e}} + M_{\text{R}}} \cdot \frac{E_\gamma}{c} \right)^2$$
$$= 2\frac{m_{\text{e}}M_{\text{R}}}{m_{\text{e}} + M_{\text{R}}} \left(E_\gamma + E_{\text{bind}}^{\text{i}} - E_{\text{bind}}^{\text{f}} \right) - \frac{m_{\text{e}}M_{\text{R}}}{(m_{\text{e}} + M_{\text{R}})^2} \left(\frac{E_\gamma}{c} \right)^2 , \qquad (1.80)$$

$$(P_{\text{e}x}^{\text{f}})^2 + (P_{\text{e}y}^{\text{f}})^2 + \left(P_{\text{e}z}^{\text{f}} - \frac{m_{\text{e}}}{m_{\text{e}} + M_{\text{R}}} \cdot \frac{E_\gamma}{c} \right)^2$$
$$= 2\frac{m_{\text{e}}M_{\text{R}}}{m_{\text{e}} + M_{\text{R}}} \left(E_\gamma + E_{\text{bind}}^{\text{i}} - E_{\text{bind}}^{\text{f}} \right) - \frac{m_{\text{e}}M_{\text{R}}}{(m_{\text{e}} + M_{\text{R}})^2} \left(\frac{E_\gamma}{c} \right)^2 . \qquad (1.81)$$

In the case when the photon momentum is small compared to the particle momenta in the final state, $E_\gamma/c \ll P_{\text{R}}^{\text{f}}$, it can be neglected, and one has:

$$|\boldsymbol{P}_{\text{R}}^{\text{f}}|^2 = (P_{\text{R}x}^{\text{f}})^2 + (P_{\text{R}y}^{\text{f}})^2 + (P_{\text{R}z}^{\text{f}})^2 = R^2 , \qquad (1.82)$$

$$|\boldsymbol{P}_{\text{e}}^{\text{f}}|^2 = (P_{\text{e}x}^{\text{f}})^2 + (P_{\text{e}y}^{\text{f}})^2 + (P_{\text{e}z}^{\text{f}})^2 = R^2 , \qquad (1.83)$$

$$R^2 = 2\frac{m_{\text{e}}M_{\text{R}}}{m_{\text{e}} + M_{\text{R}}} \left(E_\gamma + E_{\text{bind}}^{\text{i}} - E_{\text{bind}}^{\text{f}} \right) . \qquad (1.84)$$

Therefore, the absolute values of the recoil-ion and of the ejected electron momenta are equal, and their momentum distributions are given by spheres with the same radius R defined in (1.84) (see Fig. 1.7). Besides, as follows from (1.75)–(1.77), the momentum vectors $\boldsymbol{P}_{\text{R}}^{\text{f}}$ and $\boldsymbol{P}_{\text{e}}^{\text{f}}$ are pointing in opposite directions, i.e., they compensate each other. However, the energies received by the recoil ion and the electron from the photon are not shared equally. From the last two equations one has for the energies:

$$E_R = \frac{|\boldsymbol{P}_{\text{R}}^{\text{f}}|^2}{2M_{\text{R}}} = \frac{2m_{\text{e}}}{m_{\text{e}} + M_{\text{R}}} \left(E_\gamma + E_{\text{bind}}^{\text{i}} - E_{\text{bind}}^{\text{f}} \right) , \qquad (1.85)$$

$$E_e = \frac{|\boldsymbol{P}_{\text{e}}^{\text{f}}|^2}{2m_{\text{e}}} = \frac{2M_{\text{R}}}{m_{\text{e}} + M_{\text{R}}} \left(E_\gamma + E_{\text{bind}}^{\text{i}} - E_{\text{bind}}^{\text{f}} \right) . \qquad (1.86)$$

It is seen that the energy transfer to the recoil ion is $m_{\text{e}}/M_{\text{R}}$ times smaller than that to the ejected electron.

If now one takes into account the photon momentum P_γ, then two differences arise compared to the previous case (see (1.82)–(1.84)). First, the center of the spherical momentum distribution will shift on the z-axis by

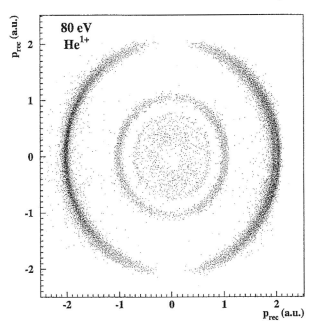

Fig. 1.7. Momentum distribution of He$^+$ recoil ions produced in single ionization of He by 80 eV photons. In the figure, the x-axis is the direction of the electric field vector of the linearly polarized light, the y-axis is the direction of the gas–jet, and $p_{\text{rec}} = P_{\text{R}}^{\text{f}}$. The data are integrated over a momentum range of ± 0.1 a.u. in the z-direction, which is the direction of the photon beam. The *outer ring* corresponds to the He$^+$ ions in the ground state, and the *inner rings* to excited states. From [14]

approximately E_γ/c for the recoil ion and $m_{\text{e}}/M_{\text{R}} \cdot E_\gamma/c$ for the ejected electron, respectively. Second, the radii of the spheres for the ion and electron momentum distributions will shrink.

In order to obtain quantitative insight into the kinematics, let us make some estimates for reaction (1.74) for a photon energy $E_\gamma = 150$ eV assuming that the He$^+$ ion in the final channel is in its ground state. In this case $E_{\text{bind}}^{\text{i}} = -79$ eV, $E_{\text{bind}}^{\text{f}} = -54.4$ eV, $P_\gamma = 0.04$ a.u., $R^2 = 9.22$ a.u., $R = 3.04$ a.u. The change in radii of the spheres

$$\Delta R = \frac{m_{\text{e}} M_{\text{R}}}{(m_{\text{e}} + M_{\text{R}})^2} \left(\frac{E_\gamma}{c}\right)^2 \approx \frac{m_{\text{e}}}{M_{\text{R}}} \left(\frac{E_\gamma}{c}\right)^2 \tag{1.87}$$

for this photon energy is very small ($\approx 2.2 \times 10^{-7}$ a.u.). The shift of the centers of the spheres along the z-axis is: $0.99986 \cdot E_\gamma/c \approx 0.04$ a.u. for the recoil ion and $1.4 \times 10^{-4} \cdot E_\gamma/c \approx 5.4 \times 10^{-6}$ a.u. for the electron, the latter being completely negligible. The deviation of the angle between vectors $\boldsymbol{P}_{\text{R}}^{\text{f}}$ and $\boldsymbol{P}_{\text{e}}^{\text{f}}$ from $180°$ is $0.04/3.04 \approx 0.75°$. Therefore, for low-energy photons the simple equations (1.82)–(1.84) can be used.

1.4.2   Compton Effect

Compton scattering dominates the photoabsorption (1.8) at rather high photon energies. In Compton photoionization,

$$\hbar\omega_i + A \rightarrow A^{m+} + \hbar\omega_f + me^-$$

a high-energy photon may deposit only part of its energy $\hbar\omega_i$ to the ionization of m electrons, and the remaining energy $\hbar\omega_f$ is taken by another photon that is scattered in a new direction. The kinematics of Compton scattering differs significantly from that for photoabsorption (1.71) and (1.72) and can be written as:

$$\boldsymbol{P}^i_\gamma + \boldsymbol{P}^i_R = \boldsymbol{P}^f_\gamma + \boldsymbol{P}^f_R + \sum_{j=1}^{m} \boldsymbol{P}^f_{ej}, \tag{1.88}$$

$$E^i_\gamma + E^i_{bind} = E^f_\gamma + E_R + E^f_{bind} + \sum_{j=1}^{m} E_{ej}, \tag{1.89}$$

where $E^{i,f}_{bind} < 0$ is the sum of all binding energies of the electrons before and after the collision, E_R is the kinetic energy of the recoil ion, $\boldsymbol{P}_\gamma$ is the vector of the photon momentum and $E_\gamma = P_\gamma c = \hbar\omega$ is the photon energy.

Let us consider the case of single-electron Compton scattering:

$$\hbar\omega_i + A \rightarrow \hbar\omega_f + A^+ + e^-.$$

If we put, as before, $P^i_R = 0$ and neglect the electron binding energy value, we will come to the Compton formula for a photon scattering on a quasifree electron. Thus, under these assumptions, the atomic nucleus will play the role of a spectator particle only. However, for the Compton scattering on the bound electrons, the three-body momentum balance has to be considered, assuming that the bound electron has the initial momentum distribution given by its Compton profile $\boldsymbol{P}^{in}_e$. Then, the momentum conservation law is written as

$$\boldsymbol{P}^i_\gamma - \boldsymbol{P}^f_\gamma - \boldsymbol{P}^f_R = \boldsymbol{P}^f_e, \tag{1.90}$$

$$\boldsymbol{P}^f_R = -\boldsymbol{P}^{in}_e. \tag{1.91}$$

Therefore, in Compton photoionization, the recoil-ion momentum reflects the *electron* initial-state momentum distribution $\boldsymbol{P}^{in}_e$. The first measurements of He^+ and He^{2+} recoil-ion momentum distributions were carried out in [15] in ionization of He atoms by 8.8 keV photons. There, in addition to the kinematic sphere for photoabsorption with a radius of $R \approx 30$ a.u. ((1.82) and Fig. 1.7), a narrow recoil-ion momentum distribution centered around $P^f_R = 0$ and resulted from Compton scattering, was observed (see also Chap. 14 for details).

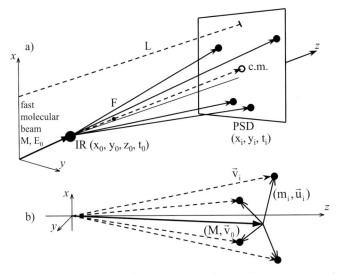

Fig. 1.8. Scheme of a fast-beam translational spectrometer for the investigation of molecular many-body dissociation reactions. (**a**) The incident fast beam of molecules with mass M and energy E_0 dissociates at the position (x_0, y_0, z_0) close to the origin and at time t_0 for negligibly short lifetime of the excited dissociative state (the molecular beam is directed along the z-axis). A fraction of undissociated molecules is labeled by F and usually is stopped by a cup. The reaction fragments are detected in the x, y-plane in coincidence by a time- and a position-sensitive multihit detector (PSD). The positions (x_i, y_i) and the arrival times t_i of the reaction products on the plane $z = L$ are measured. (**b**) Linear superposition of the center-of-mass velocities $\boldsymbol{u}_i$ of $\mathcal{N}$ fragments with masses m_i relative to the velocity $\boldsymbol{v}_0$ of the center-of-mass system. From [16]

1.5   Particle Kinematics: Fragmentation of Molecules

1.5.1   Many-Body Dissociation of Fast Molecular Beams

To initialize fragmentation of molecules, one should first excite the molecules by laser radiation, dissociative charge transfer or by atomic collisions. The final step is the dissociation into $\mathcal{N}$ fragments with masses m_i and velocities $\boldsymbol{u}_i = (u_{ix}, u_{iy}, u_{iz})$, $i = 1, 2, \ldots, \mathcal{N}$ in the center-of-mass system as displayed in Fig. 1.8 for a fast-beam translational spectrometer[1]. IR in the figure denotes the interaction region at the position (x_0, y_0, z_0) and at time t_0. The lateral extension of IR is defined by the initial molecular-beam diameter, and the longitudinal extension is determined by the details of the experiment, e.g., the diameter of the laser beam, the size of the collision cell or by the lifetime of the initial molecular state.

---

[1] In this section, the results of the work [16] and Chap. 4 are used. We omit the index f for velocities in the final state.

The center-of-mass of the molecular beam moves with the velocity $\boldsymbol{v}_0 = (v_{0x}, v_{0y}, v_{0z})$ in the laboratory frame and, due to the small divergence of the fast beam, the transverse components are supposed to be smaller than the longitudinal one:

$$v_{0y} \ll v_{0z}, \quad v_{0x} \ll v_{0z}. \tag{1.92}$$

The part of the initial molecular beam that does not dissociate is indicated by F and is usually blocked by a beam stopper. The reaction fragments are detected in the plane $z = L$ from the origin by a time- and position-sensitive multihit detector (PSD). The absolute flight times of the fragments are given by $t_i - t_0$, where t_0 is the time of molecular dissociation. In most cases, however, t_0 is not known and only the time differences $t_i - t_1$ are measured by the detector, where t_1 is the time when the first fragment hits the detector plane.

If the dissociation takes place at position (x_0, y_0, z_0) and at time t_0, one can write $3\mathcal{N}$ equations of motion for the $\mathcal{N}$ fragments in the form:

$$(v_{0x} + u_{ix})(t_i - t_0) + x_0 = x_i, \tag{1.93}$$

$$(v_{0y} + u_{iy})(t_i - t_0) + y_0 = y_i, \tag{1.94}$$

$$(v_{0z} + u_{iz})(t_i - t_0) + z_0 = L, \tag{1.95}$$

$$i = 1, 2, \ldots, \mathcal{N}.$$

The required velocity vectors $\boldsymbol{u}_i$ can be obtained from this linear set of equations provided other quantities are measurable.

Under *ideal* experimental conditions, the time t_0 of fragmentation is known exactly and

$$x_0 = y_0 = z_0 = 0, \tag{1.96}$$

$$v_{0x} = v_{0y} = 0, \quad v_{0z} = \sqrt{\frac{2E_0}{M}}, \tag{1.97}$$

i.e., the position of fragmentation coincides with the origin, the divergence and energy spread of the primary beam are negligible. Here, M is the mass of the molecules in the primary beam. Under these conditions, one has the solution of the system (1.93)–(1.95):

$$u_{ix} = x_i/(t_i - t_0), \tag{1.98}$$

$$u_{iy} = y_i/(t_i - t_0), \tag{1.99}$$

$$u_{iz} = L/(t_i - t_0) - \sqrt{\frac{2E_0}{M}}. \tag{1.100}$$

To determine the fragment momenta and energies, one has to know the masses m_i of the fragments. The m_i-values, in turn, can be obtained from the momentum and mass conservation laws, which can be written as a system of linear equations:

$$
\begin{pmatrix}
u_{1x} & u_{2x} & \cdots & u_{\mathcal{N}x} \\
u_{1y} & u_{2y} & \cdots & u_{\mathcal{N}y} \\
u_{1z} & u_{2z} & \cdots & u_{\mathcal{N}z} \\
1 & 1 & \cdots & 1
\end{pmatrix}
\begin{pmatrix}
m_1 \\ m_2 \\ \cdots \\ m_{\mathcal{N}}
\end{pmatrix}
=
\begin{pmatrix}
0 \\ 0 \\ 0 \\ M
\end{pmatrix}.
\tag{1.101}
$$

The system of equations (1.101) together with (1.98)–(1.100) allows one to determine the masses m_i of up to $\mathcal{N} = 4$ fragments. If $\mathcal{N} < 4$, the system (1.101) is overdetermined, and the additional information can be used for a consistency check or for reducing the number of equations. If the number of fragments $\mathcal{N} > 4$, there is no unique solution of the system (1.101). In this case, the finite number of permutations a for possible fragment masses $\boldsymbol{m}_{ia}$ is introduced, and for each fragmentation reaction of the parent molecule (which is a finite number), the permutation a is determined from the system (1.101).

Under *real* experimental conditions, the incident molecular beam has a finite divergence, the condition (1.96) is not satisfied but the approximation for the longitudinal velocity component v_{0z} in (1.97) is still a good approximation. The time of fragmentation t_0 can be estimated from:

$$
t_0 = \frac{1}{\mathcal{N}} \sum_{i=1}^{N} t_i - \frac{L}{v_{0z}},
\tag{1.102}
$$

and then, from different permutations a, one searches for the solutions $\tilde{\boldsymbol{u}}_{ia}$ closest to the momentum conservation law in the center-of-mass system

$$
\sum_i m_{ia}\tilde{u}_{ix} = b_x(a),
$$
$$
\sum_i m_{ia}\tilde{u}_{iy} = b_y(a),
\tag{1.103}
$$
$$
\sum_i m_{ia}\tilde{u}_{iz} = b_z(a),
$$

where the right-hand side of the system should be close to zero, $b_x = b_y = b_z = 0$, which corresponds to the ideal case.

In the system (1.93)–(1.95) with the $3\mathcal{N}$ equations of motion, one has $3\mathcal{N} + 6$ unknown quantities: the components of the vector $\boldsymbol{u}_i$, the point of dissociation (x_0, y_0, z_0), the transverse molecular-beam components v_{0x}, v_{0y} and the fragmentation time t_0. The system (1.101) gives three additional relations provided the masses m_i are chosen correctly. The remaining three unknown variables are found from the physical assumptions.

In the crossed-beam experiments, the length of the interaction region is much smaller than the length L and, therefore, one usually put $z_0 = 0$. Two other variables are found from two possible approximations:

1. the lateral extension of the fast molecular beam is neglected, i.e. one puts $x_0 = y_0 = 0$,

2. the beam divergence is neglected, i.e., $v_{0x} = v_{0y} = 0$. Applying one of these approximations, the velocity components $\boldsymbol{u}_i$ can be found for an arbitrary number of molecular fragments $\mathcal{N}$. Among the equations of the system (1.93)–(1.95), the solution of (1.93) for the longitudinal components u_{iz} is the most complicated one. However, it can be found using the *iteration* procedure (see below).

1.5.2   Longitudinal Fragment Velocity Components in the Approximation $z_0 = 0$

Putting $z_0 = 0$ and taking into account that in the center-of-mass system the fragment velocities u_i are typically much smaller than the velocity of the incident molecular beam, i.e.,

$$u_i \ll v_{0z}, \tag{1.104}$$

one can obtain from (1.95) the arrival–time difference between the i–th and the *first* fragments:

$$t_i - t_1 = L \frac{u_{1z} - u_{iz}}{(v_{0z} + u_{iz})(v_{0z} + u_{1z})}. \tag{1.105}$$

The corresponding velocity difference is given by

$$d_{iz} \equiv u_{1z} - u_{iz} = (t_i - t_1)(v_{0z} + u_{iz})(v_{0z} + u_{1z})/L, \tag{1.106}$$
$$i = 2,\ldots,\mathcal{N},$$

which has a weak dependence on the u_{iz} values because of the condition (1.104). The substitution of (1.106) into momentum and mass conservation equations (1.101) gives

$$u_{1z} = \sum_{i=2}^{\mathcal{N}} m_i d_{iz}/M, \quad u_{iz} = u_{1z} - d_{iz}, \quad i = 2,...,\mathcal{N}. \tag{1.107}$$

Now we have two systems of (1.106) and (1.107) to find the z-components of the molecular fragments u_{iz}, $i \geq 2$ from known values of the flight length L, the longitudinal primary beam velocity v_{0z} and the time differences $t_i - t_1$. These equations are solved using the iteration method by substituting the k-th iteration, $u_{iz}^{(k)}$, with the $(k-1)$-th one and, finally, for $k = 1$ with the initial one $u_{iz}^{(0)} = 0$. With parameters typical for the fast-beam experiment, the series of $u_{iz}^{(k)}$ converges.

1.5.3   Transverse Fragment Velocity Components in the Approximation of Zero-Beam Extension

In the approximation of zero extension of the beam

$$x_0 = y_0 = z_0 = 0, \tag{1.108}$$

one can obtain the transverse velocity components u_{ix} and u_{iy} knowing the longitudinal ones u_{iz}. Using (1.108), one has from (1.93)–(1.95):

$$v_{0x} + u_{ix} = x_i/\tilde{t}_i, \tag{1.109}$$
$$v_{0y} + u_{iy} = y_i/\tilde{t}_i, \tag{1.110}$$
$$\tilde{t}_i \equiv (t_i - t_0) = L/(v_{0z} + u_{iz}). \tag{1.111}$$

From the momentum conservation law (1.101) one has:

$$v_{0x} = \sum_i m_i x_i/(\tilde{t}_i M), \tag{1.112}$$

$$v_{0y} = \sum_i m_i y_i/(\tilde{t}_i M), \tag{1.113}$$

which gives the fragment velocity components in the center-of-mass frame:

$$u_{ix} = x_i/\tilde{t}_i - \sum_i m_i x_i/(\tilde{t}_i M), \tag{1.114}$$

$$u_{iy} = y_i/\tilde{t}_i - \sum_i m_i y_i/(\tilde{t}_i M). \tag{1.115}$$

1.5.4   Transverse Fragment Velocity Components in the Approximation of Zero–Beam Divergence

In the approximation of a zero beam divergence but finite interaction region, i.e.,

$$v_{0x} = v_{0y} = 0, \tag{1.116}$$
$$x_0 \neq 0, \quad y_0 \neq 0, \tag{1.117}$$

the transverse velocity components in the center-of-mass frame can also be obtained quite easily from (1.93) and (1.94):

$$u_{ix} = (x_i - x_0)/\tilde{t}_i, \tag{1.118}$$
$$u_{iy} = (y_i - y_0)/\tilde{t}_i, \tag{1.119}$$

where $\tilde{t}_i$ is defined in (1.111).

1.5.5   Fragmentation into Two Particles with Equal Masses

In the case of molecular dissociation into two particles ($\mathcal{N} = 2$), the fragment velocity components can be found, provided some additional assumptions are made. For two fragments with masses m_1 and m_2, the momentum conservation equations ((1.101)) give for the z-components of the velocities:

$$u_{1z} = m_2 d_z/M, \quad u_{2z} = -m_1 d_z/M, \quad M = m_1 + m_2, \tag{1.120}$$

where $\boldsymbol{d}$ denotes the relative velocity

$$\boldsymbol{d} = \boldsymbol{u}_1 - \boldsymbol{u}_2 \,. \tag{1.121}$$

Using approximations

$$v_{0y} = 0 \,, \quad t_1 - t_0 \approx L/v_{0z} \,, \tag{1.122}$$

one obtains from (1.94) for the mass ratio:

$$m_2/m_1 = -\frac{u_{1y}}{u_{2y}} \approx \left| \frac{y_1}{y_2} \left(1 - \frac{v_{0z}\Delta T}{L} \right) \right| \,, \quad \Delta T = t_2 - t_1 \,. \tag{1.123}$$

Equation (1.106) together with (1.120) gives the quadratic equation for the velocity difference d_z/v_{0z}:

$$\begin{aligned}
\frac{d_z}{v_{0z}} &= -\frac{\Delta T \cdot v_{0z}}{L} \left(1 + \frac{u_{1x}}{v_{0z}} \right) \left(1 + \frac{u_{2z}}{v_{0z}} \right) \\
&= \frac{\Delta T \cdot v_{0z}}{L} \left[1 + \frac{m_2 - m_1}{M} \cdot \frac{d_z}{v_{0z}} - \frac{m_1 m_2}{M^2} \left(\frac{d_z}{v_{0z}} \right)^2 \right] \,,
\end{aligned} \tag{1.124}$$

which can be solved directly or iteratively relative to d_z.

In the approximation of the vanishing extension of the interaction region (1.108), one obtains the transverse velocity v_{0y}

$$v_{0y} = (m_1 y_1 + m_2 y_2)\frac{v_{0z}}{ML} + (y_1 - y_2)d_z \frac{m_1 m_2}{M^2} \,, \tag{1.125}$$

and the d_y-component, respectively:

$$d_y = (y_1 - y_2)\frac{v_{0z}}{L} + (m_1 y_2 + m_2 y_1)\frac{d_z}{ML} \,. \tag{1.126}$$

In the case of vanishing beam divergence (1.116), similar equations for the velocity–difference d-components can be easily obtained, and the difference $y_1 - y_2$ can be found from (1.111), without determination of y_0 and v_{0y} that gives

$$y_1 - y_2 = d_y v_{0z} \Delta T / d_z \,. \tag{1.127}$$

In the case of molecular dissociation into two particles, the result (1.124) leads to the same expression for the kinetic energy release (KER) as was obtained in [17]. Actually, the KER value is defined as

$$W_{\text{KER}} = \frac{1}{2}\mu d^2 = \frac{1}{2}\frac{m_1 m_2}{m_1 + m_2} d^2 = \frac{m_1 m_2}{2M}(d_x^2 + d_y^2 + d_z^2) \,. \tag{1.128}$$

Making consequent iterations $d_z^{(k)}$ in (1.124)

$$\begin{aligned}
&d_z^{(0)}/v_{0z} = 0 \,, \\
&d_z^{(1)}/v_{0z} = \frac{v_{0z}\Delta T}{L} \,, \\
&d_z^{(2)}/v_{0z} = \frac{v_{0z}\Delta T}{L} \left[1 + \frac{m_2 - m_1}{M}\frac{v_{0z}\Delta T}{L} - \frac{m_1 m_2}{M^2} \left(\frac{v_{0z}\Delta T}{L} \right)^2 \right] \,,
\end{aligned} \tag{1.129}$$

one obtains

$$W_{\text{KER}} = \frac{m_1 m_2}{2ML} v_{0z}^2 \left(\Delta x^2 + \Delta y^2 + (v_{0z}\Delta T)^2 \right),$$

$$\times \left(1 + \frac{m_2 - m_1}{M} \frac{v_{0z}\Delta T}{L} + \ldots \right)^2 \tag{1.130}$$

$$\approx E_0 \frac{m_1 m_2}{M^2 L^2} \left(\Delta x^2 + \Delta y^2 + (v_{0z}\Delta T)^2 \right) \left(1 + 2\frac{m_2 - m_1}{M} \frac{v_{0z}\Delta T}{L} \right).$$

$$\Delta y = y_1 - y_2, \quad \Delta x = x_1 - x_2, \quad E_0 = Mv_{0z}^2/2. \tag{1.131}$$

Equation (1.130) corresponds to the equation for the kinetic energy release given in [17].

In the case of molecular dissociation into three fragments, $\mathcal{N} = 3$, with three equal masses $m_i = M/3$, this calculation technique was successfully applied for the three-body decay of the triatomic hydrogen molecule H_3 [18].

References

1. J. Ullrich, R. Moshammer, R. Dörner, O. Jagutzki, V. Mergel, H. Schmidt-Böcking, L. Spielberger: J. Phys. B **30**, 2917 (1997)
2. R. Dörner, V. Mergel, O. Jagutzki, L. Spielberger, J. Ullrich, R. Moshammer, H. Schmidt-Böcking: Phys. Rep. **330**, 95 (2000)
3. R. Dörner, Th. Weber, M. Weckenbrock, A. Staudte, M. Hattass, R. Moshammer, J. Ullrich, H. Schmidt-Böcking: *Multiple Ionization in Strong Laser Fields*, Adv. At. Mol. Opt. Phys. **48**, 1 (2003)
4. V. Mergel: PhD Thesis, Universität Frankfurt (Shaker, Aachen 1996)
5. D. Fischer, B. Feuerstein, R.D. DuBois, R. Moshammer, J.R. Crespo López-Urrutia, I. Draganic, H. Lörch, A.N. Perumal, J. Ullrich: J. Phys. B **35**, 1369 (2002)
6. M-C. Buchet-Poulizak, P.O. Bogdanovich, É.J. Knystautas: J. Phys. B **34**, 233 (2001)
7. J.D. Gillaspy (ed): *Trapping Highly Charged Ions: Fundamentals and Applications* (Nova Science Publishers, New York 2001)
8. HITRAP project: http://www-linux.gsi.de/~hitrap/
9. M. Schulz, R. Moshammer, W. Schmitt, H. Kollmus, R. Mann, S. Hagmann, R.E. Olson, J. Ullruch: J. Phys. B **32**, L557 (1999)
10. M. Unverzagt, R. Moshammer, W. Schmitt, R.E. Olson, P. Jardin, V. Mergel, J.Ullrich, H. Schmidt-Böcking: Phys. Rev. Lett. **76**, 1043 (1996)
11. R. Moshammer, J. Ullrich, H. Kollmus, W. Schmitt, M. Unverzagt, H. Schmidt-Böcking, C.J. Wood, R.E. Olson: Phys. Rev. A **56**, 1351 (1997)
12. C.J. Wood, R.E. Olson: J. Phys. B **29**, L257 (1996)
13. T. Vogt: Diploma Thesis (Johann-Wolfgang-Goethe Universität, Frankfurt 1996)
14. R. Dörner, V. Mergel, L. Spielberger, M. Achler, Kh. Khayyat, T. Vogt, H. Bräuning, O. Jagutzki, T. Weber, R. Moshammer, M. Unverzagt, W. Schmitt, H. Khemliche, M.H. Prior, C.L. Cocke, J. Feagin, R.E. Olson, H. Schmidt-Böcking: Nucl. Instrum. Methods B **124**, 225 (1997)

15. L. Spielberger, O. Jagutzki, R. Dörner, J. Ullrich, U. Meyer, V. Mergel, M. Unverzagt, M. Damrau, T. Vogt, I. Ali, Kh. Khayyat, D. Bahr, H.G. Schmidt, R. Frahm, H. Schmidt-Böcking: Phys. Rev. Lett. **74**, 4615 (1995)
16. M. Beckert, U. Müller: Eur. Phys. J. D **12**, 303 (2000)
17. D.P. De Bruin, J. Los: Rev. Sci. Instrum. **53** 1020 (1982)
18. U. Müller, Th. Eckert, M. Braun, H. Helm: Phys. Rev. Lett. **83**, 2718 (1999)

2 Recoil-Ion Momentum Spectroscopy and "Reaction Microscopes"

R. Moshammer, D. Fischer, and H. Kollmus

2.1 Introduction

The rapid and still ongoing development of recoil-ion momentum spectroscopy (RIMS) during the last ten years can undoubtedly be viewed as an experimental breakthrough for the investigation of any kind of atomic reaction dynamics. Whenever atoms or simple molecules interact with electrons, ions or photons, the concept of high–resolution recoil-ion measurements resulted in additional and complementary information compared to the traditional electron spectroscopy methods, and in some cases even kinematically complete data sets could be collected for the very first time. State–of–the–art high-resolution recoil-ion momentum spectrometers evolved through numerous technical developments like, e.g., the implementation of cold supersonic-jet targets, the use of well defined electric extraction fields for recoil ions as well as for electrons and the rapid progress in charged particle detection techniques. Among them the use of supersonic jets to produce well localized and internally cold targets (COLd Target Recoil-Ion Momentum Spectroscopy COLTRIMS) can be viewed as the most important ingredient. They allowed one to achieve a recoil-ion momentum resolution far below 1 a.u. (atomic unit) that would be impossible with room-temperature targets due to the thermal motion (the momentum spread of room-temperature helium atoms is about 3.7 a.u.). Another decisive development was the invention of completely novel and extremely efficient electron–imaging concepts. In combination with COLTRIMS they enabled the detection of recoil ions and electrons in coincidence and opened up a whole area of kinematically complete atomic–reaction studies.

   In this chapter, we discuss the working principle of these modern spectrometers and try to give some guidance for construction. We briefly discuss their range of applications, but we leave out a detailed historical overview (for this the reader is referred to [1–3]). Our aim is to introduce the reader to the basics of COLTRIMS and Reaction Microscopes.

2.2 Imaging Spectrometers for Ions

A recoil-ion momentum spectrometer is a high–precision device designed to measure recoil momenta corresponding to kinetic energies of the order of a few meV with μeV resolution. In general, the recoil-ion momentum range of interest depends on the reaction channel to be studied. For example, in photoionization at photon energies just above the ionization threshold the recoil momentum will be extremely small, it essentially reflects the ejected electron momentum. In violent collisions with highly charged ions or when Coulomb explosion of molecules is being studied, the ionic target fragments sometimes achieve kinetic energies in the eV range. From most atomic reactions, however, the created target ions emerge with typical momenta of 1 a.u. or lower (one atomic unit in momentum is the mean translational momentum of an electron bound in the ground state of a hydrogen atom). This corresponds to a kinetic energy of only 1.8 meV for a helium target.

In order to measure the magnitude and the direction of the recoil-ion momentum vector, an experimental arrangement is needed that performs an unambiguous mapping of these quantities onto experimentally accessible observables. One might think of using energy or momentum dispersive spectrometers like those extensively used in traditional electron spectroscopy. But, in spite of their excellent resolution, they suffer from transmission efficiency and, thus, they are inadequate for coincidence measurements. The simplest device to fulfil the requirements of sufficient momentum resolution and large angular and momentum acceptance would be a position–sensitive detector placed at some distance in a field–free environment viewing the point–like reaction volume. Then, simply from the position and time-of-flight (TOF) information of each detected ion, the trajectory can be reconstructed and the initial momentum vector can be calculated. The TOF has to be measured with respect to a trigger signal that uniquely defines the time of interaction of a projectile with a single target atom. To do so, either a pulsed beam of projectiles (synchrotron radiation, pulsed lasers, bunched electron or ion beams) has to be used, or single projectiles of a continuous beam have to be detected with a time–sensitive detector after the collision. In fact, spectrometers of this type have been applied in the pioneering era of recoil-ion momentum spectroscopy [4]. But, this simple projection technique has some unacceptable drawbacks. The recoil-ion angular acceptance is limited and no direct information about the ionic charge is obtained. Moreover, unavoidable contact potentials can cause major problems. These difficulties are largely circumvented by applying a constant and weak homogeneous electric field to push the ions towards the detector [5]. In this way, the angular and momentum acceptance is considerably increased and as long as the voltages applied at the spectrometer are much larger than typical contact potentials their influence and that of external stray fields is very weak. A further very important advantage is that with field extraction the TOF yields information not only about the ion momentum but, in addition, it allows one to determine the ion species because the TOF depends also on the ratio of ion mass over charge. In conclusion, a standard RIMS Spectrometer consists of a region with a homogeneous electric extraction field where recoil ions emerging from

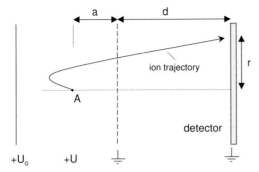

Fig. 2.1. Schematics of a recoil-ion momentum spectrometer

an ideally point–like source volume are accelerated. Afterwards they pass a field-free drift region, for reasons that will be discussed later, before they are detected by means of a position–sensitive ion detector.

To discuss the working principle in more detail let us consider a recoil-spectrometer as shown schematically in Fig. 2.1. Let us assume that at the position A target atoms of mass M are ionized due to any type of atomic reaction. During this interaction they acquire a small initial kinetic energy of $E_\parallel = P_\parallel^2/(2M)$, where $E_\parallel$ and $P_\parallel$ are the ion energy and momentum, respectively, along the spectrometer axis. This is the quantity we aim to measure. To extract the ions in a defined way a homogeneous electric field is superimposed in the reaction zone by applying, for example, a potential between a conductive back-plate and a grid or mesh. The resulting effective potential between the source point A and the grid is assumed to be U. Due to acceleration in the field the recoil ions gain an additional kinetic energy of qU, where q is the ion charge, before they pass through the grid. After traveling through a drift region (length d) they are registered in a position-sensitive detector. The recoil-ion time-of-flight as a function of their small initial energy $E_\parallel$ along the spectrometer axis is given by

$$t_{+/-}(E_\parallel) = f \cdot \sqrt{M} \cdot \left[\frac{2a}{\sqrt{E_\parallel + qU} \pm \sqrt{E_\parallel}} + \frac{d}{\sqrt{E_\parallel + qU}} \right], \qquad (2.1)$$

where the "+" sign stands for those ions that are emitted in the forward direction, i.e., toward the detector and the "−" sign for those with an initial backward velocity. Equation (2.1) yields the TOF in ns, where the ion mass M is expressed in amu (atomic mass units $1\,\mathrm{amu} = \frac{1}{12}\mathrm{C}_6^{12}$), distances in cm, energies in eV and the factor $f = 719.7 \times \sqrt{\mathrm{eV} \cdot \mathrm{ns/cm}}$ results from the conversion of units. The two terms in (2.1) correspond to the time the ions spend in the acceleration and drift region, respectively.

Usually, the extraction potential qU is much larger than $E_\parallel$ and therefore different ion species and/or charge states appear as well separated peaks in the TOF spectrum. Then, and if the recoil ions emerge with equal probability

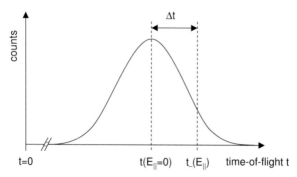

Fig. 2.2. Schematic recoil-ion time-of-flight distribution (TOF peak)

in the forward and backward direction, the position of the corresponding peak is given by

$$t_0 = t(E_\parallel = 0) = f \sqrt{M/qU} \cdot (2a + d) \tag{2.2}$$

and its shape reflects the initial energy or momentum distribution, respectively. This is true as long as $qU \gg E_\parallel$. In this case a linear expansion of $t(E_\parallel)$ (2.1) around t_0 can be used to determine the initial energy. For large initial kinetic energies, i.e. when $E_\parallel$ becomes comparable to qU, this approximation can cause unacceptable distortions. This happens most likely when energetic ions from molecular Coulomb-explosion or electrons are imaged. In this case (2.1) has to be solved numerically to extract $E_\parallel$ as a function of time. For all atomic reactions, however, where slow ions emerge, it can be easily shown that those arriving with an time offset Δt with respect to t_0 had a initial kinetic energy of

$$E_\parallel = \frac{1}{4M} \cdot \left(\frac{q \cdot U \cdot \Delta t}{a \cdot f} \right)^2. \tag{2.3}$$

This follows directly from solving $2\Delta t = (t_-(E_\parallel) - t_+(E_\parallel))$ using (2.1). With (2.3) it is possible to assign to each channel in the TOF spectrum the corresponding recoil-ion momentum (P in a.u., qU in eV, a in cm and Δt in ns)

$$P_\parallel = 8.042 \cdot 10^{-3} \times \frac{q \cdot U \cdot \Delta t}{a}. \tag{2.4}$$

For flight times shorter than t_0 the initial ion–velocity vector was pointing towards the detector and *vice versa*. Interestingly, this relation is independent of the particle mass, it is valid for ions as well as electrons. Moreover, only the knowledge about the electric field strength U/a at the source point is required to calculate the momentum component along the spectrometer axis and no further information about the overall geometry of the spectrometer is needed. Thus, with a time resolution of 1 ns in the TOF measurement under otherwise ideal conditions, a momentum resolution below 0.01 a.u. is achievable when a field of 1 V/cm is used for extraction.

2.2.1 Time Focusing

The above discussion is based on the assumption of a point–like source volume. This is fairly unrealistic. An extended reaction zone over a certain width Δa has two consequences. First, the ions start at different potentials and secondly, they travel over different acceleration distances $a + \Delta a$. According to (2.1) both effects give rise to an effective time jitter δt as a result of the target extension of

$$\frac{\delta t}{t} = \frac{2a - d}{4a + 2d} \cdot \frac{\Delta a}{a} . \tag{2.5}$$

Thus, for a spectrometer without drift ($d = 0$) where He^+ ions are extracted by a field of $1\,V/cm$, a source volume size of only $1\,mm$ results in a time jitter of about $45\,ns$. According to (2.4) this corresponds to a momentum uncertainty of more than $0.3\,a.u.$, demonstrating that a device with pure extraction only is inappropriate for momentum spectroscopy. On the other hand, if a drift path with a length of $d = 2a$ is introduced then the influence of the target size can be eliminated completely (2.5), at least to first order. This is the so-called *time-focusing* condition for spectrometers with one acceleration and one drift region [6].

A similar condition can be found for other combinations of several acceleration and drift regions. A further conclusion arises from (2.5), namely that the influence of the source volume extension on the momentum resolution is reduced if the geometrical size of the spectrometer is increased. This is not surprising, but it should be taken into account in the design of time-of-flight spectrometers. In the construction of spectrometers it is worth spending effort in order to really fulfil the time-focusing condition. If the drift path of our example spectrometer ($10\,V$ extraction over $10\,cm$ plus $20\,cm$ drift) were too short by only $1\,cm$ then the momentum resolution would be limited to be not better than $0.025\,a.u.$ An even stronger influence arises from a not well adapted acceleration distance, which is sometimes difficult to determine experimentally. It requires an exact knowledge of the location where the projectile beam interacts with the target-gas beam. Spectrometers with more than one acceleration region represent an alternative because they allow one to adjust the time-focusing condition via the applied voltages even for unknown or shifted target positions.

2.2.2 Reconstruction of Momentum Components

The recoil-ion longitudinal momentum, i.e., the component along the electric field axis, is unambiguously related with its time-of-flight. An example is shown in Fig. 2.3 for ionization of Ar in intense laser fields. The ions were extracted with a field of $1\,V/cm$ using a spectrometer with time focusing. In this specific case the Ar^+ ion has to balance the momentum of the ionized electron. This follows from momentum conservation for a negligibly small momentum carried by the absorbed photons. The result of the conversion from

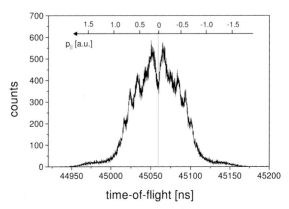

Fig. 2.3. TOF distribution of Ar^+ ions for single ionization with strong 25–fs laser pulses (at a light intensity in the upper 10^{13} W/cm^2 range). The laser polarization direction is parallel to the TOF axis. The appearance of individual peaks reflects the absorption of single 1.5–eV photons above the ionization threshold

TOF into $P_\parallel$ is shown in the inset of Fig. 2.3 demonstrating that structures in the order of 0.1 a.u. are clearly resolved. This corresponds to a kinetic energy of only 1.8 µeV. The two missing momentum components perpendicular to the field vector can be determined from the position of impact on the ion detector. Thus, all three components of the recoil-ion momentum vector are determined by a coincident measurement of the two position coordinates and the time-of-flight for each ion.

In the following, the reconstruction of the transverse momentum $P_\perp$ will be discussed. For simplicity we first assume again a point–like source and try to calculate the trajectory of an ion that is emitted in an arbitrary direction with an energy E. To stay with our notation, we split this energy into two parts, $E_\parallel$ and $E_\perp$, where $E_\perp = P_\perp^2/(2M)$ is the initial kinetic energy in the direction perpendicular to the electric field vector. Ions with $E_\perp = 0$ would hit the detector exactly in the center, while in general they are registered with a certain displacement r (see Fig. 2.1), which is given by

$$r = \frac{1}{f} \cdot \frac{\sqrt{E_\perp}}{\sqrt{M}} \cdot t(E_\parallel)\,, \tag{2.6}$$

where M is given in amu, E in eV and t in ns, r in cm. The radial deflection r at the detector depends on both the transverse and the longitudinal momentum, resulting in a very bulky expression if (2.1) were inserted. A considerable simplification is achieved if the mean flight time $t(E_\parallel = 0)$ of the specific ion is inserted instead of $t(E_\parallel)$. This approximation is very well justified as long as the width Δt of the ion TOF peak is much smaller than the mean flight time or, in other words, as long as the kinetic energy gained due to the extraction is significantly larger than the initial recoil-ion energy. With (2.2) we get

$$r = (2a + d) \cdot \sqrt{\frac{E_\perp}{qU}} \,. \tag{2.7}$$

From this it follows immediately (P being in a.u., M in amu, qU in eV, and distances in cm)

$$P_\perp = 11.6 \cdot \frac{r}{(2a + d)} \cdot \sqrt{qU \cdot M} \,. \tag{2.8}$$

Hence, from the two spatial coordinates measured with the position–sensitive detector the recoil-ion momentum components in the plane of the detector surface are obtained. Together with the TOF information the full ion momentum vector can be reconstructed.

The achievable momentum resolution depends on the position resolution Δr of the detector and, for more realistic situations, on the extension of the target zone. In many situations the latter is dominating. For our example spectrometer with time-focusing geometry and 10 eV extraction over a =10 cm a target extension of $\Delta r = 1$ mm causes a transverse momentum uncertainty of $\Delta P_\perp = 0.18$ a.u. for He ions. Keeping the concept of an homogeneous extraction field, the resolution can be improved only at the expense of a reduced transverse momentum acceptance either by lowering the applied voltage or by increasing the spectrometer size.

2.2.3 Spectrometers with Position–Focusing

To circumvent the problem of limited resolution due to the reaction volume extension, which is in the mm range even if well collimated supersonic gas–jet targets are used, spectrometers with so–called position focusing have been developed [7,8]. They focus ions starting at different positions onto a single spot on the detector, while the displacement on the detector is still proportional to the initial momentum. Thus, the resolution is not limited by the source extension but by the imaging properties of the spectrometer. To achieve position focusing, a weak electrostatic lens is implemented in the acceleration region, preferably as close as possible to the reaction zone. Using a stack of ring electrodes to generate the extraction field, either a homogeneous or a lensing field can be realized by applying appropriate potentials to the electrodes. However, the implementation of a lens in the extraction region modifies the focusing properties in the third, the time-of-flight direction. To obtain both position and time focusing, a longer drift tube compared to spectrometers with homogeneous extraction is required. A proper tuning of the spectrometer geometry and field parameters is necessary to ensure that the focal points for the time-of-flight and for the spatial focusing coincide exactly at the detector. Spectrometers with position focusing have been used mainly to study electron-capture reactions in ion–atom collisions where the recoiling target ions acquire different but discrete momenta parallel to the beam direction corresponding to different final electronic states of the captured electron [8]. To resolve these lines in the momentum spectrum a high resolution is decisive.

2.2.4   Electric–Field Distortions and Calibration

Besides the source volume extension and the finite position and time resolution, otherwise idealized conditions have been assumed up to this point. In real spectrometers, distortions due to electric stray and fringe fields, contact potentials, imperfect geometry, and external magnetic fields have to be considered. Unavoidable and sometimes severe distortions are produced when grids with large gap widths are used to separate the extraction part from the drift region. The openings of the grid act as individual lenses and even small deflections of passing ions can result in magnified images of the grid structure at the detector destroying the unique assignment of ion momentum and position of detection. Moreover, field penetration through grids into the field-free drift tube can cause a significant change of the spectrometer imaging properties and a breakdown of the time-focusing condition. This is hardly avoidable because after the drift part a region of very high field strength is required to postaccelerate the ions up to several keV in front of the detector in order to achieve a reasonable detection efficiency. Depending on the gap width of the grid used to shield the drift part from the postacceleration region a more or less prolonged drift path is required to compensate the effect of field penetration. Grids with small gaps are favored but, on the other hand, very fine grids suffer from low open areas and therefore reduced of the ion transmission. Grids with a wire spacing between 50 to 250 µm represent a reasonable compromise. In any case, due to the lensing effect the wire spacing represents an upper limit of the achievable position resolution of the detector.

In some situations it is not trivial to assign the point of zero transverse momentum to a certain position coordinate on the detector. One can either calibrate the spectrometer with well–known reactions, like electron capture, or one chooses a orientation of the spectrometer in such a way that the extraction is parallel to a symmetry axis of the physical process to be studied. In the latter case, the center of gravity of the ion position distribution corresponds to zero momentum. Possible outstanding axes of atomic reactions are the polarization direction in photoionization or the beam axis for charged–particle impact. In general, it is advantageous to make use of these symmetries, because it facilitates interpretation and cross checking of the experimental data. After having determined the zero point the spectra have to be calibrated.

This can be done in several ways. In particular, spectrometers with position focusing have to be calibrated either by relying on calculated calibrations (using, e.g., the SIMION program) or via capture measurements. In most cases, and in particular for well–known spectrometer geometries and when homogeneous fields are used, it is precise enough (within a few per cent) to apply the calculated dependence of initial momentum and position for calibration.

2.3 Target Preparation

A prerequisite for high-resolution recoil-ion momentum spectroscopy is a cold-atomic or molecular-gas target because in many cases, depending on the physical process to be studied, the recoil-ion momenta are of the order of or even smaller than the thermal momentum spread at room tempera-ture. Moreover, and as already discussed, a small target size is indispensable for momentum spectroscopy with imaging spectrometers. In almost all exist-ing devices both requirements are realized with supersonic expansion of the target gas to form a well localized and cold-atomic beam.

2.3.1 Supersonic Jets

When a gas is pressed through a small nozzle into a reservoir at low pressure it is accelerated to supersonic speed at the exit of the nozzle and an atomic beam is formed as long as the pressure ratio is larger than about two (for more detailed information see [9]). During this adiabatic expansion the free enthalpy $H = 5/2k_\mathrm{B}T_0$ (k_B being the Boltzmann constant and T_0 the gas temperature before expansion) of an ideal gas is converted into a directed motion at the expense of the internal temperature. Under ideal conditions no internal atomic motion is left over in a frame where the atoms (with mass M) move with a momentum of

$$p_\mathrm{jet} = \sqrt{5k_\mathrm{B}T_0M}\,. \tag{2.9}$$

For helium gas at room temperature we get $P_\mathrm{jet} = 5.9$ a.u. corresponding to a translational kinetic energy of 65 meV. In practice, these ideal values are not reachable but some internal temperature is left in the gas–jet. This is expressed by the speed ratio S, the ratio of the translational jet velocity to the thermal velocity spread

$$S = \sqrt{\frac{5T_0}{2T}}\,, \tag{2.10}$$

where T is the internal temperature after expansion. The speed ratio depends on the gas species, the gas temperature, and on the product of the driving pressure and the nozzle diameter $p_0 \cdot d$.

Reasonable gas-jet properties are obtained for $p_0 \cdot d > 1$ torr cm. Below this value the regime of supersonic flow behind the nozzle extends over a distance of only a few nozzle diameters, which is usually not sufficient to extract a supersonic beam. For $p_0 \cdot d > 10$ torr·cm clustering effects are cer-tainly a concern for pure gases at room temperature. In this respect, helium

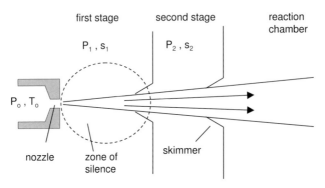

Fig. 2.4. Schematic representation of a two-stage supersonic gas–jet

is an exception and much higher driving pressures can be applied resulting in large speed ratios or low internal jet temperatures, respectively. To produce a helium gas-jet with a speed ratio of $S = 50$, corresponding to an internal jet temperature of $T = 0.3\,\mathrm{K}$, by expanding the gas at $T_0 = 300\,\mathrm{K}$ through a $30\,\mu\mathrm{m}$–diameter nozzle a driving pressure of about 25 bar is required. Under similar conditions but with precooling of the gas before expansion the jet velocity is reduced and considerably larger S values can be obtained. In practice, however, the achievable jet performance is mainly limited by the pumping speed in the first stage. One or several stages with differential pumping are used to handle the enormous gas-load and to maintain a good vacuum in the reaction chamber. To extract a geometrically well defined supersonic beam a small "skimmer" aperture is placed behind the nozzle and between the different pumping stages (Fig. 2.4). The most critical values in the design of a gas–jet are the pumping speed in the first stage and the nozzle-skimmer distance. At the nozzle exit a zone of supersonic flow bordered by a compression or shock-front is formed. A high quality gas–jet can be cut out if the edge of the skimmer immerses into this so-called "zone-of-silence", which typically extends over a range of several mm up to cm, strongly depending on the background gas pressure p_1 in the first stage. The performance is quite insensitive to the nozzle–skimmer distance and to the skimmer shape as long as a high vacuum of $p_1 < 10^{-3}$ torr is maintained. To achieve this, however, huge pumps are required. For our He-jet example either a $10000\,\mathrm{l/s}$ pump would be needed or the driving pressure has to be reduced equivalently, resulting in a lower speed ratio and lower beam intensity. The pressure rise (in torr) in the first stage can be estimated from the balance of pumping speed s_1 (in l/s) and gas throughput by

$$s_1 p_1 = C \left(\frac{300}{T_0}\right)^{3/2} p_0 d^2 , \qquad (2.11)$$

where C (in $\mathrm{l/cm^2/s}$) is a gas–dependent factor ranging from about $C = 10$ for heavy gases up to $C \approx 50$ for He and molecular hydrogen. Accordingly,

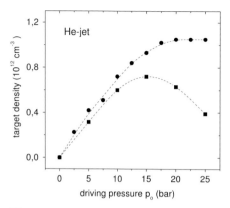

Fig. 2.5. Density of a He gas-jet for a pumping speed of 500 l/s (*circles*) and 250 l/s (*squares*) in the first stage. The nozzle diameter is 30 μm

a background pressure in the 10^{-2} torr range is achievable with a modern 500 l/s turbo pump. In this case, the zone-of-silence typically extends only over a few mm and the skimmer shape and, in particular, the nozzle–skimmer distance becomes decisive. At lower pumping speeds, and therefore higher background pressures, the zone-of-silence extension is reduced further. However, only if the skimmer tip dives into the zone–of–silence a well defined and cold supersonic beam can be extracted. The influence of the pumping speed on the jet density is demonstrated in Fig. 2.5. The gas density at the target position, i.e., 10 cm downstream from the nozzle, was measured as a function of the He driving pressure for different pumping speeds but with a fixed nozzle–skimmer distance [10]. The latter was optimized for the 500 l/s pump. The degrading of the jet performance at high pressures with the smaller pump could have been avoided by reducing the nozzle–skimmer distance slightly. Therefore, the possibility to easily adjust the nozzle position should be foreseen in the construction of a jet device, in order to obtain the best performance for different gas species and driving pressures.

For most applications a dense and narrow atomic beam is required, which is crossed with a projectile beam in the interaction chamber. Without too much effort supersonic gas beams with a particle density in the order of 10^{11} to 10^{12} cm^{-3} at a distance of about 10 cm away from the nozzle are reachable. It is obvious that this number decreases with the inverse square of the distance between the nozzle and the interaction zone. What finally determines the event rate in crossed-beam experiments is the target-area density. This increases with the geometrical size of the gas beam. The latter is defined by the nozzle–skimmer geometry, i.e., the skimmer orifice and the nozzle–skimmer distance. A small interaction volume, which is given by the overlap between the projectile and target beam, can be obtained with narrow gas beams resulting in a high momentum resolution in combination with RIMS. If the event rate is a concern, it might be necessary to work with broader gas beams in combination with a time– and position–focusing recoil spectrometer to maintain a good recoil-ion momentum resolution. Though supersonic expansion reduces the target temperature by up to a factor of 1000, or even more if precooling is used, the remnant target momentum spread certainly sets a lower limit in the achievable ion–momentum resolution. In the direction parallel to the gas-jet velocity the momentum spread is given by the intrinsic jet-temperature. For a helium jet with a speed ratio of $S = 50$ starting at room-temperature this momentum uncertainty is about 0.13 a.u. The perpendicular momentum spread is much smaller, it is determined by the jet velocity and the divergence of the atomic beam. If one or several skimmers are used to cut out a He-jet that has a diameter of 1 mm at a distance 10 cm away from the nozzle then this momentum spread is below 0.06 a.u.

Over the last five years supersonic gas-jets have become the standard devices to produce cold and localized atomic or molecular targets in connection with recoil-ion momentum spectroscopy (COLTRIMS). With this method, a large variety of different gases is accessible ranging from helium or other noble gases up to any type of molecular gas. Cooling of the gas before expansion is useful for helium or molecular hydrogen to reduce the internal gas-jet temperature and to reach the optimum momentum resolution, however, it can lead to the formation of clusters or liquid droplets if heavy gases are used. One should also be aware that the internal momentum spread increases with the mass M of the gas even if the same final jet temperature is obtained. Thus, the generation of a gas–jet with a low intrinsic momentum spread of $\Delta P < 0.5$ a.u. along the jet direction is definitely difficult to achieve for gases heavier than argon.

2.3.2   Atomic Traps (MOTRIMS)

A significant further reduction of the target temperature is achievable when laser-cooled atoms trapped in a magneto-optical-trap (MOT) are used as a target. In the first pioneering experiment, Wolf and Helm [11] measured recoil-ion energies of photoionized Rb atoms extracted from a MOT with very high resolution. Basically, the same method has been used very recently [12–14] to study single–electron capture reactions in keV ion collisions with atoms trapped in a MOT. In these first experiments an unprecedented resolution in the recoil-ion momentum was achieved by taking benefit from the sub-mK intrinsic temperature of the target. For recoiling Rb^+ ions produced in capture reactions with keV Cs^+ projectiles Flechard and coworkers [14] obtained a momentum resolution of $\Delta P = 0.03$ a.u., which means that the recoil-ion velocity has been measured with a sensitivity of below 1 m/s. Certainly, this value is not the ultimate resolution set by the gas–cloud temperature of typically 100 μK. It would correspond to a momentum resolution of $\Delta P = 0.003$ a.u. for lithium and $\Delta P = 0.01$ a.u. for rubidium.

With typical densities in the range of several 10^{10} cm^{-3} and radii of 1 mm or below, the atomic clouds trapped in MOT are ideally suited as targets for

RIMS (MOTRIMS). Whenever an ultimate momentum resolution is desirable like, e.g., in electron–capture reactions where the recoil-ion momentum gives direct access to the electronic sates populated in the projectile ion, a MOT target is superior to any gas–jet. Another important aspect is that the choice of possible targets for RIMS is considerably widened. All alkali and earth alkaline atoms are easily trapped in a MOT, but they are hardly produced and cooled in a supersonic gas-jet. These atoms are, in some situations, of particular interest because they represent single active electron targets and therefore have many features in common with atomic hydrogen, the simplest atomic target that is not directly accessible to COLTRIMS because efficient cooling methods are not at hand. Further interesting options are laser–excited atoms and the possibility to probe the occupation of these excited states in a time-resolved manner via capture reactions [15] monitored with RIMS.

2.4   Position–Sensitive Detectors

For recoil-ion momentum spectroscopy large-area position-sensitive detectors with good position resolution (typically 0.1 mm or better) and fast timing signals (below 1 ns) are essential. Multichannel plate (MCP) detectors with either chevron (two plates) or z-stack (three plates) configuration and typical plate diameters of 50, 80, or 120 mm are used. The working principle of an MCP–detector is illustrated in Fig. 2.6. The detector has a typical detection efficiency for charged particles of about 60% mainly determined by the open area. This optimum efficiency is obtained for ions that hit the detector surface with a kinetic energy of about 2 keV or more. For detection of electrons the efficiency is best at energies between 100 and 300 eV. For position encoding the emerging electron cloud created by avalanche amplification in the MCP channels is analyzed by means of wedge and strip [16] or delay-line anodes [17] at the backside of the detector.

2.4.1   Wedge and Strip Anodes

The working principle of a position-sensitive detector with a wedge and strip anode is the following. The amplified electron cloud, containing about 10^5–10^6 electrons, is accelerated onto a segmented anode consisting of electrically separated areas with a wedge and strip structure at a typical periodicity of 1.5 mm (Fig. 2.7). Since the area of the wedges and the stripes depends linearly on the x– and y–position, respectively, the pulse heights of the signals picked up at both electrodes are proportional to the position of the cloud centroid, as long as the cloud covers an area larger than one period of the anode structure. This is ensured by letting the electrons expand during their travel over a distance of several mm from the MCP to the anode. In a more refined anode concept, the wedge and strip structure is evaporated onto the backside of a glass plate. The front side is covered with a thin Ge layer that collects

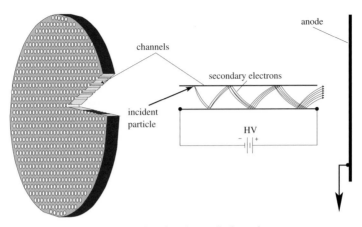

Fig. 2.6. Working principle of a channel-plate detector

the electron cloud [18]. In this case the segments pick up the image charge of the Ge layer while the widening of this image charge is adjusted by a proper selection for the Ge layer resistivity and the thickness of the glass plate. This concept has the advantage that the detector can be operated in strong magnetic field environments. In any case, to extract the position information the pulse heights of the wedge and stripe segments have to be normalized to the collected total charge, which varies from event to event. It is given by the pulse–height sum of all electrodes, the wedge, the strip, and a meander structure that fills the area between the first two. All three signals are amplified in separate charge-sensitive preamplifiers and subsequent spectroscopy amplifiers and recorded by analog-to-digital converters. The event-by-event normalization and calibration procedure is usually done by software. The position information is obtained from the measured pulse heights reflecting the amount of charge collected by the wedges Q_W, the stripes Q_S and the meander Q_M, respectively

$$X \propto \frac{Q_S}{Q_S + Q_W + Q_M}, \quad Y \propto \frac{Q_W}{Q_S + Q_W + Q_M}. \tag{2.12}$$

For a correct imaging, the amplifiers in the individual channels have to be adjusted properly and if great demands are put on linearity then the capacitive cross talk between the electrodes has to be considered. This correction can be done by software. In addition, a high gain of the MCP detector is important for a good position resolution, because the main limiting factor is the signal-to-noise ratio of the signals. With standard amplifiers a position resolution of 0.1 mm or better is easily achievable for a 50 mm–diameter anode. For timing purposes, a fast signal can be picked up either at the frontside or at the backside of the MCP stack and a time resolution well below 1 ns is easily achievable. Since several µs are required for signal integration and pulse shaping the maximum countrate is limited to about 100 kHz.

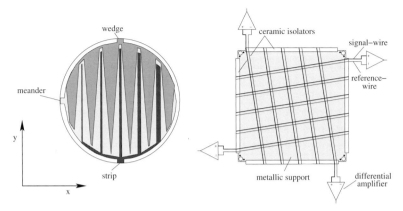

Fig. 2.7. *Left*: Structure of a wedge and strip anode with a strongly reduced number of periods to illustrate the x– and y–dependence of the electrically separated areas. *Right*: Two cable-pairs wrapped in multiple turns around a metallic support form a delay-line anode. With ceramic holders the electrical isolation between cables and support and among the cables is ensured

2.4.2 Delay-Line Anodes

The position readout with a delay-line anode combines both high count rate capabilities and good position resolution. In this case, the position encoding is done by a measurement of the time a signal needs to travel along a wire that is wound in many loops around a rectangular support with a wire spacing in the order of 0.5 mm [17]. The time difference between the signals arriving at both ends of the wire is proportional to the position coordinate where the charge cloud has been collected (Fig. 2.7). The finally achievable position resolution is not limited by the wire spacing because, similar to wedge and strip anodes, a cloud centroid averaging occurs also with delay-line anodes. For the second spatial coordinate another wire is wound perpendicular to the first one. In practice, a pair of wires forming a Lecher-cable is used for each direction. Since both wires of the Lecher-cable pick up a completely identical noise via capacitive coupling with the surrounding, their signals differ only if a real charge is deposited on one but not on the other wire. To ensure that only one wire collects the electrons from the cloud it is biased with a more positive potential and the second one is basically used to monitor the noise. The ends of the Lecher-cables are fed into fast differential amplifiers to subtract the noise from the signal and to enable a precise determination of the time difference between both ends of the Lecher-cable. Fast discriminators and time-to-digital converters (TDC) are used for further signal processing and readout. From the collected timing information the x– and y–coordinates can be calculated by

$$X = (t_{x1} - t_{x2})v_{\text{signal}}, \quad Y = (t_{y1} - t_{y2})v_{\text{signal}}, \tag{2.13}$$

where $(t_{x1} - t_{x2})$ is the time difference between the signals arriving at both ends of the wire wound in the x-direction. The difference $(t_{y1} - t_{y2})$ is the corresponding value for the y-direction. Though the signals travel almost with the speed of light along the wire the essential signal velocity v_{signal} is much smaller. It is determined by the time a signal needs to pass the anode from one side to the other, which is about 70 ns for a delay-line anode of 80 mm square. The position resolution is related to the precision obtained for the determination of the time difference between the signals collected at both ends of the delay-line wire. It can be measured very accurately and position resolutions of 0.1 mm are routinely achieved with delay-line anodes [19]. According to (2.13) a time zero or the absolute time of arrival of a particle on the detector is not needed to extract the position information. Usually the absolute time is defined by the timing signal of the MCP stack. With respect to this time base the time sums $(t_{x1} + t_{x2})$ and $(t_{y1} + t_{y2})$ are constant and equal to the transition time from end of the wire to the other. This additional information can be used for consistency checks or, as will be discussed afterwards, for multihit purposes.

2.4.3 Multiple–Hit Detection

In most applications the delay-line concept is superior to position encoding based on charge division like, e.g., wedge and strip and resistive anodes, due to several reasons. In general, the most important advantage is the very fast readout allowing countrates of several MHz limited mainly by the performance of the MCP and of the electronics. Another advantage is that large–area delay-line anodes can be produced easily without loss of position resolution, whereas a simple scaling up of a wedge and strip structure results in a proportional degrading of the absolute resolution. Only for very long wires, the signal dispersion becomes a matter of concern. Regarding COLTRIMS and, in particular, electron momentum imaging, an essential advantage is the ability to register the time and position information of multiple hits, or showers of particles arriving at the detector within a very short time interval.

The relevant timescale is set by the total delay time of the anode wires, which is between 30 and 100 ns depending on the size of the detector. As long as the time separation between two subsequent hits is larger than this total delay, no particular arrangements are needed to record their position and time, except for an electronic device for handling several consecutive signals (like, e.g., a multistop TDC). In this case, the five signals (four from the anode plus one from the MCP) of the first particle arrive first and those of the second hit always arrive later. This mode of operation with a dead-time of the order of 50 ns already enables a large variety of experimental studies, like, e.g., imaging of ionic fragments from molecular Coulomb-explosions where the number of accepted fragments is limited only by the performance of the TDC. The detector dead-time can be reduced further if more advanced signal assignment and data recovering is used. As already mentioned, the detector delivers five timing signals for each registered particle, but only three are needed to extract the position and the absolute time. It is basically this redundancy that allows reconstruction of the positions of double or multiple-hit event where the time separation of consecutive hits is considerably shorter than the total delay time of the anode. In this case, if two particles hit the detector at very different positions, it may happen that some signals from the delay-line are not in the right order (a signal from the second particle may arrive first) or some may even be missing because of signal overlapping. Using the fact that the time-sum of signals belonging together is constant and known, the correct signal ordering can be recovered and the corresponding particle positions can be determined. Moreover, if a signal is completely missing, the spatial positions are still accessible, because a signal on one end of the delay-line wire is sufficient to determine the corresponding coordinate. Only when the signals from both ends of one delay-line wire are missing, the position information is lost. Using this scheme of reordering and data recovery, the actual detector dead-time is reduced to zero as long as the x- and y-coordinates of subsequent particles differ by more than the corresponding pulse-pair resolution of the detector system, which is mainly limited by the signal widths and the pulse-pair resolution of the TDC. The latter is usually in the 5–10 ns range resulting in a cross-shaped dead-zone of some mm width in the position difference of two subsequent hits.

Based on either a complete or partial reconstruction of delay-line signals several experiments have been performed for imaging of up to three electrons emerging from a single breakup of atoms or molecules in collisions with photons, electrons or ions. In these experiments, with typical electron flight times of some 10 ns, the residual detector dead-time turned out to be not a severe limitation since in many cases the physics hinders electrons arriving exactly at the same time at identical positions on the detector. Nevertheless, the leftover dead-time sometimes complicates the interpretation of experimental data because a certain part of the electron kinematics is always cut out. Thus, a detector concept without any restrictions concerning the multihit capability would be desirable.

Such a delay-line anode, having practically no dead-time at all, has been developed by Jagutzki and coworkers [20]. The basic idea is to further increase the number of redundant signals delivered by the anode. This has been achieved by taking three instead of two wire-pair layers and by winding them not perpendicular to each other but with an angle of 60° around an hexagonal–shaped support. Thus, for each hit one signal from the MCP plus six from the anode are obtained, whereas only three of them are needed for timing and position encoding. The additional signals can be used for cross checks and data recovering. With this newly developed delay-line concept the dead-time is reduced further allowing acceptance of events where two particles hit the detector at the same time at almost identical positions without loss of information. It can be anticipated that the hexagonal anode will become the ideal tool for a large variety of applications where high demands are put on the multihit processing in combination with good position resolution at comparably low system complexity.

2.5 Imaging Spectrometers for Electrons

For most atomic reactions the detection of the target ion only yields insufficient information to characterize the reaction completely. Whenever more than two particles emerge, the desire to obtain the kinematically complete information requires the determination of all momenta for all outgoing particles in coincidence and for each single event. To achieve this, it is sufficient to measure the momentum vectors of all target fragments, but not the projectile because the momentum transferred from the projectile to the target can be deduced using momentum conservation. This strategy has been applied in various charged–particle-atom collision experiments (see [1–3]). The task is somewhat less demanding in the case of photon–induced ionization reactions because there the photon is absorbed and simply not present in the final state. Therefore, the complete reaction kinematics is obtained by registration of all charged fragments except one. In general, however, one or several electrons plus at least one or even more ionic fragments are emitted, which requires the detection of ions and electrons in coincidence with high efficiency covering a large range of the final state momentum space to obtain maximum information.

Combined recoil-ion electron momentum spectrometers have been developed within the last ten years that fulfil, to a large extent, these requirements and that enabled kinematically complete studies of atomic collision reactions with up to five outgoing particles like triple ionization of Ne in collisions with heavy ions [21]. For the majority of atomic breakup reactions the assumption holds that electrons and ions emerge with momenta of comparable magnitude. Therefore, as a result of the ion–to–electron mass ratio, the kinetic energies of electrons exceeds those of typical recoil ions by many orders (they are in the eV range) making it much harder to collect electrons with high efficiency. The radial displacement on the detector scales with the square-root of the kinetic energy (2.7). Basically, two concepts have been developed to achieve a high acceptance together with a good resolution for both electrons and ions in coincidence. Both approaches have in common that particle imaging is done by accelerating electrons and ions in the same electric field into opposite directions onto two position–sensitive detectors. Because only the low–energy part of the electron continuum is accessible with these devices another concept has been introduced taking advantage of the fact that the electron trajectories are easily modified by small magnetic fields. With proper combined electric and magnetic fields the motion is effectively confined in space. In this way extremely versatile spectrometers have been constructed, the so-called *Reac-*

tion Microscopes, allowing one to adjust the resolution and the acceptance in the electron and ion branch individually. Both approaches, direct imaging and magnetic guiding of electrons, will be discussed in the next sections.

2.5.1 Direct Imaging of Electrons

The conceptually most obvious scheme for coincident imaging of electrons and ions is to place a second position–sensitive detector opposite to the ion detector. The homogeneous electric field pushes them onto their individual detectors. In this case the same equations as those for the ions can be applied to reconstruct the electron momentum vectors from their time-of-flight and position of detection. However, because the initial electron energies easily approaches the kinetic energy they gain in the extraction field, no approximate equations should be used to calculate neither the longitudinal momentum from the measured TOF nor the transverse energy from the position information. In fact, the full equations (2.1) and (2.6) must be solved for $E_\perp$ and $E_\parallel$, respectively. The achievable resolution is, by analogy to the recoil-ion branch, limited mainly by the source extension, the detector position resolution and the precision of the TOF measurement. The finite target temperature, which contributes considerably to the ion resolution, does not impose an additional uncertainty on the electron momentum. It results only in a negligibly small electron momentum spread because of the large atom–to–electron mass ratio. As a consequence, unprecedented energy resolutions for electrons are achievable. Using our example spectrometer with time-focusing geometry and 10 V extraction over 10 cm to image electrons emitted from an extended target ($\Delta r = 1\,\mathrm{mm}$), one can easily achieve an energy resolution as small as $\Delta E_\perp = 62\,\mu\mathrm{eV}$ for the transverse direction and $\Delta E_\parallel \approx 1.3\,\mathrm{meV}$ in the TOF-direction (time resolution $\Delta t = 1\,\mathrm{ns}$). But the accepted maximum electron energy in the transverse direction is of the order of only some ten meV (less than 100 meV for an 80–mm diameter detector). For many applications this is too small.

The acceptance can be increased considerably by placing the detector closer to the target zone and/or by increasing the extraction voltage. This leads to smaller electron flight times and, therefore, to a lower momentum resolution in the TOF direction. Moreover, a direct consequence of an increased extraction field is a slightly reduced momentum resolution for the recoil ions. Nevertheless, direct electron imaging in coincidence with recoil-spectroscopy is ideally suited to investigate atomic reactions where electrons with small or very small continuum energies are generated. Placing the electron detector very close to the reaction volume (of the order of 1–2 cm) allows detection of electrons with 4π solid angle for energies in the few eV range. Such spectrometers have been used to study electron emission in low–energy ion–atom collisions [22,23] and photoionization close to threshold [24].

2.5.2   Reaction Microscopes: Magnetic Guiding of Electrons

To circumvent the limitations associated with direct imaging and to increase the electron–energy acceptance for 4π collection, maintaining at the same time the full resolution in the recoil-ion branch, novel spectrometers have been developed where a solenoidal magnetic field parallel to the electric field is superimposed [25]. This magnetic field acts over the whole flight path and it forces electrons to move on spiral trajectories from the reaction volume to the detector. In practice, the magnetic field is generated by a pair of large Helmholtz coils placed outside the vacuum chamber (Fig. 2.8). Typical coil diameters range from 50 cm up to 2 m, depending on the geometrical extension of the electron flight path. In order to avoid field distortions no magnetic materials and ideally only those with a small magnetic susceptibility should be used for construction of a Reaction Microscope. A schematic setup is shown in Fig. 2.8. It basically consists of an electric extraction field with two subsequent drift paths and position–sensitive detectors on either side. The extraction field is generated by means of potential rings in connection with a voltage divider or between two ceramic plates covered with a resistive layer [25]. The target point is defined by the crossover of the supersonic gas–jet and the particle beam.

Though the electron motion is strongly modified by the magnetic field it is still possible to reconstruct both the electron trajectory and the initial momentum vector from the two position informations and the measured time-of-flight. For a homogeneous magnetic field parallel to the electric extraction field, the longitudinal (i.e. along the TOF direction) motion of a charged particle is not changed ((2.1) is still valid). In the transverse direction (perpendicular to the field axis), however, the electron travels along a circle with radius $R = P_\perp/(qB)$ where the time T for one turn is given by the inverse of the cyclotron frequency $\omega = qB/m = 2\pi/T$, with B the magnetic

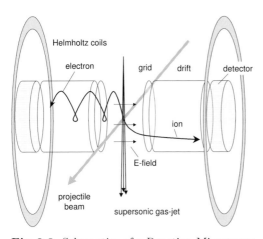

Fig. 2.8. Schematics of a Reaction Microscope

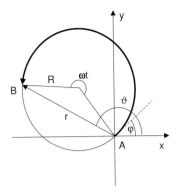

Fig. 2.9. Projection of an electron trajectory onto the detector surface. The magnetic field is perpendicular to the paper plane (see text)

field strength, m and q the electron mass and charge, respectively. In useful units these equations are

$$\omega = 9.65 \times 10^{-6} \times qB/m \quad \text{and} \quad R = 12.39 \times P_\perp/(qB)\,, \tag{2.14}$$

with ω in ns^{-1}, q and P in atomic units, B in Gauss, and m in amu. With a magnetic field of only 10 G, all trajectories of electrons with energies up to 100 eV are confined to a cylinder with a radius of 3.3 cm, independent of the electric field strength and independent of the spectrometer geometry. Hence, an electron detector with a diameter of 8 cm is sufficient to achieve a 4π collection efficiency for electron energies of more than 100 eV. Under these conditions the revolution time for electrons is 35 ns and more than 250 µs for He ions. Thus, it is in the range of or even smaller than typical electron flight times. On the other hand, the cyclotron time for ions is much longer than usual ion flight times, indicating that their trajectories are only weakly affected by the magnetic field. It basically results in a slight rotation of the ion image on the recoil detector. In most cases it is sufficient to compensate this by rotating the whole ion–position distribution by the corresponding angle. In conclusion, by changing the magnetic field strength the accepted maximum transverse momentum of electrons is adjustable without affecting the recoil-ion imaging. It is essentially this electron–zooming option that makes Reaction Microscopes extremely versatile and considerably extended the range of applications for coincident electron–ion imaging (see also [2]).

2.5.3 Reconstruction of Electron Momenta

Below we discuss how to calculate the initial electron–momentum components from the position and time information. Whereas the longitudinal electron momentum solely and unambiguously follows from the TOF, both position and time are required to reconstruct the two transverse momentum components. To illustrate this we consider the projection of an electron trajectory onto the electron detector surface, i.e., onto a plane perpendicular to the magnetic field axis (Fig. 2.9). The electron is emitted with a transverse momentum $P_\perp$ at the origin (point A) under a certain angle φ with respect to the positive x-axis. On its way to the electron detector it travels on a spiral trajectory, i.e., along a circle, before it hits the detector at B with a certain displacement r. For a given magnetic field the radius R of the cyclotron motion is a direct measure of $P_\perp$ (see (2.14)) while the arc-angle ωt depends only on the electron time-of-flight. From simple geometrical considerations (Fig. 2.9) follows that

$$ R = \frac{r}{2|\sin(\omega t/2)|} . \tag{2.15} $$

Thus, the magnitude of the transverse momentum $P_\perp$ can be calculated from the position of detection, (r, ϑ) in cylindrical coordinates, and the measured electron TOF t using (2.14) and (2.15). With the initial angle of emission φ, which follows immediately from

$$ \varphi = \vartheta - \omega t/2, \tag{2.16} $$

all three components of the electron momentum vector are determined.

Up to now we restricted the discussion to cases where the electron TOF is shorter than the cyclotron time or, in other words, where the electron passes through less than one complete turn. In many practical situations, however, the electrons perform several full turns N (up to ten) of the cyclotron motion. If this happens then ωt has to be replaced by $\omega t - N \cdot 2\pi$ where N is the next lowest integer of the ratio $(\omega t)/(2\pi)$. But still, the assignment of measured quantities and initial momentum is unique, as long as the denominator in (2.15) is larger than zero. The only prerequisite for momentum reconstruction is the knowledge of the magnetic field strength and direction, respectively. Whenever electrons perform exactly one or several complete turns they hit the detector at the origin (point A in Fig. 2.9) independent of their initial transverse momentum or, in other words, all electrons with flight times equal to a multiple integer of T (the inverse cyclotron frequency) are focused onto the same spot on the detector and no momentum information is obtained. These specific cases appear as nodal points when the radial displacement on the electron detector is plotted versus the electron time-of-flight (Fig. 2.10). Though the momentum information is lost at these nodes, however, they deliver important and valuable information. First they serve as a good control of the experimental conditions and, more importantly, they allow a very precise and intrinsic determination of the magnetic field via a measurement of the inverse cyclotron frequency. This is given by the time-distance of two nodes. Whenever important information about the physical process is masked by these nodes, which might happen, the magnetic field can be changed slightly resulting in a corresponding time-shift of the nodal points.

For a certain electron TOF the radial displacement r on the detector is proportional to the transverse momentum. Thus, the momentum resolution

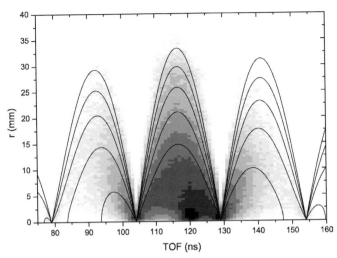

Fig. 2.10. The radial displacement r of electrons on the detector versus their time-of-flight for single ionization of H_2 by 6–MeV proton impact. The magnetic field is about 14 G corresponding to a cyclotron revolution time of $T = 25$ ns. The *solid lines* are the calculated position and time dependences for isotropically emitted electrons with kinetic energies between 10 and 50 eV in steps of 10 eV. The point (TOF $= 120$ ns, $r = 0$) corresponds to the origin ($E = 0$) on the electron energy scale

in the transverse direction is given by $\Delta P_\perp = P_\perp \Delta r / r$, where Δr is the effective position uncertainty resulting from both, the detector resolution and the target size. For an effective position resolution of 1 mm the transverse momentum of an electron that hits the detector at $r = 4$ cm with $P_\perp = 1$ a.u. is determined with an uncertainty of $\Delta P_\perp = 0.025$ a.u. This value should be regarded as a lower limit because a finite time resolution, which enters into the determination of the *sine*-term in (2.15), as well as magnetic–field distortions also contribute. As a consequence, a definite determination of the electron–momentum resolution is difficult to perform. For practical purposes and if high resolution is required, it is best to check the actual resolution by using electrons with a well–defined and sharp kinetic energy arising from known atomic processes like, e.g., Auger-electron emission or single photoionization.

2.6 New Developments

In this chapter, a few technical developments related to recoil-ion and electron momentum spectroscopy that are still ongoing and might play a certain role in future applications are listed and briefly discussed. Though this list is quite incomplete it represents the still continuing evolution of many-particle imaging devices to investigate atomic and molecular reactions.

To some extent, the limited number of target species accessible to COLTR-IMS is a constraint. This is mainly because both conditions, high target density and extremely low temperature, are difficult to combine if one thinks of targets like, for example, highly excited atoms, metastable helium, atomic hydrogen, vibrationally cold molecules or others. Very recently, the Frankfurt group succeeded in producing a supersonic jet of dense and internally cold metastable He atoms [26]. A high–pressure discharge burning between the entrance and exit surface of a 50–μm nozzle collisionally excites He while the following adiabatic expansion into the vacuum reduces the internal temperature. After a subsequent Stern–Gerlach magnet the beam can be used for experiments with oriented and cold metastable helium atoms.

If complex targets are used, for example molecules or clusters, the number of particles ejected from a breakup reaction can be quite large. The desire to detect them all increases the demands on the detectors concerning their multihit capability. The hexagonal delay-line for position encoding fulfils the requirement of almost negligible dead-time as long as channel-plates with small or medium diameters are used. With increasing detector sizes, however, the cable length and therewith the dispersion of signals travelling along the cables increases, resulting in a slowing down of the detector readout. On the other hand, with ongoing development of integrated electronic circuits, concerning speed and compactness, large–area pixel anodes with ultrafast readout of individual pixels are in the realm of possibility. Presently, work is in progress to develop such an anode where one thousand or more pixels are individually read out using highly integrated electronics mounted onboard directly at the anode. It is hoped that in the near future large active area detectors with improved multihit capability and high position resolution will be available.

The flexibility of imaging spectrometers can be increased further if the traditional concept of using only static electric and magnetic fields for projection and guiding of charged particles is skipped. Obviously, the application of pulsed or otherwise time-dependent fields can help to adapt the spectrometer performance to specific problems. To study, for example, molecular fragmentation maintaining at the same time a high resolution for the emitted electrons, a pulsed extraction field can be used. The initially small field is ramped up considerably on a sub-μs timescale whenever an electron hits the detector, because even eV ions do not travel over a large distance in the time the electrons require to reach the detector. Thus, ionic fragments experience the large acceleration field ensuring a 4π detection efficiency for them. This method has been successfully applied for the first time to investigate double photoionization of molecular hydrogen [27]. However, it is applicable for a large range of molecular breakup reactions and in most cases it is advantageous compared to a static field extraction.

An electric field between two grids placed at an appropriate position in the ion or electron flight path can serve as a switchable potential barrier. By gating the applied voltage in synchronization with an external time reference, like the timing signal of a projectile detector, specific user-defined parts of either the electron or the recoil-ion TOF spectrum can be masked out. Such gated switches in the recoil-ion branch have been used to investigate electron-transfer reactions in fast proton–helium collisions [28]. The huge amount of recoil ions from unwanted physical reactions, which would overload the detector, could be effectively suppressed by many orders of magnitude. Recently, a similar technique has been used for high-resolution electron–capture measurements. There, the recoil ions have been extracted exactly parallel to the incoming beam direction. A pulsed electric field acting perpendicular to the ion trajectory within the drift path kicked the recoil ions onto the detector, which was mounted beneath the projectile beam axis [29]. These are only two examples where gated switches or deflectors have been used. In general, the possibilities opened up with pulsed electric fields are by far not yet exhausted.

References

1. C.L. Cocke, R.E. Olson: Phys. Rep. **205**, 155 (1991)
2. J. Ullrich, R. Moshammer, R. Dörner, O. Jagutzki, V. Mergel, H. Schmidt-Böcking, L. Spielberger: J. Phys. B **30**, 2917 (1997)
3. R. Dörner, V. Mergel, O. Jagutzki, L. Spielberger, J. Ullrich, R. Moshammer, H. Schmidt-Böcking: Phys. Rep. **330**, 96 (2000)
4. J. Ullrich, H. Schmidt-Böcking: Phys. Lett. A **125**, 193 (1987)
5. R. Ali, V. Frohne, C.L. Cocke, M. Stöckli, S. Cheng, M.L.A. Raphaelian: Phys. Rev. Lett. **69**, 2491 (1992)
6. W.C. Wiley, I.H. McLaren: Rev. Sci. Instrum. **26**, 1150 (1955)
7. R. Dörner, H. Bräuning, J.M. Feagin, V. Mergel, O. Jagutzki, L. Spielberger, T. Vogt, H. Khemliche, M.H. Prior, J. Ullrich, C.L. Cocke, H. Schmidt-Böcking: Phys. Rev. A **57**, 1074(1998)
8. M.A. Abdallah, W. Wolff, H.E. Wolff, E.Y. Kamber, M. Stöckli, C.L. Cocke: Phys. Rev. A **58**, 2911 (1998)
9. R. Miller, in: Atomic and Molecular Beam Methods, Vol. **14**, eds. G. Scoles, D. Bassi, U. Buck, D. Laine (Oxford University Press, New York 1988)
10. D. Fischer: Diploma Thesis, University Freiburg (2000)
11. S. Wolf, H. Helm: Phys. Rev. A **62**, 043408 (2000)
12. M. van der Poel, C.V. Nielsen, M.A. Gearba, N. Andersen: Phys. Rev. Lett. **87**, 123201 (2001)
13. J. W. Turkstra, R. Hoekstra, S. Knoop, D. Meyer, R. Morgenstern, R. E. Olson: Phys. Rev. Lett. **87**, 123202 (2001)
14. X. Flechard, H. Nguyen, E. Wells, I. Ben-Itzhak, B.D. DePaola: Phys. Rev. Lett. **87**, 123203 (2001)
15. B. DePaola et al.: Proceedings HCI 2002 (to be published in Nucl. Instrum. Methods B)
16. A. Martin, P. Jenlinsky, M. Lampton, R.F. Malina: Rev. Sci. Instrum. **52**, 1067 (1981)
17. S.E. Sobottka, M.B. Williams: IEEE Trans. Nucl. Sci. **35**, 348 (1988)

58   R. Moshammer et al.

18. G. Battistone, P. Campa, V. Chiarella, U. Denni, E. Iarocci, G. Nicoletti: Nucl. Instrum. Methods **202**, 459 (1982)
19. I. Ali, R. Dörner, O. Jagutzki, S. Nüttgens, V. Mergel, L. Spielberger, Kh. Khayyat, T. Vogt, H. Bräuning, K. Ullmann, R. Moshammer, J. Ullrich, S. Hagmann, K.-O. Groeneveld, C.L. Cocke, H. Schmidt-Böcking: Nucl. Instrum. Methods B **149**, 490 (1999)
20. O. Jagutzki, A. Cerezo, A. Czasch, R. Dörner, M. Hattaß, M. Huang, V. Mergel, U. Spillmann, K. Ullmann-Pfleger, T. Weber, H. Schmidt-Böcking, G. Smith: IEEE Conference Proceedings (NSS), San Diego (2002), to appear
21. M. Schulz, R. Moshammer, W. Schmitt, H. Kollmus, R. Mann, S. Hagmann, R.E. Olson, J. Ullrich: Phys. Rev. A **61**, 2270 (2000)
22. R. Dörner, H. Khemliche, M.H. Prior, C.L. Cocke, J.A. Gary, R.E. Olson, V. Mergel, J. Ullrich, H. Schmidt-Böcking: Phys. Rev. Lett. **77**, 4520 (1996)
23. E. Edgü-Fry, C.L. Cocke, E. Sidky, C.D. Lin, M. Abdallah, J. Phys. B **35**, 2603 (2002)
24. R. Dörner, J.M. Feagin, C.L. Cocke, H. Bräuning, O. Jagutzki, M. Jung, E.P. Kanter, H. Khemliche, S. Kravis, V. Mergel, M.H. Prior, H. Schmidt-Böcking, L. Spielberger, J. Ullrich, M. Unverzagt, T. Vogt: Phys. Rev. Lett. **77**, 1024 (1996)
25. R. Moshammer, M. Unverzagt, W. Schmitt, J. Ullrich, H. Schmidt-Böcking: Nucl. Instrum. Methods B **108**, 425 (1996)
26. T. Jahnke et al.: http://hsbpc1.ikf.physik.uni-frankfurt.de/plasmajet/plasmajet.html
27. R. Dörner, H. Bräuning, O. Jagutzki, V. Mergel, M. Achler, R. Moshammer, J.M. Feagin, T. Osipov, A. Bräuning-Demian, L. Spielberger, J.H. McGuire, M.H. Prior, N. Berrah, J.D. Bozek, C.L. Cocke, H. Schmidt-Böcking: Phys. Rev. Lett. **81**, 5776 (1998)
28. H.T. Schmidt, A. Fardi, R. Schuch, S. H.Schwartz, H. Zettergren, and H. Cederquist, L. Bagge, H. Danared, A. Källberg, J. Jensen, and K.-G. Rensfelt, V. Mergel, L. Schmidt, and H. Schmidt-Böcking, C. L. Cocke: Phys. Rev. Lett. **89**, 163201 (2002)
29. D. Fischer, B. Feuerstein, R.D. DuBois, R. Moshammer, J.R. Crespo Lopez-Urrutia, I. Draganic, H. Lörch, A.N. Perumal, J. Ullrich: J. Phys. B **35** 1369 (2002)

3 Multiparticle Imaging of Fast Molecular Ion Beams

D. Zajfman, D. Schwalm, and A. Wolf

3.1 Introduction

For many years, two-dimensional (2D) and three-dimensional (3D) fragment imaging techniques have been successfully used in the study of molecular structure [1] and for the study of the dynamics of various molecular dissociation processes, such as photodissociation [2], dissociative recombination [3], atom–molecule collision-induced dissociation [4], and dissociative charge exchange [5]. For fast molecular ion beams (in the present context, fast means kinetic energies in the range of keV to several MeV), the basic experimental scheme includes the induced dissociation of a single molecule from the beam, and the fully correlated measurement of the asymptotic velocity vectors of the outgoing atomic and molecular fragments. If the initial velocity of the molecule is large, then all the fragments will be projected into a cone defined by the ratio of their transverse velocities and the initial beam velocity. In such a case, the transverse velocities are deduced from the 2D position on the surface of a position–sensitive detector, while the longitudinal velocities can be derived from the (relative) time of arrival at the detector. The specific physical information provided by the images depends on the particular dissociation process. In general, one obtains information about the initial molecular quantum state prior to the dissociation, the final state of the fragments and about the dynamics of the reaction, such as angular dependence, kinetic energy release or potential curves. In this chapter, we will describe the basic requirements and techniques used for the detection of the fragments, with emphasis on the optical methods based on CCD technology, and present a simple example of 2D imaging.

3.2 Basic Concepts

3.2.1 Distances and Times: Order of Magnitude

Figure 3.1 represents a typical situation where a fast homonuclear diatomic molecule (or molecular ion), with kinetic energy E_0 dissociates at a distance L from a detector. As the molecule dissociates, both fragments gain some velocity in the center-of-mass frame.

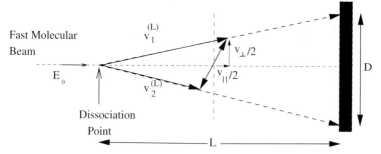

Fig. 3.1. Schematic view of the dissociation of a molecule from a fast diatomic molecular beam. $v_1^{(L)}$ and $v_2^{(L)}$ are the velocities of the fragments in the laboratory frame of reference

If we denote the beam velocity as v_0, then the relative distance D between the two fragments after traveling the distance L is given by (see Fig. 3.1)

$$D = \frac{1}{2}v_\perp \left[\frac{L}{v_0 + (1/2)v_\parallel} + \frac{L}{v_0 - (1/2)v_\parallel} \right],\tag{3.1}$$

where $v_\perp$ and $v_\parallel$ are the perpendicular and parallel components of the mutual velocity of the fragments, relative to the beam direction. Because the initial velocity of the beam v_0 is much larger than the parallel components of the relative velocity $v_\parallel$, one can reduce the previous equation to

$$D = L\frac{v_\perp}{v_0}.\tag{3.2}$$

With the same approximation, the flight–time difference between the two particles is given by

$$\Delta t = t_2 - t_1 = L\frac{v_\parallel}{v_0^2},\tag{3.3}$$

where t_1 and t_2 are the times-of-flight from the dissociation point to the detector for the two particles. Thus, the kinetic energy release E_k in the center-of-mass, which is often the value of interest, can be expressed as a function of D and Δt only:

$$E_k = \frac{E_0}{4L^2}\left[(v_0\Delta t)^2 + D^2\right].\tag{3.4}$$

The general formula for E_k for a heteronuclear molecule with mass $M = m_1 + m_2$ is

$$E_k = \left(\frac{E_0}{L^2}\right)\left(\frac{m_1 m_2}{M^2}\right)\left[(v_0\Delta t)^2 + D^2\right]\left(1 + 2\frac{m_1 - m_2}{M}\frac{v_0\Delta t}{L}\right).\tag{3.5}$$

For a polyatomic molecule, when more than two fragments are produced in the process of dissociation, the relevant quantities are the velocities of each fragment in the center-of-mass, and not only the total kinetic energy release. The Cartesian components of these velocities v_x^i, v_y^i, and v_z^i can be easily expressed as a function of the impact coordinates x^i and y^i for each particle i on the surface of the detector and by the time-of-flight t^i relative to the center-of-mass time of flight t_{cm} (assuming that the particles are identified)

$$v_x^i = v_0 \frac{x^i}{L} \qquad (3.6)$$

$$v_y^i = v_0 \frac{y^i}{L} \qquad (3.7)$$

$$v_z^i = v_0 \frac{(t_{cm} - t^i)}{t_{cm}} . \qquad (3.8)$$

Let us now estimate the typical relative distance and time of arrival between the different fragments. As an example, we suppose that a H_2^+ molecular ion ($M = 2$) moving with a kinetic energy of $E_0 = 1\,\text{MeV}$ dissociates into two protons after the electron has been rapidly stripped via a collision with a very thin foil (a process known as foil–induced Coulomb–explosion). In such a case, since the average internuclear distance between the two protons at the dissociation point is $1.06\,\text{Å}$, the energy release is $E_k \sim 15\,\text{eV}$. Based on (3.2) and (3.3), one obtains for the maximum relative distance, on the surface of the detector located at a distance of $L = 2\,\text{m}$ from the dissociation point, a value of $D_{max} = 15\,\text{mm}$, and the maximum flight–time difference is $\Delta t_{max} = 1.6\,\text{ns}$. It is interesting to point out that while the relative distance (3.2) scales with the inverse of the beam velocity, the time difference between two fragments is inversely proportional to the square of the beam velocity. As the time resolution of a 3D detector is usually the weakest point, decreasing the beam kinetic energy can improve the relative time resolution, without requiring the use of a much larger detector. Overall, a resolution of $100\,\mu\text{m}$ will result in a relative error that is better than 1% for D. On the other hand, a resolution of $100\,\text{ps}$ (which is about the present limit) yields a relative error of about 7% on Δt. Hence, in most cases, the time resolution will generally be the limiting factor as far as resolution is concerned. Other factors, such as the spread in the initial beam energy, are found to be negligible in most cases.

3.2.2 Three-Dimensional vs. Two-Dimensional Imaging

From the above description of the dissociation kinematics, it is clear that in order to measure the final velocity of each of the fragments, the detector should be able to measure the position and relative time of arrival of each particle of a given molecule, one event at the time. This is called three-dimensional (3D) imaging, while 2D imaging refers to the measurement of the position only, i.e., the transversal velocity components only. It is interesting to point out that, in the case of 3D imaging, only the relative time of arrival is needed (in addition to the spatial transverse coordinates), and not the absolute time-of-flight from the dissociation point to the detector. Since 2D imaging is much simpler than 3D imaging, it is important to consider whether a 2D imaging system yields sufficient information on the dissociation process, so that a 3D imaging detector is not required. This is often the case if the kinetic energy release in the dissociation process is a superposition of well–defined values (such as in dissociative recombination of photodissociation) which are defined by the initial state of the molecule before dissociation and the possible final states of the fragments. However, this is, in general, limited to dissociation processes yielding two or three fragments only, otherwise the reconstruction process becomes difficult because of the convolution with the different possible orientations of the molecule prior to dissociation.

3.3   Detector Concepts and Development

Various schemes for the measurement of the asymptotic velocities for each of the fragments of a dissociating molecule have been designed. Most modern imaging detectors use a microchannel plate (MCP) for the detection of the incoming particles. These devices have the ability, when used with a suitable anode, to yield excellent position and time information for the impact of particles over a wide range of energies. MCPs are basically electron amplifiers that create an electron avalanche through narrow channels close to the position where the particle hits the surface. A single MCP creates about 10^4 electrons per impact. Better amplification can be obtained by using a two–stage geometry (Chevron) (about 10^6 electrons per impact) or even a three–stage (Z-stack). By detecting these electrons on an anode located behind the MCP, one can easily measure the time as well as the position of impact of a single particle. Position resolution for the MCP itself (i.e., not including the anode) is of the order of $40\,\mu$m, and time resolution of the order of $100\,$ps can be achieved. The detection efficiency of a bare MCP is of the order of 50%. The most important part of such a detector for the purpose of imaging, is the design of the anode. Among the various schemes are the resistive anode [6], the segmented anode [7], the delay–line anode [8], and the phosphor screen coupled to a CCD camera together with a transparent or optical anode for the time measurement [9].

During the last decade, imaging systems based on the use of a phosphor screen as an anode for the MCP, have gained considerable popularity. The reason is that this method allows for the measurement of the position of a large number of particles from a single event, with good efficiency, thus making the design of a 2D imaging system relatively simple. Here we concentrate mainly on this type of detectors.

3.3.1 Optical Detection

The use of CCD cameras for recording the position of the particle hit on the surface of the detector has been around for about 10 years [9]. In this method, the cloud of electrons exiting the MCP channels are accelerated toward a metalized phosphor screen located behind the MCP. The impact of these electrons on the phosphor produces a visible light spot. The position of these spots can then be recorded by a standard CCD camera, coupled to a frame grabber [9,10], with a resolution that is basically limited by the MCP and phosphor screen. Resolution of better than 100 μm is possible. One of the advantages of the phosphor screen–CCD system is that if the region of interest (the area on the screen where the events hit the detector) is smaller than the MCP, one can use a simple lens to focus the whole CCD area on this region, thus increasing the number of pixels per unit of length. In any case, with a standard CCD camera with more than 512 pixels per line and a 40–mm diameter MCP, the resolution is better than 100 μm per pixel.

A problem arises for the measurement of correlated events (i.e., fragments originating from the same molecule) as the integration time of a frame (for a standard 50 Hz camera) is 20 ms, so that all events occurring during this time are recorded on the same frame. Thus, in order to record the position of only correlated events, one needs either to generate the events in phase with the camera timing cycle (one event per frame) [9,10], or to "switch off" the detector once an event has been detected [3].

A major advantage of CCD readout is that the number of simultaneous hits that can be measured on the detector is practically unlimited, allowing very complex 2D images to be recorded (the final efficiency of the MCP practically limits this number). The analysis of the recorded image can be easily performed using commercially available frame grabbers, or dedicated hardware, which are able to store only the positions (x, y) and intensities of pixels above a preprogrammed threshold [9,10]. This also has the advantage of producing highly contrasted images that are not affected by the background noise of the MCP (and the dark noise of the CCD). On the other hand, CCD camera are "slow" devices, when compared to the electronic readout of other anodes, although fast cameras, with up to 1000 frames/s (continuous) are now available [11], and have been used in experiments [12].

In order to create 3D images, the time of arrival of each particle must be recorded as well. This has been done by adding a transparent segmented anode on the phosphor screen [10], or by using a photomultiplier with a segmented anode, which is optically coupled to the phosphor screen [13] (see Fig. 3.2), or by using two cameras and a fast electronic shutter [14] (see Fig. 3.3). In the first two cases, each of the anode segments is read by a separated amplifier and associated digitizing electronics, hence allowing the measurement of several particles without any dead-time. The number of anodes must be large enough to minimize the probability that two particles hit the same strip. This is solved in the two–camera system, where only images

64   D. Zajfman et al.

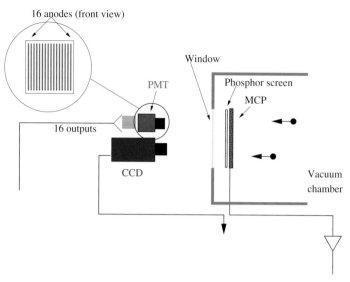

Fig. 3.2. Optical–detection scheme with a CCD camera and a segmented–anode photomultiplier

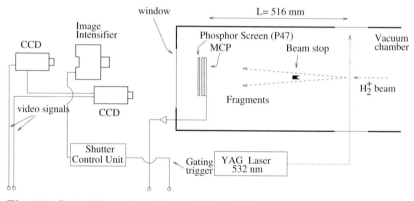

Fig. 3.3. Optical detection scheme with two CCD camera and a fast optical shutter

are analyzed, and the time information is obtained from the ratio of the pixel intensities [14].

3.3.2 Electrical Detection

The use of electrical detection (resistive anode, delay–lines) has its own advantages. The most important one is the possibility to measure with high countrate (in the range of 10 to 100 kHz). However, some of these detectors have – in general – a finite dead-time making the measurement of two simultaneous hits impossible (usually, within 10 ns) on the surface of the detector.

For a slow beam (keV) this is not a serious problem, as the maximum time between two impacts is generally longer. However, for beams of a few MeV, the time difference between two successive impacts is at most a few nanoseconds, making these detectors unsuitable for the purpose of three-dimensional imaging. Segmented anode detectors, where each anode is read out using a dedicated channel, have no dead-time (as long as two particles do not hit the same anode) and can work with high countrate. However, the electronics required, if the number of separated anode is large, can be quite voluminous. A hybrid system, where the position is read out through imaging of a phosphor screen with the help of a phosphor camera and the timing by a series of wires stretched on the surface of the phosphor screen, has also been built [9,10], and is used in the Coulomb–explosion experiments described in Chap. 22.

3.4   Image Reconstruction

When a 3D image of a dissociation event is recorded, usually very little reconstruction is needed, as the transformation to the "molecular coordinates" can be done in a straightforward manner. On the other hand, when 2D images are recorded, integration over the molecular–axis orientation has to be taken into account, and the expected measured distributions are more complicated (but the experimental setup is simpler). In the following we treat the cases when two or three particles are produced by the dissociation process, and the fragments are measured using a 2D–imaging detector.

3.4.1   Two–Body Fragmentation

It is easier to understand the expected relative distribution between two fragments, produced in a dissociation events if we assume that the energy release is a series of "delta functions" as applies, for example, in dissociative recombination or photodissociation.

Let us assume that, due to a specific interaction, a fast (kinetic energy E_0) diatomic molecule AB dissociates and that several final states of the fragments are possible, yielding n possible combination that are correlated to a specific kinetic energies release $E_{k,i}$ ($i = 1, ..., n$) in the center–of–mass. If $E_0 \gg E_{k,i}$ for all i, then after dissociation, the two fragments move forward and will start to separate from each other (see Fig. 3.1). If a detector is set up at a distance L from the dissociation point, then the *projected* distance D between the two neutral fragments on the plane of the detector is given by energy and momentum conservation (see (3.2)) as

$$D = L\delta_i \sin \theta, \tag{3.9}$$

$$\delta_i = \frac{(m_A + m_B)}{\sqrt{m_A m_B}} \sqrt{\frac{E_{k,i}}{E_0}}, \tag{3.10}$$

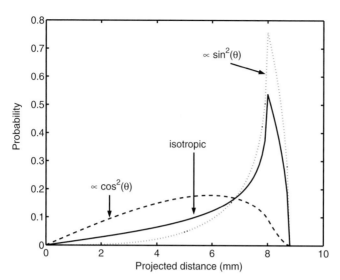

Fig. 3.4. Typical shape of the projected-distance distribution for various initial angular anisotropies. *Solid line*: isotropic distribution; *dashed line*: $(\cos^2 \theta)$-like distribution and *dotted line*: $(\sin^2 \theta)$-like distribution

where θ is the initial angle of the molecular internuclear axis relative to the beam direction and m_A and m_B are the masses of the fragments A and B, respectively. Since the angle θ is unknown, we have to integrate over all possible dissociation angles. In this case, an assumption about the distribution of θ (anisotropy) is required. As an example, for an isotropic distribution, the projected distribution is given by

$$P_{i,L}(D) = \begin{cases} (D/\delta_i^2)\,(1/L\Gamma) & \text{for} \quad 0 \le D \le \delta_i L\,, \\ 0 & \text{otherwise}, \end{cases} \tag{3.11}$$

where

$$\Gamma = \sqrt{L^2 - \frac{D^2}{\delta_i}}\,. \tag{3.12}$$

Figure 3.4 shows the typical shape of such a distribution for a single value of the kinetic energy release, and taking into account that the dissociation is not localized at a single point, but rather extended over a certain distance along the direction of motion, to represent the conditions available in storage rings, where the overlap between the molecular ion beam and the electron beam occurs over a length of 1.5 m. Also shown in this figure are the distributions for two initial angular anisotropies. When more than one final state is possible, the overall distribution is the sum over all the possible $P(i, L)$ (see (3.11)). In general, if the difference between the various kinetic energies is large enough, the branching ratio between the various channels can be obtained from the projected distribution, and full 3D imaging is not necessary. Additional distribution for various common angular anisotropies, and extended dissociation sources can be found in [15].

3.4.2   Three-Body Channel

When a molecule dissociates into three fragments with a total kinetic energy release E_k, this energy can be shared between the dissociating fragments in many ways, within the restrictions from momentum conservation. For simplicity, if we assume that all fragments have the same mass (such as in the three-body dissociation of H_3^+), this restriction limits the kinetic energy of any single fragment atom to $(2/3)E_k$. The energy sharing between the fragments uniquely determines the relative directions of the three momenta and hence the dissociation geometry, which is specified by the shape of the triangle spanned by the momentum vectors of the three fragments in the comoving frame of reference. In order to obtain information about the correlation between the motion of the three fragments (the kinematical correlation), the coordinates have to be such that a uniform density in these coordinates indicates the absence of correlations, i.e., that the phase space is filled randomly under no other constraints than total momentum conservation. The so-called two-dimensional Dalitz plot [16,17] is an excellent choice for this, and its coordinates are defined as

$$\eta_1 = (E_2 - E_1)/\sqrt{3}E_k,$$
$$\eta_2 = (2E_3 - E_2 - E_1)/3E_k \qquad (3.13)$$

with

$$E_k = E_1 + E_2 + E_3 \,, \qquad (3.14)$$

where E_i are the single-particle kinetic energies in the center-of-mass frame. Using the 3D–detection method, it is relatively simple to build such a representation from the accumulated data. However, as the 2D–imaging technique records distances in the detector plane, transverse to the beam velocity, only transverse momenta and energies are measured. It is still possible to produce a "transversal" Dalitz plot, using the transverse distances between the fragments R, as measured on the surface of the detector:

$$Q_1 = (R_2^2 - R_1^2)/\sqrt{3}R^2 \,,$$
$$Q_2 = (2R_3^2 - R_2^2 - R_1^2)/3R^2 \,, \qquad (3.15)$$

where

$$R_i^2 = (Lv_{\perp i}/v_0)^2 = 3L^2 E_{\perp i}/E_0 \,, \qquad (3.16)$$

where $v_{\perp i}$ and $E_{\perp i}$ are the transverse velocities and energies of the fragments. Similar to the (η_1, η_2) representation, each point in the (Q_1, Q_2) plane represents a single geometry of the three impact positions on the detector plane,

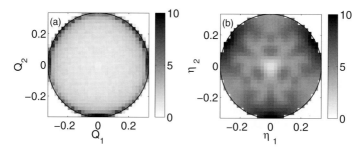

Fig. 3.5. (a): Projected Dalitz plot and (b) reconstructed Dalitz plot for the three hydrogen atoms produced by the dissociative recombination of H_3^+

which is, in fact, the projection of the (η_1, η_2) geometry in the dissociation plane onto the detector plane. However, there is no unique correspondence between the geometry in the dissociation plane and the projected geometry. Each specific dissociation geometry is transformed into a variety of different projected geometries, depending on the orientation of the dissociation plane. Nevertheless, it is possible to perform meaningful back-transformations of measured distributions in the (Q_1, Q_2) coordinates to momentum–geometry distributions in the proper (η_1, η_2) Dalitz representation [12,17]. Figure 3.5a shows, as an example, the measured transverse (Q_1, Q_2) Dalitz plot for the dissociative recombination of H_3^+ with low–energy electrons [12,17]. In order to reconstruct the original Dalitz plot, in the (η_1, η_2) coordinates, a Monte Carlo reconstruction method is used. The method, which is described in detail in [12], uses the fact that the transformations from the (η_1, η_2) to the (Q_1, Q_2) depends only on the geometry of the experiments (assuming that the angular distribution of the dissociation plane is isotropic). Thus, the deconvolution from the projected coordinates can be carried out using these transformation as response functions. The final results are shown in Fig. 3.5b. In these coordinates, each point represents a different kinematical correlation between the three particles, which is mapped out in Fig. 3.6.

The six-fold symmetry under exchange of the fragment indices is marked by the three reflection axes (dashed lines). The circle drawn at $(\eta_1^2 + \eta_2^2)^{1/2} = 1/3$ gives the limitation imposed by energy and momentum conservation. A comparison between Figs. 3.5b and 3.6 shows that the three hydrogen fragments tend to be aligned on a straight line. More details can be found in [12,17].

3.5   Conclusion and Outlook

The imaging of correlated fragments resulting from molecular dissociation is a powerful method taht observation of the "internal" mechanism of the fragmentation process, the molecular structure, and the correlation between the

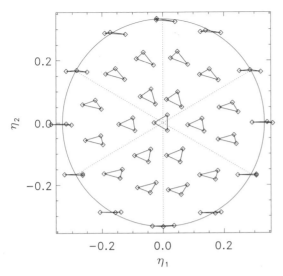

Fig. 3.6. Geometrical interpretation of the Dalitz plot using the coordinates from (3.13). The triangular shape corresponding to the respective coordinates is plotted for a sample of points in the (η_1, η_2) plane. The *dashed lines* and the *circular boundary* are discussed in the text

various degrees of freedom. Although the method has been used for about two decades, most of the work was carried out on two-body dissociation, except for the Coulomb–explosion, where up to five particles have been measured in co-incidence [18]. The availability of new detection schemes [14] should enhance the possibility for many–particle correlated measurements. Such measurements have already been performed with slow particles, which are produced by the fragmentation of atomic or molecular targets [19], when the time delay between the particles is large. For a fast beam, this time delay is of the order of few nanoseconds, and detectors with true ability to measure several particles with zero dead-time and high efficiency are needed.

References

1. Z. Vager, R. Naaman, E.P. Kanter: Nature **244**, 426 (1989)
2. A.J.R. Heck, D.W. Chandler: Annu. Rev. Phys. Chem. **46**, 335 (1995)
3. D. Zajfman, Z. Amitay, C. Broude, P. Forck, B. Seidel, M. Grieser, D. Habs, D. Schwalm, A. Wolf: Phys. Rev. Lett. **75**, 814 (1995)
4. V. Horvat, O. Heber, R.L. Watson, R. Parameswaran, J.M. Blackadar: Nucl. Instrum. Methods **B99**, 94 (1995)
5. W.J. van der Zande, W. Koot, D.P. de Bruijn: Phys. Rev. Lett. **57**, 1219 (1986)
6. C. Firmani, E. Ruiz, C.W. Carlson, M. Lampton, F. Paresce: Rev. Sci. Instrum. **53**, 570 (1982)
7. C. Martin, P. Jelinsky, M. Lampton, R.F. Malina, H.O. Anger: Rev. Sci. Instrum. **52** 1067 (1981)

8. RoentDek GmbH, Germany
9. D. Kella et al.: Nucl. Instrum. Methods **A329**, 440 (1993)
10. R. Wester et al.: Nucl. Instrum. Methods **A413**, 379 (1998)
11. Dalsa Inc., Canada
12. D. Strasser, L. Lammich, H. Kreckel, S. Krohn, M. Lange, A. Naaman, D. Schwalm, A. Wolf, D. Zajfman: Phys. Rev. A **66**, 032719 (2002)
13. Z. Amitay, D. Zajfman: Rev. Sci. Instrum. **68**, 1387 (1997)
14. D. Strasser et al.: Rev. Sci. Instrum. **71**, 3092 (2000)
15. Z. Amitay, D. Zajfman, P. Forck, U. Hechtfischer, B. Seidel, M. Grieser, D. Habs, D. Schwalm, A. Wolf: Phys. Rev. A **54**, 4032 (1996)
16. R.H. Dalitz: Philos. Mag. **44**, 1068 (1953)
17. D. Strasser, L. Lammich, S. Krohn, M. Lange, H. Kreckel, J. Levin, D. Schwalm, Z. Vager, R. Wester, A. Wolf, D. Zajfman: Phys. Rev. Lett. **86**, 779 (2001)
18. Z. Vager, D. Zajfman, T. Graber, E.P. Kanter: Phys. Rev. Lett. **71**, 4319 (1993); D. Kella, Z. Vager: J. Chem. Phys. **102**, 8424 (1995)
19. R. Dörner et al.: Phys. Rep. **330**, 95 (2000)

4 Neutral–Atom Imaging Techniques

U. Müller and H. Helm

4.1 Introduction

The dissociation of a molecule into neutral products plays an important role in astrophysics and plasma physics. In the case of larger molecules, dissociation into three or more massive fragments may constitute a major decay pathway [1–5]. Experiments to understand such processes in full detail constitute a substantial experimental challenge. The parent molecule has to be prepared in a quantum–mechanically well-characterized initial state. All fragments have to be detected in coincidence and a kinematically complete analysis of the final state has to be achieved.

Fast-beam techniques are elegant methods to study the formation of neutral fragments. The dissociation products are moving with a few-keV translational energy in the laboratory frame and release secondary electrons when they interact with solid surfaces. Therefore, fast neutrals can be detected by microchannel plates and their positions and arrival times can be determined with time- and position-sensitive anodes. Translational spectroscopy using time- and position-sensitive detectors was pioneered by de Bruin and Los [6] and applied to study dissociative charge transfer of H_2^+ and N_2^+ [7]. The production of fast neutral molecular beams by photodetachment of negative ions [8] has been pursued by groups in Berkeley [9] and San Diego [10,11]. Laser excitation of fast metastable molecules allows one to prepare the initial state in a well-defined quantum state. This has been widely applied to study molecular two-body breakup [12–15].

To investigate dissociation into three and more particles, fast–beam techniques are the only methods which allow a kinematically complete analysis of the final state. A multihit coincidence detector is required for measuring the arrival times and positions of all fragments with high accuracy and low dead–times. In the following, we first present a fast-beam translational spectrometer recently built in Freiburg University [1]. Then, we describe the multihit readout, and the data reduction to determine the fragment momentum vectors in the center-of-mass (cm) frame. We discuss suitable projections of the multidimensional photodissociation cross sections to gain insight into the decay dynamics.

4.2   Fast–Beam Apparatus

Figure 4.1 shows the Freiburg fast-beam photofragment spectrometer for the detection of molecular many-body fragmentation. Molecular ions are created in an ion source (I), accelerated to a translational energy of several keV (Acc), mass selected (MF), and focused by an Einzel Lens (EL). A small fraction of the ions is neutralized by charge transfer. After the charge-transfer cell (CT), the unreacted ions are removed by an electric field (ID), and the products of dissociative charge transfer are stopped by the aperture (Ap). A fast, well–collimated beam of neutral molecules is created propagating along the x-direction in the laboratory frame.

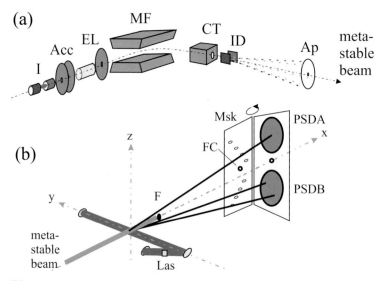

Fig. 4.1. Freiburg fast–beam photofragment spectrometer. The preparation of the molecular fast beam is shown schematically in (**a**). The laser excitation and detection of the neutral products is shown in (**b**). Legend: I: ion source, Acc: acceleration field, EL: Einzel lens, MF: mass filter, CT: charge transfer, ID: ion deflector, Ap: aperture, Las: standing–wave intracavity laser, F: beam stop, PSDA, PSDB: time- and position-sensitive multihit detectors, Msk: precision mask with hole pattern for detector calibration, FC: Faraday cup

The dissociation into n fragments is initiated in a primary process such as laser excitation or atomic collision. The undissociated part of the parent neutral molecular beam is intercepted by a miniature beam flag (F). The fragments separate from each other due to their kinetic energy in the cm frame. They are recorded in coincidence by a time- and position-sensitive multihit detector (PSDA, PSDB) after a free flight of about 150 cm. The positions and the arrival–time differences of the fragments in the plane of the detector are individually measured.

In laser-excitation experiments, a home-built standing–wave dye laser pumped by an argon-ion laser is used. To take advantage of its high intracavity power, the neutral beam is crossed with the laser beam inside the cavity. The vacuum chamber is equipped with Brewster windows to minimize the cavity losses. The laser wavelength is controlled by a birefringent filter that is tuned by a stepping motor. For operation at a fixed wavelength, an intracavity etalon is inserted to reduce the laser bandwidth to about $0.1\,\mathrm{cm}^{-1}$. The laser beam can be turned on and off by an intracavity shutter operated under computer control. This allows us to collect the events from metastable decay of the molecules separately from laser–induced fragmentation.

4.3   Detector Requirements and Specifications

A sophisticated time– and position–sensitive detection system is required to measure the products of a multiparticle dissociation process in coincidence. The sensitive detector area should be comparatively large ($\approx 10\,\mathrm{cm}$ outside diameter) to achieve a high geometric collection efficiency. To cover a significant part of the phase space of fragmentation configurations, we can discriminate consecutive fragment impacts with temporal and spatial separations as low as $20\,\mathrm{ns}$ and $1\,\mathrm{mm}$, respectively. The impact coordinates in the detector plane and the arrival–time differences are determined with accuracies of $100\,\mu\mathrm{m}$ and $100\,\mathrm{ps}$, respectively.

Common to all detector concepts in use are multichannel plates (MCP) that multiply secondary electrons produced by the impact of fast neutrals, photons, or charged particles. They achieve amplifications of 10^6–10^8, retaining the position and time information of the original impact with accuracies of $25\,\mu\mathrm{m}$ and a few ps, respectively. Several methods can be considered to read out the position and time information. Position–sensitive anodes based on charge division (resistive anodes [16,17], capacitor chains [6,18,19], wedge-and-strip anodes [20]) achieve good spatial resolution but have comparatively long recovery times (several microseconds). Such anodes are not suitable for multihit applications, unless the sensitive area is subdivided and covered by several independent detector systems, which greatly increases the hardware expenses. Using multielement anodes with individual readout circuits for each element, excellent multihit capabilities have been demonstrated [21] but the spatial resolution was only about $1\,\mathrm{mm}$. An imaging detector that combines a phosphor screen and a CCD camera with multianode photomultipliers has been developed by Amitay and Zajfman [22]. This detector offers multihit capabilities and good time- and position resolution, but the total event rate is limited by the relatively low camera readout rate ($50\,\mathrm{frames/s}$).

4.4   Multihit Methods and Readout System

The time- and position-sensitive multihit detector developed in Freiburg University [1] consists of two units. Figure 4.2 shows one of them schematically. Two Z-stacks of multichannel plates (MCP) with 50 mm outside diameter (44 mm diameter active area) are mounted in the (y, z)-plane of Fig. 4.1, centered at distances of 35 mm from the x-axis. The electron charge cloud from the MCP is collected by a position–sensitive delay–line anode (PSA) separated by 15 mm from the back surface of the MCP. The shaping element (SE) is biased appropriately to minimize distortions of the extraction field at the boundary of the MCP. Position–sensitive delay–line anodes were originally introduced by Sobottka and Williams [23]. The delay–line anodes in our apparatus were developed by Jagutzki et al. [24].

As shown in Fig. 4.2, two waveguides are formed by pairs of parallel copper wires. They are wrapped in two layers around a conductive base plate, held in place and separated from each other and the base plate by ceramic spacers. A bias voltage between the wire pairs leads to an imbalance of the amount of charge collected by the wires. Electromagnetic pulses due to the charge impact propagate to the terminals of the waveguides, which are capacitively

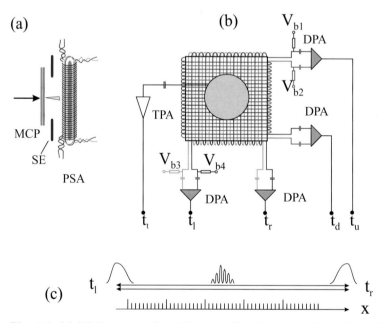

Fig. 4.2. Multihit time- and position-sensitive detector unit. Schematic side view (**a**), front view (**b**). The principle of the position measurement is shown in (**c**). Legend: MCP: Multichannel plates, SE: shaping electrode, PSA: position–sensitive delay–line anode, TPA: fast preamplifier for timing signal, DPA: differential preamplifiers for position signals

coupled to differential amplifiers (DPA). The delay–line anode transforms the position coordinate of the electron cloud into arrival–time coordinates of electromagnetic pulses. The principle of the position measurement is visualized in Fig. 4.2c for one dimension.

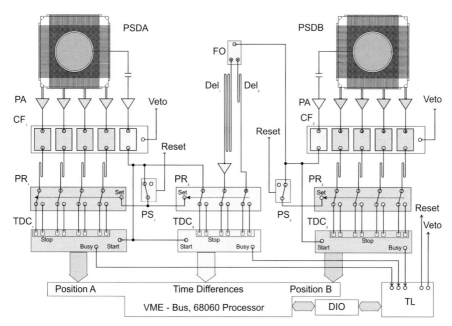

Fig. 4.3. Readout circuit of multihit time- and position-sensitive detector. Legend: PSDA, PSDB: time- and position–sensitive delay–line detector units, PA: preamplifiers, CF: constant-fraction discriminators, FO: pulse amplifier (fanout), Del: cable delays, PR: pulse routers, PS: pulse selectors, TDC: time-to-digital converters, TL: trigger logic, DIO: digital input/output interface

The computer-controlled readout system for the time– and position–sensitive detectors is given schematically in Fig. 4.3. The output pulses of the differential- and the timing amplifiers (PA) are converted to NIM-standard pulses by constant-fraction (CF) discriminators. To measure the pulse arrival times, we use three time-to-digital converters (TDC) with 8 channels each. TDC_1 and TDC_2 measure the position information of detectors PSDA and PSDB, respectively. TDC_3 measures the time differences between events. The time resolution can be hardware selected in the range between 25 ps and 125 ps. For the detection of multiple hits on both detector units, pulse routers (PR in Fig. 4.3) are provided for all position and timing signals. A description of the pulse routers (PR) developed in Freiburg can be found in [25]. The conversion activity of the TDCs is synchronized by a trigger logic

(TL). The control of the experiment, the data readout, and the preanalysis are performed by a dedicated processor.

To calibrate the position scale, a stainless steel mask (Msk) with a precisely machined hole pattern is placed in front of the detector. The mask is mounted on a hinge and can be placed and removed by a rotary feedthrough under vacuum. The timing TDC is calibrated with a precision time–interval counter. This calibration information is stored and allows us to transform online the binary raw data of the TDC into the hit positions and arrival time differences of the fragments in the laboratory frame.

4.5 Data–Reduction Algorithms

In this section, we discuss data–reduction algorithms to extract the physically relevant information, namely the momenta of the individual fragments in the cm frame from the measured positions and arrival–time differences of the fragments.

4.5.1 Two-Body Decay

For two-body decay, Newton diagrams are useful to visualize the relation between impact coordinates (y_i, z_i) in the laboratory frame and fragment momenta u_i in the cm frame. We introduce the mass ratio $Z = m_2/m_1$ and,

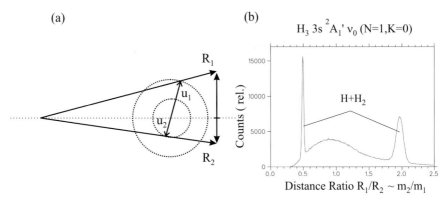

Fig. 4.4. Newton diagram to visualize the relation between the fragment velocities in the cm frame and their impact coordinates (**a**). Distance ratio of fragment pairs produced by photofragmentation of $H_3 3s \,^2A_1'$ (**b**)

using momentum conservation, we find the following relation

$$Z = -\frac{u_{1\perp}}{u_{2\perp}} \approx \left| \frac{R_1}{R_2} \right| \tag{4.1}$$

with the distances $R_i^2 = y_i^2 + z_i^2$ of the fragment hits from the neutral beam axis. The distance ratio allows us to distinguish between different two-body decay channels and to assign the fragment masses. In Fig. 4.4b, the distance ratio is shown for laser photofragmentation of $H_3 3s\ ^2A_1'$. The peaks at 0.5 and 2.0 result from $H+H_2$ fragment pairs. The continuous distribution is produced by three-body decay into three hydrogen atoms $H+H+H$.

Approximate formulas to calculate the kinetic energy release W were developed by de Bruin and Los [6]

$$W \approx E_0 \frac{m_1 m_2}{M^2\ L^2} \left(R^2 + (v_{0x}\Delta T)^2 \right) \left(1 + 2\frac{m_2 - m_1}{M}\frac{v_{0x}\Delta T}{L} \right) . \tag{4.2}$$

Input data are the kinetic energy E_0, velocity v_{0x}, and mass M of the parent molecule, the flight length L, and the spatial R and temporal T distances between the fragment hits at the detector.

4.5.2 Many-Body Decay

For three and more fragments, a geometrical interpretation of the relation between the cm momenta and the measured coordinates and arrival times becomes extremely tedious and confusing. An algebraic solution for multi-particle fragmentation with an arbitrary number n of fragments has been developed by Beckert and Müller [26] (see also Chap. 1). We have to invert the transformation from the velocity space indicated in Fig. 4.5 to the laboratory frame shown in Fig. 4.1b.

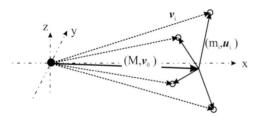

Fig. 4.5. Linear superposition of the center-of-mass velocities $\boldsymbol{u}_i = (u_{ix}, u_{iy}, u_{iz})$ of n fragments with masses m_i with the velocity $\boldsymbol{v}_0 = (v_{0x}, v_{0y}, v_{0z})$ of the center-of-mass motion

We consider the fragmentation of a parent molecule moving with a velocity $\boldsymbol{v}_0 = (v_{0x}, v_{0y}, v_{0z})$ in the laboratory frame into products with cm velocity vectors $\boldsymbol{u}_i = (u_{ix}, u_{iy}, u_{iz})$. Dissociation is taking place at position (x_0, y_0, z_0). The positions and arrival times (y_i, z_i, t_i) of the fragments measured by the PSD in the plane $x = L$ are given by the $3n$ equations of motion

$$(v_{0x} + u_{ix})(t_i - t_0) + x_0 = L, \tag{4.3a}$$
$$(v_{0y} + u_{iy})(t_i - t_0) + y_0 = y_i, \tag{4.3b}$$
$$(v_{0z} + u_{iz})(t_i - t_0) + z_0 = z_i.. \tag{4.3c}$$

The objective is to invert (4.3a–4.3c) and to determine the vectors $\boldsymbol{u}_i$ from the measured quantities. To discuss the procedure under simplified conditions, we start with an ideal experimental setup: the time of fragmentation t_0 is known and the position of fragmentation coincides with the origin of the laboratory coordinate system. Divergence and energy spread of the primary beam vanish. The momentum of the center–of–mass remains unchanged during the dissociation process. As a consequence, the longitudinal component of the cm-velocity is known from the kinetic energy E_0 and the mass M of the parent molecules. The solution of (4.3a–4.3c) is straightforward:

$$x_0 = y_0 = z_0 = 0; \quad v_{0y} = v_{0z} = 0; \quad v_{0x} = \sqrt{2E_0/M};$$

$$u_{ix} = L/(t_i - t_0) - v_{0x}, \tag{4.4a}$$

$$u_{iy} = y_i/(t_i - t_0), \tag{4.4b}$$

$$u_{iz} = z_i/(t_i - t_0). \tag{4.4c}$$

In a real experiment, the inversion is more complicated. The primary beam has a finite divergence, the position of fragmentation shows a distribution over the interaction region, and only differences between the fragment arrival times are measured. The equation system (4.3a–4.3c) is underdetermined by seven relations. Typically, $v_{0x} \approx \sqrt{2E_0/M}$ is still a good approximation. In order to make use of momentum conservation, which gives three additional relations, the masses m_i have to be known. Algorithms to determine the masses m_i for a finite number of fragmenation patterns based on approximate solutions of (4.4a–4.4c) can be found in Beckert and Müller [26]. With the assumption that the size of the interaction region is small compared to the dimensions of the spectrometer ($x_0 = y_0 = z_0 = 0$), we remove three additional unknown quantities and (4.3a–4.3c) can be solved, in principle. A closed formula, however, would be extremely complicated or even impossible. The difficult part is (4.3a) for the longitudinal fragment velocity components u_{ix}.

Keeping in mind that the fragment velocities in the cm frame $\boldsymbol{u}_i$ are typically much smaller than the velocity v_{0x} of the center-of-mass, we have developed an iterative solution of (4.3a). For $x_0 = 0$, we find the arrival–time differences between the i–th and the first fragment $t_i - t_1$:

$$t_i - t_1 = L \cdot \frac{u_{1x} - u_{ix}}{(v_{0x} + u_{ix})(v_{0x} + u_{1x})}. \tag{4.5}$$

Equation (4.5) establishes $(n-1)$ relations for the velocity differences

$$d_{ix} \equiv u_{1x} - u_{ix} = (t_i - t_1)(v_{0x} + u_{ix})(v_{0x} + u_{1x})/L \quad i = 2 \ldots n, \tag{4.6}$$

which only weakly depend on the values u_{ix}. Momentum conservation $\sum_i \boldsymbol{u}_i = 0$ for the x-component allows us to determine the u_{ix}

$$u_{1x} = \sum_{i=2}^{n} m_i d_{ix}/M \quad u_{ix} = u_{1x} - d_{ix} \quad i = 2 \ldots n \tag{4.7}$$

from the d_{ix}. Equations (4.6 and 4.7) constitute a fixed-point equation

$$\boldsymbol{u}_x = T(\boldsymbol{u}_x)\,, \tag{4.8}$$

which can be solved iteratively. With only three iteration steps, the numerical accuracy of the new algorithm [26] is far better than the experimental uncertainty of the raw data. The y- and z-components are easily found by back-substitution.

Three-body decay with equal fragment masses and two-body decay are special cases of this general procedure. In these cases, redundant information can be used to determine the transverse components of the cm velocity, which serves as an additional consistency check to suppress false coincidences. The new algorithm is also superior to the frequently used approximate formula (4.2) for two-body decay. An algorithm to evaluate three-body decay from fragment pairs has been derived by Galster et al. [27].

4.6   Projection of the Multi-Dimensional Cross Sections

In kinematically complete investigations, highly dimensional cross sections are measured. We seek suitable projections that reveal insight into the physics of the fragmentation process.

4.6.1   Two-Body Decay

Because of momentum conservation, $m_1\boldsymbol{u}_1 + m_2\boldsymbol{u}_2 = 0$, the momentum vectors of the two fragments have three degrees of freedom. A suitable projection is visualized in Fig. 4.6. A measure of the vector length is the total kinetic energy release $W = m_1\boldsymbol{u}_1^2/2 + m_2\boldsymbol{u}_2^2/2$ of the dissociation process. The orientation of the momentum vector is best described by the polar angle θ_f and the azimuth ϕ_f. If the electric field $\boldsymbol{E}$ of the linearly polarized excitation laser is

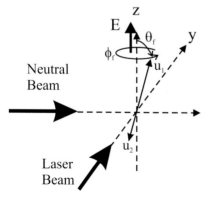

Fig. 4.6. Projection of the three–fold DCS of two-body decay

perfectly parallel to the z-axis, the distribution of ϕ_f is isotropic for symmetry reasons. The distribution of θ_f contains the information of the amount of mechanical alignment of the fragments induced by the nonisotropic excitation process.

4.6.2   Three-Body Decay

The distribution of the fragment momenta in a three-body decay has six degrees of freedom. A suitable parametrization is visualized in Fig. 4.7. Because of momentum conservation, the $\boldsymbol{u}_i$ are contained in a plane. We define a new coordinate system (x', y', z') by the normal vector on this plane (z'-axis) and the direction of the largest momentum vector observed (x'-axis). Three Euler angles (ψ, θ, ϕ) describe the orientation of the (x', y', z')-coordinate system within the laboratory reference system (x, y, z). In the case of laser excitation with $\boldsymbol{E}\|z$, the distribution of ϕ will be isotropic. The distributions of ψ and θ reflect the alignment of the molecular system due to the the laser excitation. For the remaining three parameters describing the arrangement of the three momenta in the (x', y')-plane we may choose the absolute values of the momenta $p_i = m\,|\boldsymbol{u}_i|$ or the individual fragment energies $\varepsilon_i = mu_i^2/2$. We use the total kinetic energy $W = \sum_1^3 m_i u_i^2/2$ and two parameters showing the correlation among the fragment momenta. These could be two angles between the momentum vectors. We could also plot the (x', y') momentum vector components of the fragment with the smallest energy normalized to the absolute momentum of the fastest fragment. We prefer to use a Dalitz plot [28]. To obtain a Dalitz plot we plot for each event $(\varepsilon_3/W - 1/3)$ vs. $((\varepsilon_2 - \varepsilon_1)/(W \times \sqrt{3}))$, which is basically the energy of the third fragment vs. the energy difference between the first and the second one. Energy and

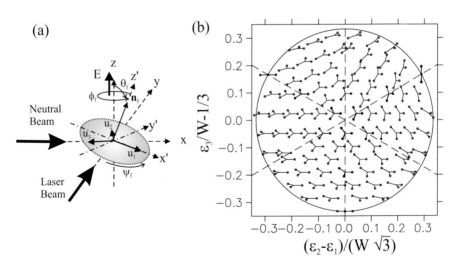

(a)

(b)

Fig. 4.7. Projection of the six–fold DCS of three-body decay

momentum conservation require that the data points in this plot lie inside a circle with radius $1/3$, centered at the origin. As shown by the symbols in Fig. 4.7b, a unique correspondence exists between the configuration of the fragment momenta and the point in the plot. In a Dalitz plot, the phase–space density is conserved, meaning that a fragmentation process with a matrix element independent of the configuration leads to a homogeneous distribution in the space of the kinematically allowed region. Preferred fragmention pathways can be immediately recognized from the point density in such a plot.

References

1. U. Müller, T. Eckert, M. Braun, and H. Helm: Phys. Rev. Lett. **83**, 2718 (1999)
2. C. Maul, T. Haas, and K.H. Gericke: J. Phys. Chem. **101**, 6619 (1997); C. Maul and K.H. Gericke: Int. Rev. Phys. Chem. **16**, 1 (1997)
3. Y. Tanaka, M. Kawaski, Y. Matsumi, H. Futsiwara, T. Ishiwata, L.J. Rogers, R.N. Dixon, and M.N.R. Ashfold: J. Chem. Phys. **109**, 1315 (1998)
4. J.J. Lin, D.W. Hwang, Y.T. Lee, and X. Yang: J. Chem. Phys. **108**, 10061 (1998)
5. M. Lange, O. Pfaff, U. Müller, and R. Brenn: Chem. Phys. **230**, 117 (1998)
6. D.P. de Bruin and J. Los: Rev. Sci. Instrum. **53**, 1020 (1982)
7. A.B. van der Kamp, L.D.A. Siebbeles, and W.J. van der Zande: J. Chem. Phys. **101**, 9271 (1994)
8. L.D. Gardner, M.M. Graff, and J.L. Kohl: Rev. Sci. Instrum. **57**, 177 (1986)
9. R.E. Continetti, D.R. Cyr, D.L. Osborn, D.J. Leahy, and D.M. Neumark: J. Chem. Phys. **99**, 2616 (1993)
10. K.A. Hanold, A.K. Luong, and R.E. Continetti: J. Chem. Phys. **109**, 9215 (1998)
11. K.A. Hanold, A.K. Luong, T.G. Clements, and R.E. Continetti: Rev. Sci. Instrum. **70**, 2268 (1999)
12. H. Helm and P.C. Cosby: J. Chem. Phys. **90**, 4208 (1989)
13. C.W. Walter, P.C. Cosby, and H. Helm: J. Chem. Phys. **99**, 3553 (1993)
14. P.C. Cosby and H. Helm: Phys. Rev. Lett. **61**, 298 (1988)
15. U. Müller and P.C. Cosby: J. Chem. Phys. **105**, 3532 (1996)
16. S.H. Courtney and W.L. Wilson: Rev. Sci. Instrum. **62**, 2100 (1991)
17. G. Leclerc, J.-B. Ozenne, J.-P. Corbeil, and L. Sanche: Rev. Sci. Instrum. **62**, 2997 (1991)
18. H. Helm and P.C. Cosby: J. Chem. Phys. **86**, 6813 (1987)
19. T. Mizogawa, M. Sato, M. Yoshino, Y. Itoh, and Y. Awaya: Nucl. Instrum. Methods A **387**, 395 (1997)
20. C. Martin, P. Jelinsky, M. Lampton, R.F. Malina, and H.O. Anger: Rev. Sci. Instrum. **52**, 1067 (1981)
21. K. Beckord, J. Becker, U. Werner, and H.O. Lutz: J. Phys. B **27**, L585 (1994)
22. Z. Amitay and D. Zajfman: Rev. Sci. Instrum. **68**, 1387 (1997)
23. S.E. Sobottka and M.B. Williams: IEEE Trans. Nucl. Sci. **35**, 348 (1988)
24. O. Jagutzki, V. Mergel, K. Ullmann-Pfleger, L. Spielberger, U. Meyer, and H. Schmidt-Böcking: Proc. SPIE **3764**, 61 (1999)
25. M. Braun, M. Beckert, and U. Müller: Rev. Sci. Instrum. **71**, 4535 (2000)

82   U. Müller and H. Helm

26. M. Beckert and U. Müller: Eur. Phys. J. D **12**, 303 (2000)
27. U. Galster, P. Kaminski, H. Helm, and U. Müller: Eur. Phys. J. D. **17**, 307 (2001)
28. R.H. Dalitz: Philos. Mag. **44**, 1068 (1953); Ann. Rev. Nucl. Sci. **13**, 339 (1963)

5 Collisional Breakup in Coulomb Systems

T.N. Rescigno and C.W. McCurdy

5.1 Introduction

Atomic–collision theorists have struggled to understand the details of the simplest problem in collisional ionization – the electron-impact ionization of atomic hydrogen – since the formulation of the problem some forty years ago by Peterkop [1] and by Rudge and Seaton [2]. In fact, it is only within the last few years that this problem has been reduced to "practical computation", meaning one has a formalism and the associated numerical algorithms that permit the calculation, with currently available computing capability, of the relevant physical quantities to any accuracy that can be tested by experiment. That fact has been demonstrated, for example, in a series of papers [3–8] applying the ideas of exterior complex scaling of electronic coordinates to the electron-impact ionization of the hydrogen atom. Other methods, including convergent close-coupling [9–11], the R-matrix pseudostates method [12], hyperspherical close-coupling [13,14], and time-dependent close-coupling [15] have also been applied to aspects of this problem with great success. For the related problem of double-photoionization of helium, the hyperspherical R-matrix method [16] has been used with great success.

The central difficulty that impeded progress on the problem of three-body breakup in Coulomb systems, particularly for the collisional breakup or "e,2e" problem (as opposed to double photoionization), is the cumbersome asymptotic form of the scattering wave function that the formal theory of ionization imposes. The appropriate boundary condition for ionization, deduced by Peterkop [1] and Rudge and Seaton [2], is

$$\Psi_{\text{ion}}^{+}(\boldsymbol{r}_1, \boldsymbol{r}_2) \underset{\rho \to \infty}{\longrightarrow} -f_{\text{i}}(\hat{r}_1, \hat{r}_2, \alpha)\sqrt{\frac{\text{i}\kappa^3}{\rho^5}} \exp\{\text{i}[\kappa\rho + \frac{\zeta(\hat{r}_1, \hat{r}_2, \alpha)}{\kappa}\ln(2\kappa\rho)]\}, (5.1)$$

where f_{i} is the ionization amplitude and the hyperspherical coordinates are defined by $\rho = (r_1^2 + r_2^2)^{1/2}$ with $\alpha = \tan^{-1}(r_1/r_2)$, and κ is related to the total energy by $E = \kappa^2/2$. The most obvious difficulty in applying this boundary condition is that the coefficient $\zeta(\hat{r}_1, \hat{r}_2, \alpha)$ of the logarithmic phase depends on the distances and ejection angles of both electrons. However, worse yet is the fact that (5.1) is not separable in spherical coordinates, and is therefore much more cumbersome to apply to numerical calculations, which are perforce done in that coordinate system. As a consequence, no one has yet applied (5.1) to the numerical solution of the Schrödinger equation for the ionization problem.

The formal theory of ionization poses another challenge to computation as well, and that is that the ordinary expression for evaluating the amplitude, starting from the scattering wave function that solves the Schrödinger equation, does not apply, because defined in the usual way it would have an infinite phase associated with integrating an expression with logarithmic phases over an infinite volume. Instead, the amplitude is given by [1,2,17]

$$f(\boldsymbol{k}_1, \boldsymbol{k}_2) = -(2\pi)^{5/2} e^{i\Delta(\boldsymbol{k}_1, \boldsymbol{k}_2)}$$
$$\times \int \int \Psi^+ (H - E) \phi(-\boldsymbol{k}_1, z_1) \phi(-\boldsymbol{k}_2, z_2) \mathrm{d}\boldsymbol{r}_1 \mathrm{d}\boldsymbol{r}_2, \qquad (5.2)$$

with *effective charges* in the one-body Coulomb functions, $\phi(-\boldsymbol{k}, z)$ depending on both the energy and direction of ejection of each electron,

$$\frac{z_1}{k_1} + \frac{z_2}{k_2} = \frac{1}{k_1} + \frac{1}{k_2} - \frac{1}{|\mathbf{k}_1 - \mathbf{k}_2|}, \qquad (5.3)$$

and with

$$\Delta(\boldsymbol{k}_1, \boldsymbol{k}_2) = 2[(z_1/k_1)\ln(k_1/\kappa) + (z_2/k_2)\ln(k_2/\kappa)]. \qquad (5.4)$$

Both of these difficulties were ultimately overcome by the successful methods for treating the electron-impact ionization problem. The first of them, the asymptotic form in (5.1), was the central issue addressed by the exterior complex scaling (ECS) method, which is the principal subject of this chapter. The second of them, the Coulomb breakup amplitude formula in (5.2) and its attendant numerical pathologies, required a reformulation and the observation that numerical computations on a finite volume can be at most affected by a finite overall phase that leaves physical observables unchanged.

5.2 Exterior Complex Scaling: Circumventing Asymptotic Boundary Conditions

The ECS method owes its origins to the long history of complex scaling methods in atomic and molecular physics, which in turn are based on a very simple observation about the behavior of solutions of the Schrödinger equation when viewed as functions of complex variables. A purely outgoing wave, $\exp(ikr)$, with $k > 0$, becomes exponentially decaying when the coordinate, r, is scaled into the upper half complex plane, $\exp(ikre^{i\eta}) \to 0$ as $r \to \infty$. The first step in the ECS formalism, therefore, is to isolate the outgoing or "scattered wave" portion of the full scattering wave function. To that end, we partition the full wave function into an initial unperturbed state, Φ_0, and a scattered wave, $\Psi_{\rm sc}$, which contains only outgoing waves in all channels:

$$\Psi^{(+)} = \Psi_{\rm sc} + \Phi_0. \qquad (5.5)$$

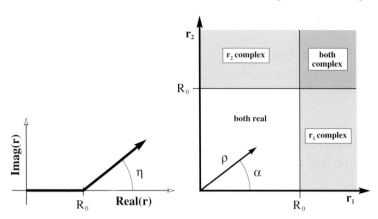

Fig. 5.1. *Left panel:* illustration of the ECS contour rotated into the upper half of the complex r-plane beyond R_0. *Right panel:* Depiction of exterior complex scaling for two radial coordinates

For a two-electron, problem, e.g., electron–hydrogen atom scattering, Φ_0 can be written, for singlet (upper sign) or triplet (lower sign) spin coupling,

$$\Phi_0 = \frac{1}{\sqrt{2k_0}}(e^{i\mathbf{k}_0\cdot\mathbf{r}_1}\varphi_0(\mathbf{r}_2) \pm e^{i\mathbf{k}_0\cdot\mathbf{r}_2}\varphi_0(\mathbf{r}_1)), \tag{5.6}$$

where $\mathbf{k}_0$ is the incident electron momentum, and φ_0 is the initial state of the atom. The scattered wave then satisfies the driven Schrödinger equation for a particular initial condition,

$$(E - H)\Psi_{\mathrm{sc}} = (H - E)\Phi_0. \tag{5.7}$$

Complex scaling reduces the Coulomb boundary condition for breakup, with its complicated logarithmic phases, to the trivial condition that $\Psi_{\mathrm{sc}}(\mathbf{r}_1, \mathbf{r}_2)$ vanishes at infinity. The subtlety in the (e,2e) problem is that we must extract the physics of breakup from Ψ_{sc} in a region in which the coordinates on which it depends are real, so we need to apply the complex scaling transformation only when either of the coordinates of the two electrons are greater than some radius, R_0. The ECS transformation that does this was invented and investigated in the context of electron–scattering resonances with only one electron in the continuum [18,19]; its adaptation to the (e,2e) problem is shown in Fig. 5.1. Specifically, under ECS, the radial coordinates of the electrons are transformed under the mapping:

$$r \to \begin{cases} r, & r < R_0, \\ R_0 + (r - R_0)e^{i\eta}, & r \ge R_0. \end{cases} \tag{5.8}$$

Because Ψ_{sc} contains only outgoing waves, which decay exponentially on the complex part of the exterior scaling contour, (5.7) can be solved by applying only the boundary condition that Ψ_{sc} vanishes at large distances. On the real part of the contour, Ψ_{sc} is the correct physical wave function from which all scattering information can, in principle, be extracted, provided it is extracted in the region of real coordinates.

There is one final subtlety: because (5.6) contains plane waves, which diverge under complex scaling, interaction potentials must be truncated at large distances, but only on the right–hand side of (5.7) [20].

5.3   Scattered-Wave Formalism: Options for Computing the Wave Function

The method of exterior complex scaling and its applications have been developed in a series of papers [3–8,21–23], which from the outset divided the solution of the problem of electron-impact ionization into two discrete steps:

1. Compute the scattering wave function without recourse to the explicit three-body asymptotic form by applying exterior complex scaling to the solution of a discretized representation of the Schrödinger equation.
2. Extract differential and total ionization cross sections from the wave function by either "interrogating" it to compute the scattered flux, or using it in an integral expression for the breakup amplitudes.

It is worth noting that these two problems, namely the computation of the scattered wave function and the subsequent extraction of the scattering information, are only distinct steps in approaches where the wave function is computed by a method that is independent of the asymptotic matching condition that defines the scattering amplitudes. In that sense, our approach is similar to time-dependent methods that track a wavepacket through the collision from initial to final states and then attempt to extract cross sections by analyzing the exiting wavepacket. It is also the case with both approaches that the extraction step, while far from being the most computationally intensive part of the overall calculation, presents many formal difficulties, particularly in the case of multielectron targets, and is the subject of much current research.

5.3.1   Time-Independent Approach: Linear Equations

To solve the scattered–wave Schrödinger equation, (5.7), we must specify the underlying representation. For all the ECS calculations to date, Ψ_{sc} is first expanded in coupled spherical harmonics of the angular coordinates of the two electrons: $\mathcal{Y}_{l_1,l_2}^{L0}(\hat{\boldsymbol{r}}_1,\hat{\boldsymbol{r}}_2)$

$$\Psi_{\mathrm{sc}}(\boldsymbol{r}_1,\boldsymbol{r}_2) = \sum_{L,l_1,l_2} \Psi_{l_1,l_2}^{L}(r_1,r_2)\mathcal{Y}_{l_1,l_2}^{L0}(\hat{\boldsymbol{r}}_1,\hat{\boldsymbol{r}}_2) \tag{5.9}$$

thereby allowing the conversion of (5.7), the driven Schrödinger equation, to a set of coupled equations for the two-particle radial functions, $\Psi_{l_1,l_2}^L(r_1,r_2)$:

$$\left(E - \hat{H}_{l_1}(r_1) - \hat{H}_{l_2}(r_2)\right)\psi_{l_1 l_2}^L(r_1,r_2)$$

$$-\sum_{l_1',l_2'}\langle l_1 l_2 \| l_1' l_2'\rangle_L \psi_{l_1',l_2'}^L(r_1,r_2) = \chi_{l_1 l_2}^L(r_1,r_2) \tag{5.10}$$

where the radial coupling potentials, $\langle l_1 l_2 \| l_1' l_2'\rangle_L$, are obtained by taking matrix elements of $\frac{1}{|r_1-r_2|}$ between two coupled spherical harmonics [6]. The inhomogeneous terms, $\chi_{l_1 l_2}^L$, arise from the partial-wave expansion of the right-hand side of (5.7).

The set of coupled equations is converted to a system of linear equations by choosing some discretization method for representing the exterior-scaled radial functions. In our earlier studies of e–H ionization, the coupled radial equations were solved on a complex two-dimensional grid using seven-point finite difference approximations to the second derivatives. A typical calculation might have ~ 450 points in each radial dimension, and for a given total angular momentum, L, have of the order of 24 (l_1, l_2) angular momentum pairs. The time–consuming step of the calculation, now a modest computation on a massively parallel supercomputer, is the solution of sparse linear equations of the order of five million. To accomplish this, we used an iterative algorithm specifically tailored to the problem at hand. The eigenvalue spectrum of a complex-scaled Hamiltonian is such that no known iterative algorithm will converge to solution without preconditioning. Therefore, finding a suitable preconditioner for the coupled equations is a necessity. The set of *uncoupled* radial equations, defined by setting $\langle l_1 l_2 \| l_1' l_2'\rangle_L = 0$ for all $(l_1', l_2') \neq (l_1, l_2)$ in (5.10), have numerical properties similar to the coupled equations, but require solving linear systems only as large as the total number of radial grid points. We have found solutions of the uncoupled equations, which can be obtained by using a direct sparse solver [24], to be a suitable preconditioner for solving the coupled equations.

The coupled equations can be solved using discretization schemes that are more efficient that high-order finite difference. In our current efforts, we use a combined finite element and discrete–variable representation (DVR), which is the most efficient numerical representation developed to date [25]. In that representation a DVR using Lobatto shape functions is constructed inside each finite element. Continuity of the wave function is enforced at the boundaries of the finite elements, and one of those boundaries is always chosen to lie at R_0. The finite-element DVR representation of the one-dimensional kinetic energy, $-\frac{1}{2}\partial^2/\partial r^2$, has a blocked structure, but the potentials, both $v(r)$ and $V(r_i, r_j)$, are diagonal in the coordinates of each electron. Since the underlying grid points are connected to a Gauss quadrature rule, the total number of points required is considerably smaller than what would be required with finite difference to achieve the same level of accuracy.

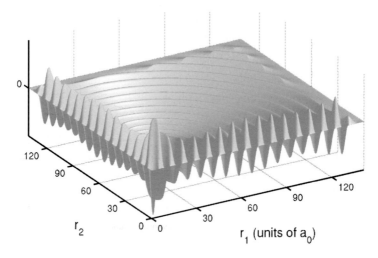

Fig. 5.2. Real part of a representative radial function for electron–hydrogen scattering at 17.6 eV incident energy. Vertical axis is $\mathrm{Re}(\Psi)$ and the two horizontal axes are r_1 and r_2 with origins at the rear left corner. $\Psi_{l_1,l_2}^{L}(r_1,r_2)$ is shown for singlet spin, $L = 2$ and $l_1 = l_2 = 1$

For the e–H problem, one of the many calculated radial functions contributing to Ψ_{sc} is shown in Fig. 5.2. In that figure one can see the outgoing flux in the discrete, inelastic channels going out near the axes, while the ionization flux goes out for large r_1 and r_2 in the structures resembling ripples from a pebble dropped in a pond.

5.3.2   Time-Dependent Approach: Wavepacket Propagation

A straightforwad extension of the method outlined in the previous section to a two-electron target atom, for example, to the case of e–He ionization, would require the solution of coupled linear equations in three radial dimensions. The most troublesome aspect of such an undertaking, aside from the computer resources that would be required, is the fact that even the *uncoupled* equations would likely require an iterative method of solution and finding a suitable preconditioner would be difficult. In contrast to time-independent methods, explicit time-dependent methods, which involve propagating a wavepacket on a multidimensional grid, have scaling properties that allow their application to three-electron systems. We have recently described a step toward a complete algorithm for solving the three-electron breakup problem by combining the idea of time propagation with that of using exterior complex scaling to solve a driven Schrödinger equation [26].

Exterior complex scaling had previously been explored as a method for solving the time-dependent Schrödinger equation [27,28]. The critical observation in that work is that if the wave packet $\Psi(\mathbf{r};t)$ contains only outgoing waves, then

$$\Psi(R(\mathbf{r});t) \to 0, \quad r \to \infty \tag{5.11}$$

on the complex contour for all times, t. Since the physical solution of the driven Schrödinger equation in (5.7) has only outgoing waves, that point is the key to the time-dependent formulation of the ECS method for the present problem.

Formally, the solution of (5.7) we seek is

$$\Psi_{\mathrm{sc}}(\mathbf{r}_1,\mathbf{r}_2,\ldots) = G^{(+)}(H-E)\Phi_0(\mathbf{r}_1,\mathbf{r}_2,\ldots), \tag{5.12}$$

with $G^{(+)}$ the Green's function with outgoing wave boundary conditions,

$$G^{(+)} = (E - H + \mathrm{i}\epsilon)^{-1}. \tag{5.13}$$

We can also write $G^{(+)}$ formally as

$$G^{(+)} = \frac{1}{\mathrm{i}} \int_0^\infty \mathrm{e}^{\mathrm{i}(E+\mathrm{i}\epsilon)t} \mathrm{e}^{-\mathrm{i}Ht} \mathrm{d}t. \tag{5.14}$$

Note that the r.h.s. of (5.7) satisfies

$$(H-E)\Phi_0(\mathbf{r}_1,\mathbf{r}_2,\ldots) \to 0, \quad r_i \to \infty \tag{5.15}$$

because Φ_0 is asymptotically an eigenfunction of H. In fact $(H-E)\Phi_0$ has the range of the interaction potential. In general, therefore, we can define a square-integrable wave packet, $\chi(\mathbf{r}_1,\mathbf{r}_2,\ldots;t)$, by

$$\chi(\mathbf{r}_1,\mathbf{r}_2,\ldots;t) = \mathrm{e}^{-\mathrm{i}Ht}(H-E)\Phi_0(\mathbf{r}_1,\mathbf{r}_2,\ldots). \tag{5.16}$$

Now if we apply the exterior scaling transformation to this equation and define the exterior scaled Hamiltonian by,

$$H \to H_{\mathrm{ECS}} = H(R(\mathbf{r}_1),R(\mathbf{r}_2),\ldots), \tag{5.17}$$

where the scaling applies only to the radial coordinates, the wave–packet then becomes

$$\chi(R(\mathbf{r}_1),R(\mathbf{r}_2),\ldots;t) = \mathrm{e}^{-\mathrm{i}H_{\mathrm{ECS}}t}(H_{\mathrm{ECS}}-E)$$
$$\times \Phi_0(R(\mathbf{r}_1),R(\mathbf{r}_2),\ldots). \tag{5.18}$$

This packet has two important properties,

$$\chi(R(\mathbf{r}_1),R(\mathbf{r}_2),\ldots;t) \to 0, \quad r_i \to \infty, \tag{5.19}$$

and

$$\chi(R(\mathbf{r}_1),R(\mathbf{r}_2),\ldots;t) \to 0, \quad t \to \infty. \tag{5.20}$$

Therefore we can write Ψ_{sc} simply as the Fourier transform of the wavepacket

$$\Psi_{sc} = \frac{1}{i} \int_0^\infty e^{iEt} \chi(t) dt, \tag{5.21}$$

and the $+i\epsilon$ in (5.14) is unnecessary.

Equation (5.21) provides a numerical representation of Ψ_{sc}, provided we can propagate $\chi(0) = (H - E)\Phi_0$ on the ECS contour in two or three dimensions. In numerical experiments we have found that the class of numerical propagators that are unitary for Hermitian Hamiltonians, i.e., before the ECS transformation is made, are generally stable for the corresponding exterior scaled Hamiltonians. For example, in one dimension it has been shown [28] that the Cranck–Nicolson propagator, for time step Δt,

$$e^{-iH\Delta t} = (1 + iH\Delta t/2)^{-1}(1 - iH\Delta t/2) + \mathcal{O}((\Delta t)^3), \tag{5.22}$$

works well, as does the two-dimensional version of this propagator.

The motivation behind the time-dependent approach is the development of a method that scales favorably with particle number, so it should not involve solutions of linear equations representing multiple dimensions at each time step. In our current work, we are using a simple version of the split operator approach [29], in which we first write,

$$H = \sum_{i=1}^d H_0(r_i) + \sum_{i>j=1}^d V(r_i, r_j), \tag{5.23}$$

where $H_0(r)$ is the one-body Hamiltonian and $V(r_i, r_j)$ is the two-body interaction potential, and then approximate the propagator by,

$$e^{-iH\Delta t} \approx e^{-i\sum_{i>j} V(r_i,r_j)\Delta t/2} \left[\prod_{i=1}^d e^{-iH_0(r_i)\Delta t}\right] e^{-i\sum_{i>j} V(r_i,r_j)\Delta t/2}. \tag{5.24}$$

With either finite difference or DVR, the potentials are diagonal in the coordinates of each electron. The operators, $\exp(-iH_0(r_i)\Delta t)$, can be represented by an $N \times N$ matrix, where N is the number of grid points in one dimension, that need be computed only once. It is straightforward to show [26] that, for a problem with d dimensions, the entire propagator requires $\mathcal{O}(2N^d) + \mathcal{O}(dN^{d+1})$ operations per time step. The scaling advantage of the time-dependent approach as outlined above is then that of N^{d+1} versus N^{2d} for the time-independent approach.

5.4 Extraction of Physical Cross Sections

With the scattering wave function in hand we are faced with the problem of extracting the information it contains about elastic, discrete inelastic, and ionization channels. A complete theoretical treatment of electron-impact ionization must necessarily include a prescription for calculating differential cross sections that give detailed information about the energies and angles of ejection of both electrons. Unlike the representation of the wave function in an ordinary atomic close-coupling calculation, its numerical representation in this approach gives no immediate indication of how to separate those contributions.

5.4.1 Flux-Operator Approach

The first ECS calculations on electron-impact ionization of hydrogen were performed by simply computing a variant of the quantum–mechanical flux through a surface that lies within the volume of coordinate space where both coordinates are real. The continuum of ionization final states is described by flux through a hypersphere of radius ρ_0 in the limit $\rho_0 \to \infty$. To this end, we define a generalized, dimensionless flux $f_{\rho_0}^{(\mathrm{ion})}$

$$
f_{\rho_0}^{(\mathrm{ion})}(\alpha, \hat{r}_1, \hat{r}_2) \equiv \mathrm{Im}\Big[k_i \rho \left(r_1 r_2 \Psi_{\mathrm{ion}}^+(\boldsymbol{r}_1, \boldsymbol{r}_2) \right)^\star
$$
$$
\times \frac{\mathrm{d}}{\mathrm{d}\rho} \left(r_1 r_2 \Psi_{\mathrm{ion}}^+(\boldsymbol{r}_1, \boldsymbol{r}_2) \right) \Big]\Big|_{\rho=\rho_0} \tag{5.25}
$$

evaluated at a hyperradius ρ_0. Since the hyperspherical angle α parametrizes the momentum distribution between the two electrons as $\rho_0 \to \infty$, we can express the total ionization cross section as an integral of $f_{\rho_0}^{(\mathrm{ion})}$, in the limit $\rho_0 \to \infty$, over α and the angular coordinates of both electrons:

$$
\sigma_{\mathrm{ion}} = \frac{1}{k_i^2} \int\limits_0^{\pi/2} \int\limits_{4\pi} \int\limits_{4\pi} f_{\rho_0}^{(\mathrm{ion})}(\alpha, \hat{r}_1, \hat{r}_2)\, d\hat{r}_1 d\hat{r}_2 d\alpha \Bigg|_{\rho_0 \to \infty} . \tag{5.26}
$$

Thus, the $\rho_0 \to \infty$ limit of the flux leads directly to a differential cross section for ionization. To compute the scattered flux, we assemble Ψ_{sc}^+ and $\frac{\mathrm{d}}{\mathrm{d}\rho}\Psi_{\mathrm{sc}}^+$ from all its partial wave components:

$$
f_{\rho_0}(\alpha, \hat{r}_1, \hat{r}_2) =
$$
$$
i\Bigg\{ k_i \rho \sum_{\substack{L',l_1',l_2' \\ L,l_1,l_2}} \left(\psi_{l_1'l_2'}^{L'}\right)^\star \frac{\mathrm{d}}{\mathrm{d}\rho}\left(\psi_{l_1 l_2}^{L}\right) \left(\mathcal{Y}_{l_1',l_2'}^{L'0}(\hat{r}_1,\hat{r}_2)\right)^\star \mathcal{Y}_{l_1,l_2}^{L0}(\hat{r}_1,\hat{r}_2) \Bigg\}\Bigg|_{\rho=\rho_0} . \tag{5.27}
$$

The flux operator approach, while conceptually straightforward, is computationally difficult. For purely geometrical reasons, the calculation of the asymptotic flux can require calculations well beyond the range of the potentials, even in the case of short-ranged interactions. Indeed, by inserting the asymptotic form for Ψ_{ion}^{+} from (5.1) into (5.25) we find that the ionization flux approaches its asymptotic limit as $\frac{1}{\rho}$, i.e., for large ρ_0

$$f_{\rho}^{(\text{ion})}(\alpha, \hat{r}_1, \hat{r}_2) \; = \; f_{\infty}^{(\text{ion})}(\alpha, \hat{r}_1, \hat{r}_2) + \mathcal{O}\left(\frac{1}{\rho_0}\right). \tag{5.28}$$

The calculated flux must therefore be numerically extrapolated to infinite ρ_0 to obtain physical results.

A more serious problem, evident in the plot of the radial wave function shown in Fig. 5.2, is that there are regions of space (near the axes) where the "ionization wave" overlaps the discrete two-body channels. The fact that the latter contaminate the ionization flux again forces one to employ grids large enough to allow the physical region inhabited only by the ionization portion of the scattered wave to be distinguishable from the parts that describe discrete two-body channels. The angular range in α subtended by the flux due to a discrete channel is $\sin^{-1}(\Delta/\rho_0)$ where Δ is the distance over which the target state is appreciably different from zero. Thus, as ρ_0 increases, contamination of the ionization flux from discrete channels is confined to smaller regions of α. In the true $\rho_0 \to \infty$ limit, the discrete channels' contributions to the flux become delta functions at $\alpha = 0$ and $\alpha = 90^o$ and equality in (5.28) holds except for infinitesimally small regions near the edges. In practice, the contamination of the ionization flux by discrete channels on finite grids limits the flux-extrapolation procedure in its ability to describe ionization when a single electron carries most of the available energy. Our early calculations of singly differential cross sections (SDCS) [5] were limited to cases where one electron carried no more than about 75% of the total energy.

The most detailed information about ionization is contained in the so-called triply differential cross section (TDCS), which measures the energy and angles of the two outgoing electrons. The calculated TDCS for electron–hydrogen ionization at 17.6 eV incident energy is compared with the absolute experimental measurements of Röder et al. [30] in Fig. 5.3. The results are shown for the coplanar symmetric experimental geometry (which means that the incident electron and both exiting electrons lie in a plane, *and* the two exiting electrons have equal energy), with a fixed angle between the exiting electrons. There has been some question about the internormalization of these measurements with others done by holding the direction of one exiting electron fixed, but previously unpublished results in Röder's thesis have recently resolved that discrepancy [31]. From 19.6 eV to 30 eV only relative measurements are available, but excellent agreement with them is attained in ECS calculations, as is demonstrated in Fig. 5.4.

5.4.2 Formal Rearrangement Theory and Scattering Amplitudes for Three-Body Breakup

While the straightforward evaluation of quantum–mechanical flux has the appeal that it corresponds to the most basic formal definition of the cross

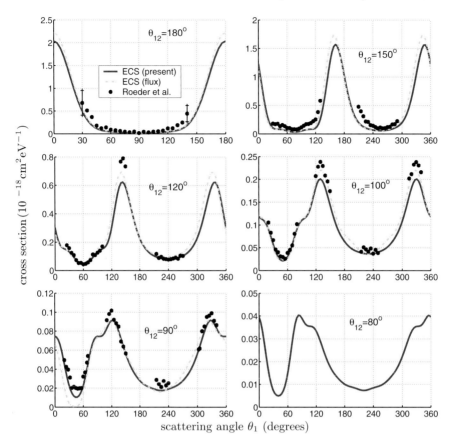

Fig. 5.3. Equal-energy sharing, coplanar TDCS for electron–hydrogen ionization at 17.6 eV incident energy shown for geometries with θ_{12} fixed. Experimental data are absolute measurements of Röder et al. with 40% error bars. *Dark solid curves*: integral expression for breakup amplitude, *lighter curves*: flux extrapolation

sections, it is not as efficient, even for simple inelastic scattering, as the calculation of scattering amplitudes via matrix elements that depend only on the range of the interaction potential. It is therefore advantageous, both computationally and theoretically, to confront the dilemma posed by the formal theory in (5.2)–(5.4).

The question of how to formulate a procedure for extracting breakup amplitudes from a wave function that is only known numerically on a finite grid was addressed in a series of recent papers [8,22,23]. In the first of these studies [22] we showed that, even in cases that involve only short-ranged potentials, some formally correct integral expressions for the breakup amplitudes can

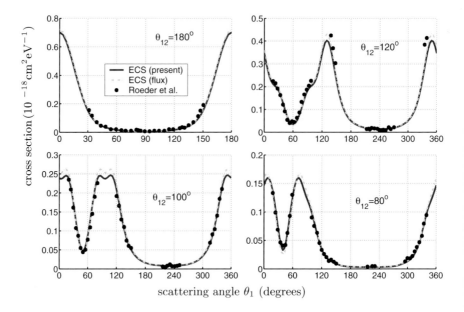

Fig. 5.4. TDCS for 25 eV incident energy. Normalization factor to convert measured values of Röder et al. from arbitrary units is 0.16

yield numerically unstable or poorly convergent results. For example, the expression

$$f = \langle \boldsymbol{p}_1, \boldsymbol{p}_2 | V | \Psi^+ \rangle, \tag{5.29}$$

where the final state is just a product of plane waves, while providing a formally correct breakup amplitude for short-ranged potentials, was found to be numerically unstable. The instability can be traced to "free–free" overlap terms that arise from discrete two-body channels in the scattered wave function. On an infinite grid, these terms are proportional to momentum–conserving delta functions, which therefore contribute nothing to the breakup amplitude, but on a finite grid, they are a source of numerical error. A practical solution was found by using formal rearrangement theory to express the amplitude in terms of distorted waves. A series of formal manipulations, combined with Green's theorem, allows us to express the breakup amplitude as a surface integral:

$$f = \frac{1}{2} \int_S (\phi_{k_1}^{(+)} \phi_{k_2}^{(+)} \nabla \Psi_{\text{sc}}^+ - \Psi_{\text{sc}}^+ \nabla \phi_{k_1}^{(+)} \phi_{k_2}^{(+)}) \cdot d\hat{\mathbf{S}}, \tag{5.30}$$

where the functions $\phi_k^{(+)}$ are distorted waves derived from the one-body terms in the interaction potential [22].

For Coulomb problems, the obvious extension is to employ Coulomb functions as distorted waves in (5.30). This is, however, at odds with the formal theory, which states that the integral expression in (5.30) will have a divergent phase unless the Coulomb functions are chosen with effective charges that satisfy (5.3). But the use of effective charges other than unity in the Coulomb functions that define the final state have the unfortunate property of destroying their orthogonality to the bound states of the hydrogen atom. We showed in [23] that, on a finite volume, the effect of using Coulomb functions with $Z = 1$ in computing the ionization amplitudes is merely to introduce an inconsequential overall phase that has no effect on the cross section. It is the application of the integral formula, together with the ECS method, to ionization of hydrogen that has given the most accurate description of the complete dynamics to date [8] and which does, in fact, "reduce the problem to practical computation". The TDCS results obtained from the integral amplitudes, which are also shown in Figs. 5.3 and 5.4, attest to the accuracy of the approach and also validate the fundamental correctness of the earlier flux–extrapolation approach.

The magnitudes and shapes of the singly differential cross sections at low energies give a particularly compelling demonstration of the accuracy of the ECS approach and make a satisfying connection with the semiclassical theories that have been applied to the threshold behavior of the ionization process. Figure 5.5 compares the SDCS computed by flux extrapolation and from integral amplitudes at incident energies from 15.4 eV (only 2 eV above the ionization threshold) to 54.4 eV. At lower energies the flux and integral formula methods for computing the SDCS disagree by as much as 10%, because the extrapolation of the flux becomes increasingly difficult as the energy is lowered. However, no such difficulty affects the integral expression in (5.30). At very low energies the SDCS is almost flat and almost constant as a function of incident energy. If it were flat and constant it would correspond to a linear threshold law for the total cross section. In semiclassical calculations at the Wannier geometry with electrons exiting in opposite directions, Rost [32] predicted qualitatively the subtle departures from flatness as the SDCS turns from a "smile" at high energies to a nearly flat shape at energies near threshold.

5.5   Multielectron Targets

In electron-impact collisions involving more complicated atoms, even for a two-electron target, there are ionization processses that cannot occur for a one-electron atom: excitation-ionization, excitation-autoionization, and double ionization. In excitation-ionization, the atom is singly ionized and the residual ion is left in an excited state. In excitation-autoionization, the target is first excited to an autoionizing state that can then decay into the ionization continuum in a process that competes with direct ionization at

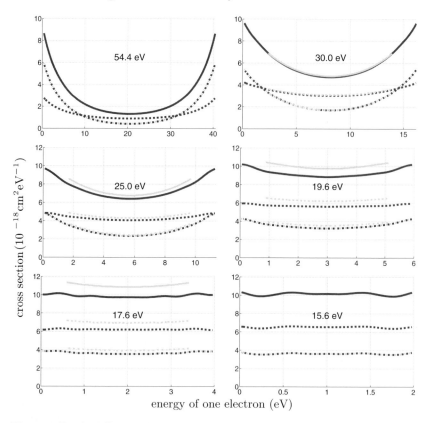

Fig. 5.5. Singly differential cross sections for e–H scattering at the collision energies indicated. Individual components for singlet (*dashed line*) and triplet (*dot-dash line*) are shown. Where applicable, results based on flux-extrapolation are shown in *light gray*

the same energy. Double ionization is the (e,3e) process in which there are three free electrons in the final state.

While we are still far from the goal of carrying out a fully *ab initio* treatment of electron-impact ionization with a multielectron target, there have already been a few proof-of-principle demonstrations involving model problems [33,26] that treat ionization of atomic targets with "two active electrons". These preliminary studies have served to establish the fact that time-dependent approaches certainly have the scaling properties that allow their application to three-electron systems and signal the emergence of a new level of sophistication in ionization studies that will go beyond currently available methods that treat multielectron atoms with frozen-core, one-electron models.

5.5.1   Asymptotic Subtraction

The extraction of ionization cross sections from a numerical representation of the scattered wave on a finite grid is substantially more difficult with a multielectron target than with a one-electron target. If one attempts to compute the ionization cross sections from an integral expression for the breakup amplitudes, one finds that in contrast to the one-electron target case, the use of distorted waves alone is not sufficient to eliminate numerical instabilities caused by discrete two-body channel terms in the scattered wave and additional steps are required to obtain a viable formula. To see why this is the case, we need only consider the asymptotic form of the scattered wave function for the case of a two-electron target at energies where both single ionization and two-body channels are open. For simplicity, we consider a case with no angular momentum. Asymptotically, there will be two-body terms in the scattered wave of the form $f_n e^{ik_n r_1} \chi_n(r_2, r_3)$, where $\chi_n(r_2, r_3)$ is a two-electron target bound state and f_n is the corresponding excitation amplitude, as well as an ionization term of the form $f_{\text{ion}}^m e^{ik\rho_{12}} \varphi_m(r_3)/\sqrt{\rho_{12}}$, where $\rho_{12} = \sqrt{r_1^2 + r_2^2}$ and φ_m is a bound state of the residual ion.

Now suppose we attempt to compute the single–ionization amplitude from an expression

$$f(k_1, k_2) = \int_S \Big(\phi_{k_1}(r_1)\phi_{k_2}(r_2)\varphi_n(r_3)\nabla\psi_{\text{sc}}^+(r_1, r_2, r_3)$$

$$-\psi_{\text{sc}}^+(r_1, r_2, r_3)\nabla\phi_{k_1}(r_1)\phi_{k_2}(r_2)\varphi_n(r_3) \Big) \cdot \hat{\boldsymbol{n}}\mathrm{d}S, \qquad (5.31)$$

which is an obvious generalization of (5.30) for a two-electron target. Since there is no orthogonality relation between the distorted waves and the *two-body* bound states, the two-body terms in the scattered wave will again give rise to overlaps between free functions in (5.31), which render it numerically unstable. One way to remedy this is first to evaluate the two-body amplitudes from the formula

$$f_n = 2\langle \sin(k_n r_1)\chi_n(r_2, r_3)|E - T - V_1|\Psi_{\text{sc}}\rangle, \qquad (5.32)$$

since there are no formal or numerical problems associated with evaluation of (5.32). We can then construct an "asymptotically subtracted" scattered wave

$$\Psi_{\text{sc}}^{\text{proj}} = \Psi_{\text{sc}} - \sum_n (f_n/k_n)e^{ik_n r_1}\chi_n(r_2, r_3), \qquad (5.33)$$

which removes the two-body channels from the asymptotic scattered wave. If we use $\Psi_{\text{sc}}^{\text{proj}}$ in (5.31), then there is, in principle, no contamination of the ionization amplitude from two-body channels and the surface integral extracts the ionization amplitude just as it does in the case of a one-electron target.

Table 5.1. Integrated cross sections for model 3D problem discussed in text

Incident energy	σ_{elastic}	σ_{ion}	$\sigma_{\text{elastic}} + \sigma_{\text{ion}}$	σ_{optical}	σ_{flux}
7 eV	17.238	3.173	20.411	19.681	20.237
9 eV	11.578	2.592	14.170	14.144	14.330
11 eV	8.568	2.176	10.744	10.708	10.883

When the collision energy moves above the threshold for double ioniza-tion, the asymptotic scattered wave will also contain a term proportional to $e^{iK\rho}/\rho$, where $\rho = \sqrt{r_1^2 + r_2^2 + r_3^2}$ and $E = K^2/2$. This term will again cause difficulties in the integral expression for the *single*–ionization amplitude. For-tunately, the following integral can be used to compute the amplitude for double ionization:

$$f_{\text{ion}}^{\text{double}}(k_1, k_2, k_3) = \langle \phi_{k_1} \phi_{k_2} \phi_{k_3} | E - T - V_1 | \Psi_{\text{sc}}^{\text{proj}} \rangle, \tag{5.34}$$

where $E = k_1^2/2 + k_2^2/2 + k_3^2/2$. If the distorted waves are chosen to be eigenstates of the one-body potential, then orthogonality between the dis-torted waves and the one-body bound states φ_m prevents the asymptotic single-ionization terms in $\Psi_{\text{sc}}^{\text{proj}}$ from causing any numerical problems. One can then extend the definition of $\Psi_{\text{sc}}^{\text{proj}}$ to include the double–ionization term,

$$\Psi_{\text{sc}}^{\text{proj}'} = \Psi_{\text{sc}} - \sum_n (f_n/k_n) e^{ik_n r_1} \chi_n(r_2, r_3) - f_{\text{ion}}^{\text{double}} e^{iK\rho}/\rho, \tag{5.35}$$

before using (5.31) to compute the amplitudes for single ionization. By fol-lowing these steps, we can, in principle, compute all the scattering amplitudes of interest in the case of a two-electron target.

We have tested these ideas in a model 3-electron problem that involves only exponentially bound one- and two-body potentials [26]. The potential strengths were chosen so that the target 'atom' and 'ion' each bind a single state, so the only channels possible are elastic scattering and breakup. The scattered waves were computed on a three-dimensional radial grid by the time-dependent version of ECS outlined in Sect. 5.2. Figure 5.6 plots the real part of the scattered wave for a fixed value of r_3, before and after subtraction of the elastic channel, at an incident energy of 11 eV. The elastic two-body component, clearly visible in the unsubtracted scattered wave near the r_1 and r_2 axes, are effectively removed by the asymptotic subtraction scheme adopted.

In Table 5.1, we show the elastic scattering cross sections together with the total ionization cross sections computed by integrating the SDCS, the lat-ter computed from the asymptotically subtracted scattered wave. The sum of these two quantities is the total cross section, which can be evaluated independently from the optical theorem or from the total flux. The differ-ence between these quantities gives some indication of the overall numerical accuracy of the results.

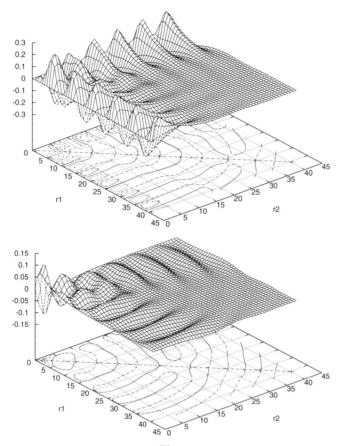

Fig. 5.6. Scattered wave, $\psi_{sc}^{(3)}$, for model 3D problem discussed in text, before (*upper plot*) and after (*lower plot*) asymptotic subtraction of elastic channel. All plotted quantities are in atomic units. The real part of $\psi_{sc}^{(3)}$ is plotted as a function of r_1 and r_2 with r_3 fixed at 0.1 Bohr radius. The incident electron energy is 11 eV

5.6   Conclusion

Theoretical and computational advances over the past few years have brought us to a point where, for the simplest (e,2e) problems, it is accurate to say that the problem has been "reduced to practical computation". For such simple systems, it will shortly become a routine matter to computationally explore all aspects of collisional breakup, including noncoplanar geometries together with unequal energy sharing only a few volts above threshold. Despite this progress, there are still questions to be answered, even in systems of only three charged particles. One notable problem yet to be solved is that of positron–impact ionization, where ionization competes with positronium formation. For collisional ionization of multielectron atoms, there are still many details to be worked out and there are still open questions about what will ultimately prove to be the best way to extract ionization cross sections from the wave functions once they are available. Despite the challenges that remain, we are confident that benchmark calculations on the electron–helium system, similar to those that now exist for the electron–hydrogen system, will appear in the next few years.

Acknowledgments

This work was performed under the auspices of the U.S. Department of Energy by the University of California Lawrence Berkeley National Laboratory and Lawrence Livermore National Laboratory under contract numbers DE-AC03-76F00098 and W-7405-Eng-48, respectively. The work was supported by the US DOE Office of Basic Energy Science, Division of Chemical Sciences, and computations were performed on the computers of the National Energy Research Scientific Computing Center.

References

1. R.K. Peterkop: Opt. Spectrosc. **13**, 87 (1962)
2. M.R.H. Rudge and M.J. Seaton: Proc. Roy. Soc. A **283**, 262 (1965)
3. C.W. McCurdy, T.N. Rescigno, and D. Byrum: Phys. Rev. A **56**, 1958 (1997)
4. C.W. McCurdy and T.N. Rescigno: Phys. Rev. A **56**, R4369 (1997)
5. T.N. Rescigno, M. Baertschy, W.A. Isaacs, and C.W. McCurdy: Science **286**, 2474 (1999)
6. M. Baertschy, T.N. Rescigno, W.A. Isaacs, X. Li, and C.W. McCurdy; Phys. Rev. A **63**, 022712 (2000)
7. W.A. Isaacs, M. Baertschy, C.W. McCurdy, and T.N. Rescigno: Phys. Rev. A **63**, 030704R (2001)
8. M. Baertschy, T.N. Rescigno, and C.W. McCurdy: Phys. Rev. A **64**, 0022709 (2001)
9. I. Bray and A.T. Stelbovics: Phys. Rev. Lett. **69**, 53 (1992)
10. I. Bray: Phys. Rev. Lett. **78**, 4721 (1997)
11. I. Bray: J. Phys. B **33**, 581 (2000)
12. K. Bartschat and I. Bray: J. Phys. B **29**, L577 (1996)
13. D. Kato and S. Watanabe: Phys. Rev. Lett. **74**, 2443 (1995)
14. N. Miyashita, D. Kato, and S. Watanabe: Phys. Rev. A **59**, 4385 (1999)
15. M.S. Pindzola and F. Robicheaux: Phys. Rev. A **54**, 2142 (1996)
16. P. Selles, L. Malegat, and A.K. Kazansky: Phys. Rev. A **65**, 032711 (2002)
17. M.R.H. Rudge: Rev. Mod. Phys. **40**, 564 (1968)
18. B. Simon: Phys. Lett. A **71**, 211 (1979)
19. C.A. Nicolaides and D. R.Beck: Phys. Lett. A **65**, 11 (1978)
20. T.N. Rescigno, M. Baertschy, D. Byrum, and C.W. McCurdy: Phys. Rev. A **55**, 4253 (1997)
21. M. Baertschy, T.N. Rescigno, W.A. Isaacs, and C.W. McCurdy: Phys. Rev. A **60**(R13) (1999)
22. C.W. McCurdy and T.N. Rescigno: Phys. Rev. A **62**, 032712 (2000)

23. C.W. McCurdy, D.A. Horner, and T.N. Rescigno: Phys. Rev. A **63**, 022711 (2001)
24. J.W. Demmel, S.C. Eisenstat, J.R. Gilbert, X.S. Li, and J.W.H. Liu: SIAM J. Matrix Analysis Applic. **20**, 720 (1999)
25. T.N. Rescigno and C.W. McCurdy: Phys. Rev. A **62**, 032706 (2000)
26. C.W. McCurdy, D.A. Horner, and T.N. Rescigno: Phys. Rev. A **65**, 042714 (2002)
27. C.W. McCurdy and C.K. Stroud: Comput. Phys. Commun. **63**, 323 (1991)
28. C.W. McCurdy, C.K. Stroud, and M. Wisinski: Phys. Rev. A **43**, 5980 (1991)
29. M.D. Feit, J.D.F. Jr., and A. Steiger: J. Comput. Phys. **47**, 412 (1982)
30. J. Röder, J. Rasch, K. Jung, C.T. Whelan, H. Ehrhardt, R. Allan, and H. Walters: Phys. Rev. A **53**, 225 (1996); J. Röder, H. Ehrhardt, C. Pan, A.F. Starace, I. Bray, and D. Fursa: Phys. Rev. Lett. **79**, 1666 (1997)
31. J. Röder, M. Baertschy, and I. Bray: Phys. Rev. A **67**, 010702 (2003)
32. J.M. Rost: Phys. Rev. Lett. **72**, 1998 (1994)
33. M.S. Pindzola, D. Mitnik, and F. Robicheaux: Phys. Rev. A **59**, 4390 (1999)

6 Hyperspherical $\mathcal{R}$-Matrix with Semiclassical Outgoing Waves

L. Malegat, P. Selles, and A.K. Kazansky

6.1 The Double-Electronic Continuum Problem

The 'γ, 2e' process, where one photon pulls two electrons out of an atom or a molecule, and the 'e, 2e' process, in which one incident electron knocks one electron out of the target, both lead to a final state composed of two electrons in the continuum of an ionic core. The description of this final state sets a very difficult problem to theoreticians.

The reason for this is best understood if one considers the leading members of these two families of processes, namely the double photoionization (DPI) of He and the electron-impact ionization of H, in the case where the excess energy above the threshold divides itself in a balanced way between the two outgoing electrons. In these two cases, the three particles carry comparable charges and are separated by comparable distances, so that the three Coulomb pair-interactions in the system are of the same order of magnitude. This defines a strongly correlated motion. The important point is that, due to the long range of Coulomb interactions, this three-particle motion remains strongly correlated up to very large interparticle distances.

This property puts the description of this system beyond the scope of the standard methods of either quantum chemistry or collision theory. Actually, quantum chemistry has developed sophisticated treatments of the electron-correlation problem, but within the framework of *finite–range bound states*. On the other side, collision theory considers the case of *a single particle in the continuum*. When this particle is far away from the target, the particle–target interaction takes on a very simple form no matter how complex the target is, as it can be approximated by the lowest term in its expansion in inverse powers of the particle–target distance r. As a result, one is able to solve the resulting two-body asymptotic problem exactly and thoroughly: this means that a complete basis of exact solutions can be found. The Coulomb spherical waves form such a basis in the case of a charged target, and so do the free spherical waves in the case of a neutral target. This knowledge of the solution of the asymptotic problem turns out to be the cornerstone of standard collision theory: actually, the concepts of phase shifts, or, more generally, of $\boldsymbol{K}$, $\boldsymbol{S}$ or $\boldsymbol{T}$ matrices, are defined with respect to this solution. When a second electron is excited into the continuum, the asymptotic problem defined by large values of the electron–target distances r_1 and r_2 is a real three-body problem that cannot be reduced to a pair of two-body problems due to the long-range electron–electron interaction in the continuum. The full exact solution of this problem is unknown. Accordingly, neither the standard form of collision theory nor any obvious generalization of it apply to the double continuum problem. This is well illustrated by the lasting difficulties encountered by the convergent close–coupling (CCC) [1–4] theory in describing 'γ, 2e' or 'e, 2e' processes. Actually, *the treatment of this three-particle correlated motion extending over all space requires a specific approach.*

Recently, three new approaches to this challenging problem have appeared. They have in common to turn their back to the standard collision theory: the authors do not resign themselves to applying the two-body boundary condition attached to this method to the three-body problem of interest; they do not attempt to formulate an explicit three-body boundary condition either; they simply work out methods where the need for an explicit boundary condition is circumvented. The most obvious way to attain this end is to use a time-dependent approach as in the time dependent close coupling (TDCC) method [5–7]. However, this goal can also be achieved within the framework of a stationary approach. This is demonstrated in another chapter of this book as to the external complex scaling (ECS) method [8–11]. Here, we will show how the hyperspherical $\mathcal{R}$-matrix with semiclassical outgoing waves (HRM-SOW) method [12,13] manages to complete this task. A detailed account of the method has been given in [13], along with a wide sample of results. Here, we do not want the reader to get bogged down in technical details or an abundance of cross sections. Our concern instead is to focus on the physical specificities of the approach and to illustrate them by a limited set of examples, in as pedagogical a way as possible.

6.2   The 'γ, 2e' Case: a Stationary Formulation

The HRM-SOW method has been developed in the framework of the He double photoionization problem, but it can be easily adapted to the electron-impact ionization problem. In this chapter, we consider the 'γ, 2e' case, which lends itself to a simpler presentation. The equation that is central to the method is then best introduced starting from the time-dependent Schrödinger equation

$$i\frac{\partial}{\partial t}\Psi(\boldsymbol{r_1},\boldsymbol{r_2},t) = \left(H_0(\boldsymbol{r_1},\boldsymbol{r_2}) + \frac{1}{2}\mathbf{E_0}.\boldsymbol{d}\,e^{-i\omega t}\right)\Psi(\boldsymbol{r_1},\boldsymbol{r_2},t) \qquad (6.1)$$

involving the field-free Hamiltonian H_0 and the atom–field interaction expressed in the resonant dipole approximation that is relevant to the case of a synchrotron source. $\Psi(\boldsymbol{r_1},\boldsymbol{r_2},t)$ is the two-electron wavepacket, $\boldsymbol{d}$ is the dipole moment of the atom, ω is the frequency of the incident radiation and $\boldsymbol{E_0}$ is its associated electric field. Note that the study of the plain case where the incident radiation is 100% linearly polarized provides a sufficient basis for the treatment of more complicated situations where the polarization state of the light is characterized by a full set of nonzero Stokes parameters. We then solve (6.1) within the framework of perturbation theory, looking for a solution in the form

$$\Psi(\boldsymbol{r_1}, \boldsymbol{r_2}, t) = \mathrm{e}^{-\mathrm{i}E_0 t}\Psi_0(\boldsymbol{r_1}, \boldsymbol{r_2}) + \mathrm{e}^{-\mathrm{i}(E_0+\omega)t}\Psi_1(\boldsymbol{r_1}, \boldsymbol{r_2}) \qquad (6.2)$$

of a superposition of the He ground state Ψ_0 of energy E_0 and the state that is reached after absorption of one photon, weighted by the appropriate time-dependent phase factors. Let us stress that the perturbative treatment here applies to the atom–field interaction and not to the electron–electron interaction, as in the usual "perturbative treatments" of the 'e, 2e' and 'γ, 2e' processes. At photon energies above the double–ionization threshold, a large number of competing channels are open, including single ionization to the ground and excited states of He$^+$ as well as double ionization. The photoabsorption state Ψ_1 contains the full information regarding all these coupled one-photon processes. Putting (6.2) into (6.1) yields the following stationary equation

$$(H_0 - E)\,\Psi_1(\boldsymbol{r_1}, \boldsymbol{r_2}) = -\frac{1}{2}\mathbf{E_0}.\boldsymbol{d}\,\Psi_0(\boldsymbol{r_1}, \boldsymbol{r_2}) \qquad (6.3)$$

for Ψ_1, where $E = E_0 + \omega$. The inhomogeneous term on the right–hand side acts as a source term that feeds the photoabsorption state Ψ_1. It is proportional to Ψ_0 – hence of short range – and to the atom–field interaction operator – the explicit expression of which depends on the gauge used to describe the electromagnetic field. The purpose of the H$\mathcal{R}$M-SOW method is to solve (6.3) for outgoing waves boundary conditions in order to get Ψ_1 and to extract from it the information it contains regarding the single- and, especially, the double-ionization process.

6.3   Outline of the Three-Step H$\mathcal{R}$M-SOW Resolution Scheme

As one can infer from its name, the H$\mathcal{R}$M-SOW method relies upon hyperspherical coordinates, consisting of the hyperspherical radius $R = \sqrt{r_1^2 + r_2^2}$ that measures the overall size of the system, and five angles, denoted collectively by Ω_5 for convenience. Ω_5 includes the spherical angles $\theta_1, \varphi_1, \theta_2, \varphi_2$ specifying the directions of the two electrons with respect to the main axis of polarization of the incident radiation, and the angle $\alpha = \tan^{-1}(r_1/r_2)$ that describes the radial correlation between the two electrons.

The choice of (R, α) instead of (r_1, r_2) to define the radial distances of the electrons has many advantages. First, one variable instead of two goes to ∞, which is favorable from a numerical point of view. Also, the calculation of the wavefunction flux through a hypersphere $R = \text{constant}$, which is closely related to the cross sections, is very easy in the first case since the wave function is obtained directly on the hypersphere. In the second case, by contrast, a two-dimensional interpolation procedure is needed to get the values of the wave function on the hypersphere from its values on a two-dimensional $r_1 \times r_2$ mesh. Finally, single ionization to the various ionic states and double ionization, which are all accounted for by Ψ_1, can be disentangled based on the values of the angle α when $R \to \infty$. Two cases can occur in this limit actually. Either r_1 and r_2 tend to ∞, having a nonzero finite ratio associated to $\alpha \in]0, \pi/2[$, which corresponds to double ionization. Or r_1 or r_2 tends to ∞ so that their ratio is either null or infinite, leading to $\alpha = 0$ or $\pi/2$, which corresponds to single ionization. In the double–ionization case, in addition, r_1/r_2 tends towards the ratio of the speeds of the two electrons, that is to say towards the square root of the energy ratio E_1/E_2, so that the value of α is related to the sharing of the energy between the two continuum electrons.

In practice, at any large but finite R, single ionization is confined within angular sectors of width $\Delta\alpha = \sin^{-1}(r_{n_{max}}/R)$ around $\alpha = 0$ and $\Delta\alpha = \pi/2 - \cos^{-1}(r_{n_{max}}/R)$ around $\alpha = \pi/2$ where $r_{n_{max}} \simeq n_{max}^2$ a.u. is the mean value of r in the highest Rydberg state of the ion that is significantly excited in the one-photon absorption process considered. As $n_{max} = 5$ seems a realistic upper limit from the experimental literature on the subject [14], we obtain $\Delta\alpha \simeq 0.025$ rad at $R = 10^3$ a.u., which is already a strong confinement. In other words, if we manage to get the solution Ψ_1 at large enough distances R, we will be able to extract the double ionization contribution from the full signal using the value of α. On the other hand, this strong confinement will preclude the resolution of the various single–ionization channels using this technique.

The other basic feature of the method is the splitting of configuration space into two regions depending on the value of the hyperspherical radius. In the inner region, defined by $R < R_0$, we use a quantum approach based on $\mathcal{R}$-matrix and adiabatic partial waves expansion techniques. In the complementary external region, we retain a quantum treatment for all angles, while assuming that the R-motion is regular enough to be treated semiclassically.

We then proceed in three steps. First, we extract Ψ_1 on the hypersphere $R = R_0$. Next, we propagate Ψ_1 from R_0 to a very large hyperradius denoted R_{max}. Finally, we extract the cross sections from their definition by computing the corresponding flux of Ψ_1 through the hypersphere $R = R_{max}$.

Before explaining each of these steps in more detail, let us recast (6.3) into a more appropriate form by premultiplying it by the factor $R^{5/2} \sin 2\alpha$ and by introducing the reduced function $\overline{\Psi}_1(R; \Omega_5) = R^{5/2} \sin 2\alpha \, \Psi_1(R; \Omega_5)$ to get

$$\left(\overline{H}_0 - E\right) \overline{\Psi}_1(R; \Omega_5) = \overline{S}, \tag{6.4}$$

where $\overline{H}_0$ and $\overline{S}$ are the modified Hamiltonian and source term, respectively.

6.4   First Step: Extraction of the Solution at R_0

The extraction procedure is based on three relations which have different origins and status, two of them being exact, and the other approximate. Below we present each of them and discuss the validity of the entire procedure.

6.4.1 $\mathcal{R}$-Matrix Relation

The first relation is the $\mathcal{R}$-matrix one that can be deduced from (6.4) by a standard sequence of formal manipulations including: (i) the introduction of the additional equation

$$\left(\overline{H}_0 + B - \varepsilon_k\right) \Phi_k = 0, \quad R \leq R_0, \tag{6.5}$$

which is made hermitic over the finite inner region by the addition of the appropriate Bloch operator B; (ii) the introduction of partial wave expansions of both $\overline{\Psi}_1$ and the Φ_k,

$$\overline{\Psi}_1(R; \Omega_5) = \sum_i f_i(R) x_i(\Omega_5) \quad \text{and} \quad \Phi_k(R; \Omega_5) = \sum_i g_i^k(R) x_i(\Omega_5), \tag{6.6}$$

involving some appropriate orthonormal angular basis set $x_i(\Omega_5)$; (iii) the left multiplication of (6.4) by Φ_k^* and of (6.5) by $\overline{\Psi}_1^*$, the subtraction of the two equations obtained, and the integration of the resulting relation over the finite inner region with the help of (6.6). One thus obtains the linear relationship

$$\boldsymbol{f}'(R_0) = \boldsymbol{\mathcal{R}}(R_0)\,\boldsymbol{f}(R_0) + \boldsymbol{\mathcal{V}}(R_0) \tag{6.7}$$

between the vector $\boldsymbol{f}$, having as components the hyperradial channel functions f_i, and the vector $\boldsymbol{f}'$ having as components their derivatives, at $R = R_0$. The matrix $\boldsymbol{\mathcal{R}}$ is constructed from the eigenvalues ε_k of (6.5) and from the hyperradial channel functions $g_i^k(R_0)$ associated with the corresponding eigenvectors Φ_k. The vector $\boldsymbol{\mathcal{V}}$ involves, in addition to these elements, the projections of the source term in (6.4) on the Φ_k. Both $\boldsymbol{\mathcal{R}}$ and $\boldsymbol{\mathcal{V}}$ depend analytically on the total energy E, which is a well–known advantage of the $\mathcal{R}$-matrix approach.

6.4.2   Local Properties of the Adiabatic Channels

The second relation we use is obtained by considering the expansion

$$\overline{\Psi}_1(R; \Omega_5) = \sum_\lambda F_\lambda(R)\, X_\lambda(R; \Omega_5) \tag{6.8}$$

of $\overline{\overline{\Psi}}_1$ on the adiabatic angular basis $X_\lambda(R;\Omega_5)$ that diagonalizes $\overline{H}_0$ *at fixed* R. We note $\overline{\overline{H}}_0$ the resulting operator, where the differential term in R has been suppressed. We then have

$$\left(\overline{\overline{H}}_0(R;\Omega_5) - E_\lambda(R)\right) X_\lambda(R;\Omega_5) = 0 \,. \tag{6.9}$$

The eigenvectors of (6.9) can be viewed as *locally adapted* angular basis vectors depending parametrically on R. The associated eigenvalues are referred to as *adiabatic potentials*. If we put (6.8) into (6.4), and project onto each adiabatic angular basis vector, we obtain the set of coupled differential equations for the hyperradial channel functions

$$-\frac{1}{2}\frac{\mathrm{d}^2}{\mathrm{d}R^2}F_\lambda(R) - \sum_{\lambda'}\langle X_\lambda|\frac{\partial X_{\lambda'}}{\partial R}\rangle F_{\lambda'}(R) - \frac{1}{2}\sum_{\lambda'}\langle X_\lambda|\frac{\partial^2 X_{\lambda'}}{\partial R^2}\rangle F_{\lambda'}(R)$$
$$- (E - E_\lambda(R))\, F_\lambda(R) = \langle X_\lambda|\,\overline{S}\rangle \,, \tag{6.10}$$

where the brackets stand for integration over the angular variables Ω_5. This system simplifies drastically if we restrict it to a very small interval surrounding R_0, *provided R_0 is taken large enough*. Actually, we can then neglect the short-range right–hand side, as well as the variation of the adiabatic potentials and adiabatic angular vectors with R, hence omitting the so-called nonadiabatic couplings involving the radial derivatives of the adiabatic angular vectors, and putting $E_\lambda(R) = E_\lambda(R_0)$. We then obtain the system of uncoupled differential equations

$$\frac{\mathrm{d}^2}{\mathrm{d}R^2}F_\lambda(R) + p_\lambda^2(R_0)\, F_\lambda(R) \simeq 0 \,, \quad \text{with} \quad p_\lambda^2(R_0) = 2\,(E - E_\lambda(R_0)) \,, \tag{6.11}$$

the solution of which is given by ingoing and outgoing waves associated with the channel-dependent local momenta $p_\lambda(R_0)$. The physically acceptable solutions are given by outgoing waves, and their derivatives satisfy

$$\boldsymbol{F}'(R_0) \simeq \mathrm{i}\,\boldsymbol{p}(R_0)\,\boldsymbol{F}(R_0) \,, \tag{6.12}$$

where the components of the vectors $\boldsymbol{F}$ and $\boldsymbol{F}'$ are the adiabatic hyperradial channel functions and their derivatives at R_0, while the diagonal elements of the diagonal matrix $\boldsymbol{p}$ are the channel momenta defined in (6.11).

6.4.3   Frame Transformation

The third relation we use is the frame transformation

$$\boldsymbol{X}(R_0;\Omega_5) = \boldsymbol{x}(\Omega_5)\,\boldsymbol{U}(R_0) \tag{6.13}$$

that connects the two angular basis sets that we have introduced above. The corresponding sets of hyperradial channel functions are then related by

$$\boldsymbol{F}(R_0) = \boldsymbol{U}^\dagger(R_0)\,\boldsymbol{f}(R_0) \,. \tag{6.14}$$

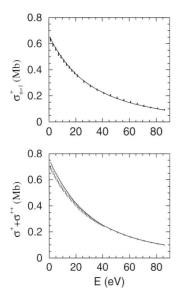

Fig. 6.1. *Bottom panel*: total photoionization cross section of He, in Mb, versus the excess energy above the double-ionization threshold, in eV. *Dots* with error bars: experiment [15]; *Solid lines*: H$\mathcal{R}$M-SOW calculations [13]. *Top panel*: total cross section for single ionization of He without excitation, in Mb, versus the excess energy above the double–ionization threshold, in eV. *Dots with error bars*: experimental data compilation [14]; *Solid line*: H$\mathcal{R}$M-SOW calculations [13]

6.4.4   Discussion

The solution can then be extracted by putting together the $\mathcal{R}$-matrix relation (6.7) and the property (6.12) of the adiabatic channel functions given the frame transformation (6.14). Unlike (6.7) and (6.14), which are exact, (6.12) is approximate. It relies upon many assumptions regarding the physical situation at R_0. First, the source term is considered as negligible, which is very likely to be valid given its exponentially decaying behavior. Secondly, the R-dependence of the adiabatic potentials and vectors is assumed to be weak. And thirdly, the solution is expected to take on the form of an outgoing wave according to its asymptotic behavior. Note that this third assumption will probably hold if the second does insofar as smooth potentials are not likely to reflect outgoing waves. The key point then lies in the smoothness of the R-dependence of the quantities that control the dynamics.

One way to check the validity of these hypotheses is to compute the total outgoing flux of the resulting $\Psi_1(R_0; \Omega_5)$ through the hypersphere $R = R_0$ and to divide it by the incoming photon flux in order to get the total ionization cross section $\sigma^+ + \sigma^{++}$. The bottom panel of Fig. 6.1 shows that this quantity, calculated for He using $R_0 = 10$ a.u., is in good agreement with the measurements over a wide energy range. This provides strong support to the extraction procedure proposed.

Interestingly, the lowest adiabatic energy $E_{\lambda=0}(R_0)$ is very close to the energy of one free electron a distance R_0 away from the He$^+$ ion in its ground state. This suggests we should identify the lowest adiabatic channel in (6.8) with the lowest single–ionization channel. The single–ionization cross section to the ground ionic state $\sigma_{n=1}^+$ obtained acccordingly when one normalizes the flux associated to the lowest adiabatic channel by the incoming photon flux is represented on the top panel of Fig. 6.1. It agrees very well with the experimental data over the same energy range as considered before for $\sigma^+ + \sigma^{++}$. The next adiabatic channels, however, cannot be attributed to the next ionization channels. This is not surprising since the *mean* radius of the first excited state of the ion is of the order of 4 a.u., which means that this state hardly fits within the inner hypersphere. This means that, at the relatively short distance $R_0 = 10$ a.u. considered, the lowest single–ionization channel is already decoupled from the highest ionization channels, which, by contrast, are still tightly coupled together. This identification between adiabatic channels and ionization channels is thus very partial. Yet we shall see in the following that it plays an important role in the theory.

Let us compare now the present approach with standard collision theory. First, the $\mathcal{R}$-matrix relation (6.7) differs from the $\mathcal{R}$-matrix relation of collision theory in two respects: (i) because the radial coordinate in (6.7) is the hyperradius R instead of the radius vector r of the *single* continuum particle of collision theory; and (ii) because (6.7) is inhomogeneous due to the source term in (6.4). Secondly, whatever the approach used, be it hyperspherical or standard $\mathcal{R}$-matrix, the solution can only be obtained if the $\mathcal{R}$-matrix relation is complemented with a second independent relation involving the channel functions and their derivatives. In the standard approach, this second relation is obtained from the known general form of the solution in the asymptotic region. In the present hyperspherical approach, where this general form is unknown, it is the property (6.12) of the adiabatic channel functions that provides the required alternative.

6.5 Second Step:
Propagation of the Solution from R_0 to $R_{\max}$

The only cross sections that we have been able to extract from the solution obtained at $R_0 = 10$ a.u. are $\sigma^+ + \sigma^{++}$ and $\sigma_{n=1}^+$. Propagating the solution beyond R_0 to much larger distances will now allow us to access the double-photoionization cross sections, which are the main object of our interest.

6.5.1 Defining a Semiclassical Treatment of the R-motion

In the external region $R \geq R_0$, we assume that the short-range inhomogeneous term in (6.4) can be neglected, and that the R-motion is smooth enough to be treated semiclassically. Accordingly, we look for the solution of the homogeneous counterpart of (6.4) in the form

$$\overline{\Psi}_1(R;\Omega_5) \simeq \frac{1}{\sqrt{p(R)}}\, \mathrm{e}^{\mathrm{i}\int_{R_0}^{R} p(R')\,\mathrm{d}R'} \times \widetilde{\Psi}_1(R;\Omega_5) \tag{6.15}$$

of the product of a hyperradial semiclassical *outgoing* wave, associated with the local momentum $p(R)$, by a reduced function of the angles that depends slowly on R. Actually, we assume that the second–order partial derivative of $\widetilde{\Psi}_1$ with respect to R is zero. Note that (6.12) and (6.15) have similar conditions of validity, which should be fulfilled if R_0 is taken large enough.

The definition of the local momentum $p(R)$ is a rather subtle task. It is based on the assumption that the R-motion, at the energy E, is governed by the Coulomb potential associated with the local effective charge $Z_{\mathrm{eff}}(R)$ so that

$$p(R) = \sqrt{2\left(E + \frac{Z_{\mathrm{eff}}(R)}{R}\right)}. \tag{6.16}$$

The effective charge is then determined by the following one-parameter rational interpolation between the Wannier charge $Z_{\mathrm{W}} = (4Z - 1)/\sqrt{2}$ – with Z the nuclear charge – that holds at very large R [16], and the charge Z_0 that is obtained by expressing the conservation of the 'norm' of the wave function *on* the hypersphere $R = R_0$ and of the associated total outgoing flux through this hypersphere:

$$Z_{\mathrm{eff}}(R) = Z_0\left(1 - \frac{\eta(R-R_0)^2}{1+\eta(R-R_0)^2}\right) + Z_{\mathrm{W}}\left(\frac{\eta(R-R_0)^2}{1+\eta(R-R_0)^2}\right). \tag{6.17}$$

In the first implementation of our approach, we defined Z_0 from the conservation properties of $\overline{\Psi}_1$ at $R_0 = 10$ a.u., where $\overline{\Psi}_1$ was given by (6.8) with *all open* adiabatic channels $\lambda \geq 0$ included. The value obtained turned out to be markedly different from Z_{W} which made the cross sections dependent on the parameter η that controls the rate of merging from the initial value at R_0 to the final value at $R \gg R_0$. This dependence hardly affected the shapes of the cross sections but it did modify their absolute values by as much as a few tens of per cent. We then realized that, since the lowest adiabatic channel corresponding to the lowest single–ionization channel was already decoupled from the bulk of the wave function at R_0, it was enough to propagate the restriction of $\overline{\Psi}_1$ to the open channels $\lambda > 0$. The value of Z_0 obtained in this way was very close to Z_{W} so that the independence of the results with respect to η was fully restored.

This can of course appear as a lucky occurrence, especially if one analyzes the R_0-dependence of Z_0 on the basis of the expression

$$Z_0 = R_0 \times \left[\frac{1}{2}\left(\frac{\sum\limits_{\lambda>0} p_\lambda(R_0)\,|F_\lambda(R_0)|^2}{\sum\limits_{\lambda>0} |F_\lambda(R_0)|^2}\right)^2 - E\right], \tag{6.18}$$

which follows from its definition – the summations over λ being restricted to open channels with $\lambda > 0$. Actually, the dominant open channels are the lowest ones, associated with negative adiabatic energies $E_\lambda(R)$, which increase with R_0 and tend to zero as R tends to ∞. The corresponding momenta $p_\lambda(R_0)$ accordingly decrease with R_0, so that the related $|F_\lambda(R_0)|^2$ increase to ensure conservation of the flux given by the numerator of the fraction in the expression of Z_0. The quantity within the square brackets in (6.18) then decreases with R_0. As a result, Z_0 increases with R_0 more slowly than linearly, despite the multiplicative factor R_0 in (6.18).

One can thus anticipate that increasing the size of the inner region, as required in certain dynamical situations, would lead to values of Z_0 increasingly higher than Z_W resulting in a renewed dependence of our approach on the parameter η. We think that this fear is groundless for two reasons: First, because the increase of Z_0 with R_0 is relatively slow, as explained above, and, secondly, because we anticipate that as R_0 increases, the ionization channels to He^+ $n \geq 2$ merge into the adiabatic channels with $\lambda \geq 1$ and decouple from the bulk of the wave function, so that their contribution to (6.18) can be omitted, leading back to values of Z_0 close to Z_W. We obtained a first confirmation of this process while performing He DPI calculations with an inner–region hyperradius $R_0 = 20$ a.u. At this distance, the energies of the second to fourth adiabatic channels become close to those of one electron a distance R_0 away from He^+ in the excited state $n = 2$, the degeneracy of which is partially removed by the Stark effect induced by this continuum electron. A definite identification would, however, be premature since, for instance, the equal spacing of the Stark levels is not well verified yet, and also, more importantly, the cross section deduced from the flux associated with these three channels does not agree with the available experimental data to better than 10% in the 10–80 eV energy range. In other words, at $R_0 = 20$ a.u., the decoupling of the He^+ $n = 2$ excitation-ionization channel is not fully achieved yet. Even so, it is clearly in progress. This provides a first confirmation of the process we have described above. (Note also that at $R_0 = 20$ a.u., (6.18) yields $Z_0 \simeq 7$, a value that is still close enough to $Z_W \simeq 5$ to avoid any significant η-dependence of the results.)

The stability of the matching procedure defined by (6.15)–(6.18) is also supported by a recent study of the DPI of Be [17] using a model core potential and an inner region of intermediate size $R_0 = 15$ a.u. In that case again, (6.18) yields a value of Z_0 close enough to Z_W to ensure the independence of our results with respect to the parameter η.

This increasing experience leads us to characterize our approach as parameter-free.

6.5.2 Solving the Resulting R-Propagation Equation

Now if we put the ansatz (6.15) into (6.4), neglecting the inhomogeneous term in (6.4) and the second–order partial derivative of $\tilde{\Psi}_1$ with respect to R, as mentioned above, we get a partial differential equation of first order over R for the unknown $\widetilde{\Psi}_1$. The change of variable $R \to \tau$ defined by $R\,p(R)\,d\tau = dR$ then allows one to recast this equation into the standard form

$$i\frac{\partial}{\partial \tau}\widetilde{\Psi}_1(\tau; \Omega_5) = \left[R(\tau)\,\overline{\overline{H}}_0\,(R(\tau); \Omega_5)\right]\widetilde{\Psi}_1(\tau; \Omega_5) \tag{6.19}$$

of a propagation equation. We then propagate the solution known at R_0 towards larger distances using a unitary algorithm based on the combination of the split-operator method with the joint use of Crank–Nicholson and exponential propagators. The procedure used is so stable and so efficient that it allows propagation up to distances as large as millions of a.u., or even larger within reasonable times.

6.6   Third Step: Extraction of the DPI Cross Sections at R_{max}

As pointed out in Sect. 6.2, the photoabsorption state Ψ_1 accounts for all processes that are energetically allowed, given the energy of the single absorbed photon. When this energy is above the DPI threshold, the dynamics of the system is characterized by the entanglement of the dominant single–ionization channels to the various states of the ion with the weak double ionization channel. This makes the extraction of the cross sections a nontrivial task.

6.6.1   Definition of the DPI Cross Sections

The DPI cross sections we are most interested in are the so-called "triply differential cross sections" (TDCSs) denoted $d^{(3)}\sigma/d\Omega_1 d\Omega_2 dE_1$. These quantities contain the most detailed information about the DPI process, because they are differential with respect to all the kinematical parameters of the final state: the solid angles Ω_1 and Ω_2, in which the two electrons are emitted, and the energy of one electron – the energy of the other being determined by the energy conservation relation $E = E_1 + E_2$. The TDCSs are measured in the so-called 'complete' experiments where the two particles are detected in coincidence and their momentum vectors are measured simultaneously -the momentum vector of the third particle being obtained, if needed, by momentum conservation.. They are defined by normalizing the flux of $\Psi_1(R; \Omega_1, \Omega_2, \alpha = \sin^{-1}\sqrt{E_1/E})$ through an elementary five-dimensional surface on a hypersphere of very large hyperradius R by the incoming photon flux $\mathcal{F}_{\mathrm{phot}}$, yielding

$$\frac{d^{(3)}\sigma}{d\Omega_1 d\Omega_2 dE_1} = \frac{1}{2E\mathcal{F}_{\mathrm{phot}}}$$

$$\times \operatorname{Im}\left(\lim_{R\to\infty} R^5 \sin 2\alpha\left(\Psi_1^*\frac{\partial \Psi_1}{\partial R}\right)\bigg|_{\alpha=\sin^{-1}\sqrt{E_1/E}}\right). \tag{6.20}$$

'Less-differential' cross sections can be defined from the TDCSs by integrating over one, two, or three variables. The interest has focused on those that depend on the characteristics of only one electron and can accordingly be measured in noncoincidence experiments. They are the doubly differential cross section (DDCS)

$$\frac{d^{(2)}\sigma}{d\Omega_1 dE_1} = \int d\Omega_2 \frac{d^{(3)}\sigma}{d\Omega_1 d\Omega_2 dE_1} , \qquad (6.21)$$

the singly differential cross section (SDCS)

$$\frac{d^{(1)}\sigma}{dE_1} = \int d\Omega_1 \int d\Omega_2 \frac{d^{(3)}\sigma}{d\Omega_1 d\Omega_2 dE_1} , \qquad (6.22)$$

which reflects the sharing of the excess energy above threshold between the two electrons, as well as the integrated cross section (ICS) for double photoionization

$$\sigma^{++} = \int_{0^+}^{E/2} dE_1 \frac{d^{(1)}\sigma}{dE_1} , \qquad (6.23)$$

where the symbol 0^+ at the lower energy-integration limit recalls that integration has to exclude the singular point noted $E_1 = 0$, which corresponds to $\alpha = 0$, i.e., to single ionization, the integration being restricted in addition to half the available energy interval $]0, E[$ to avoid double counting of the same physical event.

6.6.2   The Flux-Based Extraction: A Specificity of H$\mathcal{R}$M-SOW

The possibility of evaluating the cross sections accurately from their definition based on the flux of the wave function is specific to the H$\mathcal{R}$M-SOW method. It is a consequence of this method's ability to propagate the solution up to macroscopic distances of the order of the millimeters. By contrast, the largest distance at which the solution is obtained in the ECS or TDCC methods we referred to in Sect. 6.1 is of the order of hundreds of a.u. Accordingly, these methods rely upon alternative extraction procedures of the cross sections. In the TDCC method [6,7], the wavepacket at the final time is projected onto an approximate representation of the double–ionization channel. In the ECS method, the cross sections were initially computed by extrapolating the flux obtained at $R_0 \simeq 100$ a.u. to larger distances based on its assumed R^{-1} dependence [8,9]. However, the difficulties encountered at low excess energy above the threshold, or for very asymmetric sharings of this excess energy between the two electrons, led the authors to abandon this technique. The cross sections are now obtained from an expression of the transition amplitude that is formally valid only for short–range potentials [10,11].

Actually, the main reason for the failure of the flux-based methods at moderate distances is that the two-body and three–body channels are still

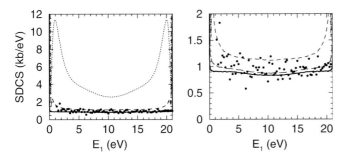

Fig. 6.2. SDCS for He DPI at 21 eV above threshold, in kb/eV, versus the energy of one electron, in eV. The *left* and *right* panels display the same measurements but on different vertical scales. *Full circles*: measurements [18]; *Dotted lines*: HRM-SOW calculations [13] at $R_0 = 10$ a.u. *Dashed lines*: same at $R_0 = 100$ a.u. *Long-dashed lines*: same at $R_0 = 1000$ a.u. *Thick solid lines*: same at $R_0 = 10^5$ a.u. where convergence is reached

tightly coupled there. Returning to the crude analysis we gave in Sect. 6.3 based on the upper limit $n_{\max} = 5$ for the excited Rydberg states of the ion, we can evaluate to about $\Delta\alpha \simeq 0.25$ rad at $R = 100$ a.u. the extension of the angular sectors located around $\alpha = 0$ and $\pi/2$, which are associated to single–continuum channels At this distance, significant α-angular sectors are thus still contaminated by two-body processes. This lasting coupling of the two types of processes is well illustrated on Fig. 6.2 in the case of He DPI. Evaluations of (6.22) are given for increasing values of R and compared with experiment. The curve obtained at $R = 10$ a.u. looks like a basin surrounded by two large peaks. As R increases, the basin flattens, widens, and goes down, while the peaks go up and shrink to ensure conservation of the total ionization cross section represented by half the area below the curve. Beyond $R = 1000$ a.u., the side peaks can no longer be distinguished from the vertical axis at the scale of the figure. In other words, in the course of the expansion of the system, its density of probability leaks out of the 'Wannier' ridge at $\alpha = \pi/4$ – which would lead to double escape, into the potential wells at $\alpha = 0$ and $\pi/2$ – associated with single ionization. The 30% lowering of the central plateau when passing from $R = 100$ a.u. to $R = 1000$ a.u., which is visible on the right panel of Fig. 6.2, can be compared with the 10% decrease of the 19.6 eV SDCS for electron-impact ionization of H reported in [11] when switching from the flux method applied at $R = 100$ a.u. to the amplitude method that seems more suitable at these moderate distances.

The amplitude (resp. projection)-based extraction methods used in ECS (resp. TDCC) introduce matrix elements involving the scattered wave (resp. the final wavepacket) and a product of two Coulomb waves. This product is orthogonal to the single–continuum channels that are then projected out of the scattered wave (resp. final wavepacket). This elimination of the single–ionization channels is achieved, however, at the expense of reintroducing into

the theory an approximate asymptotic representation of the double continuum. On the contrary, H$\mathcal{R}$M-SOW lets the physical interactions act freely
up to the point where the decoupling between two- and three-body channels
occurs naturally. As such, it is the only method that does not rely at all upon
any two-body asymptotic condition.

6.6.3   Structure of the Outgoing Flux at Intermediate Distances

A subsequent advantage of H$\mathcal{R}$M-SOW is that one can follow the redistribution of the outgoing flux between the different open channels during the
expansion of the system, as expressions like (6.20)–(6.22) can be evaluated at
any R in the external region. Figure 6.3, for instance, illustrates the formation
of a selected set of TDCSs. The evolution with increasing R is dominated by
two main effects: (i) a drastic decrease of the magnitude of the signal, which
reflects the loss of flux into the competing single–ionization channels essentially – but also into alternative double–ionization channels associated with
different kinematical arrangements; and (ii) a concentration of the signal into
the half–plane opposite to the ejection direction of the first electron, which
reflects the progressive development of electron–electron repulsion in the continuum. As expected, it is for the most asymmetric energy sharings that the
R-evolution is the slowest.

6.7   Extracting the H$\mathcal{R}$M-SOW
Single Ionization Cross Sections

Let us now summarize the status of H$\mathcal{R}$M-SOW with respect to the extraction
of the various cross sections associated with photoabsorption above the DPI
threshold.

   We have seen that the DPI cross sections can be extracted directly from a
trivial flux calculation at the macroscopic distances reached by the method.
This is due to the very strong confinement of the single–ionization channels
in the vicinity of the singular points $\alpha = 0$ and $\pi/2$. This confinement, on
the other hand, implies that the various single–ionization channels cannot
be resolved, so that the flux method does not apply to the extraction of the
single–ionization cross sections.

   The latter are extracted accordingly using a completely different approach. In Sect. 6.4, we analyzed the expansion of the solution at $R_0 = 10$ a.u.
into adiabatic partial waves and found that the lowest adiabatic channel corresponds to single ionization to the ground ionic state. In Sect. 6.5, we have
shown that this type of identification could be pushed further if the size of
the inner region is increased significantly. The increase of the inner region
however will quickly run into the available computer's limits. As a result, a
precise evaluation of the He ionization-excitation cross sections to the levels
$n \geq 3$ of the ion does not seem realistic using this technique.

The solution will therefore be to project Ψ_1 onto the single–ionization channels at large R, which can be described *exactly* by the product of a continuum Coulomb wave with the charge $Z - 1 = 1$ and a bound hydrogenic state. This will be the subject of future work.

6.8   Comments on the Current State of the Arts in the Field

The best way to test a new DPI theory is to compare its predictions with absolute measurements of the most informative quantity, i.e., the He TDCS. But absolute measurements of this TDCS are very rare. Three measurements are available at 0.1 eV above threshold [20], but at the moment, this energy region is beyond the scope of current theories. Fortunately, a very complete set of TDCSs at 20 eV above threshold was produced in 1998 using the COLTRIMS technique [19]. All recent DPI theories (CCC, TDCC, H$\mathcal{R}$M-SOW) have been tested succesfully against these measurements. Regarding H$\mathcal{R}$M-SOW, a comprehensive comparison with the COLTRIMS TDCSs and with various DDCSs, SDCSs, and ICSs obtained in non-coincidence experiments has been presented in [13]. The agreement obtained *on the absolute scale* establishes the relevance of the method. In addition, the *gauge-independence* property, which is essential to any theory, but proves difficult to obtain in general, is attained to an excellent precision.

A consequence of all these recent successes of the theory is a widespreading feeling that the double– continuum problem is now under full control. To conclude this contribution, we would like to present an example that makes this opinion appear untimely.

The statistics of the COLTRIMS TDCSs measurements is not very high, as one can see from Fig. 6.3. Accordingly, the agreement in shape between experiment and the various theories cannot be appreciated with a high precision. Other coincidence spectrometers, which do not ensure 4π-detection of all particles and hence do not provide absolute values of the cross sections, may, however, yield better-resolved TDCSs with smaller error bars, which constitute important test cases for the theory. In the course of our H$\mathcal{R}$M-SOW calculations of these precise but relative TDCSs, our attention has been attracted to two problems, which are illustrated by Fig. 6.4.

First, a rapid comparison between the thick solid line (H$\mathcal{R}$M-SOW), the dotted line (CCC), and the dashed line (TDCC) on the left graph of Fig. 6.4 shows that there are significant discrepancies between the H$\mathcal{R}$M-SOW and CCC theories, on the one hand, and the TDCC theory, on the other side, regarding the absolute values of the cross sections. According to our experience, this problem is not specific to the particular case considered in Fig. 6.4, but occurs repeatedly. This disagreement must be resolved to allow experimentalists to put their relative measurements on an absolute scale with confidence.

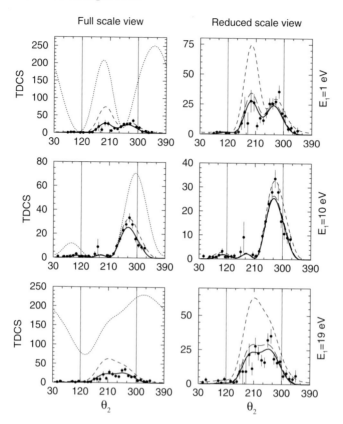

Fig. 6.3. TDCS for the DPI of He at 20 eV above threshold in b(eV)$^{-1}$(sr)$^{-2}$. The two electrons are in the plane perpendicular to the linearly polarized photon beam, the electron labelled '1' is emitted at 30° of the electric field, and the TDCS is plotted as a function of the azimuthal angle θ_2 of electron '2' measured in degrees. Each line in the array of graphs corresponds to a fixed energy of electron '1': from top to bottom, $E_1 = 1$ eV, 10 eV, 19 eV. The *left* and *right columns* display the same data but on different vertical scales. *Full circles with error bars*: measurements [19]. The conventions for the lines are the same as in Fig. 6.2. The angular range between the two vertical lines at $\theta_2 = 120°$ and 300° corresponds to emission of the second electron in the half–plane opposite to the direction of the first electron

The same graph makes another problem appear that concerns the shape of the TDCS. The measured TDCS shows three peaks of equal magnitude. This shape is not reproduced by any of the available theories, which all generate a central peak with more or less pronounced symmetric shoulders. If one discards the possibility of an experimental artefact, one is led to acknowledge that the dynamical situation where the fast electron is emitted along the electric field is particularly demanding towards theory.

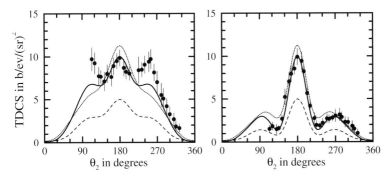

Fig. 6.4. TDCS for the DPI of He at 40 eV above threshold in $b(eV)^{-1}(sr)^{-2}$. The two electrons are in the plane perpendicular to the linearly polarized photon beam, the electron labelled '1' is emitted at $0°$ of the electric field with an energy $E_1 = 35\,eV$ (*left panel*) or $5\,eV$ (*right panel*), and the TDCS is plotted as a function of the azimuthal angle θ_2 of electron '2' measured in degrees. *Dots with error bars*: measurements [21]. Thick solid line: H$\mathcal{R}$M-SOW calculations. *Dotted line*: CCC calculations [21]. *Dashed line*: TDCC calculations [7]

We propose to interpret the specificity of this kinematics in terms of the conflict between the electric field, which drives the two electrons in the same direction, and the Coulomb repulsion, which pushes them apart in opposite directions. Actually, this conflict is made most acute when one electron is ejected along the electric field, since electronic correlations in this case act *exactly* against the field. This conflict should also be sensitive to the energy sharing between the two electrons. At equal sharing, the antagonism between the singlet and odd character of the electronic pair wave function for antiparallel emission results in a node at $\theta_2 = 0°$ that smears the peculiarities of the TDCS shape and reduces it to a simple two–lobe structure that is easily reproduced by all theories. For very unequal energy sharing, electronic correlations weaken, and so does the intensity of their conflict with the electric field. We thus expect this antagonism to be the strongest for intermediate energy sharings. In that case, the distribution of the slowest electron is likely to be the most affected by the resulting very subtle and unstable dynamics. The right panel of Fig. 6.4, where the agreement in shape between all theories and experiment is much better, confirms this expectation.

Our conclusion is that still more effort should be put in the near future in the definite establishment of absolute scales for the cross sections as well as in the study of critical dynamics of which we have given an example. A still more fundamental challenge in the field is the development of reliable computational schemes for energies less than 1 eV above the threshold, where the leaking of the system from the Wannier ridge $\alpha = \pi/4$ has to be taken into account very precisely. Another set of problems arise with the consideration of the simplest molecular systems (H_2, D_2) at small energies, when the Coulomb-explosion can influence the receding electrons. We believe that the H$\mathcal{R}$M-SOW approach has a potential for consideration of these problems.

Acknowledgements

We thank A. Kheifets and J. Colgan for communicating the original files of their results to us. A.K. Kazansky acknowledges a partial support from RFFI via grant 02-02-16586, P. Selles and L. Malegat acknowledge the support of the CNRS computer center IDRIS (Orsay, France) through the project 021485.

References

1. I. Bray, A.T. Stelbovics: Phys. Rev. A **46**, 6995 (1992)
2. I. Bray, D.V. Fursa: Phys. Rev. A **54**, 2991 (1996)
3. I. Bray: Phys. Rev. Lett. **78**, 4721 (1997)
4. A. Kheifets, I. Bray: J. Phys. B **31**, L447 (1998)
5. M.S. Pindzola, F. Robicheaux: Phys. Rev. A **57**, 318 (1998)
6. J. Colgan, M.S. Pindzola, F. Robicheaux: J. Phys. B **34**, L457 (2001)
7. J. Colgan, M.S. Pindzola: Phys. Rev. A **65**, 032729 (2002)
8. T.N. Rescigno, M. Baertschy, W.A. Isaacs, C.W. McCurdy: Science **286**, 2474 (1999)
9. M. Baertschy, T.N. Rescigno, W.A. Isaacs, X. Li, C.W. McCurdy: Phys. Rev. A **63**, 022712 (2001)
10. C.W. McCurdy, D.A. Horner, T.N. Rescigno: Phys. Rev. A **63**, 022711 (2001)
11. M. Baertschy, T.N. Rescigno, C.W. McCurdy: Phys. Rev. A **64**, 022709 (2001)
12. L. Malegat, P. Selles, A. Kazansky: Phys. Rev. Lett. **85**, 4450 (2000)
13. P. Selles, L. Malegat, A.K. Kazansky: Phys. Rev. A **65**, 032711 (2002)
14. J.M. Bizau, F.J. Wuillemier: J. Electron Spectrosc. Relat. Phenom. **71**, 205 (1995)
15. J.A.R. Samson, Z.X. He, L. Yin, G.N. Haddad: J. Phys. B **27**, 887 (1994)
16. G.H. Wannier: Phys. Rev. **90**, 817 (1953)
17. F. Citrini, L. Malegat, P. Selles, A.K. Kazansky: Phys. Rev. A **67**, 042709 (2003)
18. R. Wehlitz, F. Heiser, O. Hemmers, B. Langer, A. Menzel, U. Becker: Phys. Rev. Lett. **67**, 3764 (1991)
19. H. Bräuning, R. Dörner, C.L. Cocke, M.H. Prior, B. Krässig, A.S. Kheifets, I. Bray, A. Bräunig-Demiam, K. Carnes, S. Dreuil, V. Mergel, P. Richard, J. Ullrich, H. Schmidt-Böcking: J. Phys. B **31**, 5149 (1998)
20. A. Huetz, J. Mazeau: Phys. Rev. Lett. **85**, 530 (2000)
21. P. Bolognesi, R. Camilloni, M. Coreno, G. Turri, J. Berakdar, A.S. Kheifets, L. Avaldi: J. Phys. B **34**, 3193 (2001)

7 Convergent Close-Coupling Approach to Electron–Atom Collisions

I. Bray and A.T. Stelbovics

7.1 Introduction

The convergent close-coupling (CCC) method was developed in order to resolve the long-standing discrepancy between two consistent experiments and all available theories for 2p excitation of atomic hydrogen [1]. The method was unable to resolve this discrepancy, but subsequent experiments [2,3] found much more in favor of theory than the previous experiments. There have been a number of reviews of the applications of the CCC theory with the most recent one being by Bray et al. [4]. The method has been extended to ionization [5], resulting in some controversy [6,7] that required further explanation [8,9]. Our own confidence in the ability of the CCC method to reproduce electron–hydrogen fully differential ionization cross sections was shaken by the less than satisfactory agreement with experiment [10]. However, this turned out to be primarily due to insufficient computational resources available at the time [11]. Consequently, we are now confident that the CCC method is able to solve the e–H, γ–He, and e–He (within the frozen-core model) collision systems at all energies with one or two outgoing electrons. We shall attempt to explain here the underlying foundations as clearly as possible. The example of the S-wave model will be used to demonstrate the method. A published program is available that shows the workings of the method discussed here [12]. We will finish by concentrating on the application of the method to fully differential ionization processes.

7.2 Electron–Hydrogen Collisions

We assume atomic units throughout unless stated otherwise. The atomic hydrogen Hamiltonian H_2 (the subscript indicates electron space) we write as

$$H_2 = K_2 + V_2 \,, \tag{7.1}$$

where K is the free–electron Hamiltonian and V is the attractive Coulomb potential. The full Hamiltonian H of the electron–hydrogen system is then

$$H = H_1 + H_2 + V_{12} \,, \tag{7.2}$$

where V_{12} is the electron–electron potential. The total energy E and wave function of the system $|\Psi_i^{(S+)}\rangle$ satisfy the Schrödinger equation

$$(E - H)|\Psi_i^{(S+)}\rangle = 0\,, \tag{7.3}$$

where S is the total electron spin and the $+$ superscript denotes outgoing spherical–wave boundary conditions.

7.2.1 Structure

In the CCC approach we first diagonalize H_2 using an orthogonal Laguerre basis

$$\xi_{kl}(r) = \left(\frac{\lambda_l(k-1)!}{(2l+1+k)!} \right)^{1/2} (\lambda_l r)^{l+1} \exp(-\lambda_l r/2) L_{k-1}^{2l+2}(\lambda_l r)\,, \tag{7.4}$$

where the $L_{k-1}^{2l+2}(\lambda_l r)$ are the associated Laguerre polynomials, and k ranges from 1 to the basis size N_l. This results in (pseudo) target states $|\phi_n^{(N)}\rangle$ with energies $\epsilon_n^{(N)}$, which for each l satisfy

$$\langle \phi_m^{(N)}|H_2|\phi_n^{(N)}\rangle = \delta_{mn}\epsilon_n^{(N)}\,, \tag{7.5}$$

where we use generic N to indicate the dependence on the basis size.

Given a fixed λ_l, as the basis size N_l is increased the negative-energy states converge to the true discrete eigenstates, while the positive-energy states yield an increasingly dense discretization of the continuum. In Fig. 7.1 we give results of some of the wave functions generated in realistic calculations. For states with principal quantum number n the true exponential falloff is $-r/n$, yet we see that even for $n = 5$ the true eigenstate is very well described by the $N = 40$ calculation with an exponential falloff being $-r/2$ or even $-r$. The smaller $N = 20$ calculation succeeds only for $\lambda = 1$.

We now turn our attention to the two positive-energy states. To compare the true eigenstate $\chi_e(r)$ with the corresponding pseudostate $\phi_n^{(N)}(r)$ of energy $\epsilon_n^{(N)} = e$ we plotted $\langle\phi_n^{(N)}|\chi_e\rangle\phi_n^{(N)}(r)$. At 1 eV $\chi_e(r)$ is well reproduced by the $(N, \lambda) = (40, 1)$ state out to 100 a.u., while the $(20,1)$ does so only out to 50 a.u. or so. Doubling the λ substantially decreases the range on which the pseudostates describes the corresponding $\chi_e(r)$. However, it is important to note that in going from the $(20,1)$ to the $(40,2)$ calculation the description of $\chi_e(r)$ improves. In other words, doubling the basis size more than compensates for the negative effect of λ doubling.

The 2–eV case shows that the often–expressed idea that λ determines a box size for the atom of fixed radius is not quite accurate. The accuracy of the pseudostate description of the corresponding $\chi_e(r)$ clearly diminishes with energy. As the pseudostate energy increases, its useful range lessens because the pseudostate oscillations have a decreasing wavelegth and a fixed basis approximates a similar number of them, independent of energy.

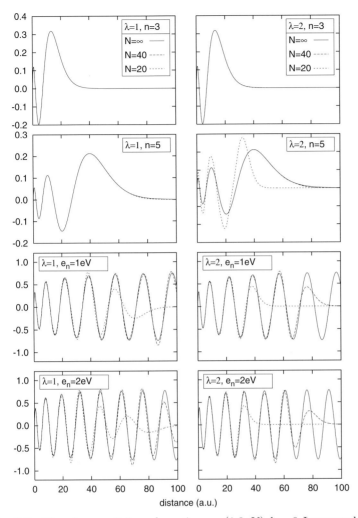

Fig. 7.1. Discrete $n = 3, 5$, and continuous (1,2 eV) $l = 0$ Laguerre-based states evaluated with specified λ and N. The eigenstates are denoted with $N = \infty$

7.2.2   Scattering

The total wave function $|\Psi_i^{(S+)}\rangle$ is written as an explicitly symmetrized expansion over the discretised target states[1]

$$|\Psi_i^{(S+)}\rangle \approx |\Psi_i^{(NS+)}\rangle = \left(1 + (-1)^S P_r\right) | \sum_{n=1}^{N} \phi_n^{(N)} f_{ni}^{(NS)} \rangle, \qquad (7.6)$$

---

[1] In the limit $N \to \infty$ the exact total wave function is reproduced owing to the completeness of the Laguerre basis from which the target states are constructed.

where the unknown one-electron functions $|f_{ni}^{(S+)}\rangle$ need to be found. The above expansion ensures the required symmetry of $|\Psi_i^{(S+)}\rangle$, but at the price of a non-unique definition of the $|f_{ni}^{(NS)}\rangle$. For example, with $N = n = S = 1$ (7.6) becomes

$$|\Psi_i^{(NS+)}\rangle = |\phi_n^{(N)} f_{ni}^{(NS)}\rangle - |f_{ni}^{(NS)} \phi_n^{(N)}\rangle, \tag{7.7}$$

and then $|f_{ni}^{(NS)}\rangle + c|\phi_n^{(N)}\rangle$ also satisfies (7.7) for an arbitrary constant c.

Though, in principle, nonuniqueness of the $|f\rangle$ is not a problem so long as (7.6) is satisfied, in practice computational methods are adversely affected. To solve this problem we first note that the nonuniqueness is only in the space spanned by the $|\phi_n^{(N)}\rangle$, and without loss of generality may write two solutions of (7.6) for any i as

$$|f_n^{(NS)}\rangle = |h_n^{(NS)}\rangle + \sum_{n'=1}^{N} c_{nn'}^{(NS)}|\phi_n^{(N)}\rangle \tag{7.8}$$

and

$$|F_n^{(NS)}\rangle = |h_n^{(NS)}\rangle + \sum_{n'=1}^{N} d_{nn'}^{(NS)}|\phi_n^{(N)}\rangle, \tag{7.9}$$

where $\langle\phi_n^{(N)}|h_{n'}^{(NS)}\rangle = 0$ for all n and n'. In writing the above two equations we concentrate the problem to the nonuniqueness of coefficients $c_{nn'}^{(NS)}$ and $d_{nn'}^{(NS)}$ when satisfying (7.6). By considering $\langle\phi_n^{(N)}\phi_{n'}^{(N)}|\Psi_i^{(NS+)}\rangle$ for the two solutions (7.8) and (7.9) we have the relation

$$c_{nn'}^{(NS)} + (-1)^S c_{n'n}^{(NS)} = d_{nn'}^{(NS)} + (-1)^S d_{n'n}^{(NS)}, \tag{7.10}$$

which may be readily satisfied with $c_{nn'}^{(NS)} \neq d_{nn'}^{(NS)}$. To fix this problem we impose the condition that solutions $|f_n^{(NS)}\rangle$ of (7.6) must satisfy

$$\langle\phi_n^{(N)}|f_{n'}^{(NS)}\rangle = (-1)^S\langle\phi_{n'}^{(N)}|f_n^{(NS)}\rangle. \tag{7.11}$$

This condition leads to $c_{nn'}^{(NS)} = (-1)^S c_{n'n}^{(NS)}$ and $d_{nn'}^{(NS)} = (-1)^S d_{n'n}^{(NS)}$, and hence $c_{nn'}^{(NS)} = d_{nn'}^{(NS)}$. Note that (7.11) does not affect $|h_n^{(NS)}\rangle$, and only ensures a unique description in the space spanned by the target states $|\phi_{n'}^{(N)}\rangle$. Now, with the extra condition (7.11), (7.6) provides for a unique expansion of the total wave function that has the required symmetry conditions.

We next derive the close-coupling equations for the description of electron–hydrogen collisions. Utilizing the Bubnov–Galerkin principle [13] it follows that minimizing the error of the approximate trial solution (7.6) applied to (7.3) leads to a system of N coupled equations for the single–electron functions $f_{ni}^{(NS)}(\boldsymbol{r}_1)$,

$$\langle\phi_{n'}^{(N)}|(E - H)(1 + (-1)^S P_r)| \sum_{n=1}^{N} \phi_n^{(N)} f_{ni}^{(NS)}\rangle = 0, \tag{7.12}$$

for all $1 \leq n' \leq N$. Since we are seeking solutions for the f that correspond to outgoing scattered waves leaving the target in the ground or an excited state, we define the asymptotic Hamiltonian $K = K_1 + H_2$. The L^2 expansion of the target space ensures that, at large distances, the projectile sees a neutral target, albeit always in a negative- or a positive-energy discrete state. Consequently, we write (7.12) as

$$\langle\phi_{n'}^{(N)}|(E - K_1 - H_2)|\sum_{n=1}^{N}\phi_n^{(N)}f_{ni}^{(NS)}\rangle = (E - K_1 - \epsilon_{n'}^{(N)})|f_{n'i}^{(NS)}\rangle$$

$$= \langle\phi_{n'}^{(N)}|V_1 + V_{12} + (H - E)(-1)^S P_r|\sum_{n=1}^{N}\phi_n^{(N)}f_{ni}^{(NS)}\rangle$$

$$\equiv \langle\phi_{n'}^{(N)}|V^{(S)}|\sum_{n=1}^{N}\phi_n^{(N)}f_{ni}^{(NS)}\rangle, \qquad (7.13)$$

and hence applying the ougoing scattering boundary conditions we convert the coupled Schrödinger equation into a Lippmann–Schwinger equation

$$|f_{n'i}^{(NS)}\rangle = \delta_{n'i}|\boldsymbol{k}_{n'}\rangle + \frac{1}{E + i0 - K_1 - \epsilon_{n'}^{(N)}}\langle\phi_{n'}^{(N)}|V^{(S)}|\sum_{n=1}^{N}\phi_n^{(N)}f_{ni}^{(NS)}\rangle$$

$$= \delta_{n'i}|\boldsymbol{k}_{n'}\rangle + \int d^3k\frac{|\boldsymbol{k}\rangle\langle\boldsymbol{k}\phi_{n'}^{(N)}|V^{(S)}|\sum_{n=1}^{N}\phi_n^{(N)}f_{ni}^{(NS)}\rangle}{E + i0 - k^2/2 - \epsilon_{n'}^{(N)}}, \qquad (7.14)$$

where $E = k_{n'}^2/2 + \epsilon_{n'}^{(N)}$ on the energy shell.

Rather than solving for the one-electron functions f we premultiply (7.14) by $\langle\boldsymbol{k}_f\phi_f^{(N)}|V^{(S)}|\sum_{n'=1}^{N}\phi_{n'}^{(N)}$ to form coupled integral equations for the T matrix

$$\langle\boldsymbol{k}_f\phi_f^{(N)}|T^{(NS)}|\phi_i^{(N)}\boldsymbol{k}_i\rangle = \langle\boldsymbol{k}_f\phi_f^{(N)}|V^{(S)}|\phi_i^{(N)}\boldsymbol{k}_i\rangle$$

$$+ \sum_{n=1}^{N}\int d^3k\frac{\langle\boldsymbol{k}_f\phi_f^{(N)}|V^{(S)}|\phi_n^{(N)}\boldsymbol{k}\rangle\langle\boldsymbol{k}\phi_n^{(N)}|T^{(NS)}|\phi_i^{(N)}\boldsymbol{k}_i\rangle}{E + i0 - k^2/2 - \epsilon_{n'}^{(N)}}, \qquad (7.15)$$

where

$$\langle\boldsymbol{k}_f\phi_f^{(N)}|T^{(NS)}|\phi_i^{(N)}\boldsymbol{k}_i\rangle = \langle\boldsymbol{k}_f\phi_f^{(N)}|V^{(S)}|\sum_{n=1}^{N}\phi_n^{(N)}f_{ni}^{(NS)}\rangle$$

$$= \langle\boldsymbol{k}_f\phi_f|(H - E)(1 + (-1)^S P_r)|\Psi_i^{(NS+)}\rangle, \qquad (7.16)$$

so long as $\boldsymbol{k}_f$ is on the energy shell. The derivation of the above relation took no account of the nonuniqueness problem. A complete discussion of its treatment and the general structure of the resulting unique equations was given by Stelbovics [14]. For our calculations, the following practical approach is used.

To implement the condition (7.11) we modify the potential $V^{(S)}$ in the following way. Let us consider the total–energy–dependent part

$$E\langle \boldsymbol{k}\phi_{n'}^{(N)}|P_r|\sum_{n=1}^{N}\phi_n^{(N)}f_{ni}^{(NS)}\rangle = E\sum_{n=1}^{N}\langle \boldsymbol{k}|\phi_n^{(N)}\rangle\langle\phi_{n'}^{(N)}|f_{ni}^{(NS)}\rangle \qquad (7.17)$$

$$= (-1)^S E\sum_{n=1}^{N}\langle \boldsymbol{k}|\phi_n^{(N)}\rangle\langle\phi_n^{(N)}|f_{n'i}^{(NS)}\rangle$$

$$= (-1)^S E\langle \boldsymbol{k}\phi_{n'}^{(N)}|I_1^{(N)}|\sum_{n=1}^{N}\phi_n^{(N)}f_{ni}^{(NS)}\rangle,$$

where $I_1^{(N)} = \sum_{n=1}^{N}|\phi_n^{(N)}\rangle\langle\phi_n^{(N)}|$ operates in projectile space. Thus, we may simply replace the term $(-1)^S EP_r$ in (7.13) with $EI_1^{(N)}$ and then the nonuniqueness problem is solved. In practice, we introduce an arbitrary constant θ and replace $(-1)^S EP_r$ with $\theta EI_1^{(N)}+(1-\theta)(-1)^S EP_r$, which implements the solution to the nonuniqueness problem for arbitrary nonzero θ. The potential operator $V^{(S)}$ then gains an explicit dependence upon both N and θ

$$V^{(NS)}(\theta) = V_1 + V_{12} - \theta EI_1^{(N)} + (H - E(1 - \theta))(-1)^S P_r. \qquad (7.18)$$

Variation of θ allows for a numerical check of the stability of the calculated T matrix, which must be independent of θ off or on the energy shell, i.e.,

$$\langle \boldsymbol{k}\phi_{\mathrm{f}}^{(N)}|V^{(NS)}(\theta)|\Psi_i^{(NS+)}\rangle = \langle \boldsymbol{k}\phi_{\mathrm{f}}^{(N)}|V^{(S)}|\Psi_i^{(NS+)}\rangle$$
$$= \langle \boldsymbol{k}\phi_{\mathrm{f}}^{(N)}|T^{(NS)}|\phi_i^{(N)}\boldsymbol{k}_i\rangle. \qquad (7.19)$$

The above amplitude is for the excitation of initial target state $\phi_i^{(N)}$ to final state $\phi_{\mathrm{f}}^{(N)}$ by an electron of initial momentum $\boldsymbol{k}_i$ and final momentum $\boldsymbol{k}_{\mathrm{f}}$. It may be directly applied to discrete transitions so long as N is sufficiently large so that $\phi_{\mathrm{f}}^{(N)}$ and $\phi_i^{(N)}$ are true discrete eigenstates. The close-coupling method evolved with just such transitions in mind. However, we may also consider ionizing processes in the following way. The true transition amplitude is approximated as

$$\langle \Phi_f|H - E|\Psi_i^{(S+)}\rangle \approx \langle \boldsymbol{k}_f\boldsymbol{q}_f^{(-)}|I_2^{(N)}(H - E)(1 + (-1)^S P_r)I_2^{(N)}|\Psi_i^{(S+)}\rangle$$
$$= \langle \boldsymbol{q}_f^{(-)}|\phi_f^{(N)}\rangle\langle \boldsymbol{k}_f\phi_f^{(N)}|T^{(NS)}|\phi_i^{(N)}\boldsymbol{k}_i\rangle, \qquad (7.20)$$

where $\boldsymbol{q}_f$ is a positive-energy eigenstate of the target Hamiltonian H_2 of energy $q_f^2/2 = \epsilon_f^{(N)}$ and satisfies $\langle \boldsymbol{q}_f^{(-)}|\phi_n^{(N)}\rangle = \delta_{fn}\langle \boldsymbol{q}_f^{(-)}|\phi_f^{(N)}\rangle$. Here we assume the same asymptotic Hamiltonian K_1 as used for discrete excitation. The square-integrable nature of the target expansion states justifies this.

The solution method of (7.15) has been discussed in detail by Bray and Stelbovics [1]. Briefly, (7.15) is expanded in partial waves and is rewritten in terms of the real K matrix, for computational efficiency, and the complex

T matrix is reconstructed after solution for K. For each partial wave the K-matrix equation is

$$\langle k_{\rm f}\phi_{\rm f}^{(N)}|K^{(NS)}|\phi_{\rm i}^{(N)}k_{\rm i}\rangle = \langle k_{\rm f}\phi_{\rm f}^{(N)}|V^{(NS)}|\phi_{\rm i}^{(N)}k_{\rm i}\rangle \qquad (7.21)$$
$$+ \sum_{n=1}^{N}\mathcal{P}\int_{0}^{\infty}dk k^{2}\frac{\langle k_{\rm f}\phi_{\rm f}^{(N)}|V^{(NS)}|\phi_{n}^{(N)}k\rangle\langle k\phi_{n}^{(N)}|K^{(NS)}|\phi_{\rm i}^{(N)}k_{\rm i}\rangle}{E-k^{2}/2-\epsilon_{n'}^{(N)}},$$

where $\mathcal{P}$ denotes a principal value type integral. The integral over k is performed using a suitable Gauss-type quadrature with the singularity placed in an interval containing an even number of points. To solve (7.21) $k_{\rm f}$ is allowed to range over the same quadrature points k and may be written as

$$K_{\rm fi}^{(NS)} = V_{\rm fi}^{(NS)} + \sum_{n}w_{n}V_{\rm fn}^{(NS)}K_{\rm ni}^{(NS)}, \qquad (7.22)$$

where the w_{n} contain the Green's function and the Gaussian weights. The resultant linear matrix equation for K is solved using LAPACK routines [15]. The required T is then found from

$$K_{\rm fi}^{(NS)} = \sum_{n_{o}}T_{{\rm f}n_{o}}^{(NS)}\left(\delta_{n_{o}{\rm i}} + i\pi k_{n_{o}}K_{n_{o}{\rm i}}^{(NS)}\right), \qquad (7.23)$$

where the sum ranges over open states only. The partial cross sections $\sigma_{\rm fi}^{(NS)}$, in a.u., are given by

$$\sigma_{\rm fi}^{(NS)} = \pi\frac{k_{\rm f}}{k_{\rm i}}|\langle k_{\rm f}\phi_{\rm f}^{(N)}|T^{(NS)}|\phi_{\rm i}^{(N)}k_{\rm i}\rangle|^{2}, \qquad (7.24)$$

with summation over all the partial waves yielding the full result.

For a given initial state i we define the (partial) total ionization cross section $\sigma_{\rm ion}^{(NS)}$ as

$$\sigma_{\rm ion}^{(NS)} = \sum_{\epsilon_{\rm f}^{(N)}>0}\sigma_{\rm fi}^{(NS)}. \qquad (7.25)$$

From this we define the (partial) singly differential cross section (SDCS), which governs the distribution of electrons going out with the secondary energy $0 < e < E$, from

$$\sigma_{\rm ion}^{(NS)} = \sum_{\epsilon_{\rm f}^{(N)}>0}\sigma_{\rm fi}^{(NS)}$$
$$\approx \int_{0}^{\sqrt{2E}}dq_{f}q_{f}^{2}|\langle q_{f}^{(-)}|\phi_{\rm f}^{(N)}\rangle|^{2}\sigma_{\rm fi}^{(NS)}$$
$$\equiv \int_{0}^{E}de\frac{d\sigma^{(NS)}}{de}(e). \qquad (7.26)$$

There are some points to note about the above. The integrand is only available at the discrete energy points of the pseudostates $\epsilon_f^{(N)} = q_f^2/2$, and so some interpolation is required to obtain the SDCS on a continuous domain. The integration range ends at the total energy E, and yet cannot have any double counting since the close-coupling formalism is unitary and the total cross section is the same whether obtained via the optical theorem or as a sum over all open states.

7.2.3   S-Wave Model: Proof–of–Principle

Having defined the CCC formalism, the fundamental questions are whether solutions of (7.15) converge with increasing N, and if so, is the convergence to the right answer? To answer these questions, without any loss of generality, we may consider the S-wave model, where only states of zero orbital angular momentum are included. This model, often called the Temkin–Poet model in the case of e–H scattering was first considered by Temkin [17] and then Poet [18] who gave highly accurate results for some transitions and incident energies. Most recently, Jones and Stelbovics [19] applied similar techniques to yield highly accurate results for $n \le 6$ transitions, total and singly differential ionization cross sections. In this model, the effect of the target continuum and the symmetry of the total wave function are fully treated. The former is particularly problematic since the derivation of the coupled equations utilizing the true target eigenstates leads to a class of V matrix elements in (7.15) that are distributions, thereby not allowing for numerical solution. These occur for the free–free class of channel potentials [20].

In Fig. 7.2 we compare the 30-state CCC calculations [16] with the recent and most accurate finite difference method [19]. We see excellent agreement between the two calculations. At the time when the CCC calculations were first published [21] we had no idea about the accuracy of the method for total ionization cross sections (TICS). Here we see that it is quite accurate. The same has proved to be the case for the full e–H problem [22].

Figure 7.2 demonstrates the utility of the CCC method for discrete transitions. We consider TICS as a discrete transition because via the optical theorem it may be obtained using only cross sections for discrete channels. What about the energy distribution within a TICS for a given total energy $E > 0$? This is the so-called singly differential cross section (SDCS).

As soon as we consider the ionization amplitude (7.20) we are faced with a problem. For a given $q < k$ we have two independent ionization amplitudes

$$f^{(NS)}(\boldsymbol{q}_{\mathrm f}, \boldsymbol{k}_{\mathrm f}) = \langle \boldsymbol{q}_{\mathrm f}^{(-)}|\phi_{\mathrm f}^{(N)}\rangle \langle \boldsymbol{k}_{\mathrm f}\phi_{\mathrm f}^{(N)}|T^{(NS)}|\phi_{\mathrm i}^{(N)}\boldsymbol{k}_{\mathrm i}\rangle \qquad (7.27)$$

and

$$f^{(NS)}(\boldsymbol{k}_{\mathrm f'}, \boldsymbol{q}_{\mathrm f'}) = \langle \boldsymbol{k}_{\mathrm f'}^{(-)}|\phi_{\mathrm f'}^{(N)}\rangle \langle \boldsymbol{q}_{\mathrm f'}\phi_{\mathrm f'}^{(N)}|T^{(NS)}|\phi_{\mathrm i}^{(N)}\boldsymbol{k}_{\mathrm i}\rangle , \qquad (7.28)$$

where $\boldsymbol{q}_{\mathrm f'} = \boldsymbol{q}_{\mathrm f}$ and $\boldsymbol{k}_{\mathrm f'} = \boldsymbol{k}_{\mathrm f}$. We know that they are independent because we obtain TICS from (7.26), and there is no double counting. What are we

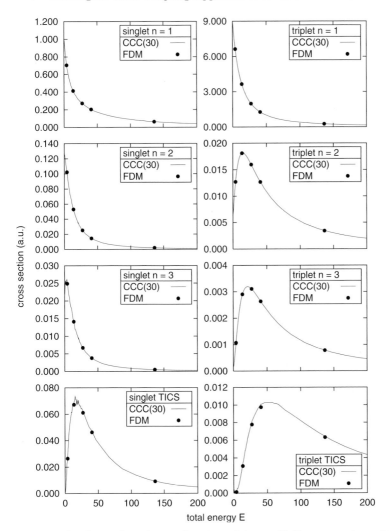

Fig. 7.2. Singlet and triplet, with spin weights, e-H discrete and total ionization cross section for the S-wave model. The CCC calculations are due to Bray and Stelbovics [16], while the finite difference method (FDM) is due to Jones and Stelbovics [19]

to do with the two independent amplitudes for the same process? Following a numerical study, we suggested that the CCC-calculated SDCS should converge with increasing N to zero for $q > k$ [23]. In other words, the ionization amplitude (7.20) converges to zero for $q_f > k_f$, and then the integral endpoint in (7.26) effectively becomes $E/2$, as desired.

To expand on the idea we now have the benefit of accurate SDCS calculated using the FDM [19]. In Fig. 7.3 we consider the SDCS at three energies,

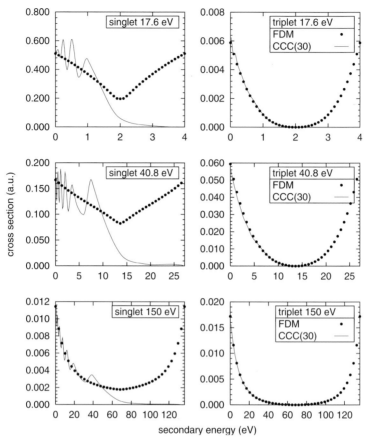

Fig. 7.3. Singlet and triplet, with spin weights, e–H ionization singly differential cross section for the S-wave model. The CCC calculations have been performed to compare specifically with those published by Jones and Stelbovics [19] using the finite difference method (FDM)

and for the two total spins separately. Considering the triplet case first we see excellent agreement between the CCC calculation and the FDM theory, but only for the secondary energy range $0 < e \leq E/2$. For greater secondary energies the CCC calculation yields effectively zero cross sections, whereas the FDM yields the required symmetric result. However, in the FDM theory TICS is obtained from an integral from zero to $E/2$ and hence there is a satisfactory one-to-one correlation between the CCC- and FDM-calculated amplitudes.

Now let us consider the singlet case. Here, the CCC-calculated SDCS is rather oscillatory, though about the correct result for $0 < e < E/2$, and is very small for $E/2 < e < E$. A careful analysis of the singlet case [24] revealed that at $e = E/2$ the CCC SDCS is about one quarter of the true result, and hence that the close-coupling approximation is behaving like a Fourier expansion of a step function with convergence of the underlying amplitudes to half the step height. The results presented in Fig. 7.3 are consistent with this interpretation. If the size of the step is zero then the CCC results yields an accurate smooth cross section for $q_f \leq k_f$. As the size of the step increases the CCC results for $q_f \ll k_f$ increasingly oscillate about the true result with the cross section at $q_f = k_f$ being one quarter the true value. In these cases a relatively accurate result may still be obtained from the CCC calculation if we take into account how Fourier expansions of step functions behave. For example, the $q_f = k_f$ CCC amplitude needs doubling to give the height of the step, and an integral-preserving averaging procedure of some kind may be devised to yield the expected to be smooth cross sections.

This completes our discussion of the S-wave model. It is very helpful in simply explaining how the close-coupling approach to electron–atom collisions works. This understanding does not apply to just our own CCC implementation of the close-coupling method, but to the many other implementations also.

7.2.4   Full Calculations

In the case of discrete excitation the directly obtained amplitude (7.19) should be used when forming comparisons with experiment. However, in the case of general ionization the situation is more complicated. From Fig. 7.3 we see that finite CCC calculations have difficulty in obtaining accurate results whenever the SDCS($E/2$) is significant, which occurs for low to intermediate energies in full calculations. Let us consider the special case of $q_f = k_f$ first. In full e–H calculations [10] the underlying amplitudes were found to satisfy the symmetry property

$$f^{(NS)}(\boldsymbol{q}, \boldsymbol{k}) = (-1)^S f^{(NS)}(\boldsymbol{k}, \boldsymbol{q}) + \delta^{(NS)}(\boldsymbol{q}, \boldsymbol{k}), \qquad (7.29)$$

where δ was small. Then, in forming the (fully) triply differential cross sections (TDCS) the effective doubling of the raw amplitudes may be achieved by noting that

$$|f^{(NS)}(\boldsymbol{q}, \boldsymbol{k}) + (-1)^S f^{(NS)}(\boldsymbol{k}, \boldsymbol{q})|^2 = 2\left(|f^{(NS)}(\boldsymbol{q}, \boldsymbol{k})|^2 + |f^{(NS)}(\boldsymbol{k}, \boldsymbol{q})|^2\right)$$
$$-|\delta^{(NS)}(\boldsymbol{q}, \boldsymbol{k})|^2. \qquad (7.30)$$

In the case of e-He calculations the situation is similar, but the symmetries are more complicated [25], explaining the observed comparison between theory and experiment. Thus, we have the curious situation that CCC solves this special case entirely *ab initio*.

The incoherent combination of the amplitudes on the RHS of (7.30), without the factor of two, was suggested by Bray and Fursa [5] in order to preserve unitarity when the energy integration ends at $E/2$. The factor of two at the single point of $q_f = k_f$ does not affect this. The coherent combination on the LHS of (7.30), derived by Stelbovics [24] for $q_f = k_f$, if applied generally preserves unitarity only if the amplitudes are zero for $q_f > k_f$. To obtain reasonably accurate TDCS for $q_f < k_f$ a semiempirical estimate of the underlying step-function SDCS is required to smooth the unphysical oscillations of the kind evident in Fig. 7.3, see for example, Plottke and Bray [20].

An extensive body of literature exists for applications of the CCC method to electron–atom collisions. The most recent review [4] explained the discrepancies between the earlier CCC-calculated TDCS and experiment [10] by reference to the 25–eV case. Here we discuss briefly the case of 17.6–eV e–H ionization with 2–eV outgoing electrons, see Bray [11] for more details. The 17.6–eV absolute experimental data have been recently revised [26], and are now much more internally consistent than was the case earlier [10]. In addition, the problem of e–H ionization has already been solved numerically using the exterior complex scaling (ECS) theory by Rescigno et al. [27], with the complete set of calculations yielding unprecedented agreement with experiment [28].

The CCC calculations are controlled by the Laguerre basis (7.4) parameters λ_l, N_l, and the maximum l value of the target-space $l_{\max}$. The original 17.6 eV CCC calculations [10] used $N_l = 20 - l$, $\lambda_l = (0.814, 0.828, 0.855, 0.891, 0.936, 0.988)$, with the variation ensuring that the $n = 12 - l$ pseudostate energy was exactly 2 eV for each $0 \leq l \leq l_{\max} = 5$. The energies on either side of the 2–eV state are around 1 and 3 eV. Given the rapid fall at 2 eV from the higher to lower energies, as can be seen in Fig. 7.3, being able to avoid interpolation at $E/2$ seemed an advantage. Taking $N_l = 20 - l$ was done to keep the calculations tractable within the then–existing computational resources.

At low energies we require the continuum-like pseudostates to go out accurately as far as possible, while always ensuring that the resulting scattering calculation is not too big. From Fig. 7.1 we see that a good way to extend the continuum-like pseudostates is to increase both N_l and λ_l simultaneously. The increase of λ_l is counter intuitive, but is necessary to ensure a commensurately consistent discretization of both the discrete subspace and the open continuum. With substantially larger N_l the open continuum is much more densely populated reducing any errors associated with interpolation of the complex scattering amplitudes. In Fig. 7.4 we compare experiment with ECS and a new CCC calculation. The latter was obtained by taking $N_l = 50 - l$ with $\lambda_l = 2$.

The substantially larger N_l and fixed λ_l have resulted in the CCC calculation yielding excellent absolute agreement with the ECS calculation and the experiment. Note that the experiment has been uniformly multiplied by 0.7 for best overall visual fit to theory, which is within the $\pm 40\%$ stated experimental uncertainty [29]. The situation is much improved on the comparison presented previously [10], where the CCC theory was substantially too low for

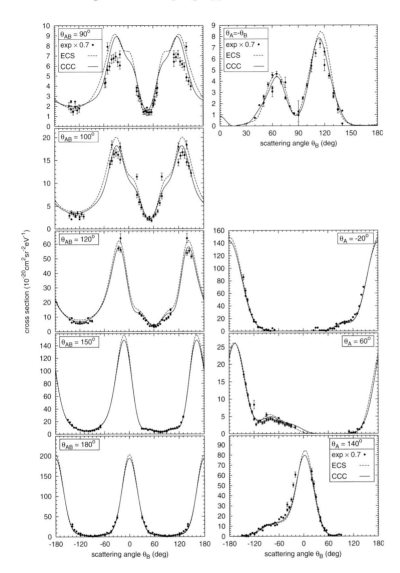

Fig. 7.4. Coplanar triply differential cross sections, in the specified fixed-θ_{AB}, fixed-θ_A and symmetric ($\theta_A = -\theta_B$) geometries, for the 17.6–eV electron-impact ioniza-tion of atomic hydrogen with 2–eV outgoing electrons. The experimental data, with $\pm 40\%$ uncertainty in absolute value, are the full set presented by Röder et al. [26]. The ECS calculation is due to Baertschy et al. [28]. The CCC calculation is described in the text

the geometries with the larger values of θ_{AB}, the angle between the two detectors. Generally the agreement between the two theories is excellent, with the only exception being at the small angles of the $\theta_A = -\theta_B$ geometry. Here, the CCC result has a visible cross section rather than the expected near–zero one. The problem stems from the interpolation of the calculated scattering amplitudes, which upon combination to form the cross section should completely cancel. Finally, the difference between the two amplitude combinations in (7.30) is generally within the thickness of the presented CCC curve.

7.3   Conclusions and Future Directions

We have presented the CCC formalism and demonstrated how it works in the complexity extremes of the simple S-wave model and the fully differential ionization with equal-energy outgoing electrons. The model has been invaluable for the testing and the understanding of how the close-coupling method works. In our opinion it has effectively fulfilled the promise of a "complete scattering theory" [30] in the sense that a single large-enough calculation, at any incident electron energy, is able to give accurate amplitudes for all atomic excitation and ionization processes of practical interest. The rate of convergence with the basis size varies substantially depending on the process of interest and kinematical arrangements. The key issue is that the formalism is able to yield the true scattering amplitudes. We do note that as yet the step-function hypothesis [23] is still not formally proven, but appears consistent with all available numerical and analytical work.

Now that we have a complete understanding of the CCC method the next challenge is very clear. It is to expand the technique to more and more complicated atomic systems. Thus far, helium has been treated in the frozen-core approximation [31], which has proved sufficiently accurate for one-electron transitions. Relaxing this approximation to allow for ionization with excitation and double–ionization processes is a major challenge currently being undertaken. In addition, treatment of relativistic effects in heavier systems and multi-electron atoms are under investigation.

References

1. I. Bray, A.T. Stelbovics: Phys. Rev. A **46**, 6995 (1992)
2. H. Yalim, D. Cvejanovic, A. Crowe: Phys. Rev. Lett. **79**, 2951 (1997)
3. R.W. O'Neill, P.J.M. van der Burgt, D. Dziczek, P. Bowe, S. Chwirot, J.A. Slevin: Phys. Rev. Lett. **80**, 1630 (1998)
4. I. Bray, D.V. Fursa, A.S. Kheifets, A.T. Stelbovics: J. Phys. B **35**, R117 (2002)
5. I. Bray, D.V. Fursa: Phys. Rev. A **54**, 2991 (1996)
6. G. Bencze, C. Chandler: Phys. Rev. A **59**, 3129 (1999)
7. V.L. Shablov, V.A. Bilyk, Y.V. Popov: Phys. Rev. A **65**, 042719 (2002)
8. I. Bray: Phys. Rev. A **59**, 3133 (1999)

9. I. Bray, A.T. Stelbovics: Phys. Rev. A **66**, 036701 (2002)
10. I. Bray: J. Phys. B **33**, 581 (2000)
11. I. Bray: Phys. Rev. Lett. **89**, 273201 (2002)
12. I. Bray, A.T. Stelbovics: in *Computational Atomic Physics*, ed. K. Bartschat (Springer, Berlin, Heidelberg, New York 1996) pp. 161–180
13. S.G. Mikhlin: *Variational Methods in Mathematical Physics* (Pergamon, Oxford 1964)
14. A.T. Stelbovics: Phys. Rev. A **41**, 2536 (1990)
15. E. Anderson, Z. Bai, C. Bischof, J. Demmel, J. Dongarra, J.D. Croz, A. Greenbaum, S. Hammarling, A. McKenney, S. Ostrouchov, et al.: *LAPACK Users's guide* (Society for Industrial and Applied Mathematics, Philadelphia 1992)
16. I. Bray, A.T. Stelbovics: At. Data Nucl. Data Tables **58**, 67 (1994)
17. A. Temkin: Phys. Rev. **126**, 130 (1962)
18. R. Poet: J. Phys. B **11**, 3081 (1978)
19. S. Jones, A.T. Stelbovics: Phys. Rev. A **66**, 032717 (2002)
20. C. Plottke, I. Bray: J. Phys. B **33**, L71 (2000)
21. I. Bray, A.T. Stelbovics: Phys. Rev. Lett. **69**, 53, (1992)
22. I. Bray, A.T. Stelbovics: Phys. Rev. Lett. **70**, 746 (1993)
23. I. Bray: Phys. Rev. Lett. **78**, 4721 (1997)
24. A.T. Stelbovics: Phys. Rev. Lett. **83**, 1570 (1999)
25. I. Bray, D.V. Fursa, A.T. Stelbovics: Phys. Rev. A **63**, 040702 (2001)
26. J. Röder, M. Baertschy, I. Bray: Phys. Rev. A **67** 010702 (2003)
27. T.N. Rescigno, M. Baertschy, W.A. Isaacs, C.W. McCurdy: Science **286**, 2474 (1999)
28. M. Baertschy, T.N. Rescigno, C.W. McCurdy: Phys. Rev. A **64**, 022709 (2001)
29. J. Röder, H. Ehrhardt, C. Pan, A.F. Starace, I. Bray, D.V. Fursa: Phys. Rev. Lett. **79**, 1666 (1997)
30. I. Bray, D.V. Fursa: Phys. Rev. Lett. **76**, 2674 (1996)
31. D.V. Fursa, I. Bray: Phys. Rev. A **52**, 1279 (1995)

8 Close-Coupling Approach to Multiple–Atomic Ionization

A.S. Kheifets

8.1 Introduction: Photoionization vs. Electron-Ion Scattering

Photoionization is a collision process where a quantum of electromagnetic energy is absorbed by an atom and one or several atomic electrons are ejected into the continuum. In the independent electron approximation, at most practical photon energies, the electromagnetic field can only couple to a single electron. Changing the quantum state of the second or further electrons can only take place via *many-electron correlation*. Here, this correlation is understood broadly as the ability of atomic electrons to change, without any external field or interaction, a well-defined set of individual quantum numbers. Uncorrelated electrons in the ground state are labeled by the principle, angular, and magnetic quantum numbers n, l, m, respectively. The ground-state correlation make these quantum numbers only approximate. For instance, the $1s^2$ pair of electrons in the ground–state helium atom can find itself, with a finite probability, in the $2s^2$, $2p^2$ and higher excited states. In an atomic continuum, an uncorrelated electron is labeled by its energy E and momentum $\mathbf{k}$. However, due to elastic or inelastic scattering on other atomic electrons, one or both of these quantum numbers can change.

For the reason outlined above, double or many-fold photoionization of an atom by a single photon is driven entirely by electron correlations, both in the ground state and the photoelectron continuum. The ground–state correlation is a static process that can be relatively easily described by various configuration-interaction schemes. A dynamic correlation in the photoelectron continuum is much more difficult to describe theoretically. Due to the long range of the Coulomb interaction, the continuum correlation cannot be treated perturbatively and a nonperturbative approach has to be applied. A most appealing and physically transparent method to treat such a correlation is the close-coupling method, when the final–state wave function is constructed as a linear combination of the channel wave functions each of which being a symmetrized product of one-electron orbitals. In this sense, the close-coupling method is a generalization of the configuration-interaction method for bound states into the states with one or few electrons in the continuum.

A two-electron continuum state can be treated very efficiently by the convergent close-coupling (CCC) method in which one of the electrons is represented by a complete set of discrete (positive and negative energy) pseudostates while the second electron is treated as a true continuum state. Details of the CCC method, as applied to electron–atom scattering, are outlined in Chap. 7. Application of the CCC method to double ionization is straightforward. Consider, for instance, double photoionization (DPI) of helium, the simplest two-electron atomic system. After the first electron is ejected into the continuum following absorption of a photon, the system of interest evolves through elastic and inelastic scattering of the photoelectron on the He^+ ion. Electron scattering on the He^+ ion is very similar to that of e–H scattering as the wave functions for He^+ belong to the same isoelectronic sequence as atomic hydrogen (H-like ions) with the nuclear charge $Z = 2$. Another minor difference is that the target has an asymptotic charge requiring the use of Coulomb waves (including bound states) for the description of the projectile rather than plane waves. These are relatively simple modifications, and the first application of the CCC method to e–He^+ scattering was able to describe the total ionization cross section quantitatively [1].

With this realization, a complete theory of DPI on He and similar two-electron targets can be constructed, as was demonstrated by Kheifets and Bray in a series of publications [2–6]. DPI of more complex many-electron targets like alkaline-earth atoms can also be described in a similar way by taking the asumption of a frozen core of electrons [7]. A related process of double ionization of helium by fast charged–particle impact is very similar to DPI and can be treated accurately by the CCC method [8].

In this chapter, we perform a formal derivation of the CCC theory as applied to multiple atomic ionization and present a collection of results in comparison with the latest experimental data. In conclusion, we discuss possible generalization of the method to more complex atomic targets and ionization mechanisms.

8.2   Two-Electron Photoionization

In this section, we consider DPI and a related process of photoionization with simultaneous excitation of the second electron into one of the discrete nl states. Both processes involve transition of two atomic electrons and shall be treated on a similar theoretical ground.

8.2.1   Shake-Off and Two-Step Mechanisms

In our formalism we consider the two-electron photoionization as a two-stage process. Stage one (left diagram of Fig. 8.1) is a single ionization that is followed by electron-impact excitation or ionization of the resultant He^+ ion (right diagram of Fig. 8.1). Here, a thin straight line with an arrow indicates

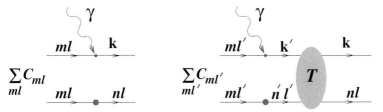

Fig. 8.1. Graphical representation of the shake-off (left) and the two-step (right) mechanisms of two-electron atomic photoionization process

an electron state in the Coulomb field of the nucleus, the wavy line shows a photon. The filled oval represents the interelectron interaction in all orders of the perturbation theory (the so-called T-matrix theory). The dot indicates an overlap of the non-orthogonal one-electron orbitals of the initial and final states of the target.

Taken alone, the left diagram represents the so-called *shake-off* mechanism in which the second electron is excited or ionized because of a sudden change of the atomic potential due to ejection of the first electron. The right diagram depicts a true *two-step* mechanism in which the first ejected electron is scattered inelastically on the ion. The relative contributions of these two mechanisms depends on the photon energy. The shake-off mechanism requires a certain excess energy above the ionization threshold that enables the first ejected electron to leave the atom quickly enough to produce a sudden change of the atomic field. For this reason one would not expect this mechanism to be dominant near the double–ionization threshold. On the contrary, when the excess energy is large, the two ejected electrons leave the atom with a highly asymmetric energy sharing. In this case one can neglect the interaction of the fast–ejected electron with the ion and the whole yield of the DPI process would be due to the shake-off mechanism. More elaborate separation of the shake-off and two-step mechanisms is given in [9].

8.2.2   Ground-State Correlation and Gauge Invariance

The ground–state correlation is shown in Fig. 8.1 as a superposition of various two-electron configurations. This corresponds to a ground–state wave function

$$\Phi_0(\boldsymbol{r}_1, \boldsymbol{r}_2) = \sum_{nl} \frac{C_{nl}}{\sqrt{(2l+1)}} \sum_m (-1)^{l-m} \phi_{nlm}(\boldsymbol{r}_1)\, \phi_{nl-m}(\boldsymbol{r}_2), \qquad (8.1)$$

where the sum is taken over a few lowest nl orbitals coupled to the ^1S term. The most optimal choice of the one-electron orbitals in (8.1) is made by employing a multi-configuration Hartree–Fock (MCHF) method. This excludes nondiagonal components from the sum and provides a fast convergence. For example, in the helium atom, a 7-term expansion with s, p, and d orbitals recovers 95% of the correlational energy, which is defined as a difference between the exact and noncorrelated, Hartree–Fock ground–state energies.

Another form of a highly correlated ground state for a two-electron system is a Hylleraas expansion over the powers of the three parameters: $u = |\boldsymbol{r}_1 - \boldsymbol{r}_2|$, $s = r_1 + r_2$, and $t = r_1 - r_2$.

$$\Phi_0 = Ne^{-zs} \sum_{ijk} a_{ijk}\, u^i s^j t^k \,. \tag{8.2}$$

A Hylleraas-type wave function is much more rapidly convergent and can deliver an unparalleled accuracy in terms of the ground–state energy. However, it is not separable and it is, therefore, much more difficult to deal with in practical computations.

Convergence, in terms of energy is not the only indication of the accuracy of the ground–state wave function. More revealing is the so-called *gauge invariance* of the photoionization cross section. The photoionization cross section, as a function of the photon energy w, corresponding to a particular bound electron state j (degenerate with magnetic sublevels m_j) is given by [10]:[1]

$$\sigma_j(w) = \frac{4\pi^2}{wc} \sum_{m_j} \int d^3k \, |\langle \Psi_j^{(-)}(\boldsymbol{k})|\mathcal{D}|\Phi_0\rangle|^2 \, \delta(w + E_0 - k^2/2 - \epsilon_j) \,. \tag{8.3}$$

The dipole electromagnetic operator $\mathcal{D}$ for the light polarized along the z-direction can be written in one of the following three forms commonly known as length, velocity, and acceleration gauges [10]:

$$\mathcal{D}^r = w(z_1 + z_2)\,, \quad \mathcal{D}^\nabla = \nabla_{z_1} + \nabla_{z_2}\,, \quad \mathcal{D}^{\dot{V}} = \frac{2}{w}\left(\frac{z_1}{r_1^3} + \frac{z_2}{r_2^3}\right)\,. \tag{8.4}$$

If the initial Φ_0 and the final $\Psi_j^{(-)}(\boldsymbol{k})$ state wave functions in (8.3) are exact the three forms of the dipole operator would provide identical cross sections. Numerical wave functions (8.1) and (8.2) provide deviating cross sections, the length form being most sensitive as it enhances large distances from the origin that contribute little to the ground–state energy. Only the most accurate 20-term Hylleraas expansion fixes this problem, providing essentially the gauge-insensitive cross section [3].

8.2.3   Total Cross Sections for Ionization-Excitation and Double Photoionization of Helium Isoelectronic Sequence

For the final state of the two-electron system consisting of one bound and one continuum electrons, we use a close-coupling expansion:

$$|\Psi_j^{(-)}(\boldsymbol{k})\rangle = |j\boldsymbol{k}^{(-)}\rangle + \sum_i \int\!\!\!\!\!\!\sum d^3p \, \frac{\langle \boldsymbol{p}^{(+)}i|T|j\boldsymbol{k}^{(-)}\rangle}{E - \varepsilon_p - \epsilon_i + i\delta}|i\boldsymbol{p}^{(+)}\rangle \,, \tag{8.5}$$

---

[1] System of atomic units is used with $\hbar = m_e = e = 1$.

where $\cos\theta_{12} = \hat{\mathbf{k}}_1 \cdot \hat{\mathbf{k}}_2$. This immediately takes us to the final expression [29]

$$\sigma_0 = C\left|(\cos\theta_1 + \cos\theta_2)a_g(E_1, E_2) + (\cos\theta_1 - \cos\theta_2)a_u(E_1, E_2)\right|^2, (8.12)$$

where the symmetric and antisymmetric (gerade-g and ungerade-u) DPI amplitudes are

$$a_{\substack{g\\u}} = \frac{1}{4\pi}\sum_{l=0}^{\infty}\frac{(-1)^l}{\sqrt{l+1}}\left[P'_{l+1}(\cos\theta_{12}) \mp P'_l(\cos\theta_{12})\right] D^{\pm}_{ll+1}(E_1, E_2). \quad (8.13)$$

For the special case of equal energy sharing $E_1 = E_2$, the antisymmetric amplitude vanishes and a simple Gaussian ansatz can be applied to the symmetric amplitudes. At low excess energy, close to the double–ionization threshold, this ansatz is predicted by Wannier-type quasiclassical theories [30–32]. Numerically, it was demonstrated that the Gaussian parametrization holds at quite high excess energies, at least up to 60 eV [33]. Cvejanović and Reddish [34] went one step further and suggested an empirical Gaussian parametrization for the antisymmetric amplitude as well. They introduced the so-called *practical parameterization* when the whole set of the DPI TDCS at given photoelectron energies E_1, E_2 can be described by a small set (three or, more generally, four) adjustable parameters.

Here we demonstrate how this parametrization can be applied within the CCC model [36]. We write the complex amplitudes a_g, a_u as $a_g = A_g \exp(i\delta_g)$, $a_u = A_u \exp(i\delta_u)$, where $A_g, A_u, \delta_g, \delta_u$ are real. We introduce the Gaussian ansatz for the real amplitudes A_g, A_u:

$$A_g = b_g \exp\left[-2\ln 2\frac{(\pi - \theta_{12})^2}{\Gamma_g^2}\right]$$

$$A_u = b_u \exp\left[-2\ln 2\frac{(\pi - \theta_{12})^2}{\Gamma_u^2}\right]. \quad (8.14)$$

The 4-parameter (4P) parameterization of Cvejanović and Reddish [34] corresponds to an assumption of a constant phase shift $\phi = \delta_g - \delta_u$ between the symmetric and antisymmetric amplitudes for all mutual angles θ_{12}. The constant phase shift ϕ, the two Gaussian widths Γ_g, Γ_u and the magnitude ratio $\eta = b_u/b_g$ are sufficient to reproduce the shape of the TDCS at a given photoelectron energies at any geometry of two-electron ejection.

In [36] the Gaussian parametrization (8.14) was applied to CCC calculated amplitudes at the excess energies of $E = 9, 20, 40$, and 60 eV. The central part of the amplitudes A_g and A_u amplitudes around $\theta_{12} = 180°$ could be well fitted with the Gaussian ansatz and the phase difference in this area was indeed almost constant, as is seen at the top panels of Fig. 8.4. At small mutual angles $\theta_{12} < 90°$ or, equivalently, $\theta_{12} > 270°$, the amplitudes noticeably deviate from the Gaussian shape, especially the antisymmetric amplitude A_u, which shows substantial "wings". In the area of the wings, the

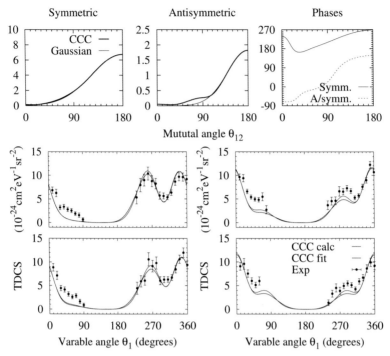

Fig. 8.4. *Top*: Symmetrized amplitudes of the helium DPI at $E = 40\,\mathrm{eV}$ and energy sharing ratio $E_1/E_2 = 7$ fitted with the Gaussian ansatz (8.14). *Bottom*: DPI TDCS of helium at the same excess energy and sharing ratio. The *arrow* on the polar plots indicates the direction of the slower ($E_2 = 5\,\mathrm{eV}$) electron. The *thick solid line* is the CCC calculation (velocity form, other gauges near identical), the *thin solid line* is the four-parameter model with the Gaussian parameters extracted from the CCC calculation Experimental data, shown by *solid circles with error bars*, are taken from [35]

phase δ_g shows some rapid variation and the phase shift ϕ no longer remains constant. However, the wings area does not contribute to the TDCS as the amplitude here is very small.

This observations lends support to the 4P parametrization of Cvejanović and Reddish [34]. We illustrate it on the bottom panel of Fig. 8.4 where we show the TDCS as a function of the variable escape angle of the fast electron ($E_1 = 35\,\mathrm{eV}$) at several selected escape angles of the slow photoelectron ($E_2 = 5\,\mathrm{eV}$). The two sets of theoretical amplitudes are shown: the CCC calculation and the Gaussian fit with the parameters extracted from the present calculation. We observe that the TDCS generated from the CCC calculated and CCC-fitted amplitudes are hardly discernible. This is despite the fact that the CCC amplitudes, especially the antisymmetric one, noticeably deviate from the Gaussian ansatz. Agreement with experiment [35] is generally good.

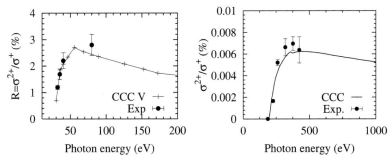

Fig. 8.5. *Left panel*: Double-to-single photoionization cross section ratio for the outer $2s^2$ shell of Be. The experiment is by Wehlitz and Whitfield [37], the CCC calculation in the velocity (V) gauge is from [7]. *Right panel*: Triple-to-single photoionization cross section ratio for Li. The experiment is by Wehlitz et al. [38], the CCC calculation (all gauges indistinguishable) is from [5]

8.2.6   Beyond Two-Electron Targets: Double Photoionization of Beryllium and Triple Photoionization of Lithium

There is another class of atomic targets, namely the alkaline-earth atoms, which can be treated, in some approximation, as two-electron systems. Indeed, a compact electron core is well separated, both in the coordinate space and in energy, from the valence ns^2 shell. At relatively small photon energies the inner core electrons can be treated as "spectators" not taking a direct part in the photoionization of the outer valence electrons. In this case the influence of the core on the valence electrons can be included via the self-consistent field and/or the polarization potential.

In [7] the CCC method was applied to the two-electron photoionization of the valence $2s^2$ shell of the beryllium atom. The static ground–state correlation in this shell was described by employing a MCHF wave function with the frozen $1s^2$ core. The dynamic correlation in the two-electron continuum is represented by a momentum–space close-coupling expansion, obtained from a CCC calculation for electron-impact ionization of Be^+.

The photon energy dependence of the total ionization-excitation and the double–ionization cross sections in Be qualitatively resembles that of He. On the left panel of Fig. 8.5 we show the ratio of the double-to-single photoionization cross sections in Be, which is in a good agreement with the experiment [7]. Despite similarity of the total cross sections, the TDCS in Be and ground–state He are quite different. Our calculations at the excess energy of $E = 20\,\text{eV}$ shared equally between the photoelectrons show considerably smaller Gaussian width in Be than in He (68° and 91°, respectively). This indicates a much stronger angular correlation in the double–ionized continuum of beryllium than that of helium.

Triple photoionization (γ,3e reaction) on lithium can also be calculated using the CCC method for a two-electron ionization. Wehlitz et al. [38] showed that $(\gamma,3e)$ on Li is well separated into the DPI of the core $1s^2$ shell and subsequent shake-off of the outer $2s$ electron. The core DPI of neutral Li is very similar to that of the Li^+ ion and can be handled by a standard CCC formalism. On the right panel of Fig. 8.5 we show the triple-to-single photoionization cross section ratio, as measured and calculated using this shake-off model. Good agreement with experiment can be achieved but the shake-off rate should be taken as that estimated by Wehlitz et al. [38].

8.3   Two-Electron Charged Particle Impact Ionization

8.3.1 $(\gamma,2e)$ and (e,3e) Reactions on Helium

When the energy of the fast charged projectile is large its interaction with the target can be treated perturbatively by employing the Born series. The lowest first and second terms in this series are shown in Fig. 8.6.

The first Born term is very similar to the two-step diagram of the DPI process shown in Fig. 8.1. It can be calculated using the same initial MCHF (8.1) or Hylleraas (8.2) ground–state wave functions and the CCC final state wave function (8.5) thus taking both the ground– and final–state correlations fully into account. The dipole operator in (8.4) should be substituted by the Born operator taken in either length or velocity form:

$$\mathcal{B}^r = \exp(i\boldsymbol{qr})\,, \quad \mathcal{B}^\nabla = \frac{1}{2\omega}\{\exp(i\boldsymbol{qr})[\boldsymbol{q}\nabla] - [\nabla\boldsymbol{q}]\exp(i\boldsymbol{qr})\}\,. \qquad (8.15)$$

Here, $\boldsymbol{q} = \boldsymbol{k}_0 - \boldsymbol{k}_1$ is the momentum transfer from the projectile to the target. Partial–wave expansion can be used for the Born operator (8.15) leading to separation of the radial and angular variables. Angular integration is performed analytically.

The first Born CCC calculation of the (e,3e) reaction on helium at the incident energy of 1 keV is presented in Fig. 8.7, together with another first Born calculation based on the three-body Coulomb waves method (known in the literature as BBK) [40], and the experiment [39]. It can be seen in Fig. 8.7 that the BBK calculation is far from the CCC calculation. In contrast, similar $(\gamma,2e)$ calculations performed with the BBK and CCC models produce very close, nearly identical results [35,41,42].

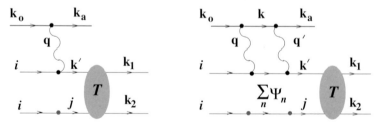

Fig. 8.6. Graphical representation of charged–particle impact double ionization. *Left*: 1st Born amplitude, *right*: 2nd Born amplitude

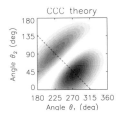
CCC theory

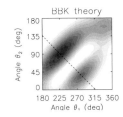
BBK theory

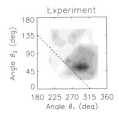
Experiment

Fig. 8.7. Contour plots of the (e,3e) FDCS at the kinematical conditions of the experiment [39]. The escape angles of the two slow ejected electrons θ_1 and θ_2 are shown on the axes. Only the quadrant of the experimentally accessible range of angles is displayed. Theoretical results are from [40]. *Left*: CCC calculation, *middle*: BBK calculation, *right*: experiment. The dashed line indicates the symmetry axis of the first Born model

8.3.2   Second Born Corrections

Incident–electron energy in (e,3e) reactions can always be taken sufficiently low so that there is a need for higher Born corrections. At 1–keV incident energy the higher Born effects are clearly visible in the experiment (see Fig. 8.7) as a deviation from the symmetry line of the momentum transfer $\boldsymbol{q}$. The second Born process is illustrated by the right diagram of Fig. 8.6. Here the projectile interacts with the target twice thus being able to eject two electrons sequentially. Direct evaluation of the second Born term is difficult. A common approximation is made to substitute a state-dependent energy denominator into the intermediate state by an average value and then to use a closure relation to perform summation over the intermediate states of the targets [43,44].

Remaining integration over the projectile momentum in the intermediate state $\boldsymbol{k}$ is made numerically. This, however, is a very inefficient computational procedure. Instead, we do a partial wave expansion of the two Born operators, as in the first Born amplitude, and restrict ourselves to the dipole contribution, which is the leading term at small momentum transfer. This approach was suggested by Franz and Altic [45]. It is justified by the fact that the second Born amplitude is proportional to $(q_0 q_1)^{-2}$ where $\boldsymbol{q}_0 = \boldsymbol{k}_0 - \boldsymbol{k}$ and $\boldsymbol{q}_1 = \boldsymbol{k} - \boldsymbol{k}_1$ are momenta transferred from the projectile to the target in the first and second collisions. However, Franz and Altic took the optical limit of the dipole term, which, we find, leads to a significant overestimation of the second Born amplitude. By properly evaluating the dipole second Born term, we otained electron impact ionization-excitation cross sections very similar to those reported in the literature [43,44]. We thus assured the correctness of our second Born calculation.

Results of our first and second calculations of the (e,3e) on helium at the excess energy of 0.5 keV are presented in Fig. 8.8. The second Born effects are significant and they take the calculated cross section closer towards the experiment.

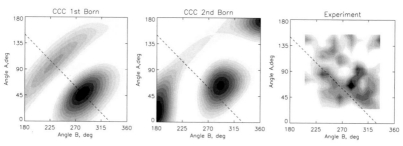

Fig. 8.8. Contour plots of the (e,3e) FDCS at the scattered electron energy $E_1 = 500\,\mathrm{eV}$ and the two ejected electrons $E_a = E_b = 11\,\mathrm{eV}$. Experiment is from Lahamam-Bennani and Duguet [46])

8.4 Conclusion: Towards Larger Dynamical Freedom

We have demonstrated that the CCC method can be successfully employed to calculate a variety of multiple–ionization processes caused by photon or electron impact. Future applications of the method will rest on its ability to treat more complex atomic targets. The Laguerre basis of the positive–energy pseudostates is not the only choice. It was demonstrated in [12] that pseudostates can be built on the spline basis that is widely used in atomic physics in a variety of applications.

More generally, multiple–ionization processes with larger numbers of particles open up a larger dynamical freedom in the sense that there are fewer constraints dictated by the rigid selection rules. This places a significantly stronger demand on accurate numerical models. For example, DPI is very much restricted by the dipole selection rule and even mediocre–quality continuum wave functions, like BBK and its clones, produce reasonable results. In contrast, for the (e,3e) reaction the dipole selection rules are relaxed and BBK calculations find themselves at variance with more accurate close-coupling results. The ultimate challenge of theory will be treating three (or even more) particles in continuum nonperturbatively. This, however, requires significant modification of existing computational methods.

References

1. I. Bray, I.E. McCarthy, J. Wigley, and A.T. Stelbovics: J. Phys. B **26**, L831 (1993)
2. A.S. Kheifets and I. Bray: Phys. Rev. A **54**, R995 (1996)
3. A.S. Kheifets and I. Bray: Phys. Rev. A **57**, 2590 (1998)
4. A.S. Kheifets and I. Bray: J. Phys. B **31**, L447 (1998)
5. A.S. Kheifets and I. Bray: Phys. Rev. A **58**, 4501 (1998)
6. A.S. Kheifets and I. Bray: Phys. Rev. Lett. **81**, 4588 (1998)
7. A.S. Kheifets and I. Bray: Phys. Rev. A **65**, 2710 (2002)
8. A.S. Kheifets et al.: J. Phys. B **32**, 5047 (1999)

9. A.S. Kheifets: J. Phys. B **34**, L247 (2001)
10. M.Y. Amusia, in *Atomic Photoeffect*, ed. by K.T. Taylor (Plenum Press, New York 1990)
11. I. Bray: Phys. Rev. A **49**, 1066 (1994)
12. A.S. Kheifets, A. Ipatov, M. Arifin, and I. Bray: Phys. Rev. A **62**, 2724 (2000)
13. A.S. Kheifets and I. Bray: Phys. Rev. A **58**, 4501 (1998)
14. R. Wehlitz et al.: J. Phys. B **30**, L51 (1997)
15. R. Dörner et al.: Phys. Rev. Lett. **76**, 2654 (1996)
16. I. Bray and A.T. Stelbovics: Phys. Rev. Lett. **70**, 746 (1993)
17. A.S. Kheifets and I. Bray: J. Phys. B **31**, L447 (1998)
18. D.A. Varshalovich, A.N. Moskalev, and V.K. Khersonskii: *Quantum Theory of Angular Momentum* (World Scientific, Singapore 1988)
19. O. Schwarzkopf and V. Schmidt: J. Phys. B **28**, 2847 (1995)
20. O. Schwarzkopf and V. Schmidt: J. Phys. B **29**, 1877 (1996)
21. O. Schwarzkopf, B. Krassig, and V. Schmidt: J. Phys. (Paris) **3**, 169 (1993)
22. V. Mergel et al.: Phys. Rev. Lett. **80**, 5301 (1998)
23. M. Achler et al.: J. Phys. B **34**, 965 (2001)
24. A.S. Kheifets and I. Bray: Phys. Rev. A **62**, 65402 (2000)
25. A.S. Kheifets and I. Bray: Phys. Rev. Lett. **81**, 4588 (1998)
26. J. Berakdar and H. Klar: Phys. Rev. Lett. **69**, 1175 (1992)
27. J. Berakdar, H. Klar, A. Huetz, and P. Selles: J. Phys. B **26**, 1463 (1993)
28. A. Kono and S. Hattori: Phys. Rev. A **29**, 2981 (1984)
29. L. Malegat, P. Selles, and A. Huetz: J. Phys. B **30**, 251 (1997)
30. A.K. Kazansky and V.N. Ostrovsky: J. Phys. B **26**, 2231 (1993)
31. J.M. Rost: Phys. Rev. A **53**, R640 (1996)
32. J.M. Feagin: J. Phys. B **29**, L551 (1996)
33. A.S. Kheifets and I. Bray: Phys. Rev. A **62**, 5402 (2000)
34. S. Cvejanović and T. Reddish: J. Phys. B **33**, 4691 (2000)
35. S. Cvejanović et al.: J. Phys. B **33**, 265 (2000)
36. A.S. Kheifets and I. Bray: Phys. Rev. A **65**, 2708 (2002)
37. R. Wehlitz and S.B. Whitfield: J. Phys. B **34**, L719 (2001)
38. R. Wehlitz et al.: Phys. Rev. Lett. **81**, 1813 (1998)
39. A. Lahmam-Bennani et al.: J. Phys. B **34**, 3073 (2001)
40. A.S. Kheifets, I. Bray, J. Berakdar, and C.D. Cappello: J. Phys. B **35**, L15 (2002)
41. C. Dawson et al.: J. Phys. B **34**, L525 (2001)
42. P. Bolognesi et al.: J. Phys. B **34**, 3193 (2001)
43. P.J. Marchalant, C.T. Whelan, and H.R.J. Walters: J. Phys. B **31**, 1141 (1998)
44. Y. Fang and K. Bartschat: J. Phys. B **34**, L19 (2001)
45. A. Franz and P.L. Altick: J. Phys. B **28**, 4639 (1995)
46. A. Lahmam-Bennani and A. Duguet: in *Correlations, Polarization and Ionization in Atomic Systems*, ed. by D. Madison and M. Schulz (AIP Press, New York 2002) pp. 96–101

9 Numerical Grid Methods

K.T. Taylor, J.S. Parker, D. Dundas, K.J. Meharg, L.R. Moore,
E.S. Smyth, and J.F. McCann

9.1 Introduction

Over recent years in Belfast we have developed numerical grid methods for
solving the two-electron time-dependent Schrödinger equation (TDSE) in its
full-dimensionality [1–4]. We have specifically had in mind during these developments the TDSE for laser-driven helium and, more recently, that for the
laser-driven hydrogen molecule. Our developments were initially prompted,
and continue to be stimulated, by the experimental interest [5–8] in these
few-electron strongly time-dependent systems. As reported elsewhere in this
volume, advances in laser technology and in detection techniques [9] steadily
increase the possibilities and refinement of experimental measurement. For
instance, angular information regarding the ionization of both electrons can
now be gained and the upcoming free-electron lasers will soon make available
unprecedented high radiation intensities in the UV to soft X-ray regimes. If
theory is to play a meaningful role, and especially a predictive one in such
circumstances, sophisticated calculational methods, typified by those we set
out to summarize below, are required.

9.2 Solution of the Time-Dependent Schroedinger Equation Using Numerical Spatial Grids and High-Order Time Propagation

In both two-electron problems of interest to us (laser-driven helium and H_2)
linearly polarized laser radiation provides one element of symmetry in the
electronic motion and thus full-dimensionality for this motion amounts to five
spatial degrees of freedom plus time. We handle the time variable by high-
accuracy techniques (Taylor series and Arnoldi propagator methods) that
will first be described below. So far as individual spatial degrees of freedom
are concerned, we find, depending on the problem, that it may in some cases
be very advantageous to use pure basis-state representations and thus avoid
grid techniques completely. But in each problem there are spatial variables
remaining that must be handled by grid techniques, and finite-difference and
discrete variable representation (DVR) methods are used selectively for these.
The second part of this subsection thus summarizes the principal points of
both of these grid methods.

9.2.1   Time Propagation using Taylor Series and Arnoldi Propagator Methods

Krylov subspace techniques were originally introduced by Lanczos and Arnoldi for calculating eigenstates of matrices. The methods are readily adapted to the integration of differential equations. We have used these methods, which we call Arnoldi propagators [10], for propagating forward in time wave function solutions to the two-electron TDSEs for both laser-driven helium and the hydrogen molecule (H_2) as well as for the much simpler one-electron TDSEs governing the laser-driven hydrogen atom and laser-driven hydrogen molecular ion (H_2^+).

Popular alternatives such as the second-order Crank–Nicholson propagator tend to be either too inaccurate or too inefficient for our applications. The Crank–Nicholson, for example, has local truncation error (LTE) per time step δt of order δt^3, (and global integration error of order δt^2). In a typical example it may exhibit a LTE of 10^{-4}. In many problems we examine, incorrect answers are obtained if LTE exceeds 10^{-9}. To reduce a Crank–Nicholson LTE from 10^{-3} to 10^{-9} requires a reduction in δt by a factor of 100.

Our primary test of the Arnoldi propagator is through comparison with the results of an independently written Taylor series propagator. The Taylor series propagator proves to be particularly reliable but is 2–30 times slower than the Arnoldi one. The Taylor series is perhaps conceptually the simplest of explicit single-step propagators. The better known Runge-Kutta integrators become identical to the Taylor series when applied to time-independent linear differential equations, $\dot{\Psi} = M\Psi$ with time-independent M, but require more evaluations of $M\Psi$ than the Taylor series if the order of the integrator is greater than four. The Taylor series is formed by repeatedly differentiating the differential equation. In the limit in which the time derivatives of H can be neglected the k-th derivative of Ψ is just $(-iH)^k\Psi$, and the Taylor series for $\Psi(t + \delta t)$ in terms of $\Psi(t)$ is:

$$\Psi(t + \delta t) = c_0\Psi(t) + c_1 H\Psi(t) + c_2 H^2\Psi(t) + \ldots + c_n H^n\Psi(t), \qquad (9.1)$$

where c_k are the Taylor series coefficients $(-i\delta t)^k/k!$, and H the Hamiltonian. The higher-order time-derivatives of $H(t)$ can easily be included at negligible cost but neglect of these terms is found to introduce no detectable error.

The Krylov subspace K_{n+1} is that spanned by the vectors $\Psi, H\Psi, \ldots, H^n\Psi$. Gram-Schmidt, with iterative refinement, is used to obtain an orthonormal set of vectors that span K_{n+1}, which we write $Q_0, Q_1, \ldots, Q_n$, where $Q_0 = \Psi/|\Psi|$. Q_k is obtained by calculating HQ_{k-1} and then orthonormalizing this vector with respect to $Q_0, \ldots, Q_{k-1}$. We write this as

$$HQ_k = h_{k+1,k}Q_{k+1} + h_{k,k}Q_k + \ldots . \qquad (9.2)$$

If we define Q to be formed from the $n + 1$ column vectors $(Q_0, Q_1, \ldots, Q_n)$, then the above equation in matrix form reduces to $h = Q^\dagger HQ$. We see then that h is the Krylov subspace Hamiltonian, (i.e., H in the space spanned by

$Q_0, Q_1, \ldots, Q_n$). It is clear from the above outline that h is calculated simultaneously with Q, at no extra cost. As Lanczos showed, the eigendecomposition of h can be used as the first step in an iterative scheme to calculate eigenvalues of H. The approach outlined above was first given by Arnoldi [10]. More recently, it has been appreciated that h (or more accurately $\tilde{H} = QhQ^\dagger$) can be used as a replacement for H in a wide variety of applications [11–15], including the integration of differential equations.

In our applications, the TDSE, replacing H with $\tilde{H}$ is beneficial because the Taylor series in $\tilde{H}$ can be summed to effectively arbitrary order in negligible time compared to that required for the calculation of $\Psi, H\Psi, \ldots, H^n\Psi$. This follows from $(QhQ^\dagger)^m = Qh^mQ^\dagger$, so that $e^{-i\tilde{H}\delta t} = Qe^{-ih\delta t}Q^\dagger$. The matrix h is typically a 19×19 tridiagonal matrix, so that its exponentiation through direct diagonalization of h is inexpensive.

The method proves attractive for four reasons. First, despite the work required to orthonormalize the vectors, the computational overhead rises linearly with n. The work is almost entirely in the calculation of $\Psi, H\Psi, \ldots, H^n\Psi$. Even on a parallel machine, we have found the calculation of h and Q to be less than 10% of the total overhead. Since the method is explicit it obviates the need for costly inversion operations. Secondly, the method may be viewed as a means of constructing a unitary propagator that is correct to order n in δt. Thirdly, the Arnoldi formulation provides a very efficient means of obtaining eigenstates of the Hamiltonian. This feature is used in the generation of the ground state of helium and in other initial states of time-dependent calculations. Finally, and perhaps most importantly, the Arnoldi propagator demonstrates at least twice the efficiency of the Taylor series and the performance ratio improves linearly as the order of the method increases. For example, suppose we set as a constraint the requirement that local truncation errors remain less than 10^{-10}. If we increase the order of the Arnoldi propagator from 8 to 12, the computational cost of each timestep increases by 50%, whereas δt can be doubled. If we increase the order of the Arnoldi propagator from 8 to 16, the computational cost of each timestep increases by a factor of 2, whereas δt can be quadrupled. By contrast, the performance of the Taylor series (and equivalently Runge-Kutta) remains constant as n is increased. In other words, if n is doubled (doubling the computational cost of each timestep), then δt can at most be doubled. At $n = 18$ the Arnoldi propagator outperforms the Taylor series by a factor of 4.5. This scaling law for the Arnoldi propagator favors the highest order possible, but storage limitations for the Q vectors limits n to between 12 and 24.

9.2.2   Grid Methods:
Finite-Difference and Discrete Variable Representation
(DVR) Methods for Spatial Variables

In both problems of interest we have five spatial variables to handle. We will see below that geometrical and center-of-force considerations demand these be handled differently in the two problems. Finite-difference and DVR are the fundamental grid methods that come into play, however, and here we outline the concepts underlying each, together with any difficulties specific to the problems in hand.

The finite-difference method is a long-established reliable technique for solving PDEs. Each spatial coordinate is discretized on a uniformly spaced grid of points. The evaluation of spatial derivatives at a specific point couples values of the function from neighbouring points. If the spatial coordinate is of infinite (or semi-infinite) range, as in our work then, since computer memory is finite, some finite upper limit must be taken. We thus consider representation over a box of finite size. If the box size is too small, the wave function can be reflected back from the boundary towards the atom or molecule. This can alter the population, hence affecting ionization rates. If the reflected wave function reaches the region near the nucleus or nuclei, it can further cause spurious peaks in the harmonic spectra, and spurious ionization rates. Reflections from the edge of the box are eliminated by a splitting technique that breaks the wave function into two parts. We illustrate this in a generic case using r as the spatial variable spanning the box over the range $[0, R_{\mathrm{max}}]$. A masking function, $M(r)$, is used to perform the splitting. M is a function that equals 1 for small electron-nuclear distances r and goes to 0 very gradually asymptotically. The two parts of the wave function, Ψ, can therefore be written, $\Psi_1 = M(r)\Psi(r)$ and $\Psi_2 = (1 - M(r))\Psi(r)$. This splitting operation is performed each timestep and ensures that the $\Psi_1 \to 0$ as $r \to R_{\mathrm{max}}$. The total population removed by the mask is accumulated in arrays that store ionization yields. The design and optimization of the mask requires considerable care, and extensive numerical simulations must be performed to characterize the optimal shape. Various shapes were investigated, but a Gaussian with the following characteristics very clearly gave the best results:

$$M(r) \quad = \quad 1.0 \qquad 0 \le r < R_{\mathrm{max}} \times \alpha \,, \tag{9.3a}$$

$$= \quad e^{-T(r)^2} \qquad R_{\mathrm{max}} \times \alpha \le r \le R_{\mathrm{max}} \,, \tag{9.3b}$$

where $T(r) = 2(r - R_{\mathrm{max}} \times \alpha)/(R_{\mathrm{max}}\sigma)$. The optimal values of α and σ were determined by numerical experiment. In our current runs, typical values are $\alpha = 0.3333$ and $\sigma = 5.0$.

The Lagrange mesh method is a grid method founded on Lagrange interpolation and Gaussian quadrature. It is a special case of the discrete variable representation method [16] that has recently been extensively applied to both time-independent problems [17,18] and time-dependent problems [19–21]. An outline of the method [22] follows. Consider a set of N differentiable mesh functions, $f_i(x)$, defined on a domain $a \le x \le b$ with $i = 1, 2, \ldots, N$. These mesh functions are chosen to satisfy the Lagrange interpolation condition

$$f_i(x_j) = \lambda_i^{-1/2}\delta_{ij} \,. \tag{9.4}$$

If the grid points $(x_1, \ldots, x_N)$ are selected to be the pivots of an n-point Gaussian quadrature rule these functions are quadrature orthogonal,

$$
\begin{aligned}
\langle f_i | f_j \rangle &\equiv \int_a^b f_i^\star(x) f_j(x) \mathrm{d}x \\
&\approx \lambda_i^{1/2} f_i^\star(x_j) = \lambda_i^{1/2} f_j(x_i) = \delta_{ij} \,,
\end{aligned}
\tag{9.5}
$$

where λ_i is the i-th weight. In order to construct the mesh functions we first consider N differentiable basis functions, $\varphi_k(x)$, defined on the domain $a \leq x \leq b$ with $k = 0, 1, \ldots, N - 1$ and that satisfy the orthonormality conditions

$$
\langle \varphi_i | \varphi_j \rangle = \delta_{ij}
\tag{9.6}
$$

exactly. Then provided

$$
\sum_{k=0}^{N-1} \varphi_k^\star(x_i) \varphi_k(x_j) = \lambda_i^{-1} \delta_{ij} \,,
\tag{9.7}
$$

we can construct the transformation

$$
f_i(x) = \lambda_i^{1/2} \sum_{k=0}^{N-1} \varphi_k^\star(x_i) \varphi_k(x) \,,
\tag{9.8}
$$

such that

$$
\lambda_i = \left[\sum_{k=0}^{N-1} |\varphi_k(x_i)|^2 \right]^{-1} \,,
\tag{9.9}
$$

satisfying (9.4) and (9.5). To show how the grid equations are obtained we apply the method to a one-dimensional, time-dependent problem. Starting from the TDSE

$$
H(x,t)\Psi(x,t) = [T + V(x,t)] \Psi(x,t) = \mathrm{i}\frac{\partial}{\partial t}\Psi(x,t) \,,
\tag{9.10}
$$

where T is the kinetic energy operator, $V(x,t)$ a potential and $\Psi(x,t)$ the wave function, we expand the wave function in the Lagrange basis

$$
\Psi(x,t) \approx \sum_{i=1}^{N} c_i(t) f_i(x) \,,
\tag{9.11}
$$

where $c_i(t)$ are expansion coefficients. Clearly

$$
c_i(t) = \int_a^b f_i^\star(x)\Psi(x,t)\mathrm{d}x \approx \lambda_i^{1/2}\Psi(x_i,t) \,.
\tag{9.12}
$$

Substituting (9.11) into (9.10), taking the Gaussian quadrature inner product of both sides with an arbitrary mesh function, $f_j(x)$, using (9.4) and (9.5) and introducing the linear scaling $y_i = hx_i$ leads to

$$\sum_{j=1}^{N} \left[\frac{1}{h^2} T_{ij} + V(y_i, t)\delta_{ij} - i\frac{\partial}{\partial t}\delta_{ij} \right] \lambda_j^{1/2} \Psi(y_j, t) = 0, \qquad (9.13)$$

where $T_{ij} = \langle f_i |T| f_j \rangle$. We see that this equation represents a set of linear equations where we need only evaluate the potential term and the wave function at the mesh points. While V is diagonal in this approximation, T is generally dense.

9.3   Mixed Finite-Difference and Basis Set Techniques for Spatial Variables in Spherical Geometry with Application to Laser-Driven Helium

Handling the electron-laser interaction in the electric dipole velocity gauge, the Hamiltonian H for laser-driven helium takes the following form

$$H = \sum_{i=1,2} \left(-\frac{1}{2}\boldsymbol{\nabla}_i^2 - \frac{Z}{r_i} + \frac{A(t)}{ic}\frac{\partial}{\partial z_i} \right) + \frac{1}{r_{12}}, \qquad (9.14)$$

where atomic units $(e = m_{\mathrm{e}} = \hbar = 1)$ have been used and $A(t)\hat{z}$ is the vector potential for the linearly polarized electric field of the laser. The single nuclear center present makes a choice of spherical geometry with all its associated algebraic advantages and selection rules very convenient for this problem. If the laser radiation is restricted to be linearly polarized and the polarization axis chosen to be the z-axis then the overall two-electron wave function is rotationally symmetric about this axis and its spatial part can be conveniently written as:

$$\Psi(\boldsymbol{r_1}, \boldsymbol{r_2}, t) = \sum_{l_1, l_2, L} \frac{1}{r_1 r_2} f_{l_1, l_2, L}(r_1, r_2, t) |l_1, l_2, L\rangle, \qquad (9.15)$$

where the $|l_1, l_2, L\rangle$ consist of one-electron spherical harmonics (with orbital angular momentum quantum numbers l_1 and l_2) vectorially coupled to yield an overall orbital angular momentum L with conserved projection $M = m_1 + m_2 = 0$ along the z-axis. The three quantum numbers l_1, l_2 and L together with the explicit radial coordinates r_1 and r_2 on the RHS of (9.15) reflect the five degrees of freedom remaining in the electronic motion.

The $|l_1, l_2, L\rangle$ form a basis set representation of the angular co-ordinates with the two radial variables r_1 and r_2 handled by finite-difference (FD) techniques over a 2-dimensional grid of up to 600×600 (but more typically 300×300) points. Each of the three operators appearing in the Hamiltonian, ∇^2, p_z, and $1/r_{12}$, is in some way r_1 and r_2 dependent and must be cast in FD form. Each of these operators presented us with special difficulties. The $1/r_{12}$ operator, for example, which we represent as the series $\sum_l r_<^l /r_>^{l+1} P_l(\cos\theta_{12})$, has a singularity at $r_1 = r_2$, which falls exactly on the diagonal elements of the lattice. It is far from clear a priori that the series representation of $1/r_{12}$ will converge to the correct value as $\delta r \to 0$ and as the number of terms in the series increases. Care must be taken to ensure that convergence is possible, to verify that the truncated series approximates $1/r_{12}$ adequately, and to estimate the truncation error. First, we consider two of the most basic parameters in the code, the FD lattice-point spacing, δr, and the timestep used in the propagation, δt. The box-size (the radius of the integration volume) together with the associated wave function splitting techniques have already been addressed above.

In intense field experiments ionizing electrons with energies well over $100\,\mathrm{eV}$ have been detected. We believe that the lattice, at a minimum, should be able to model excitations as high as $500\,\mathrm{eV}$. We have measured excitations (plane waves representing the ionized electron) at energies as high as $270\,\mathrm{eV}$ ($10\,\mathrm{a.u.}$), and consider it prudent to allow for energies as high as $500\,\mathrm{eV}$.

The question now arises whether or not the largest value of δr we are using, namely $0.333\,\mathrm{a.u.}$, is sufficiently small to meet the above requirements. To first approximation, the highest-energy plane wave supported by the lattice has a de Broglie wavelength roughly equal to twice δr, $0.666\,\mathrm{a.u.}$, which corresponds to an energy of $44.4\,\mathrm{a.u.}$ ($1200\,\mathrm{eV}$). We judge this to be adequate although far from ideal. Values of $\delta r = 0.25\,\mathrm{a.u.}$ are routinely used, especially for high photon frequencies that are less computationally demanding. As well as the constraints on δt coming from the numerical methods used, there is an upper limit set by the need to sample physical quantities on a time scale that is short compared to the period $T_{\min}$ of oscillation of the highest energy excitation, where $T_{\min} = 2\pi/\omega_{\max}$. For $E_{\max} = \hbar\omega_{\max} = 18\,\mathrm{a.u.}$ ($\sim 500\,\mathrm{eV}$) we have $T_{\min} = 0.35$. We use $\delta t = 0.036$, hence there are about 10 timesteps per period $T_{\min}$, which is adequate.

The second derivative operator, $\mathrm{d}^2/\mathrm{d}r^2$, appears as a contribution to the ∇^2 operator acting on each electron, but applies only to the functions $f_{l_1,l_2,L}(r_1,r_2)$ as defined in (9.15). The simplest FD representation for $\mathrm{d}^2/\mathrm{d}r^2$ is a 3-point formula. However, this results in a 30% error in the helium ground state. As $\delta r \to 0$, these errors should disappear. However, we are constrained to a relatively large value of δr and so the 3-point method is unacceptable.

A 5-point representation has a smaller error, and it also has the desirable characteristic that it can be tuned to give the correct derivative at the boundaries of the radial lattice, even on the states with the largest finite-differencing errors. To see how this very important property comes about consider, for simplicity, a 1-dimensional FD lattice for a radial coordinate r. The lattice is laid out so that lattice points lie at the two boundaries, $r = 0$ and $r = R_{\max}$. At the two boundaries, $r = 0$ and $r = R_{\max}$, f is 0, so f need not be explicitly calculated there. But this fact is taken into account in the calculation of derivatives of f at points near the boundary. The largest error occurs in the s states, i.e., those states in which l is zero, due to the fact that they have a nonzero derivative at the $r = 0$ boundary (called the inner boundary). If a 5-point FD rule is used to calculate d^2f/dr^2 near the inner boundary then f must be given a fictional value at $r = -\delta r$. Freedom is available to us in the choice of $f(-\delta r)$ and this enables us to improve substantially the degree to which the FD lattice can model the atom, especially the lowest energy states that play a vital role in the problems of interest. The policy that worked best in general was to set $f(-\delta r)$ in such a way that minimized the finite-differencing error in the ground state of He^+. This turns out to minimize the error in the ground state of He as well.

The structure of the FD operator at the first grid point $(r = \delta r)$ is as follows:

$$f''(\delta r) = \frac{1}{(\delta r)^2} \left[-\frac{1}{12}Cf(\delta r) - \frac{30}{12}f(\delta r) + \frac{16}{12}f(2\delta r) - \frac{1}{12}f(3\delta r) \right] \tag{9.16}$$

where C is the ratio of $f(-\delta r)$ to $f(\delta r)$, and we have used $f(0) = 0$. In the helium two-electron SE for a fixed nonzero r_1, as $r_2 \to 0$ the SE describes an atom in which electron 2, described by r_2, is unscreened by electron 1 and hence sees the full nuclear charge $Z = 2$. For fixed r_1 in the situation described above the functional form of the wave function $f_{l_1,l_2,L}(r_1, r_2)$ looks like $r_2^{l_2+1}e^{-\alpha r_2}$, where α (for bound states) is proportional to the square root of the absolute value of the energy. We see then that as $\delta r \to 0$, $C = f(-\delta r)/f(\delta r)$ approaches -1 if $l_2 = 0$ and 0 if $l_2 \neq 0$. The choice of $C = -1$ for $l_2 = 0$ is an acceptable choice, but we have found that a much better policy (in the case of $l_2 = 0$ partial waves only) is to tune C to minimize the error in the calculation of $d^2f(\delta r)/dr^2$ for the lowest-energy eigenstate of singly-ionized helium. In this case $\alpha = 2$ and $C = -\exp(4\delta r)$. Using $C = -\exp(4\delta r)$, the ground states of both singly ionized helium and neutral helium are within 1% of the true values, even with a coarse grid, $\delta r = 0.333$. The helium ground-state energy using $C = -\exp(4\delta r)$ is calculated to be -2.93 a.u., compared to a value of -2.51 a.u. with $C = -1$. The true value is -2.9037 a.u. In actual runs of the code additional corrections are performed to ensure that all bound-state energies are correct to four significant figures. The quality of the numerical results is sensitive to these states because the ground state of neutral helium is typically the initial state in time-dependent calculations and because the ground state of singly ionized helium is a major component of the screening of the ionizing electron. Normally these compact states are the most poorly represented by a FD lattice. The correction $C = -\exp(4\delta r)$ is correct for all states as $\delta r \to 0$, but is not optimal for excited states when $\delta r > 0$. However, the excited states are far more extended than the ground states, encompass more grid points, are better modeled by the lattice, and are less sensitive to the boundary point at $r = \delta r$ where C is invoked.

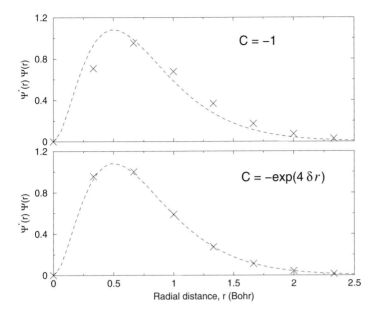

Fig. 9.1. Functional form of the He$^+$ ground-state wave function. The *dashed line* indicates the desired form. The *'X' points* indicate the values obtained at each lattice point with spacing $\delta r = 0.3333$ Bohr

Figure 9.1 shows the form of the He$^+$ ground-state wave function for the two different values of C discussed above. The choice of $C = -\exp(4\delta r)$ is clearly superior.

A very important application of the HELIUM code has been to model the full electronic response of helium at a laser wavelength of 390 nm, which corresponds to frequency-doubled Ti:Sapphire light. A comparison with experiment [23] is plotted in Fig. 9.2 for the ratio of He^{++} to He$^+$ obtained over the experimental pulse.

The calculation [24] has fully taken into account the temporal and spatial character of the laboratory laser pulse. However, the agreement obtained in the figure has only been achieved by a unilateral shift upwards by 50% of each experimental measurement of intensity. Calibration of laser intensity at high operating intensities has long been recognized by experimentalists as something notoriously difficult to achieve reliably. We believe that our computational method, by yielding reliable and accurate ionization rates for an experimentally accessible atom (helium), gives the first opportunity for high-intensity lasers to be calibrated reliably in intensity.

The calculations have uncovered many other previously unknown features of the electronic response at laser wavelengths of 200 nm and longer. Especially notable amongst these findings has been the discovery [25], subsequently confirmed by experiment [6], that the two simultaneously ionizing

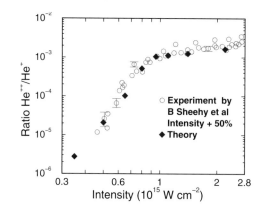

Fig. 9.2. Ratio of He^{++} to He^{+} ion yields. Experimental data points are given by *open circles* and the theoretically derived ratios by *solid diamonds*

electrons emerge on the same side of the nucleus with equal momenta. This discovery was made by subjecting the time-evolving wave function to scientific visualization analysis after first constructing an appropriate lower dimensionality cut in the five spatial dimensions. This processing of the calculated wave function, especially when a transformation to a different geometry is required, is a nontrivial task requiring a computational effort often comparable to that involved in the initial generation of the time-evolving wave function.

Given the wide wavelength range applicability of the method and the confidence in the method through its successful predictions confirmed by experiment at optical wavelengths, we have recently embarked on calculations handling laser wavelengths as low as 14 nm [26]. Wavelengths in this neighborhood are shortly to be available, at the unprecedented high focused intensities of between 5×10^{15} W/cm^2 and 5×10^{17} W/cm^2 through the free-electron laser (FEL) sources under construction. At such laser wavelengths one photon carries sufficient energy that, when shared by the two electrons, both can ionize. A fundamental question in the case of a two-electron system absorbing two photons is does each electron absorb a photon or does only one electron absorb both?

Figure 9.3 displays a plot in the radial momentum space of the two electrons of the probability density (averaged over all angular coordinates) at the end of a 46-cycle pulse at 14 nm with a ramp up and ramp down from peak intensity of 2×10^{16} W/cm^2. The figure displays two principal features. At low momentum values there is a circular arc. This corresponds to single-photon absorption with the electrons sharing the energy absorbed so that both ionize. The flatness of the arc indicates that the excess kinetic energy of escape is shared without any particular partition preferred. The second principal feature is the spot (and its symmetry required partner) at a higher momentum value. The coordinates of these spots in radial momentum space

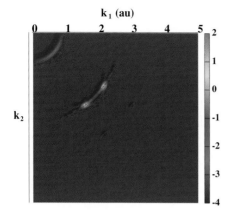

Fig. 9.3. Plot of the probability distribution $P(k_1, k_2)$ of doubly ionizing electrons in momentum space after excitation with a 46 field period laser pulse of wavelength 14 nm and of peak intensity 2.0×10^{16} W/cm^2

allow us to say that one electron has been ionized by an electron in the He ground state absorbing a photon whilst the other has been ionized from the ground state of the residual ion He$^+$.

9.4 Mixed Finite-Difference and DVR Techniques for Spatial Variables in Cylindrical Geometry with Application to Laser-Driven H$_2$

We have considered the diatomic, homonuclear H$_2$ molecule subjected to an ultrashort, linearly polarized laser pulse. The molecule is assumed to be initially in its ground state, $^1\Sigma_g^+$, and in current work, it is also assumed that the nuclei are of infinite mass separated by a *fixed* distance R. Furthermore, the polarization axis of the laser pulse is taken to be aligned along the internuclear axis. With these assumptions only electronic motion is present, and with an element of symmetry remaining about the common internuclear/laser polarization axis, this motion has five degrees of freedom. The two-center nature of the system, however, makes it much more difficult to handle than the laser-driven helium problem of identical dimensionality. In effect it is the loss of spherical symmetry anywhere in the wave function that prevents us taking advantage of the spherical algebra and associated selection rules that could be used to tremendous computational advantage in the latter problem. Moreover, since our overall goal is to also allow for nuclear motion there is no sensible development pathway based on two-center ellipsoidal coordinate systems. It is thus most appropriate to use a cylindrical polar coordinate system, with z-axis collinear with the common inter-nuclear/laser polarization axis and origin midway between the two nuclei. In such a system the position vectors of the two electrons are therefore given by

$$\boldsymbol{r}_s = \rho_s \cos\phi_s \boldsymbol{i} + \rho_s \sin\phi_s \boldsymbol{j} + z_s \boldsymbol{k} \quad s = 1,2\,, \tag{9.17}$$

so that the Hamiltonian for the H_2 molecule is given by

$$
\begin{aligned}
H = \sum_{s=1}^{2} &\left(-\frac{1}{2\mu} \left\{ \frac{\partial^2}{\partial z_s^2} + \frac{1}{\rho_s} \frac{\partial}{\partial \rho_s} \rho_s \frac{\partial}{\partial \rho_s} \right\} + V(\rho_s, z_s, R) + U(z_s, t) \right) \\
&- \frac{1}{2\mu} \left[\frac{1}{\rho_1^2} + \frac{1}{\rho_2^2} \right] \frac{\partial^2}{\partial\phi^2} + V_{\mathrm{ee}}(\boldsymbol{r}_1, \boldsymbol{r}_2) + \frac{Z_1 Z_2}{R}\,,
\end{aligned}
\tag{9.18}
$$

where $V(\rho_s, z_s, R)$ represents the Coulomb interaction of electron s with each of the nuclei

$$V(\rho_s, z_s, R) = -\frac{Z_1}{\sqrt{\rho_s^2 + \left(z_s - \frac{1}{2}R\right)^2}} - \frac{Z_2}{\sqrt{\rho_s^2 + \left(z_s + \frac{1}{2}R\right)^2}}\,, \tag{9.19}$$

R being the internuclear distance and Z_1 and Z_2 being the nuclear charges. $V_{\mathrm{ee}}(\boldsymbol{r}_1, \boldsymbol{r}_2)$ represents the electron-electron interaction

$$V_{\mathrm{ee}}(\boldsymbol{r}_1, \boldsymbol{r}_2) = \frac{1}{\sqrt{\rho_1^2 + \rho_2^2 + (z_1 - z_2)^2 - 2\rho_1\rho_2 \cos\phi}}\,, \tag{9.20}$$

while $U(z_s, t)$ represents the interaction between electron s and the laser field, and may take either length or velocity form, as described previously [4].

This choice of coordinate system also has the advantage that the five spatial degrees of freedom in the electronic wave function are manifest explicitly through dependence only on the quantities ρ_1, ρ_2, z_1, z_2, and $\phi = \phi_1 - \phi_2$. It is reasonable to use grid methods in representing the wave function in all five of these co-ordinates. However, for principally physical reasons, different grid methods are appropriate to different coordinates. The electrons are predominantly driven back and forth along the laser polarization axis and one must allow for short de Broglie wavelengths near the nuclear centers in z_1 and z_2 (with recollisional events occurring in such localities) as well as for extended spatial coverage in these coordinates. In contrast, for the ρ_1, ρ_2, and ϕ coordinates, the wave function nodal structure is much more uniformly distributed. Moreover ϕ extends over a closed interval ($[0, \pi]$ when symmetries are taken into account) and the ρ coordinates need to be handled over only comparatively small spatial extents. Note that ρ_1, ρ_2, and ϕ are required for a proper description of the exchange of angular momentum between the electrons and the laser field. A further mathematical reason counting against the use of finite difference techniques for the ρ coordinates is that appropriate scaling of these coordinates leads to nonunitary time evolution, which we wish to avoid. For these reasons we have decided on a Lagrange mesh

(DVR) treatment of the ρ_1, ρ_2, and ϕ coordinates together with the z_1 and z_2 coordinates handled by finite-difference techniques.

Full details of the Lagrange mesh treatment of the ρ_1, ρ_2, and ϕ coordinates can be found in [4,28] and we can give only a brief summary here. For the ρ coordinates the quadrature points are the zeros of the generalized Laguerre polynomial $L^1_{N_\rho}(\rho)$ defined over the range $[0, \infty)$. These quadrature points are denoted by ρ_{si}, for $s = 1, 2$ and $i = 1, \ldots, N_\rho$, where ρ_{si} indicates the i-th quadrature point of the independent variable ρ_s.

The normalized Lagrange functions, $\{g_i(\rho_s)\}$, $i = 1 \ldots, N_\rho$, $s = 1, 2$, have been obtained through use of the orthonormal set, $\varphi_k(\rho_s)$, $k = 0 \ldots, N_\rho - 1$, $s = 1, 2$, where,

$$\varphi_k(\rho_s) = \left[\frac{\Gamma(k+2)}{k!}\right]^{-1/2} \rho_s^{1/2} e^{-\rho_s/2} L^1_k(\rho_s), \tag{9.21}$$

and have been found to take the form

$$g_i(\rho_s) = \lambda_i^{-1/2} \frac{1}{\varphi'_{N_\rho}(\rho_{si})} \frac{\varphi_{N_\rho}(\rho_s)}{\rho_s - \rho_{si}}, \tag{9.22}$$

where the quadrature weights are given by

$$\lambda_i = \frac{1}{\rho_{si} \varphi'_{N_\rho}(\rho_{si})^2} \quad i = 1, \ldots, N_\rho. \tag{9.23}$$

These three sets of variables make it possible to evaluate the matrix elements of the ρ_s-dependent kinetic energy terms exactly.

In the case of the ϕ variable, we sought a set of normalized Lagrangian functions and weights, denoted $\{v_m(\phi)\}$ and $\{w_m\}$, $m = 1, \ldots, N_\phi$ respectively, defined on the interval $[0, \pi]$, together with a set of quadrature points that would enable us to evaluate the ϕ-dependent kinetic energy terms exactly. Karabulut and Sibert [27] found such a desirable set, to within a scale factor. Therefore, following this work, we have chosen quadrature points in ϕ given by

$$\phi_m = \frac{(2m - 1)}{2N}\pi \quad m = 1, \ldots, N_\phi, \tag{9.24}$$

together with normalized Lagrange functions

$$v_m(\phi) = \frac{1}{\sqrt{\pi N_\phi}}\left[1 + 2\sum_{k=1}^{N_\phi - 1} \cos(k\phi_m)\cos(k\phi)\right] \quad m = 1, \ldots, N_\phi, \tag{9.25}$$

and quadrature weights

$$w_m = \frac{\pi}{N_\phi} \quad m = 1, \ldots, N_\phi, \tag{9.26}$$

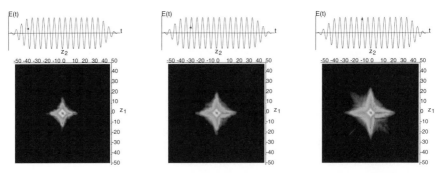

Fig. 9.4. The *dots* on the upper E-field plots indicate the instants in time for which the probability density $P(z_1, z_2, t)$ of the hydrogen molecule are portrayed, during its excitation by a 20 field period laser pulse of wavelength 20 nm and of peak intensity 2×10^{16} W/cm^2

expanding the wave function in terms of their respective normalized Lagrange functions

$$\psi(\rho_1, \rho_2, z_1, z_2, \phi, t) \approx \sum_{i=1}^{N_\rho} \sum_{k=1}^{N_\rho} \sum_{m=1}^{N_\phi} \lambda_i^{1/2} \lambda_k^{1/2} w_m^{1/2} \psi(\rho_{1i}, \rho_{2k}, z_1, z_2, \phi_m, t)$$
$$\times g_i(\rho_1) g_k(\rho_2) v_m(\phi) . \tag{9.27}$$

As illustrated in the 1D example in Sect. 9.2.2 above, substituting this form into the TDSE yields a coupled set of grid equations (with in this case $N_\rho \times N_\rho \times N_\phi$ equations in the set) involving functions that depend on the remaining coordinates z_1 and z_2. Five-point central difference formulae are used to approximate the kinetic energy terms in z_1 and z_2, as well as for the first-order derivatives in z_1 and z_2, necessary for the calculation of the laser-electron interaction in the velocity gauge.

The code is mapped over processors in such a way that a given processor handles a particular section of the (z_1, z_2) space with information on all values of the other three coordinates stored on every processor. This arrangement has the important advantage that only nearest-neighbor communication is involved, making the code readily scalable over a wide variety of parallel architectures.

Figure 9.4 displays probability density plotted against (z_1, z_2) at four indicated instants during the exposure of the molecule (with fixed equilibrium spacing $R = 1.4$ a.u.) to a laser pulse of wavelength 20 nm and 2×10^{16} W/cm^2 peak intensity. The most important feature to remark on here is the predominance in double-electron ionization occurring in the $z_1 = -z_2$ quadrants, indicating at this short wavelength simultaneous electron ionization on both sides of the nucleus, in contrast to that found for helium at longer wavelengths.

The code, H2MOL, under development for this problem awaits a Terascale facility for its full exploitation and especially to allow its extension to include internuclear motion. Some numbers bring out this point clearly. Thus if we use 500 points to map out each of z_1 and z_2; 50 points to map out each of ρ_1 and ρ_2 and another 20 points to map out $\phi = \phi_1 - \phi_2$ then recalling that a single 8-byte (complex) number – the wave function value – must be stored at all points in the 5-dimensional space we have a storage requirement of $500 \times 500 \times 50 \times 50 \times 20 \times 8 = 100$ Gbytes. A conservative 20 points brought in to map out the internuclear coordinate (again via a DVR approach) brings the memory requirement up to 2 Tbytes.

9.5   Conclusions

We have set out above some details of the methods we continue to develop at Belfast in order to reliably and accurately solve the time-dependent Schrödinger equation for laser-driven helium and the laser-driven hydrogen molecule. The methods for these systems are implemented in the computer codes HELIUM and H2MOL, respectively.

An efficient high-order time propagator is common to both HELIUM and H2MOL. This is essential for efficiency and indeed for accuracy and reliability, especially when more than just a few photons are absorbed – the situation with optical lasers.

For laser-driven helium it is important to take advantage of the single force center present – done by casting the problem in spherical polar coordinates. These coordinates as well as bringing into play strong selection rules in the dipole approximation, mean that three of the spatial degrees of freedom can be handled by basis-set techniques. This leaves only the two radial coordinates to be represented using finite-difference methods.

For laser-driven H_2 the presence of two nuclear centers makes spherical polar coordinates inappropriate. Moreover, since the eventual goal is to allow for movement of the nuclei along the laser polarization axis, we avoid ellipsoidal coordinates. Cylindrical polar coordinates prove the best choice with DVR techniques employed to handle the ρ_1 and ρ_2 co-ordinates as well as the single remaining azimuthal coordinate ϕ. The z_1 and z_2 co-ordinates, measuring electron distance from the center-of-mass along the laser polarization axis, and where the electron experiences most acceleration, are most appropriately taken care of by finite-difference methods.

We have illustrated the applications of HELIUM and H2MOL by way of sample results pertinent to optical lasers as well as to the upcoming free-electron laser sources.

References

1. E.S. Smyth, J.S. Parker, K.T. Taylor: Comput. Phys. Commun. **114**, 1 (1998)
2. J.S. Parker, L.R. Moore, E.S. Smyth, K.T. Taylor: J. Phys. B: At. Mol. Opt. Phys. **33**, 1057 (2000)
3. D. Dundas, J.F. McCann, J.S. Parker, K.T. Taylor: J. Phys. B: At. Mol. Opt. Phys. **33**, 3261 (2000)
4. D. Dundas: Phys. Rev. A **65**, 023408 (2002)
5. B. Walker, B. Sheehy, L.F. DiMauro, P. Agostini, K.J. Schafer, K.C. Kulander: Phys. Rev. Lett. **73**, 1227 (1994)
6. T.H. Weber, H. Giessen, M. Weckenbrock, G. Urbasch, A. Staudte, L. Spielberger, O. Jagutzki, V. Mergel, M. Vollmer, R. Dörner: Nature **405**, 658 (2000)
7. M.R. Thompson, M.K. Thomas, P.F. Taday, J.H. Posthumus, A.J. Langley, L.J. Frasinski, K Codling: J. Phys. B: Mol. Opt. Phys. **30**, 5755 (1997)
8. R. Lafon, J.L. Chaloupka, B. Sheehy, P.M. Paul, P. Agostini, K.C. Kulander, L.F. DiMauro: Phys. Rev. Lett. **86**, 2762 (2001)
9. B. Feuerstein, R. Moshammer, D. Fischer, A. Dorn, C.D. Schroter, J. Deipenwisch, J.R.C Lopez-Urrutia, C. Hohr, P. Neumayer, J. Ullrich, H. Rottke, C. Trump, M. Wittmann, G. Korn, W. Sandner: Phys. Rev. Lett. **87**, 043003 (2001)
10. W.E. Arnoldi: Quart. Appl. Math. **9**, 17 (1951)
11. H. Tal-Ezer, R. Kosloff: J. Chem. Phys. **81**, 3967 (1984)
12. T.J. Park, J.C. Light: J. Chem. Phys. **85**, 5870 (1986)
13. H. van der Vorst: J. Comput. Appl. Math. **18**, 249 (1987)
14. E. Gallopoulos, Y. Saad: SIAM J. Sci. Stat. Comput. **13**, 1236 (1992)
15. A. Nauts, R.E. Wyatt: Phys. Rev. Lett. **51**, 2238 (1983)
16. J.C. Light, I.P. Hamilton, J.V. Lill: J. Chem. Phys. **82**, 1400 (1985)
17. M. Hesse, D. Baye: J. Phys. B: At. Mol. Opt. Phys. **32**, 5605 (1999)
18. D. Baye, M. Hesse, J.M. Sparenberg, M. Vincke: J. Phys. B: At. Mol. Opt. Phys. **31**, 3439 (1998).
19. J.T. Muckerman, R.V. Weaver, T.A.B. Kennedy, T. Uzer. In: *Numerical Grid Methods and Their Applications to Schrödinger's Equation*, ed. by C. Cerjan (Kluwer Academic Publishers, The Netherlands 1993) pp. 89–119
20. V.S. Melezhik, D. Baye: Phys. Rev. C **59**, 3232 (1999)
21. K. Sakimoto: J. Phys. B: At. Mol. Opt. Phys. **33**, 5165 (2000)
22. D. Baye, P-H. Heenen: J. Phys. A: Math. Gen. **19**, 2041 (1986)
23. B. Sheehy, R. Lafon, M. Widmer, B. Walker, L.F. DiMauro, P.A. Agostini, K.C. Kulander: Phys. Rev. A **58**, 3942 (1998)
24. J.S. Parker, L.R. Moore, D. Dundas, K.T. Taylor: J. Phys. B: Mol. Opt. Phys. **33**, L691 (2000)
25. K.T. Taylor, J.S. Parker, D. Dundas, E.S. Smyth, S. Vivirito: Laser Phys. **9**, 98 (1999)
26. J.S. Parker, L.R. Moore, K.J. Meharg, D. Dundas, K.T. Taylor: J. Phys. B: At. Mol. Opt. Phys. **34**, L69 (2001)
27. H. Karabulut, E.L. Sibert III: J. Phys. B: At. Mol. Opt. Phys. **30**, L513 (1997)
28. D. Dundas, K.J. Meharg, J.F. McCann and K.T. Taylor: J. Phys. B: At. Mol. Opt. Phys. to be published 2002

10 S-Matrix Approach
to Intense–Field Processes
in Many-Electron Systems

F.H.M. Faisal

10.1 Introduction

The theoretical challenge for intense-field laser–atom interaction in many-electron systems arises from a combination of

1. highly nonperturbative light interaction,
2. quantum many-body problem, and
3. nonseparable Coulomb interaction between the electrons, or 'correlation' $\sum_{i \neq j} 1/r_{ij}$.

An *ab initio* systematic approximation method that has proved to be useful in this context is the so-called 'intense-field many-body S-matrix theory' (or IMST). In this approach, the usual perturbation expansion of the transition amplitude of a process is rearranged in such a way that the dominant virtual channels, when present, appear already in the leading terms of the S-matrix series. Such a rearrangement was carried out [1] by introducing *three* alternative partitions, namely, an initial, a final, *and* an intermediate partition of the total Hamiltonian of the interacting system and incorporating them systematically in the series expansion. The resulting S-matrix expansion allows one to take account of the 'transition states' (virtual fragment, or doorway states) through which a reaction of interest may proceed predominantly already in the leading terms. In contrast, the well-known 'prior' or 'post' expansions (e.g., [2]) would have to be carried out to very high (if not to infinite) orders to account for such virtual fragment states. This is because the 'prior' or the 'post' expansion is obliged to use for the intermediate propagator the *same* reference Hamiltonian that is appropriate for the initial or the final-state partition only; the IMST is characteristically free from this constraint. In this chapter we shall introduce the rearranged S-matrix theory and briefly discuss its application to a number of phenomena observed recently for ionizations of atoms and molecules in intense laser fields.

10.2 The Rearranged Many-Body S-Matrix Theory

We formulate the rearranged S-matrix theory [1,3] for a general time–dependent Hamiltonian. The results for the stationary systems reduce to a special case (with merely exponential time dependence). First, the time-dependent Schrödinger equation of the system of interest is rewritten as an integral equation (a generalized time-dependent Lippmann–Schwinger equation). Next, three different partitions of the total Hamiltonian of the system, $H(t)$, are introduced,

$$
\begin{aligned}
H(t) &= H_i^0 + V_i(t) && \text{(initial partition)} \\
&= H_f^0 + V_f(t) && \text{(final partition)} \\
&= H_0 + V_0(t) && \text{(intermediate partition)} .
\end{aligned}
\tag{10.1}
$$

The Green's functions (propagators) associated with the initial, final, and the intermediate reference Hamiltonians H_i^0, H_f^0, and H_0 are denoted by G_i^0, G_f^0, and G_0, respectively. By employing an appropriate sequence of iterations [1,3] involving the reference Green's functions in the time-dependent Lippmann–Schwinger equation, one can rewrite the total wave function in the form:

$$
\begin{aligned}
|\Psi(t)\rangle &= |\phi_i(t)\rangle + \int_{t_i}^{t} dt_1 G_f^0(t, t_1) V_i(t_1) |\phi_i(t_1)\rangle \\
&\quad + \int_{t_i}^{t} \int_{t_i}^{t_2} dt_2 dt_1 G_f^0(t, t_2) V_f(t_2) G_0(t_2, t_1) V_i(t_1) |\phi_i(t_1)\rangle \\
&\quad + \int_{t_i}^{t} \int_{t_i}^{t_3} \int_{t_i}^{t_2} dt_3 dt_2 dt_1 G_f^0(t, t_3) V_f(t_3) G_0(t_3, t_2) V_0(t_2) \\
&\quad \times G_0(t_2, t_1) V_i(t_1) |\phi_i(t_1)\rangle + \dots .
\end{aligned}
\tag{10.2}
$$

where $\phi_i(t)$ is the initial state of unperturbed system. Note that in the above expansion the wave function not only satisfies the initial condition but also is well arranged for computing the transition amplitude by projection onto any state of the final reference Hamiltonian, even when the latter is *not* identical to the initial reference Hamiltonian and/or the intermediate reference Hamiltonian.

The transition amplitude $T_{i\to f}(t)$ from any initial reference state $\phi_i(t)$ to any final reference state $\phi_f(t)$ can now be obtained by orthogonal projection from the left

$$
T_{i\to f}(t) \equiv (S-1)_{i\to f}(t) = \sum_{j=1}^{\infty} T_{i\to f}^{(j)}(t),
$$

$$
T_{i\to f}^{(1)}(t) = \int_{t_i}^{t} dt_1 \langle \phi_f(t_1)| V_i(t_1) |\phi_i(t_1)\rangle
\tag{10.3}
$$

$$
T_{i\to f}^{(2)}(t) = \int_{t_i}^{t} \int_{t_i}^{t_2} dt_2 dt_1 \langle \phi_f(t_2)| V_f(t_2) G_0(t_2, t_1) V_i(t_1) |\phi_i(t_1)\rangle
\tag{10.4}
$$

$$
\begin{aligned}
T_{i\to f}^{(3)}(t) &= \int_{t_i}^{t} \int_{t_i}^{t_3} \int_{t_i}^{t_2} dt_3 dt_2 dt_1 \langle \phi_f(t_3)| V_f(t_3) \\
&\quad \times G_0(t_3, t_2) V_0(t_2) G_0(t_2, t_1) V_i(t_1) |\phi_i(t_1)\rangle + \dots .
\end{aligned}
\tag{10.5}
$$

Note the presence of the intermediate propagator G_0 as well as the initial- and the final-state interactions, $V_i(t)$ and $V_f(t)$, already in the second leading term of the series. This form of the S-matrix expansion is appropriate not only for any 'direct process' but also for all kinds of 'exchange processes', e.g., charge-transfer, chemical reaction processes, and, as will be seen below, also for intense-field ionization processes in many-electron systems.

10.3   Applications to Intense-Field Ionization Dynamics

In the case of interactions with intense laser field the theory is referred to 'Intense-field Many-body S-matrix Theory' or IMST. For ionization processes in atoms and molecules in intense fields, the nonperturbative laser interaction with the free electron (in the final or in the virtual intermediate states) can be taken into account to all orders of the field strength, F, by using the Volkov solution, i.e., the exact solution of a free electron in a laser field (e.g., [4]) and the associated Volkov propagator [5,6]. Among other things, for example, this permits one to introduce and identify the overwhelming importance of 'transition states' of the form of $(\{|\text{Volkov}\rangle\}) \otimes (\{|\text{ion}\rangle\})$ for the so-called 'nonsequential double ionization' in intense fields ([1,3,7–9]).

To illustrate the usefulness of the theory (IMST), below we shall briefly discuss a number of applications of the theory to single ionization and double ionization of atoms, as well as ionization of diatomic and polyatomic molecules in intense laser fields, and compare the theoretical results with the recent experimental observations.

10.3.1   Intense–Field Ionization of Atoms

Using the Volkov wave function (approximately corrected for the final-state Coulomb tail) for the ionized electron and the initial-state interaction $V_i(t)$ (i.e., the interaction of the electrons of the atom with the vector potential of the laser field)

$$V_i(t) = \sum_{j=1}^{N_e} \left[-\frac{e}{mc} \boldsymbol{p}_j \cdot \boldsymbol{A}(t) + \frac{e^2}{2mc^2} A^2(t) \right] \tag{10.6}$$

in $T_{i\to f}^{(1)}(\infty)$, one obtains the following expression for the rate of single ioniza- tion of a many-electron atom [10] (given here for the *linear polarization* of the field ($e = \hbar = m_e = 1$):

$$\Gamma^{(+)} = 2\pi N_e C^2(Z, E_B, F)$$
$$\times \sum_{n=n_0}^{\infty} \int d\boldsymbol{k}_n k_n |\langle \phi^0(\boldsymbol{k}_n; \boldsymbol{r}_1)\phi_f^{(\text{ion})}(\boldsymbol{r}_2, \dots \boldsymbol{r}_{N_e})|\phi_i(\boldsymbol{r}_1, \dots \boldsymbol{r}_{N_e})\rangle|^2$$
$$\times (U_p - n\omega)^2 J_n^2(a_{\boldsymbol{k}_n}, b), \tag{10.7}$$

where $J_n(a, b)$ is a generalized Bessel function of two arguments (e.g., [4]), ϕ^0 is a plane wave, $\alpha_0 = \sqrt{I}/\omega^2$ is the quiver radius, $U_p = I/(4\omega^2)$ is the ponderomotive energy, $I = F^2/(4\pi)$ is the field intensity, $k_n^2/2 = (n\omega - E_B + U_p)$ is the kinetic energy of the ionized electron on absorption of n photons, E_B is the binding energy of the atom and $C^2 = \left(\frac{2\kappa_B E_B}{F}\right)^{2Z/\kappa_B}$ is the Coulomb correction factor (see, e.g., [7,10,11]); this factor is equal to unity for a plane–wave Volkov state. This formula is analogous to the well-known Keldysh–Faisal–Reiss (KFR) formula [12] originally obtained for electron detachment of a one-electron system.

As an example of application of the above amplitude to the ionization of a complex atom we show the results of both experimental ion yields and the corresponding theoretical results, for the Xe atom. For the atomic ground state a self-consistent field wave function is used, e.g., [13]. In Fig. 10.1 the dependence of the ionization yields as a function of field intensity is presented for a number of wavelengths and pulse durations. It can be seen immediately from the figure that the agreement between the experimental data and the theoretical calculations using $T_{i \to f}^{(1)}$ of the IMST is very satisfactory. Similar agreements have been obtained for numerous other atoms for different polarizations, and for a wide range of wavelengths and pulse durations of the laser fields. We refer the reader for these comparisons, and for the references to the experimental data to [10].

10.3.2   Recoil-Momentum Distributions for Nonsequential Double Ionization

We next consider the more challenging problem of the so-called nonsequential double ionization mentioned above. Double ionization of He and other noble gas atoms interacting with intense infrared or optical lasers (e.g., [14,15]) have shown surprisingly large ion yields that are, in fact, larger by *many* orders of magnitude than expected initially. More over the momentum distributions of the recoil (doubly charged) ions in coincidence with an ejected electron have recently been measured [16,17], which revealed previously unsuspected features. The most prominent among them are:

1. the component of the recoil momenta *parallel* to the laser polarization direction shows a prominent *double*-hump distribution with a central minimum,
2. the perpendicular component of the recoil momentum shows a single-hump distribution,
3. the parallel component distribution is very broad,
4. the perpendicular distribution is narrow, and
5. the maximum momentum transfer of the parallel component shows a sharp cutoff.

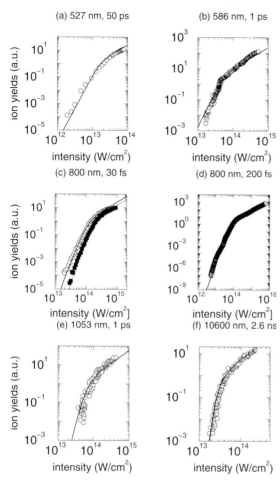

Fig. 10.1. Ionization yields of Xe atom at various wavelengths and pulse durations as a function of field intensity (**a**) 527 nm, 50 ps (**b**) 586 nm, 1 ps (**c**) 800 nm, 30 ps (**d**) 800 nm, 2000 fs (**e**) 1053 nm, 1 ps, and (**f**) 10600 nm, 2.6 ns; experiment: *circle* (linear polarization) and *solid squares* (circular polarization); theory: *solid line* (linear polarization) and *dashed line* (circular polarization); from [10]

With the help of IMST, the non-sequential double ionization process, in which the electrons are emitted directly (without the formation of singly charged ions in a first step followed by ionization of the ions), has been identified with a predominant 'correlated energy-sharing diagram' which predicted by the second term $T_{i \to f}^{(2)}$. This diagram, given in Fig. 10.2, has been shown [1,7] to automatically incorporate the classical 're-scattering' mechanism [18,19], the 'antenna' mechanism [20], as well as the effect of 'direct scattering' and the Volkov-dressing effect on both the outgoing electrons, in a unified way. One may translate the diagram analytically by explicitly specifying the two-

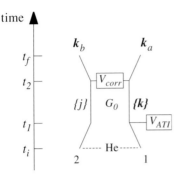

time

t_f

t_2

t_1

t_i

k_b k_a

V_{corr}

$\{j\}$ G_0 $\{k\}$

V_{ATI}

------ He ------

2   1

Fig. 10.2. 'Correlated energy-sharing diagram' from IMST for laser–induced non-sequential double ionization of He; from [9]

particle 'transition states' through the intermediate propagator appearing in the diagram 10.2 [1,9]:

$$G_0(t, t') = -\frac{\mathrm{i}}{\hbar}\theta(t - t') \tag{10.8}$$

$$\times \sum_j \frac{1}{(2\pi)^3} \int \mathrm{d}\boldsymbol{k} |\boldsymbol{k}\rangle |\phi_j^+(2)\rangle$$

$$\times \exp -\frac{\mathrm{i}}{\hbar}\int_{t'}^{t}\{[(\boldsymbol{p}_1 - e\boldsymbol{A}(\tau)/c]^2/2m + E_j\}\mathrm{d}\tau]\langle \boldsymbol{k}|\langle\phi_j^+(2)|\,,$$

where, $\theta(t - t')$ is the Heaviside theta function, $\{|\boldsymbol{k}\rangle\}$ is the complete set of plane–wave states with wave vectors $\boldsymbol{k}$, and $\{|\phi_j^+(2)\rangle\}$ is the complete set of residual ionic states; the electron–electron correlation now stands for V_f. This gives the following analytic expression for the nonsequential double–ionization amplitude [1,9]:

$$T_{i\to f}^{(2)}(\infty)|_{\mathrm{NS}} = -2\pi i \sum_N$$

$$\times \delta\left(\frac{\hbar^2 k_{\mathrm{a}}^2}{2m} + \frac{\hbar^2 k_{\mathrm{b}}^2}{2m} + E_{\mathrm{B}} + 2U_{\mathrm{p}} - N\hbar\omega\right)T^{(N)}(\boldsymbol{k}_{\mathrm{a}}, \boldsymbol{k}_{\mathrm{b}})\,, \tag{10.9}$$

where,

$$T^{(N)}(\boldsymbol{k}_{\mathrm{a}}, \boldsymbol{k}_{\mathrm{b}}) = \sum_n \sum_j \int \frac{1}{(2\pi)^3}\mathrm{d}\boldsymbol{k}$$

$$\times \langle \phi^0(\boldsymbol{k}_{\mathrm{a}}, \boldsymbol{r}_1)\phi^0(\boldsymbol{k}_{\mathrm{b}}, \boldsymbol{r}_2)|\frac{1}{r_{12}}|\phi_j^+(\boldsymbol{r}_2)\phi^0(\boldsymbol{k}, \boldsymbol{r}_1)\rangle$$

$$\times \frac{J_{N-n}\left(\boldsymbol{\alpha}_0 \cdot (\boldsymbol{k}_{\mathrm{a}} + \boldsymbol{k}_{\mathrm{b}} - \boldsymbol{k}); U_{\mathrm{p}}/(2\hbar\omega)\right) J_n\left(\boldsymbol{\alpha}_0 \cdot \boldsymbol{k}; U_{\mathrm{p}}/(2\hbar\omega)\right)}{(\hbar^2 k^2)/(2m) - E_j + E_{\mathrm{B}} + U_{\mathrm{p}} - n\hbar\omega + i0}$$

$$\times (E_j - E_{\mathrm{B}} - \frac{\hbar^2 k^2}{2m})\langle \phi_j^+(\boldsymbol{r}_2)\phi^0(\boldsymbol{k}, \boldsymbol{r}_1)|\phi_{1S}(\boldsymbol{r}_1, \boldsymbol{r}_2)\rangle \tag{10.10}$$

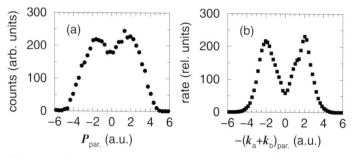

Fig. 10.3. Recoil–ion momentum distribution of He^{2+} parallel to the polarization direction, $P_{par.}$ (experiment: [16] panel **a**) and the sum-momentum of the two outgoing electrons in the opposite direction (theory: IMST [9], panel **b**); from [9]

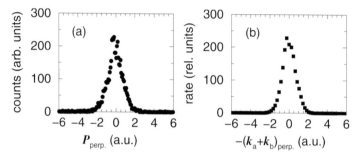

Fig. 10.4. Distribution of recoil momentum of He^{2+} ions, perpendicular to polarization axis: (**a**) experiment [16], (**b**) theory (IMST); from [9]

and $\phi^0(\mathbf{k}, \mathbf{r})$ is a plane–wave state. It involves a formidable multidimensional integral that has been approximately evaluated by a combined analytical plus Monte Carlo integrations [9]. The recoil–momentum distributions for the double ionization of He have been investigated using the above formula and compared with the observed data. The experimental data [16] for the recoil momentum of the doubly charged helium ions and the calculated distributions [9] for the sum-momentum of the ejected electrons are shown in Figs. 10.3 and 10.4.

It should be noted that the recoil momentum of the doubly charged helium ions is equivalent to the sum-momentum of the electrons under the experimental conditions [16,17]). Figures 10.3 and 10.4 correspond to the recoil–momentum distribution *parallel* and *perpendicular* to the (linear) polarization axis, respectively. They are obtained for a Ti:sapphire laser field of duration 200 fs, intensity 6.6×10^{14} W/cm^2, and $\lambda = 800$ nm, as employed by Weber et al. [16]. It can be seen from the comparison that all the physical features (see items 1 to 5 above) of the experimental distributions are consistent with the calculated distributions. For a more detailed comparison we refer the reader to [9]. We may add simply that the origin of the remarkable double-hump structure of the parallel-momentum distribution is unequivocally shown to be due to the *final*–state 'Volkov-dressing' of the two outgoing electrons [9]. Also, the *absence* of the double-hump structure in the perpendicular distribution is due to the effective *decoupling* of the field interaction with the two outgoing electrons that move in a direction *orthogonal* to the field–polarization direction. In effect, the perpendicular distribution can be thought of as due only to the momentum transfer from the atomic correlation in the intermediate state, since the field is effectively switched off on the ejection of the electrons in the perpendicular direction. Such a condition would be otherwise difficult to achieve directly in the laboratory.

IMST analysis of the coincidence spectrum of momentum distribution has been able to show that the two electrons in double ionization preferentially emerge in the *same* dierection with the same non-zero energy [21]. In fact the analysis fully corroborates the so-called 'correlated energy sharing mechanism' of nonsequential double ionization by intense infrared laser fields, and at the same time rules out the so-calld 'shake-off' mechanism of double ionization in such fields. The latter conclusion in fact is in sharp contrast to the mechanism of double ionization in weak high-frequency synchrotron radiation fields (where the shakeoff process plays a significant role).

Coincident measurements of energy distribution of the two electrons in double ionization of He have recently been measured which revealed a surprising 'excessive production' of hot electrons (e.g. [22,23]). IMST calculations [21] of the same distrubution have shown an excellent agreement with the data [22] and confirmd the phenomenon as being another consequence of the (dynamic) correlation between the electrons.

Cutoff Law of Sum-Momentum Distribution A prominent feature of the distribution of recoil momentum along the polarization direction is its large width with a sharp cutoff. Is there a rule for determining this *cutoff* momentum? How does it depend on the field parameters? IMST provides a simple analytical answer to these questions. From the explicit expression of the transition matrix element and using the properties of the Bessel functions involved, like the exponential decrease for the orders exceeding the argument, one can derive [9] the following expression for the cutoff momentum of the distribution,

$$|\boldsymbol{P}_{\text{par.}}|_{(\text{cutoff})} \approx |\,((\hbar \boldsymbol{k}_{\text{a}})_{\text{par.}} + (\hbar \boldsymbol{k}_{\text{b}})_{\text{par.}})\,|_{(\text{cutoff})}$$
$$= Re(4\sqrt{mU_{\text{p}}} + \sqrt{m(8U_{\text{p}} - E_{\text{B}})})\,,$$

where E_{B} is the binding energy. For the case of the distribution shown in Fig. 10.3, use of this formula predicts the cutoff momentum ≈ 5 a.u., in good agreement with both the experimental value, and the numerical result. Satisfactory agreement with the above cutoff formula and the available experimental values at other intensities for He and Ne has also been found [9].

10.3.3   Intense-Field Ionization of Molecules

IMST has been applied also to investigate the ionization dynamics of more complex systems of diatomic as well as polyatomic molecules. Recently, a number of unexpected results of molecular ionization in intense fields have been obtained experimentally. These phenomena have been analyzed with the help of the lowest–order IMST amplitude, which is equivalent to the well-known KFR model [12].

Ionizaion of Diatomic Molecules and N_2/O_2 Puzzle It had been observed some time ago [24–26] that the measured ionization signals of a diatomic molecule and a noble gas atom, which have comparable ionization energies, also ionize very similarly at high laser intensities. This phenomenon was often assumed to be understood in terms of the optical tunneling process [27] in which the probability of intense–field ionization depends primarily on the ionization energy of the species. Thus, for example, the ionization signals of the pair N_2 and Ar (with analogous ionization energies $E_{\mathrm{ion}}(\mathrm{Ar})/E_{\mathrm{ion}}(N_2)$ $= 15.76\,\mathrm{eV}/15.58\,\mathrm{eV} = 1.01$) showed very similar ionization yields as a function of intensity [24,26]. Recently, however, a strong deviation from this general expectation is discovered in two independent experiments using femtosecond Ti:sapphire laser pulses. Thus, Talebpour et al. [28] and Guo et al. [29] have found that although as before the ionization signal for N_2 remains comparable to that of its companion Ar atom, the signal for the O_2 molecule is greatly suppressed compared to that of its companion noble gas Xe atom ($E_{\mathrm{ion}}(\mathrm{Xe})/E_{\mathrm{ion}}(O_2) = 12.13\,\mathrm{eV}/12.07\,\mathrm{eV} = 1.005$). In fact, the reduction for O_2 ionization turns out to be more than an order of magnitude. Considerable interest has been generated by this unexpected finding.

The wave function of a homonuclear diatomic molecule can be written conveniently as LCAO-MOs (e.g., [30]):

$$\Phi(\boldsymbol{r}; \boldsymbol{R}_1, \boldsymbol{R}_2) = \sum_{i=1}^{i_{\max}} a_i \phi_i(\boldsymbol{r}, -\boldsymbol{R}/2) + b_i \phi_i(\boldsymbol{r}, \boldsymbol{R}/2). \tag{10.11}$$

Here, the atomic orbitals ϕ_i are centered at $\boldsymbol{R}_1 = -\boldsymbol{R}/2$ and $\boldsymbol{R}_2 = \boldsymbol{R}/2$, where R is the internuclear separation and a_i and b_i denote the orbital coefficients. For a bonding molecular orbital (σ_g or π_u) both sets of coefficients, a_i and b_i, have the same value *and* sign, $a_i = b_i$, while, in contrast, for an antibonding orbital (σ_u or π_g), they have the opposite sign, $a_i = -b_i$. Within the first approximation, $T_{\mathrm{i}\to\mathrm{f}}^{(1)}$, of IMST the rate of ionization (per species), in a linearly polarized laser field, for a diatomic molecule with N_e equivalent electrons in a bonding or in an antibonding valence orbital takes the form [31]:

$$\Gamma^+ = N_e \sum_{N=N_0}^{\infty} \int \mathrm{d}\hat{\boldsymbol{k}}_N \left(\frac{\mathrm{d}W^{(N)}}{\mathrm{d}\hat{\boldsymbol{k}}_N}\right)_{\mathrm{at}}$$

$$\times 4 \begin{cases} \cos^2(\boldsymbol{k}_N \cdot \boldsymbol{R}/2) & : \quad \text{bonding} \\ \sin^2(\boldsymbol{k}_N \cdot \boldsymbol{R}/2) & : \quad \text{antibonding} \end{cases} \tag{10.12}$$

where

$$\left(\frac{\mathrm{d}W^{(N)}}{\mathrm{d}\hat{\boldsymbol{k}}_N}\right)_{\mathrm{at}} = 2\pi\, C^2\, k_N (U_\mathrm{p} - N\omega)^2\, J_N^2 \left(\boldsymbol{\alpha}_0 \cdot \boldsymbol{k}_N, \frac{U_\mathrm{p}}{2\omega}\right)$$

$$\times \left|\sum_{i=1}^{P} a_i \langle \boldsymbol{k}_N | \phi_i(\boldsymbol{r})\rangle\right|^2, \tag{10.13}$$

and

$$C^2 = \left(\frac{2\kappa_{\mathrm{ion}} E_{\mathrm{ion}}}{F}\right)^{2Z/\kappa_{\mathrm{ion}}},$$

where Z is the charge state of the molecular ion ($Z = 1$ for single ionization), and F is the peak field strength of the laser. This shows that the rate essentially factorizes into that of an atom-like ('atom in molecule') part, (10.13), and a trigonometric part associated with an interference effect between the waves of the ionizing electron centered about the two nuclei. In the case of molecules with valence orbitals of antibonding symmetry, one may expect that the interference term would tend to behave destructively (while that for the bonding symmetry, constructively), since for a small argument ($\boldsymbol{k}_N \cdot \boldsymbol{R}/2$), which corresponds to the first few and dominant photon orders N, $\sin^2(\boldsymbol{k}_N \cdot \boldsymbol{R}/2) \ll 1$ and $\cos^2(\boldsymbol{k}_N \cdot \boldsymbol{R}/2) \approx 1$.

In Fig. 10.5 the experimental ion signals (upper panel) for the pairs N_2 and Ar, and O_2 and Xe, measured by Guo et al. [29], and the corresponding signals obtained from the IMST (lower panels) are presented. The experimental data clearly exhibit the phenomenon of suppressed or reduced ionization for O_2 and its absence in the case of N_2. A similar reduction of ionization of O_2 by more than an order of magnitude, compared to that for N_2 has been observed experimentally by Talebpur et al. [28] and are reproduced by IMST [31]. It should be noted that for highly electronegative molecules such as F_2 for which the role of intermediate ionic states could very well be significant to alter the results of the first order prediction and Hartree-Fock approximation of the initial unperturbed molecular state. For a more detailed discussion of the first order approximation for the total yields as well as for the effects of the symmetry–induced interference on the corresponding energy distributions (the molecular ATI spectra) we refer the reader to [31] as well as to [32], respectively.

Ionization of Polyatomic Molecules Generalization of the lowest–order IMST ionization amplitude from the diatomic to the more complex polyatomic molecules has recently been carried out [36–38]. It has then been applied to obtain the ionization yields of polyatomic hydrocarbon molecules

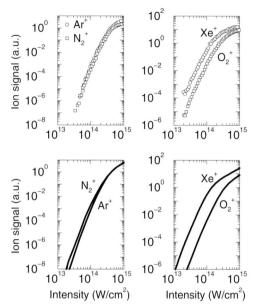

Fig. 10.5. Suppressed molecular ionization as observed by Guo et al. [29] (*upper panels*) in the case of O_2 (open-shell, antibonding, π_g-symmetry), compared to that of its companion noble gas Xe atom ($E_{\mathrm{ion}}(\mathrm{Xe})/E_{\mathrm{ion}}(O_2) = 12.13\,\mathrm{eV}/12.07\,\mathrm{eV} = 1.005$), and its absence in N_2 (closed-shell, bonding, σ_g-symmetry), compared to its companion Ar atom ($E_{\mathrm{ion}}(\mathrm{Ar})/E_{\mathrm{ion}}(N_2) = 15.76\,\mathrm{eV}/15.58\,\mathrm{eV} = 1.01$). Results of IMST calculations for the respective pairs are shown in the *lower panels*. Laser parameters (in experiments and calculations): $\lambda = 800\,\mathrm{nm}$, pulse duration, $\tau = 30\,\mathrm{fs}$; from [31]

such as C_2H_2, C_2H_4, and C_6H_6. In these calculations the ground–state wave functions for the molecules were obtained from the respective Hartree–Fock methods with default Gaussian basis sets extended by an additional diffuse s and a p function, using the GAMESS code [13]. The basic rates of ionization so calculated are then employed in the rate equation to compute the total ion yields (assuming a Gaussian laser beam (TEM$_{00}$ mode) and a Gaussian pulse profile having the same beam waist and pulse duration as that used in the experiments). Finally, the calculated yields have been orientation averaged assuming random orientations of the molecular axis with respect to the laser polarization direction.

Figure 10.6 shows the results of IMST calculations (solid lines) for ionization of C_2 (panel **a**), C_2H_2 (panel **b**) and C_2H_4 (panel **c**), in the field of a Ti:sapphire laser pulse at $\lambda = 800\,\mathrm{nm}$, and $\tau = 200\,\mathrm{fs}$ (panels **a** and **c**) or $\tau = 50\,\mathrm{fs}$ (panel **b**), respectively, along with the experimental data (circles) obtained by Cornaggia and Hering (for C_2H_2, [33]) and by Talebpour et al. (for C_2H_4, [34]). The calculated molecular ion yields are seen to be in good agreement with the experimental data. We also note that the reduction of ionization of C_2H_4 compared to that of I (its companion atom with analogous ionization potential) should show up as an increase of the saturation intensity of the former compared to that of the latter. This expectation is also in agreement with the recent experimental measurement of the saturation intensities of a set of hydrocarbons [35].

Ionization of Benzene Ionization of the highly symmetric benzene molecule, C_6H_6, presents a particularly interesting example of 'ionization suppression' phenomenon discussed above for the diatomics. In fact, a strong reduction of ionization yield of benzene compared to its companion atom Be (having comparable ionization potentials) can be shown from the IMST analysis as due to the '*anti-bonding*'–like symmetry of the molecule.

As is well known, benzene has a delocalized π-electron system (arising from the six $2p_z$-orbitals of the carbon atoms in the benzene ring), which leads to a doubly degenerate e_{1g} valence orbital (containing four electrons) and an energetically lower lying a_{2u}-orbital (with two electrons). Each of the two e_{1g} valence orbital functions has a nodal plane perpendicular to the ring of the six C atoms. One may also view the density of the active π-electrons of the carbon ring as arising from three equivalent effective homonuclear diatomic-like (C_2) molecular orbitals of antibonding (π_g) symmetry. One, therefore, expects as seen in the case of diatomic molecules that the benzene molecule too would exhibit a suppressed molecular ionization due to the destructive interference of the subwaves emerging from the three equivalent pairs of C atoms (that may be only weakly perturbed by the presence of the relatively distant pairs of H atoms). To test this expectation, calculation of the ion yields for benzene as a function of the field intensity has been

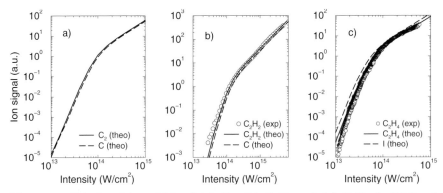

Fig. 10.6. Calculated ion yields of (**a**) C_2, (**b**) C_2H_2 and (**c**) C_2H_4 (*solid lines*) as well as those of their *atomic* companion atoms (*dashed lines*) compared with the experimental data (*open circles*), (**b**) [33], (**c**) [34]. The laser parameters are: $\lambda = 800$ nm and $\tau = 200$ fs (panel **a** and **c**), and $\lambda = 800$ nm and $\tau = 50$ fs (panel **b**); from [37]

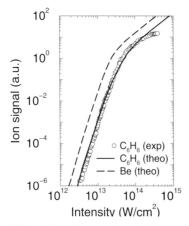

Fig. 10.7. Comparison of the predictions of the present theory for the ion yields of C_6H_6 (*solid line*) in comparison with the experimental data (circles) and that of the companion atom, Be (*dashed line*). Note the reduced ionization of benzene compared to that of Be. The experimental data are for a 200–fs Ti:sapphire laser pulse at $\lambda = 800$ nm [39]; from from [37]

carried out [37]. The results of the IMST calculations (solid line) are shown in Fig. 10.7. They are also compared in the figure with the recent experimental data obtained by Talebpour et al. [39]. The calculated ionization yields of the atomic companion Be (dashed line), with comparable ionization potential as of benzene, are also shown. Both the experimental and theoretical results are obtained for a linearly polarized Ti:sapphire laser with $\lambda = 800$ nm and pulse duration of 200 fs. As can be easily seen from the figure, the experimental data (circles) and the calculated ion yields (solid line) vs. field intensity are in very good agreement with each other. Furthermore, the comparison with the companion Be atom shows a strong reduction of the ionization signal of benzene (circles) with respect to that of Be (dashed line), confirming the theoretical expectation.

10.4   Conclusion

To summarize, in this chapter the so-called intense-field many-body S-matrix theory (IMST) is briefly presented with application to a number of recently observed laser–induced ionization phenomena in many-electron atomic *and* molecular systems; the results show the usefulness of IMST for gaining insights into the nonperturbative ionization dynamics of complex many-body systems in intense laser fields.

It is a pleasure to thank Dr. A. Becker and Dr. J. Muth-Böhm for their fruitful collaborations.

References

1. F.H.M. Faisal and A. Becker: in: *Selected Topics on Electron Physics*, ed. by D.M. Campbell and H. Kleinpoppen (New York, Plenum 1996) p. 317; Commun. Mod. Phys. D **1**, 15 (1999)
2. C. Joachain: *Quantum Collision Theory*, 3rd edn., (Amsterdam, North-Holland 1983)
3. A. Becker and F.H.M. Faisal: Opt. Express **8**, 383 (2001)
4. F.H.M. Faisal: *Theory of Multiphoton Processes* (New York, Plenum 1987), p. 11
5. F.H.M. Faisal: Phys. Lett. A **187**, 180 (1994)
6. A. Becker and F.H.M. Faisal: Phys. Rev. A **50**, 3256 (1994)
7. A. Becker and F.H.M. Faisal: J. Phys. B **29**, L197 (1996); Phys. Rev. A **59**, R1742 (1999); ibid., R3182 (1999)
8. A. Becker and F.H.M. Faisal: J. Phys. B **32**, L335 (1999)
9. A. Becker and F.H.M. Faisal: Phys. Rev. Lett. **84**, 3546 (2000)
10. A. Becker, L. Plaja, P. Moreno, M. Nurhuda and F.H.M. Faisal: Phys. Rev. A **64**, 0234008 (2001)
11. V.P. Krainov: J. Opt. Soc. Am. B **14**, 425 (1997)
12. L.V. Keldysh: Zh. Eksp. Teor. Fiz. **47**, 1945 (1964) [Sov. Phys. JETP **20**, 1307 (1965)]; F.H.M. Faisal: J. Phys. B **6**, L89 (1973); H.R. Reiss: Phys. Rev. A **22**, 1786 (1980)
13. M.W. Schmidt, K.K. Baldridge, J.A. Boatz, S.T. Elbert, M.S. Gordon, J.H. Jensen, S. Koseki, N. Matsunaga, K.A. Nguyen, S.J. Su, T.L. Windus, M. Dupuis and J.A. Montgomery: J. Comp. Chem. **14**, 1347 (1993)
14. D.N. Fittinghoff, P.R. Bolton, B. Chang and K.C. Kulander: Phys. Rev. Lett. **69**, 2642 (1992)
15. B. Walker, B. Sheehy, L.F. DiMauro, P. Agostini, K.J. Schafer and K.C. Kulander: Phys. Rev. Lett. **73**, 1227 (1994)
16. T. Weber, M. Weckenbrock, A. Staudte, L. Spielberger, O. Jagutzki, V. Mergel, F. Afaneh, G. Urbasch, M. Vollmer, H. Giessen and R. Dörner: Phys. Rev. Lett. **84**, 443 (2000)
17. R. Moshammer, B. Feuerstein, W. Schmitt, A. Dorn, C.D. Schröter and J. Ullrich: Phys. Rev. Lett. **84**, 447 (2000)
18. P.B. Corkum: Phys. Rev. Lett. **71**, 1994 (1993)
19. K.C. Kulander, J. Cooper and K.J. Schafer: Phys. Rev. A **51**, 561 (1995)
20. M.Y. Kuchiev: J. Phys. B. **28**, 5093 (1995)
21. A. Becker and F.H.M. Faisal, Phys. Rev. Lett. **89**, 193003 (2002)
22. R. Lafon et al., Phys. Rev. Lett. **86**, 2762 (2001)
23. B. Witzel, N.A. Papdogiannis and D. Charalambidis: Phys. Rev. Lett. **85**, 2265 (2001)
24. G.N. Gibson, R.R. Freeman and T.J. McIlrath: Phys. Rev. Lett. **67**, 1230 (1991)
25. S.L. Chin, Y. Liang, J.E. Decker, F.A. Ilkov and M.V. Ammosov: J. Phys. B: At. Mol. Opt. Phys. **25**, L249 (1992)
26. T.D.G. Walsh, J.E. Decker and S.L. Chin: J. Phys. B: At. Mol. Opt. Phys. **26**, L85 (1993); T.D.G. Walsh, F.A. Ilkov, J.E. Decker and S.L. Chin: J. Phys. B: At. Mol. Opt. Phys. **27**, 3767 (1993)

27. M.V. Ammosov, N.B. Delone and V.P. Krainov: Z. Eksp. Teor. Fiz. **91**, 2008 (1986); Sov. Phys. JETP **54**, 1191 (1986)
28. A. Talebpour, C-Y. Chien and S.L. Chin: J. Phys. B: At. Mol. Opt. Phys. **29**, L677 (1996)
29. C. Guo, Li M, J.P. Nibarger and G.N. Gibson: Phys. Rev. A **58**, R4271 (1998)
30. J.C. Slater: *Quantum Theory of Molecules and Solids* (New York, McGraw-Hill 1963)
31. J. Muth-Böhm, A. Becker and F.H.M. Faisal: Phys. Rev. Lett. **85**, 2280 (2000)
32. F. Grasbon, G.G. Paulus, S.L. Chin, H. Walther, J. Muth-Böhm, A. Becker and F.H.M. Faisal: Phys. Rev. A **63**, 041402 (2001)
33. C. Cornaggia and P. Hering: J. Phys. B: At. Mol. Opt. Phys. **31**, L503 (1998)
34. A. Talebpour, A.D. Bandrauk, J. Yang and S.L. Chin: Chem. Phys. Lett. **313**, 789 (1999)
35. S.M. Hankin, D.M. Villeneuve, P.B. Corkum and D.M. Rayner: Phys. Rev. Lett. **84**, 5082 (2000)
36. F.H.M. Faisal, A. Becker and J. Muth-Böhm: Laser Phys. **9**, 115 (1999)
37. J. Muth-Böhm, A. Becker, S.L. Chin and F.H.M. Faisal: Chem. Phys. Lett. **337**, 313 (2001)
38. J. Muth-Böhm, A. Becker and F.H.M. Faisal: *Proceedings of the 8th ICOMP, 1999*, ed. by L.F. DiMauro, R.R. Freeman and K.C. Kulander (AIP, New York 2000)
39. A. Talebpour, S. Larochelle and S.L. Chin: J. Phys. B: At. Mol. Opt. Phys. **31**, 2769 (1998)

11 Quantum Orbits and Laser-Induced Nonsequential Double Ionization

W. Becker, S.P. Goreslavski, R. Kopold, and S.V. Popruzhenko

11.1 Introduction

Intense laser fields affect processes that already take place in their absence or may induce others that are not possible otherwise, e.g., laser-assisted vs. laser-induced processes (for reviews, see [1–3]). Electron–atom scattering on the background of a laser field is an example of the former, single and multiple ionization of the latter. The transfer of energy between the laser field and the charged particles involved can amount to a very large number of laser photons. For the theoretical description, the presence of the laser field poses a formidable challenge (for a recent review, see [4]). Clearly, lowest-order perturbation theory is completely inadequate for the treatment of the laser–atom interaction. Consider a titanium-sapphire laser with $\hbar\omega = 1.55\,\mathrm{eV}$ and an intensity of $I = 10^{15}\,\mathrm{Wcm}^{-2}$. In this case, the so-called Keldysh parameter $\gamma = \sqrt{|E_0|/(2U_\mathrm{P})}$, which specifies the ratio of the tunneling time over the laser period, for helium has the value $\gamma = 0.45 < 1$ where E_0 denotes the ionization energy of the atom and U_P the ponderomotive energy of the laser field; see Sect. 11.3.2. In this case, envisioning the electron to enter the continuum with zero velocity via tunneling provides a valid physical picture. Accelerated by the laser field with the amplitude E, the electron will then perform a wiggling motion with the amplitude

$$\alpha = eE/(m\omega^2) = 3.95 \times 10^{-6}\sqrt{I}/\omega^2\,a_0\,,$$

where the laser intensity is given in Wcm^{-2}, the laser frequency in eV, and a_0 is the Bohr radius. For the above example, we have $\alpha = 52a_0$. Hence, the wiggling motion extends over a scale much larger than the diameter of the atom. For a numerical solution of the time-dependent Schrödinger equation, this is a major problem, since the spatial grid must be chosen large enough to allow keeping track of this motion.

In this "tunneling regime", the laser is able to couple significant energy into the atom via the "recollision mechanism"[1]. Thereby, the freed electron is accelerated by the laser field and driven back into a recollision with the ion. In this recollision, it may scatter elastically, recombine into the ground state with emission of a photon, or scatter inelastically, dislodging a second electron

---

[1] This also has been aptly dubbed the "atomic-antenna mechanism"[5].

(or more) [6,7]. The corresponding processes are referred to as high-order above-threshold ionization (ATI), high-order harmonic generation (HHG), or nonsequential double (multiple) ionization (NSDI), respectively (for reviews, see [1,8]).

The motion of the ionized electron in the continuum is largely classical. However, the electron can reach a given final state via several space–time pathways. Their contributions add up coherently in the transition amplitude and, therefore, interfere quantum mechanically. The electron enters the continuum via tunneling, which can be well described semiclassically. These facts combined lead to a picture where the S (scattering) matrix is expanded in terms of "quantum orbits": space-time orbits that are complex so that they account for the tunneling process [9–11]. They provide an illustration of Feynman's path integral in a situation where few paths are sufficient to generate the transition amplitude [12]. Quantum orbits work particularly well for low frequency and high intensity – that is, in the regime where the numerical solution of the time-dependent Schrödinger equation is most cumbersome.

Here, we shall survey the usage of quantum orbits for the description of nonsequential double ionization [13,14] and the calculation of the S-matrix elements that govern the multiply differential rates that have recently been measured via the COLTRIMS technique [15–17]; see also [18–21] for doubly charged-ion–electron correlation measurements.

11.2   The S-Matrix Element

11.2.1   Volkov Wave Functions

Calculating any process on the background of an intense laser field is, in principle, a straightforward matter. One starts from the perturbation expansion of the respective S-matrix element with respect to the various interparticle interactions in the absence of the field. Then, the laser field is taken into account by replacing the wave functions and propagators of all free charged particles by Volkov wave functions and propagators. The Volkov wave function is the exact wave function of an otherwise free charged particle in the presence of a plane-wave laser field. For a laser field that is described by a purely time-dependent vector potential $\boldsymbol{A}(t)$ it has the (length-gauge) form

$$\psi_{\boldsymbol{k}}^{(\mathrm{Vv})}(x) = (2\pi)^{-\frac{3}{2}} \mathrm{e}^{\mathrm{i}[\boldsymbol{k}-e\boldsymbol{A}(t)]\cdot\boldsymbol{r}} \exp\left(-\frac{\mathrm{i}}{2m}\int^t \mathrm{d}\tau[\boldsymbol{k}-e\boldsymbol{A}(\tau)]^2\right), \quad (11.1)$$

where $x = (\boldsymbol{r}, t)$. This is a plane wave with time-dependent velocity, corresponding to the average (drift) momentum $\boldsymbol{k}$ (see Sect. 11.3.2). If the laser field is turned off as a function of time only, the drift momentum will agree with the momentum outside the field. We have adopted the dipole approximation, that is, we ignore the spatial dependence of the laser field. This is justified provided: (i) the laser pulse is so short that the electron never experiences the spatial gradient of the field and (ii) the intensity is sufficiently low so that electrons do not reach a speed where the magnetic field has to be considered. The corresponding (time-ordered) Volkov propagator is

$$iG^{(\text{Vv})}(x,x') = \theta(t-t') \int d^3 k \psi_{\boldsymbol{k}}^{(\text{Vv})}(x) \psi_{\boldsymbol{k}}^{(\text{Vv})}(x')^* \,. \tag{11.2}$$

It has a compact analytical expression (see, e.g., [22]), which, however, we will not need.

11.2.2   The S Matrix in the Strong-Field Approximation

The straightforward procedure outlined in the preceding section is much too complicated to be actually carried out. For the case of low-frequency high-intensity laser fields, the strong-field approximation (SFA) incorporates the essential physics and leads to viable expressions. The SFA neglects the effect of the laser field on bound states and the effect of the electron–ion and the electron–electron interaction on electrons that have been promoted into continuum states. This is partly justified by the large excursion amplitude mentioned in the Introduction. In this approximation, the contribution to the S matrix that implements the recollision scenario is

$$S_{\boldsymbol{p}_1,\boldsymbol{p}_2} = \int d^4 x d^4 x' d^4 x'' \psi_{\boldsymbol{p}_1}^{(\text{Vv})}(x)^* \psi_{\boldsymbol{p}_2}^{(\text{Vv})}(x')^* \delta(t-t') V(\boldsymbol{r},\boldsymbol{r}')$$
$$\times \psi_2^{(0)}(x') G^{(\text{Vv})}(x,x'') V(\boldsymbol{r}'') \psi_1^{(0)}(x'') + (\boldsymbol{p}_1 \leftrightarrow \boldsymbol{p}_2) \,, \tag{11.3}$$

where $x \equiv (\boldsymbol{r},t)$, $x' \equiv (\boldsymbol{r}',t')$, and $x'' \equiv (\boldsymbol{r}'',t'')$. The potential $V(\boldsymbol{r})$ denotes the effective binding potential of the first electron, and $V(\boldsymbol{r},\boldsymbol{r}')$ is the electron–electron interaction. The wave functions $\psi_i^{(0)}(x) = \phi_i^{(0)}(\boldsymbol{r}) \times \exp(-iE_{0i}t)(i=1,2)$ describe, respectively, the field-free initial bound states of the first and the second electron with ionization energies $E_{02} < E_{01} < 0$. The S-matrix element (11.3) has been successfully applied to the calculation of *total* multiple-ionization rates [23–26]. The scenario implemented in the "crapola" model is close to the physical content of (11.3), too [27]. For the *differential* rates, the S-matrix element (11.3) or closely related expressions were evaluated by various groups [28–33].

   The corresponding Feynman diagram of Fig. 11.1 exhibits the physical content of the approximation (11.3) more readily than the formula: At the time t'', the first electron is ionized into the continuum. During its subsequent propagation, which is described by the Volkov propagator $G^{(\text{Vv})}(x,x'')$, it is able to collect energy from the laser field. During this stage, the effect of the binding potential is neglected. All the while, the second electron remains inactive in its bound state until, at the later time $t = t'$, the crucial electron–electron interaction occurs that is to kick out the second electron. This is mediated by the interaction potential $\delta(t-t')V(\boldsymbol{r},\boldsymbol{r}')$. Thereafter, both electrons propagate in the presence of the laser field towards the detector, without feeling the potential of the ion nor each other.

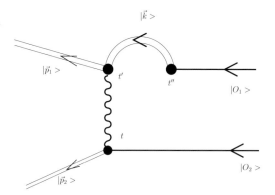

Fig. 11.1. Graphical representation of the S-matrix element (11.3) in the form of a Feynman diagram. The crucial electron–electron interaction $\delta(t - t')V(r, r')$ is represented by the *vertical wavy line*. *Double lines* represent Volkov states and propagators, *single lines* bound states in the absence of the field

In view of the Volkov propagator (11.2), the S-matrix element (11.3) treats the electron–electron interaction in the Born approximation, with Volkov states replacing the plane waves that would appear in the absence of the laser field.

11.3   Quantum Orbits

If contact potentials are chosen for the binding potential $V(r)$ as well as the electron–electron interaction potential $V(r, r')$, the S-matrix element can be reduced to a quadrature (see Sect. 11.4.2). For other potentials, in particular, for the Coulomb potential, this is not possible. If, however, the potentials can be represented in momentum space in analytical form, the saddle-point approximation provides a comparatively simple and physically appealing approach.

11.3.1   The Saddle-Point Approximation

To this end, we rewrite the S-matrix element (11.3) in the form

$$S_{\boldsymbol{p}_1, \boldsymbol{p}_2} = \int_{-\infty}^{\infty} \mathrm{d}t \int_{-\infty}^{t} \mathrm{d}t'' \int \mathrm{d}^3 \boldsymbol{k} \, \mathrm{e}^{\mathrm{i}\Phi_{\boldsymbol{p}_1, \boldsymbol{p}_2}(t, t'', \boldsymbol{k})} \langle \boldsymbol{k} - e\boldsymbol{A}(t'') | V(\boldsymbol{r}) | \phi_1^{(0)} \rangle$$

$$\times \langle \boldsymbol{p}_1 - e\boldsymbol{A}(t), \boldsymbol{p}_2 - e\boldsymbol{A}(t) | V(\boldsymbol{r}, \boldsymbol{r}') | \boldsymbol{k} - e\boldsymbol{A}(t), \phi_2^{(0)} \rangle . \qquad (11.4)$$

Here, the exponential dependence on t, t'', and $\boldsymbol{k}$, which derives from the Volkov states, the Volkov propagator, and the initial bound states $\psi_i^{(0)}(x)$, has been collected into the phase

$$\Phi_{\boldsymbol{p}_1,\boldsymbol{p}_2}(t,t'',\boldsymbol{k}) = -E_{01}t'' - E_{02}t - \frac{1}{2m}\int_{t''}^{t} \mathrm{d}\tau[\boldsymbol{k} - e\boldsymbol{A}(\tau)]^2 \tag{11.5}$$

$$+ \frac{1}{2m}\int^{t}\mathrm{d}\tau[\boldsymbol{p}_1 - e\boldsymbol{A}(\tau)]^2 + \frac{1}{2m}\int^{t}\mathrm{d}\tau[\boldsymbol{p}_2 - e\boldsymbol{A}(\tau)]^2\,.$$

The matrix elements in (11.4) of the binding potential $V(\boldsymbol{r})$ and the interaction potential $V(\boldsymbol{r},\boldsymbol{r}')$ between the indicated plane-wave states, that is, the form factors, will be discussed below.

If we regard the dependence of the form factors on the times t, t'', and the drift momentum $\boldsymbol{k}$ as slow, we conclude that, for sufficiently high laser intensity, the S-matrix element (11.4) will acquire its dominant contributions from those values of the integration variables that render the phase (11.6) stationary. They are determined by equating to zero the derivatives of this phase with respect to the three variables t, t'', and $\boldsymbol{k}$. This yields the three equations

$$[\boldsymbol{k} - e\boldsymbol{A}(t'')]^2 = 2mE_{01} = -2m|E_{01}|\,, \tag{11.6}$$

$$\int_{t''}^{t}\mathrm{d}\tau[\boldsymbol{k} - e\boldsymbol{A}(\tau)] = 0\,, \tag{11.7}$$

$$[\boldsymbol{p}_1 - e\boldsymbol{A}(t)]^2 + [\boldsymbol{p}_2 - e\boldsymbol{A}(t)]^2 = [\boldsymbol{k} - e\boldsymbol{A}(t)]^2 - 2m|E_{02}|\,. \tag{11.8}$$

These equations hold regardless of the form of the electron–electron interaction and the binding potential. They are the saddle-point equations that determine the "quantum orbits" [9–12] of intense-laser–atom physics. Note that owing to $E_{01} < 0$ their solutions will be complex. Their physical meaning will be discussed below in Sects. 11.3.2–11.3.4.

Let us denote the solutions of the saddle-point equations (11.6)–(11.8) by t_s, t''_s, and $\boldsymbol{k}_s$ ($s = 1, 2, \ldots$). Then, the S-matrix element is represented as the coherent superposition of the contributions of the individual saddle points,

$$S_{\boldsymbol{p}_1,\boldsymbol{p}_2} = \sum_{\bar{s}} A_{\bar{s}}(\boldsymbol{p}_1,\boldsymbol{p}_2)e^{i\Phi_{\boldsymbol{p}_1,\boldsymbol{p}_2}(t_{\bar{s}},t''_{\bar{s}},\boldsymbol{k}_{\bar{s}})}\,, \tag{11.9}$$

where the form factors as well as the determinant resulting from the saddle-point integration are included in the prefactor $A_s(\boldsymbol{p}_1,\boldsymbol{p}_2)$. Actually, only the subset $\{\bar{s}\} \subset \{s\}$ of the *relevant* saddle points is to be included in the sum (11.9). It consists of those saddle points that are visited by an appropriate deformation into complex space of the original real integration manifold $-\infty < t < \infty$, $-\infty < t'' \leq t$, $-\infty < k_i < \infty$ ($i = x, y, z$) in the integral (11.4). Their actual determination requires great care [11,34].

If we let $E_{01} = 0$, that is, if we ignore the fact that the electron reaches the continuum via tunneling, then the solutions of the saddle-point equations (11.6)–(11.8) are real, provided the laser field is linearly polarized and the final momenta $\boldsymbol{p}_1$ and $\boldsymbol{p}_2$ are within the classically accessible regime; see

Sect. 11.3.3. The ensuing kinematics are simple and often referred to as the "simple-man model", discussed in the following subsection. In order actually to evaluate the S matrix, one may either solve the saddle-point equations (11.6)–(11.8) numerically. In this procedure, the (real) solutions of the simple-man model can be used as starting points for an iterative solution. For above-threshold ionization, this is carried out, for example, in [11,35]. Or, alternatively, in the tunneling regime where the Keldysh parameter γ is small, one can largely proceed analytically, incorporating the imaginary parts only to lowest order [36]. This will be sketched in Sect. 11.3.5. The occurrence of complex times then can be entirely avoided.

In either case, one has to select the *relevant* saddle points, whose number is, in principle, infinite, and to decide where to terminate the sum (11.9). Usually, the pair of saddle points with the shortest travel times $\mathrm{Re}(t_s - t_s'')$ provides an excellent approximation. However, for certain intensities, a large number has to be taken into account. This phenomenon and its physical origin are discussed in Sect. 11.5.

11.3.2   The Simple-Man Model

An electron in a time-dependent laser field $\boldsymbol{E}(t)$ has the velocity

$$m\boldsymbol{v}(t) = \boldsymbol{k} - e\boldsymbol{A}(t), \qquad (11.10)$$

where $\boldsymbol{A}(t)$ is the vector potential such that $\boldsymbol{E}(t) = -\mathrm{d}\boldsymbol{A}(t)/\mathrm{d}t$. We choose the vector potential so that its average over one period of the field vanishes, $\langle \boldsymbol{A}(t) \rangle_t = 0$. Obviously, the canonical momentum $\boldsymbol{k}$ then has the physical meaning of the drift momentum inside the field. The electron's time-averaged kinetic energy is

$$\frac{m}{2}\langle \boldsymbol{v}(t)^2 \rangle_t = \frac{1}{2m}\boldsymbol{k}^2 + U_\mathrm{P} \quad \text{with} \quad U_\mathrm{P} = \frac{e^2 \langle \boldsymbol{A}(t)^2 \rangle_t}{2m}, \qquad (11.11)$$

which displays the ponderomotive energy U_P as the cycle-averaged kinetic energy of the wiggling motion. If the field sweeps over the electron as a function of time only, the canonical momentum $\boldsymbol{k}$ is a conserved quantity. When the electron leaves the pulse, the energy of the wiggling motion is lost or, more correctly, returned to the field. For laser pulses with a duration of less than about 100 fs, this is the normal situation. If, in the opposite limit, the electron leaves the pulse on one side, it experiences a spatial gradient, energy is conserved, and the energy of its wiggling motion is added to its directional motion, so that its drift momentum is increased.

If in (11.6)–(11.8) we let $E_{01} = 0$, then for a linearly polarized field the solutions are real for momenta $\boldsymbol{p}_1$ and $\boldsymbol{p}_2$ that are classically accessible (see Sect. 11.3.3). The first equation (11.6) mandates that the electron start its orbit with velocity zero at the time t''. The second equation (11.7) makes sure that at a later time $t > t''$ it returns to its starting point, and the

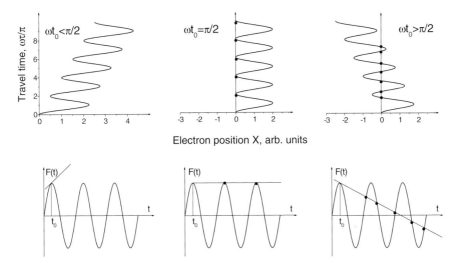

Fig. 11.2. Examples of simple-man orbits or, equivalently, the real parts of quantum orbits. All orbits start at the time t_0 from the origin with zero velocity. The field is given by the vector potential $\boldsymbol{A}(t) = A\hat{\boldsymbol{x}}\cos\omega t$. The *upper three panels* exhibit three orbits having positive, zero, and negative drift momentum $\boldsymbol{k} = e\boldsymbol{A}(t_0)$, such that the orbit returns to its starting point never, infinitely often, and a finite number of times, respectively. The *lower three panels* illustrate how to obtain the return time(s) for given start time t_0: One draws the tangent to the curve $F(t) = \int^t d\tau A(\tau)$ at the time t_0. Its intersections with $F(t)$ for $t > t_0$ specify the return times [37]. One can see that, depending on the start time t_0, there may be no, infinitely many, or a finite number of return times, respectively, from left to right

third equation (11.8) describes the instantaneous energy balance for inelastic scattering at the time t'' into the final state, which is characterized by the drift momenta $\boldsymbol{p}_1$ and $\boldsymbol{p}_2$. Equations (11.6)–(11.8) determine t, t'', and $\boldsymbol{k}$. With $E_{01} = 0$, they can have any number of real solutions between zero and infinity. This is illustrated in Fig. 11.2.

11.3.3   Classical Cutoffs

An important insight to be gained from the simple-man model is the existence of classical cutoffs. From (11.10) it follows that $\boldsymbol{k} = e\boldsymbol{A}(t'')$, where t'' is the ionization time, and therefore that $\boldsymbol{k}^2/(2m) \leq 2U_P$ for linear polarization. Moreover, since electrons are preferably ionized near the maxima of the electric field, $\boldsymbol{A}(t'')$ and, thereby, the drift momentum tends to be small.

With some more effort, it can be shown from (11.10) and (11.7) that the maximal kinetic energy of an electron at the time when it returns to its starting point (i.e., to the ion) is $3.17U_P$ [6,7]. It is important to note that such an electron returns to the ion near times when the electric field goes through zero, see Fig. 11.3. For above-threshold ionization with allowance for elastic rescattering the maximal electron energy at the detector is $10.007U_\mathrm{P}$ in this classical model [37].

For double ionization, too, $(\boldsymbol{p}_1, \boldsymbol{p}_2)$-momentum space is divided into a classically accessible and a classically inaccessible region. Classically accessible are those momenta that are, for some time t'', compatible with (11.7) and (11.8) for $\boldsymbol{k} = e\boldsymbol{A}(t'')$ and real t. That is, they are the outcome of inelastic rescattering at the time t of an electron that was "born" with zero velocity at some time t'' at the position of the ion. As a consequence of (11.7), for a linearly polarized laser field the drift momentum $\boldsymbol{k}$ is in the direction of the laser field. Let us decompose the final electron momenta into components parallel and perpendicular to the laser field, $\boldsymbol{p}_i = (\boldsymbol{p}_\|, \boldsymbol{p}_\perp)_i (i = 1, 2)$. Then, (11.8) can be rewritten as

$$[\boldsymbol{p}_{\|,1} - e\boldsymbol{A}(t)]^2 + [\boldsymbol{p}_{\|,2} - e\boldsymbol{A}(t)]^2$$
$$= [\boldsymbol{k} - e\boldsymbol{A}(t)]^2 - (2m|E_{02}| + \boldsymbol{p}_{\perp,1}^2 + \boldsymbol{p}_{\perp,2}^2). \tag{11.12}$$

where the recollision time t and the drift momentum $\boldsymbol{k}$ are determined from the ionization time t'' via (11.6) and (11.7). For each t, (11.12) defines a spherical surface in the six-dimensional $(\boldsymbol{p}_1, \boldsymbol{p}_2)$ space. For $\boldsymbol{p}_{\perp,1} = \boldsymbol{p}_{\perp,2} = 0$ this reduces to a circle in the $(p_{\|,1}, p_{\|,2})$ plane. Its interior is the classically accessible region for electrons that return at the time t. The envelope of all these spheres, when the return time t varies over one cycle of the field, is the boundary of the classical region. It can only be obtained numerically; see Fig. 11.6 for an example. Further examples can be found in [31], and [38] gives a general discussion of the constraints imposed by classical kinematics[2].

11.3.4   The Long and the Short Orbits

It is important to realize that for a given final state, that is for given $\boldsymbol{p}_1$ and $\boldsymbol{p}_2$, solutions to the saddle-point equations (11.6)–(11.8) come in pairs. This fact is crucial for numerous features of high-order harmonic generation [9]. The argument is presented in Fig. 11.3 for the case of the energy of the returning electron, using the graphical construction already employed in Fig. 11.2 and explained in the caption. For $E \geq E_{\mathrm{max}} = 3.17U_\mathrm{P}$, the saddle-point equations have no real solutions, but a pair of complex solutions continues to exist. However, only one solution of this pair is relevant, as defined in Sect. 11.3.1.

The situation is completely analogous for double ionization. For values of $\boldsymbol{p}_1$ and $\boldsymbol{p}_2$ exactly at the classical boundary discussed in Sect. 11.3.3, there is precisely one *real* simple-man orbit that leads to these momenta, while inside

---

[2] If the field at the time t is taken into consideration, then depending on the shape of the binding potential the bound electron may escape over the barrier even when the two momenta $\boldsymbol{p}_1$ and $\boldsymbol{p}_2$ are outside the classically accessible region [25]. This is instrumental for the classical model of [39].

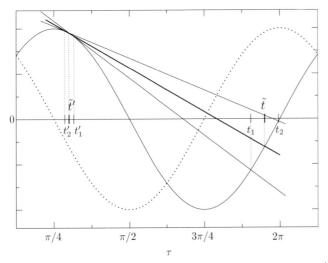

Fig. 11.3. The *solid curve* depicts the function $F(t) = \int^t d\tau A(\tau)$, which for a monochromatic field is proportional to the electric field; the *dotted curve* denotes the vector potential. The unique simple-man orbit that yields the maximal return energy of $E_{\max} \equiv 3.17U_{\mathrm{P}}$ corresponds to the tangent to $F(t)$ at the time $\tilde{t}'$ (the *thick solid line*), which intersects $F(t)$ at the return time $\tilde{t}$. Note that the electric field $E(\tilde{t})$ is near zero. Obviously, for $E < E_{\max}$, there are *two orbits* with this energy, one where the electron starts earlier (t_2') and another where it starts later (t_1') than for $E = E_{\max}$. In the present case, the first returns later (t_2) and the second earlier (t_1), so that the travel time of the first is longer than that of the second

the classical regime there are two. If complex solutions are admitted, there are always two solutions. Hence, we can state, in general, that solutions to the saddle-point equations for the specified final state come in pairs. The superposition of their contributions according to (11.9) generates a pronounced beating pattern.

11.3.5   An Analytical Approximation
to the Complex Quantum Orbits

In the tunneling regime, where $\gamma \ll 1$, the imaginary parts of the stationary points t_s, t_s'', and $\mathbf{k}_s$ are small provided the final state of the system lies inside the classically accessible region. This allows us to simplify (11.6)–(11.8) significantly. However, even in this region the imaginary parts of the stationary points must not be neglected since, otherwise, the phase in (11.9) becomes real and the origin of the first electron via tunneling is no longer adequately described. To obtain an appropriate approximation, we decompose the system (11.6)–(11.8) into real and imaginary parts. Taking into account that in the tunneling regime $\operatorname{Im}\omega t''_s \ll 1$, $\operatorname{Im}\omega t_s \ll 1$ and $\operatorname{Im}|\boldsymbol{k}_s| \ll e|\boldsymbol{A}|$ we can show that up to terms of the order of γ^3 [36]

$$\operatorname{Im}\omega t''_s = \gamma - \frac{1}{6}\gamma^3 + \mathrm{O}(\gamma^5) \quad \text{and} \quad \operatorname{Im}\omega t_s = \operatorname{Im}\boldsymbol{k}_s = 0\,. \tag{11.13}$$

In the limit where $\gamma \ll 1$, the approximation (11.13) is sufficient to keep all contributions to the phase $\Phi_{\boldsymbol{p}_1,\boldsymbol{p}_2}$ in (11.9) that are of order unity or larger. Thereafter, the integral over the (now real) intermediate momentum $\boldsymbol{k}$ can be carried out analytically, and the remaining equations for the *real* parts of the stationary points t and t'' are

$$\boldsymbol{r}_0 + e \int_{t''}^{t} \mathrm{d}\tau [\boldsymbol{A}(t'') - \boldsymbol{A}(\tau)] = 0\,, \tag{11.14}$$

$$[\boldsymbol{p}_1 - e\boldsymbol{A}(t)]^2 + [\boldsymbol{p}_2 - e\boldsymbol{A}(t)]^2 = e^2[\boldsymbol{A}(t'') - \boldsymbol{A}(t)]^2 - 2m|E_{02}|\,, \tag{11.15}$$

which do have real solutions for momenta $\boldsymbol{p}_1$ and $\boldsymbol{p}_2$ inside the classically accessible region. Comparing (11.14) and (11.15) with (11.6)–(11.8) we note that $\boldsymbol{k}$ has been replaced by $e\boldsymbol{A}(t'')$ and $E_{01} = 0$. Moreover, the first electron starts its orbit at the position $\boldsymbol{r}_0 = |E_{01}|\boldsymbol{E}(t'')/[e\boldsymbol{E}^2(t'')]$ (the "exit of the tunnel") rather than at the origin (where it still rescatters). The presence of $\boldsymbol{r}_0$ in (11.14) originates from the imaginary part of t''_s. Equations (11.14) and (11.15) are the equations of the simple-man model, improved by taking into account the exit of the tunnel at $\boldsymbol{r}_0$.

Therefore, in the tunneling regime of the first electron the five equations (11.6)–(11.8) (having complex solutions) have been reduced to the two equations (11.14) and (11.15) (having real solutions) for the start time and the return time of the active electron, which are known from the simple-man model. This procedure is valid, however, only in the case when the final state of the two electrons lies inside the classically allowed domain. In Sect. 11.3.4 we have seen that when the final momenta approach the boundary of this domain, the two saddle points of the simple-man model merge. This leads to the result of the simple-man model becoming divergent [36]. When this case is approached, the complex equations (11.6)–(11.8) have to be employed and the saddle-point approximation has to be abandoned in favor of the so-called uniform approximation [34,40]. This is no more complicated than the saddle-point approximation. It is tailored to the case where saddle points come in pairs and can be used both inside and outside the classically accessible region [34].

11.4   Results

In this section, we will display representative results of calculations based on (11.9) and obtained with the help of the semianalytical method described in Sect. 11.3.5. The results for a contact interaction between the two electrons have been obtained as explained in Sect. 11.4.2.

11.4.1   The Choice of the Electron–Electron Interaction Potential

For the electron–electron interaction $V(\boldsymbol{r}, \boldsymbol{r}')$, two different forms will be considered. The Coulomb interaction

$$V(\boldsymbol{r}, \boldsymbol{r}') \equiv V_c(\boldsymbol{r}, \boldsymbol{r}') = e^2/|\boldsymbol{r} - \boldsymbol{r}'| \tag{11.16}$$

appears to be the obvious choice. However, one has to realize that the S-matrix element (11.3) does not incorporate electron–electron repulsion in the final state nor the interaction of the returning electron and the two final-state electrons with the ion, except for the fact that the second electron is bound to the ion in the state $\psi_2^{(0)}(\boldsymbol{x}')$ up to the time $t = t'$. In reality, the bound electron can "hide" behind the ion so that the long-range Coulomb repulsion (11.16) may be partly shielded. Therefore, it may be reasonable to reinterpret $V(\boldsymbol{r}, \boldsymbol{r}')$ as an *effective* interaction potential. Then, the simplest choice is a contact interaction restricted to the position of the ion

$$V(\boldsymbol{r}, \boldsymbol{r}') \equiv V_s(\boldsymbol{r}, \boldsymbol{r}') = V_0 \delta(\boldsymbol{r} - \boldsymbol{r}') \delta(\boldsymbol{r}') . \tag{11.17}$$

The form factor of this contact potential is a constant independent of the momenta, while the form factor of the Coulomb potential (11.16) is given by [30]

$$\langle \boldsymbol{p}_1 - e\boldsymbol{A}(t), \boldsymbol{p}_2 - e\boldsymbol{A}(t) | V_c(\boldsymbol{r}, \boldsymbol{r}') | \boldsymbol{k} - e\boldsymbol{A}(t), \phi_2^{(0)} \rangle$$
$$\sim [(\boldsymbol{p}_1 - \boldsymbol{k})^2]^{-1} [\kappa_2^2 + (\boldsymbol{p}_1 + \boldsymbol{p}_2 - \boldsymbol{k} - e\boldsymbol{A}(t))^2]^{-2} + (\boldsymbol{p}_1 \leftrightarrow \boldsymbol{p}_2), \tag{11.18}$$

with $\kappa_2^2 \equiv 2m|E_{02}|$.

Let us investigate the consequences of the Coulomb form factor (11.18): In the tunneling limit, where $\gamma^2 \equiv |E_{01}|/(2U_{\mathrm{P}}) < 1$, the first ionization preferentially takes place near the maximum of the electric field, that is, when $\boldsymbol{A}(t'') \approx 0$. Then, in view of (11.6), the corresponding drift momentum $\boldsymbol{k}$ is small. Moreover, only those electrons have a sizeable kinetic energy that return (at time t) to the core near a zero of the electric field (see Sect. 11.3.2), so that $\boldsymbol{A}(t)$ is near an extremum. The form factor (11.18) will be largest if $\boldsymbol{p}_1 - \boldsymbol{k}$ (or $\boldsymbol{p}_2 - \boldsymbol{k}$) is as small as is possible, given that (11.8) does not admit $\boldsymbol{p}_1 = \boldsymbol{k}$ (or $\boldsymbol{p}_2 = \boldsymbol{k}$). If, however, $\boldsymbol{p}_1 - \boldsymbol{k}$ is small, then $\boldsymbol{p}_1$ is small. The second factor in the denominator of the form factor then favors small $\boldsymbol{p}_2 - e\boldsymbol{A}(t)$. In consequence, $\boldsymbol{p}_2$ is not small. The details depend greatly on whether the right-hand side of (11.8) is positive or not, that is, whether or not the returning electron has sufficient energy to liberate the bound one. If this energy is large, then owing to the Coulomb form factor the momentum of one electron will be near zero and the other one comparatively large, though significantly below $2\sqrt{mU_{\mathrm{p}}}$. The fact that for high-energy impact the two electrons tend to have different energies is well known from electron–impact ionization (see, e.g., [41]). If, on the other hand, the second ionization energy is comparatively large, that is if κ_2^2 in the denominator of the Coulomb form factor (11.18) is large, then the momentum dependence of the form factor plays a lesser role. In this limit, the distributions for the Coulomb potential and the contact potential will approach each other until they are centered about the momenta $|\boldsymbol{p}_{\|,1}| = |\boldsymbol{p}_{\|,2}| \approx 2\sqrt{mU_P}$. This is the center of the contact-interaction distribution for all intensities. Its simple origin is the fact that the maximal drift energy E_e that an electron released at rest in a laser field with vector potential $\boldsymbol{A}(t)$ can acquire, is $E_e \equiv \boldsymbol{p}^2/2m = 2U_P$.

In Sect. 11.4.3, we will recover these statements in the calculated momentum distributions and compare the latter with the data.

11.4.2 The Case of the Contact Electron–Electron Interaction

For the contact interaction (11.17), we do not depend on the saddle-point evaluation, since the spatial integrations in the S-matrix element (11.3) can be carried out trivially. The result is an integral over a sum of Bessel functions, which is then computed numerically. Since this is straightforward and does not depend on the occasional subtleties of the saddle-point method (see Sect. 11.5 below), the result is also useful as a benchmark, and we reproduce it here.

For a laser field with the vector potential $\boldsymbol{A}(t) = \boldsymbol{A}_0 \cos \omega t$, we have [42]:

$$
S_{\boldsymbol{p}_1,\boldsymbol{p}_2} \propto \hbar\omega \sum_n \delta\left[2U_p - E_{01} - E_{02} + \frac{1}{4m}(\boldsymbol{P}^2 + \boldsymbol{p}^2) - n\hbar\omega\right]
$$
$$
\times \sum_{l=-\infty}^{\infty} J_{n+2l}\left(\frac{e\boldsymbol{P}\cdot\boldsymbol{A}_0}{m\omega}\right)\left(\int_0^{\infty} d\tau \left(\frac{m}{i\tau}\right)^{3/2} \exp i\left(\frac{E_{01}\tau}{\hbar}\right)\right.
$$
$$
\times \left\{\exp i\left[\frac{4\eta\sin^2\frac{\tau}{2}}{\tau} - \eta\tau - l\phi(\tau)\right] J_l[\eta y(\tau)] - J_l(\eta)\right\}
$$
$$
\left. + J_l(\eta)\int_0^{\infty} d\tau \left(\frac{m}{i\tau}\right)^{3/2} \exp i\left(\frac{E_{01}\tau}{\hbar}\right)\right), \tag{11.19}
$$

where the $J_l(x)$ are Bessel functions and the real functions $y(\tau)$ and $\phi(\tau)$ are given by

$$
y(\tau)e^{i\phi(\tau)} = 1 - ie^{i\tau}\left(\frac{4\sin^2\frac{\tau}{2}}{\tau} - \sin\tau\right). \tag{11.20}
$$

The ponderomotive potential is $U_P = (eA_0)^2/4m$, and $\boldsymbol{P} \equiv \boldsymbol{p}_1 + \boldsymbol{p}_2$ and $\boldsymbol{p} \equiv \boldsymbol{p}_1 - \boldsymbol{p}_2$ denote the total and the relative momentum of the two final electrons.

The δ function in (11.19) allows the transfer of an, in principle, arbitrarily large number n of laser photons, to aid in the double-ionization process. However, photon transfers in excess of the classically allowed values will be more and more suppressed. On the technical side, it should be noted that the first integral over τ on the right-hand side of (11.19) converges absolutely at the origin, owing to the function in the curly bracket [for $\tau = 0$, we have $\phi(\tau) = 0$ and $y(\tau) = 1$], while the second integral, which is formally divergent at the origin, can be calculated analytically by analytic continuation in the fractional power of τ. Equation (11.19) underlies all of the explicit results of this chapter for the contact interaction.

11.4.3 Distribution of the Ion Momentum and the Electron Momenta

The momentum most straightforward to measure is that of the doubly charged ion [15,16]. Provided it is permissible to neglect the momenta of the laser photons, the ion momentum is the opposite of the total momentum of the two electrons, $\boldsymbol{P} \approx -(\boldsymbol{p}_1 + \boldsymbol{p}_2)$. Figure 11.4 exhibits a density plot [29,42] of the ion-momentum distribution as a function of $\boldsymbol{P}$, integrated over all relative momenta $\boldsymbol{p} = \boldsymbol{p}_1 - \boldsymbol{p}_2$, for double ionization of neon, which can be compared with the corresponding experimental data of Rottke et al. given in Chap. 18. The distribution is symmetric upon $\boldsymbol{P} \to -\boldsymbol{P}$. It has a pronounced minimum at $\boldsymbol{P} = 0$ and maxima near $\boldsymbol{P} = 4\sqrt{mU_\mathrm{P}}\hat{\boldsymbol{x}}$, as predicted by the simple-man model for two electrons each released with zero velocity near a zero of the electric field.

Later measurements were able to record, in effect, the momenta of both electrons [17,43–45]. In Fig. 11.5 we present the results of pertinent calculations for the parameters of neon and argon, comparing the consequences of adopting the contact potential (11.17) or the Coulomb potential (11.16) for the electron–electron interaction. The implications of the Coulomb form factor (11.18) as discussed above in Sect. 11.4.1 are very noticeable: in the Coulomb case, the momentum distribution has its center at a lower momentum and it trends away from the main diagonal in the $(p_{\parallel,1}, p_{\parallel,2})$ plane. An extreme example for very high intensity can be found in [31]: there, the momentum distribution has almost entirely moved into the second and the fourth quadrants.

Generally, the results calculated from the S-matrix element (11.3) agree rather well with the experimental data in neon, provided the electron–electron interaction is described by the contact interaction (11.17). Neither the Coulomb nor the contact interaction yield agreement with the data in argon, where significant populations are observed in the second and the fourth quadrants of the $(p_{\parallel,1}, p_{\parallel,2})$ distribution [17,44,45]. It is now believed that this is due to the contribution of an additional recollision-excitation mechanism [44]. Indeed, a one-dimensional model of this mechanism yields results that point in the right direction [29]; see, also, [26]. If, on the other hand, the contribution of the latter mechanism is subtracted from the data [44], the remainder agrees well with the pure rescattering results here presented. Some indication of the Coulomb effects that are so obvious in Fig. 11.5 can be found in [43,45] for small values of the transverse momenta $\boldsymbol{p}_{i,\perp}$. This conforms with the tendencies that follow from the form factor (11.18).

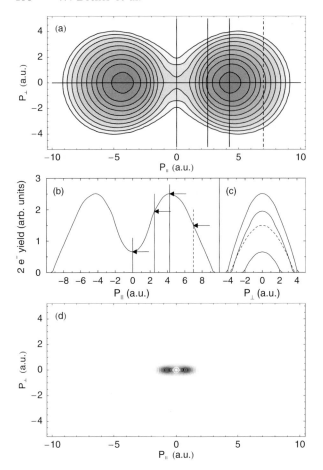

Fig. 11.4. Nonsequential double ionization of neon by a Ti:Sa laser ($\hbar\omega = 1.55\,\mathrm{eV}$) at $8 \times 10^{14}\,\mathrm{W\,cm^{-2}}$. (**a**) Distribution of the ion-momentum components $P_\parallel$ and $\boldsymbol{P}_\perp$ parallel and perpendicular to the electric field of the laser. (**b**) A cut through this distribution at $\boldsymbol{P}_\perp = \boldsymbol{0}$. (**c**) Cuts at various values of $P_\parallel$ as indicated in (**a**) and (**b**). (**d**) Momentum distribution for single ionization of neon for the same laser parameters, calculated from the KFR matrix element [46]. From [29,42]

11.5   Resonances and the Effects of Orbits with Long Travel Time

As discussed in Sect. 11.3.2, the same final state can be reached by several quantum orbits, which differ by their ionization and return times and can be ordered according to their travel times. Figure 11.2 shows examples of ionization times that, within the simple-man model, lead to a finite number or to infinitely many returns. The decomposition (11.9) of the S-matrix element in terms of the contributions of the various quantum orbits shows that all

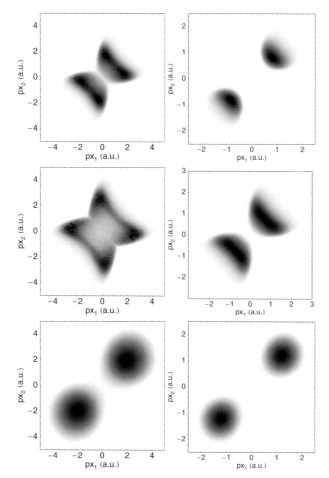

Fig. 11.5. Density plots of the distribution of the momentum components $p_{\parallel,1}$ and $p_{\parallel,2}$ parallel to the laser field, calculated from the S-matrix element (11.3) and integrated over the transverse–momentum components. The *two upper rows* are for the Coulomb interaction (11.16), the *bottom one* is for the contact interaction (11.17). The *left-hand* (*right-hand*) column is for the ionization energy of neon (argon). For neon, the laser intensities are 6, 10, and $6 \times 10^{14}\,\mathrm{W/cm^2}$ from top to bottom, for argon they are 2, 3, and $2 \times 10^{14}\,\mathrm{W/cm^2}$. The laser frequency corresponds to $\hbar\omega = 1.55\,\mathrm{eV}$. The gray scale is linear. From [32]

of them have to be added coherently. In most cases, the significance of the orbits will quickly decrease with increasing travel time. However, at certain intensities, constructive interference of many orbits occurs. For high-order ATI, experiments have shown [47–49] and theory has confirmed [48–51] that this effect is capable of raising the yields of the peaks in the rescattering plateau by up to one order of magnitude.

11.5.1 Channel Closings

For a zero-range binding potential, which does not support any excited state, such an enhancement due to constructive interference occurs when the electron can reach the ponderomotively upshifted continuum threshold from the atomic ground state by the absorption of an integer number of photons, that is, when

$$|E_0| + U_P = n\hbar\omega\,. \tag{11.21}$$

The effect is strongest for even n, less so for odd n. If the condition (11.21) is satisfied, the electron can be promoted into the continuum state with the lowest energy possible, so that no energy is left for its drift motion and its drift momentum is zero. In the simple-man picture, this corresponds to the middle panel of Fig. 11.2. Equation (11.21) defines the so-called channel closings: for an intensity just below the one given by (11.21), absorption of n photons suffices for ionization, just above a minimum of $n+1$ is required. This condition is of purely quantum-mechanical origin.

For a realistic binding potential, this picture changes only quantitatively. Strong resonant enhancements are observed when the electron is multiphoton resonant with a highly excited Rydberg state with binding energy E_R [47,48]. Very roughly, such a state can be envisioned in position space as the field-free orbit with the ponderomotive oscillation superimposed. The result is much the same as in Fig. 11.2: the electron has many recurring opportunities to rescatter. In order to account for the Rydberg state, in (11.21) one has to replace $|E_0|$ by $|E_0| - |E_R|$. In theory, the transition from a Coulomb potential to a short-range potential can readily be modeled in order to track the intensities where resonance occurs. For high-order harmonic generation, this has been done in [52]. The result is that, in the absence of Rydberg states, their role is taken over by the continuum threshold.

To the extent that NSDI is caused by the rescattering mechanism, as implied by the S-matrix element (11.3), one expects that such a constructive interference may also enhance NSDI. Figure 11.6 presents $(p_{1,\parallel}, p_{2,\parallel})$-momentum distributions just above and precisely at the 16-photon channel closing that have been calculated including the contributions of the orbits with longer travel times [53]. Indeed, right at the channel-closing intensity, an additional crescent-shaped contribution appears in the momentum distribution, which is centered on the diagonal. Note that the kinetic energy of the returning electron is smaller for the longer orbits. Hence, the term $[\mathbf{k} - e\mathbf{A}(t)]^2$ on the right-hand side of (11.12) is smaller, which leads to a smaller classically allowed domain. Figure 11.6 is calculated for the Coulomb potential (11.16); for the contact interaction (11.17) the crescent-shaped population does not change very much, but it is swamped by the contribution of the two shortest orbits.

For the case of quantum orbits with long travel time the system (11.6)–(11.8) may be solved analytically for arbitrary values of the Keldysh pa-

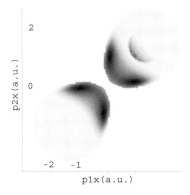

Fig. 11.6. Density plot of the distribution of the momentum components p_{1x}, p_{2x} ($\boldsymbol{p}_{i\perp} = 0$) parallel to the laser field for the case of NSDI of xenon by a Ti:Sa laser calculated in the SFA approach with Coulomb form factor (11.18) [30,31] including the contributions of trajectories with long travel times. The gray scale is linear. The *lightly shaded regions* indicate the classical domain, as defined in Sect. 11.3.3. In the first quadrant, the laser intensity corresponds to the closing of the ionization channel $n = 16$, see (11.21), where the contribution from the long orbits is maximal. The third quadrant is for the slightly higher intensity where $|E_{01}| + U_P = 16.1\omega$, and this contribution is no longer visible. ([The actual distributions are symmetric upon $(p_{1x}, p_{2x}) \to (-p_{1x}, -p_{2x})$). From [53]

rameter γ. Hence, the constructive interference at the channel closings can essentially be discussed in analytical terms [53].

Resonant–like enhancements of NSDI have also been observed in numerical simulations based on the solution of the two-electron time-dependent Schrödinger equation and have been traced to constructive interference of wavepackets launched by the ion at various times in the past [54]. In experiments, it was observed that NSDI decreases for laser pulses short enough to eliminate the contributions of the long orbits [55]. Like here, this was attributed to the fact that short pulses cut off the contributions of the long orbits. However, the explanation given was classical invoking Coulomb refocusing, and the effect is nonresonant.

11.5.2 Relation to Formal Scattering Theory

More precise investigation of the S-matrix element (11.3) shows that at the closing of an *even-n* ionization channel the first derivative of the cross sections of all rescattering processes (ATI, HHG, and NSDI) with respect to the final energy of the system diverges. At the closing of an *odd-n* channel, this is the case for the second derivative. This observation suggests that the resonant-like enhancements described above are the manifestation of the very general phenomenon known as Wigner threshold anomalies or Wigner cusps [56]. They are a consequence of the unitarity of the S matrix and occur at the precise moment when a new channel opens (or closes) as a function of some parameter. The threshold anomalies in the cross sections of elastic and inelastic processes were investigated rigorously by Baz [57] for short-range potentials and by Gailitis [58] for potentials with a Coulomb tail. However, these results are based on stationary scattering theory and cannot be directly applied to the present case.

The first observation of Wigner-type threshold peculiarities was made for a zero-range model atom in numerical simulations of ATI spectra [59] and for HHG [60]. Recently, in high-order ATI spectra based on the exact solution of the zero-range atom in a laser field, cusps at channel closings have been observed and interpreted in terms of Wigner threshold phenomena [61]. However, to the best of our knowledge, the strong-field analog of the Wigner–Baz–Gailitis theory of threshold phenomena does not yet exist.

11.6 Conclusions

Comparing S-matrix calculations as presented here with experimental data, in particular [15–17,43–45], has allowed us to identify inelastic rescattering as the dominant mechanism for nonsequential double ionization. The formalism of quantum orbits makes it possible to deal with various interaction potentials in the Born approximation and helps develop an appealing physical picture with high predictive power.

Acknowledgements

We acknowledge stimulating discussions with M. Dörr, C. Figueira de Morisson Faria, H. Rottke, and W. Sandner. This work was supported in part by Deutsche Forschungsgemeinschaft.

References

1. L.F. DiMauro, P. Agostini: Adv. At., Mol., Opt. Phys. **35**, 79 (1995)
2. M. Protopapas, C.H. Keitel, P.L. Knight: Rep. Prog. Phys. **60**, 389 (1997)
3. N.B. Delone, V.P. Krainov: Phys.–Uspekhi **41**, 469 (1998)
4. C.J. Joachain, M. Dörr, N. Kylstra: Adv. At., Mol., Opt. Phys. **42**, 225 (2000)
5. M.Y. Kuchiev: Pis'ma Zh. Eksp. Teor. Fiz. **45**, 319 (1987) [JETP Lett. **45**, 404 (1987)]
6. P.B. Corkum: Phys. Rev. Lett. **71**, 1994 (1993)
7. K.C. Kulander, K.J. Schafer, K.L. Krause: *Super-Intense Laser-Atom Physics*, Vol. 316 of NATO Advanced Study Institute Series B: Physics, ed. by B. Piraux, A. L'Huillier, K. Rzążewski (Plenum, New York, 1993) p. 95
8. W. Becker, F. Grasbon, R. Kopold, D.B. Milošević, G.G. Paulus, H. Walther: Adv. At., Mol., Opt. Phys. **48**, 35 (2002)
9. M. Lewenstein, P. Balcou, M.Y. Ivanov, P.B. Corkum: Phys. Rev. A **49**, 2117 (1994)

10. M. Lewenstein, K.C. Kulander, K.J. Schafer, P.H. Bucksbaum: Phys. Rev. A **51**, 1495 (1995)
11. R. Kopold, W. Becker, M. Kleber: Opt. Commun. **179**, 39 (2000)
12. P. Salières, B. Carré, L. Le Déroff, F. Grasbon, G.G. Paulus, H. Walther, R. Kopold, W. Becker, D.B. Milošević, A. Sanpera, M. Lewenstein: Science **292**, 902 (2001)
13. A. L'Huillier, L.A. Lompré, G. Mainfray, C. Manus: Phys. Rev. A **27**, 2503 (1983)
14. B. Walker, B. Sheehy, L.F. DiMauro, P. Agostini, K.J. Schafer, K.C. Kulander: Phys. Rev. Lett. **73**, 1227 (1994)
15. T. Weber, M. Weckenbrock, A. Staudte, L. Spielberger, O. Jagutzki, V. Mergel, F. Afaneh, G. Urbasch, M. Vollmer, H. Giessen, R. Dörner: Phys. Rev. Lett. **84**, 443 (2000)
16. R. Moshammer, B. Feuerstein, W. Schmitt, A. Dorn, C. D. Schröter, J. Ullrich, H. Rottke, C. Trump, M. Wittmann, G. Korn, K. Hoffmann, W. Sandner: Phys. Rev. Lett. **84**, 447 (2000)
17. T. Weber, H. Giessen, M. Weckenbrock, G. Urbasch, A. Staudte, L. Spielberger, O. Jagutzki, V. Mergel, M. Vollmer, R. Dörner: Nature (London) **404**, 608 (2000)
18. B. Witzel, M.A. Papadogiannis, D. Charalambidis: Phys. Rev. Lett. **85**, 2268 (2000)
19. R. Lafon, J.L. Chaloupka, B. Sheehy, P.M. Paul, P. Agostini, K.C. Kulander, L.F. DiMauro: Phys. Rev. Lett. **86**, 276 (2001)
20. G.D. Gillen, M.A. Walker, L.D. Van Woerkom: Phys. Rev. A **64**, 043413 (2001)
21. E.R. Peterson, P.H. Bucksbaum: Phys. Rev. A **64**, 053405 (2001)
22. W. Becker, S. Long, J.K. McIver: Phys. Rev. A **50**, 1540 (1994)
23. M.Y. Kuchiev: J. Phys. B **28**, 5093 (1995)
24. A. Becker, F.H.M. Faisal: J. Phys. B **29**, L197 (1996); Phys. Rev. A **59**, R1742 (1999); ibid. R3182 (1999)
25. H.W. van der Hart, K. Burnett: Phys. Rev. A **62**, 013407 (2000)
26. H.W. van der Hart: J. Phys. B **33**, L699 (2000)
27. J.B. Watson, A. Sanpera, D.G. Lappas, P.L. Knight, K. Burnett: Phys. Rev. Lett. **69**, 1884 (1997)
28. A. Becker, F.H.M. Faisal: Phys. Rev. Lett. **84**, 3546 (2000)
29. R. Kopold, W. Becker, H. Rottke, W. Sandner: Phys. Rev. Lett. **85**, 3781 (2000)
30. S.V. Popruzhenko, S.P. Goreslavskii: J. Phys. B **34**, L239 (2001)
31. S.P. Goreslavskii, S.V. Popruzhenko: Opt. Express **8**, 395 (2001)
32. S.P. Goreslavskii, S.V. Popruzhenko, R. Kopold, W. Becker: Phys. Rev. A **64**, 053402 (2001)
33. J. Chen, J. Liu, L.B. Fu, W.M. Zheng: Phys. Rev. A **63**, 011404(R) (2001); L.-B. Fu, J. Liu, J. Chen, S.-G. Chen: Phys. Rev. A **63**, 043416 (2001)
34. C. Figueira de Morisson Faria, H. Schomerus, W. Becker: Phys. Rev. A **66**, 043413 (2002); C. Figueira de Morisson Faria, W. Becker: Laser Phys. **13**, xxx (2003).
35. R. Kopold, D.B. Milošević, W. Becker: Phys. Rev. Lett. **84**, 3831 (2000)
36. S.P. Goreslavskii, S.V. Popruzhenko: Zh. Eksp. Teor. Fiz. **117**, 895 (2000) [JETP **90**, 778 (2000)]
37. G.G. Paulus, W. Becker, H. Walther: Phys. Rev. A **52**, 4043 (1995)

38. B. Feuerstein, R. Moshammer, J. Ullrich: J. Phys. B **33**, L823 (2000)
39. K. Sacha, B. Eckhardt: Phys. Rev. A **63**, 043414 (2001)
40. S.P. Goreslavski, S.V. Popruzhenko: J. Phys. B **32**, L531 (2000)
41. L.D. Landau, E.M. Lifshitz: *Quantum Mechanics (Nonrelativistic Ttheory)*, 3rd edn. (Pergamon Press, 1977) § 148
42. R. Kopold: Atomare Ionisationsdynamik in starken Laserfeldern. Doctoral dissertation, Technische Universität München (2001)
43. M. Weckenbrock, M. Hattass, A. Czasch, O. Jagutzki, L. Schmidt, T. Weber, H. Roskos, T. Löffler, M. Thomson, R. Dörner: J. Phys. B **34**, L449 (2001)
44. B. Feuerstein, R. Moshammer, D. Fischer, A. Dorn, C.D. Schröter, J. Deipenwisch, J.R. Crespo Lopez-Urrutia, C. Höhr, P. Neumayer, J. Ullrich, H. Rottke, C. Trump, M. Wittmann, G. Korn, W. Sandner: Phys. Rev. Lett. **87**, 043003 (2001)
45. R. Moshammer, B. Feuerstein, J. Crespo López-Urrutia, J. Deipenwisch, A. Dorn, D. Fischer, C. Höhr, P. Neumayer, C.D. Schröter, J. Ullrich, H. Rottke, C. Trump, M. Wittmann, G. Korn, W. Sandner: Phys. Rev. A **65**, 035401 (2002)
46. L.V. Keldysh: Zh. Eksp. Teor. Fiz. **47**, 1945 (1964) [Sov. Phys. JETP **22**, 1307 (1964)]; F.H.M. Faisal: J. Phys. B **6**, L98 (1973); H.R. Reiss: Phys. Rev. A **22**, 1786 (1980)
47. M.P. Hertlein, P.H. Bucksbaum, H.G. Muller: J. Phys. B **30**, L197 (1997); P. Hansch, M.A. Walker, L.D. Van Woerkom: Phys. Rev. A **55**, R2535 (1997)
48. M.J. Nandor, M.A. Walker, L.D. Van Woerkom, H.G. Muller: Phys. Rev. A **60**, R1771 (1999)
49. G.G. Paulus, F. Grasbon, H. Walther, R. Kopold, W. Becker: Phys. Rev. A **64**, 021401(R) (2001)
50. H.G. Muller, F.C. Kooiman: Phys. Rev. Lett. **81**, 1207 (1998)
51. R. Kopold, W. Becker, M. Kleber, G.G. Paulus: J. Phys. B **35**, 217 (2002)
52. C. Figueira de Morisson Faria, R. Kopold, W. Becker, J.M. Rost: Phys. Rev. A **65**, 023404 (2002)
53. S.V. Popruzhenko, P.A. Korneev, S.P. Goreslavski, W. Becker: Phys. Rev. Lett. **89**, 023001 (2002)
54. H.G. Muller: Opt. Express **8**, 425 (2001)
55. V.R. Bhardwaj, S.A. Aseyev, M. Mehendale, G.L. Yudin, D.M. Villeneuve, D.M. Rayner, M.Y. Ivanov, P.B. Corkum: Phys. Rev. Lett. **86**, 3522 (2001); G.L. Yudin, M.Y. Ivanov: Phys. Rev. A **63**, 033404 (2001)
56. E.P. Wigner: Phys. Rev. **73**, 1002 (1948)
57. A.I. Baz: Zh. Eksp. Teor. Fiz. **33**, 923 (1957) [Sov. Phys. JETP **7**, 709 (1958)]
58. M. Gailitis: Zh. Eksp. Teor. Fiz. **44**, 1974 (1963) [Sov. Phys. JETP **17**, 1328 (1963)]
59. F.H.M. Faisal, P. Scanzano: Phys. Rev. Lett. **68**, 2909 (1992)
60. W. Becker, S. Long, J.K. McIver: Phys. Rev. A **46**, R5334 (1992)
61. B. Borca, M.V. Frolov, N.L. Manakov, A.F. Starace: Phys. Rev. Lett. **88**, 193001 (2002)

12 Time-Dependent Density Functional Theory in Atomic Collisions

H.J. Lüdde

12.1 Introduction

Scattering processes leading to excitation or fragmentation of atomic or molecular systems are a source of information on various physical effects associated with the mutual Coulomb interaction of such many-particle systems. This chapter focuses on the discussion of a quantum–mechanical system (the electrons of the target) influenced by a classical environment (the projectile) that provides the energy that disturbs the electronic system [1]. The classical environment can, for instance, be realized by an intense laser field or an ion beam that imposes its time-dependence on the electronic subsystem and defines a typical timescale for the scattering process – femtoseconds for an electronic system exposed to a laser beam and attoseconds for heavy-particle collisions.

¿From a theoretical point of view it is the time-dependent many-electron problem that has to be solved for different external potentials representing the interaction with the classical environment. In practice, one often has to be content with a single-particle approximation providing effective single-particle equations that can be solved for any active electron at the price, however, that the true two-particle character of the electron–electron interaction has to be neglected.

It is the power of density functional theory (DFT) to provide a mathematical framework that allows one to *exactly* map the many-electron system onto a set of effective single-particle equations. This chapter gives an outline of the basic concepts behind time-dependent DFT based on a series of review articles [2–4] that document the activity in this field. For a comprehensive introduction to stationary DFT the monograph of Dreizler and Gross is specially recommended [5].

Although there is hardly an alternative to time-dependent DFT for a theoretical investigation of systems with *many* active electrons it is not always clear how to extract observables of the system to establish contact with experimental results. The second part of this chapter addresses this problem and gives an idea of its complexity.

A few applications of DFT for typical collisional situations are summarized at the end of this chapter. Atomic units are used.

12.2   Basic Concepts
of Time-Dependent Density-Functional Theory

The starting point for the theoretical description of a nonrelativistic electronic system is the time-dependent Schrödinger equation (TDSE)

$$i\partial_t \Psi(t) = \hat{H}(t)\Psi(t),\tag{12.1}$$

which determines the propagation in time of the N-particle system evolving from its initial state

$$\Psi(t_0) = \Psi_0.\tag{12.2}$$

Although $\Psi(t)$ fully describes the electronic system it is, of course, not observable. In a typical collision problem one usually measures the probabilities of finding some of the electrons in a given state or at a certain place in configuration space. These *inclusive probabilities* [6] can be expressed in terms of the *q-particle density* [7]

$$\gamma^q(\boldsymbol{x}_1,\dots\boldsymbol{x}_q,t) = \binom{N}{q}\int \mathrm{d}^4x_{q+1}\dots\mathrm{d}^4x_N\,|\Psi(\boldsymbol{x}_1,\dots\boldsymbol{x}_N,t)|^2,\tag{12.3}$$

where $\boldsymbol{x}_j = (\boldsymbol{r},s)$ denotes coordinates and spin of the j-th electron and d^4x_j indicates summation over spin and integration over space coordinates, respectively. With the definition (12.3) $\gamma^q\mathrm{d}^3r_1\dots\mathrm{d}^3r_q$ describes the inclusive probability of finding q electrons at given positions in configuration space, while the remaining $N-q$ electrons are not detected explicitly. As a special case one obtains for $q=1$ the spin-free one-particle density

$$n(\boldsymbol{r},t) = \sum_s \gamma^1(\boldsymbol{r},s,t),\tag{12.4}$$

which is the key for the understanding of DFT.

The total hamiltonian that enters the TDSE (12.1)

$$\hat{H}(t) = \hat{T} + \hat{W} + \hat{V}(t)\tag{12.5}$$

includes the kinetic energy, the mutual Coulomb repulsion between the electrons

$$\hat{T} = \sum_{j=1}^{N}\left(-\frac{1}{2}\nabla_j^2\right),$$

$$\hat{W} = \sum_{i<j=1}^{N}\frac{1}{|\boldsymbol{r}_i - \boldsymbol{r}_j|},\tag{12.6}$$

and the external potential

$$\hat{V}(t) = \sum_{j=1}^{N} v(\boldsymbol{r}_j,t),\tag{12.7}$$

which characterizes the geometry as well as the explicit time-dependence of the particular quantum system. This decomposition of the Hamiltonian into a universal, time-independent part (12.6) and a system-specific time-dependent external potential is a second important ingredient for the conceptual understanding of DFT.

Let us, thus, define a map $\mathcal{F} : v(\boldsymbol{r}, t) \rightarrow \Psi(t)$ by solving the TDSE for different external potentials but common initial state Ψ_0. As the one-particle density is uniquely determined by the time-dependent many-particle state $\Psi(t)$ this obviously defines a second map $\mathcal{G} : v(\boldsymbol{r}, t) \rightarrow n(\boldsymbol{r}, t)$. The foundation of DFT involves the proof that the map $\mathcal{G}$ is invertible, i.e., $\Psi(t)$ can be obtained as a functional of the one-particle density $\Psi(t) = \mathcal{F}\mathcal{G}^{-1} n(\boldsymbol{r}, t)$. As a consequence any observable that can be written as an expectation value $\langle \Psi(t) | \hat{O}(t) | \Psi(t) \rangle$ of a Hermitian operator would then be uniquely determined by $n(\boldsymbol{r}, t)$.

More precisely the conditions under which $\mathcal{G}$ is a 1-1 map between the external potential and the one-particle density are formulated by the Runge–Gross theorem [8]:

- For every single-particle potential $v(\boldsymbol{r}, t)$ that can be expanded into a Taylor series with respect to time around $t = t_0$ there exists a map $\mathcal{G} : v(\boldsymbol{r}, t) \rightarrow n(\boldsymbol{r}, t)$ by solving the TDSE with a fixed initial state Ψ_0. This map can be inverted up to an additive merely time-dependent function in the potential.

At this point a few remarks might be appropriate: (i) As the invertibility of the map between potentials and densities can only be shown with respect to a given initial state Ψ_0, the solution of the TDSE rigorously depends on the density *and* the initial state. Consequently, any observable of the system is a functional of n and Ψ_0. (ii) The potential as a functional of n is only determined up to a merely time-dependent function. This corresponds to an ambiguity in $\Psi(t)$ up to a time-dependent phase factor, which cancels out for any observable characterized by an operator that is free of time derivatives. (iii) The Runge–Gross theorem can be applied to any system characterized by a given interaction $\hat{W}$, in particular for $\hat{W} = 0$. This fact is used in the subsequent section, where a set of effective one-particle Schrödinger equations yielding the *exact* one-particle density is derived.

12.3 Time-Dependent Kohn–Sham Equations

The essence of DFT is the determination of the exact one-particle density n from which any many-particle observable can be derived. Let us assume we know this density and can calculate it from a set of N ficticious orbitals $\{\phi_j, j = 1, \ldots N\}$

$$n(\boldsymbol{r}, t) = \sum_{j=1}^{N} |\phi_j|^2 \,. \tag{12.8}$$

The ansatz (12.8) suggests that the orbitals fulfil a single-particle Schrödinger equation of the form

$$i\partial_t \phi_j(\boldsymbol{r}, t) = \left(\frac{-\nabla^2}{2} + v_{\mathrm{KS}}(\boldsymbol{r}, t) \right) \phi_j(\boldsymbol{r}, t) \, , j = 1, \ldots N \, , \tag{12.9}$$

where the existence of the single-particle potential v_{KS} is assumed. (This is discussed in the literature under the topic *v-representability* [9].) If this potential exists, the Runge–Gross theorem guarantees its uniqueness, i.e., there is up to an additive merely time-dependent function exactly one single-particle potential, which together with the time-dependent Kohn–Sham (KS) equations (12.9) reproduces the *exact* one-particle density n. Essentially, the KS equations represent an exact mapping of the N-electron problem onto a set of N single-particle problems. The crucial point is, however, that one does not know the KS potential explicitly.

12.3.1 Kohn–Sham Potential

Nevertheless, a few general properties of the KS potential can be established: (i) the KS potential is local in space in contrast to the exchange term in Hartree–Fock theory, (ii) by virtue of the Runge–Gross theorem the KS potential must be a unique functional of the exact density *for a given initial state* Ψ_0 and *for a given KS determinant* $\Phi_0 = \det(\phi_1, ..\phi_N)/\sqrt{N!}$. The latter condition can be largely simplified if we assume that the time-dependent electronic system evolves from a *non-degenerate ground state* of the initially undisturbed system, which, via stationary DFT, is fully determined by its corresponding density $n_0(\boldsymbol{r})$. In this case the KS potential is a unique functional of the density alone

$$v_{\mathrm{KS}}[n, \Psi_0, \Phi_0] = v_{\mathrm{KS}}[n](\boldsymbol{r}, t) \, . \tag{12.10}$$

Based on the experience with single-particle pictures one usually splits the KS potential into its classical parts – the external (Coulomb) interaction $v(\boldsymbol{r}, t)$ and the Hartree potential v_{H}, which includes the screening of the external potential due to the electrons – and a genuine quantum part v_{xc} the *exchange-correlation potential*

$$v_{\mathrm{KS}}[n](\boldsymbol{r}, t) = v(\boldsymbol{r}, t) + v_{\mathrm{H}}[n](\boldsymbol{r}, t) + v_{\mathrm{xc}}[n](\boldsymbol{r}, t) \tag{12.11}$$

$$v_{\mathrm{H}}[n](\boldsymbol{r}, t) = \int \mathrm{d}^3 r' \, \frac{n(\boldsymbol{r}', t)}{|\boldsymbol{r} - \boldsymbol{r}'|} \, .$$

However, as nothing is known about the exchange-correlation potential so far, one needs an additional property of the many-particle system to determine the KS potential explicitly.

The solution of the TDSE corresponds to a stationary point of the action integral

$$\mathcal{A} = \int_{t_0}^{t_1} \mathrm{d}t \, \langle \Psi(t) | i\partial_t - \hat{H}(t) | \Psi(t) \rangle \, , \tag{12.12}$$

which essentially should be a functional of n as Ψ is. The TDSE is then obtained by variation of $\mathcal{A}$ with respect to Ψ. Can one therefore conclude that the exact one-particle density is a stationary point of the action integral as well, thus $\delta\mathcal{A}/\delta n(\boldsymbol{r}, t) = 0$?

This is obviously not the case, as the Runge–Gross theorem predicts the functional $\Psi[n]$ only up to an arbitrary time phase. Because of the time-derivative in the Schrödinger operator $i\partial_t - \hat{H}(t)$ the action (12.12) is, in fact, a functional of n *and* the undetermined phase (for a detailed discussion on appropriate action functionals see [4] and [10]). Consequently, (12.12) is not useful as an additional source for the derivation of the KS potential.

A more pragmatic approach rests on the assumption that the time-dependence of the KS potential is only due to the time-dependence of the density, where the functional dependence on the density is taken from stationary DFT. This is called the *adiabatic* approximation. The exchange-correlation (xc) potential for a nondegenerate ground state is related to the corresponding energy on the basis of the Rayleigh–Ritz variational principle

$$v_{\mathrm{xc}}[n_0](\boldsymbol{r}) = \frac{\delta E_{\mathrm{xc}}[n]}{\delta n(\boldsymbol{r})}\Big|_{n(\boldsymbol{r})=n_0} \tag{12.13}$$
$$E_{\mathrm{xc}} = T - T_s + W - E_{\mathrm{H}},$$

where T and W are the expectation values of the interacting system, T_s denotes the kinetic energy, and E_{H} the Hartree energy of the KS system. The functional derivative (12.13) yields the functional dependence of the xc-potential on $n_0(\boldsymbol{r})$ for any particular approximation of the energy functional E_{xc}. Replacement of the ground–state density by $n(\boldsymbol{r}, t)$ is the essential idea of the adiabatic approximation.

The Local-Density Approximation

The most convenient ansatz for the xc-energy is based on the assumption that the energy functional can be locally approximated by that of the homogeneous electron gas. This *local-density approximation* (LDA) yields for the exchange part a simple analytical expression

$$v_{\mathrm{x}}^{\mathrm{LDA}} = -\frac{1}{\pi}(3\pi^2 n(\boldsymbol{r}, t))^{1/3}, \tag{12.14}$$

while the correlation part is given in terms of a parametrization [11]. The LDA (or ALDA for adiabatic LDA) should give reasonable results for systems in which the density is slowly varying both in space and time. However, it is important to note that the exchange potential (12.14) decreases exponentially, giving rise to an asymptotically incorrect compensation of the self-energy contained in the Hartree term. In atomic physics this can be cured by forcing the correct asymptotic decrease of the KS potential [12].

The Optimized-Potential Method

The problems associated with the self-energy effects can be solved more systematically on the basis of orbital–dependent density functionals. As the KS orbitals are also functionals of n one can express the xc-potential as a functional of the KS orbitals. Using the chain rule for functional differentiation one obtains

$$v_{\mathrm{xc}}(\boldsymbol{r}) = \frac{\delta E_{\mathrm{xc}}}{\delta n(\boldsymbol{r})} \qquad (12.15)$$

$$= \int \mathrm{d}^3 r' \, \frac{\delta v_{\mathrm{KS}}(\boldsymbol{r}')}{\delta n(\boldsymbol{r})} \int \mathrm{d}^3 r'' \sum_{k=1}^{N} \frac{\delta \phi_k^*(\boldsymbol{r}'')}{\delta v_{\mathrm{KS}}(\boldsymbol{r}')} \frac{\delta E_{\mathrm{xc}}[\boldsymbol{\phi}]}{\delta \phi_k^*(\boldsymbol{r}'')} + \mathrm{cc} \,,$$

with cc indicating the complex conjugate of the preceding expression. For the x-only term the exact energy functional is known

$$E_{\mathrm{x}} = -\frac{1}{2} \int \mathrm{d}^3 r \int \mathrm{d}^3 r' \sum_{k,l=1}^{N} \delta_{m_{s_k}, m_{s_l}} \frac{\phi_k^*(\boldsymbol{r})\phi_l(\boldsymbol{r}) \; \phi_l^*(\boldsymbol{r}')\phi_k(\boldsymbol{r}')}{|\boldsymbol{r} - \boldsymbol{r}'|} \,, \qquad (12.16)$$

which together with (12.15) yields an integral equation for the local exchange part of the KS potential. This scheme, originally introduced by Talman and Shadwick [13] as the *optimized (effective) potential method* (OEP → OPM) has been shown to be very successful in ground–state DFT [14] as well as in time-dependent systems. The local character of the x-potential allows for a compensation of the self-energy not only for the occupied ground–state orbitals – as is the case in Hartree–Fock theory – but also for the virtual orbitals that is of particular interest in time-dependent systems. The first attempts to include correlation [15] are very promising for ground–state problems but are yet too complex for time-dependent systems.

12.3.2 Numerical Solution of the Kohn–Sham Equations

The numerical solution of the time-dependent KS equations (12.9) is quite involved. It is the long range of the Coulomb interaction that requires (i) stable algorithms as one has to propagate atomic systems over large time scales and (ii) high accuracy to adequately account for the enormous delocalization of the electronic density. It is, thus, not surprising that the development of appropriate numerical methods is well represented in the literature. Two reviews on time-dependent methods for quantum dynamics [16] document the lively activity in this field over the last two decades. More recent developments of particular interest for ion–atom collisions include (i) lattice techniques discretizing the TDSE in configuration space [17], momentum space [18] or a combination thereof based on FFT algorithms [19], (ii) expansion methods relying on single-center [20] or two-center [21] basis sets, (iii) the hidden crossing method [22] valid for adiabatic collisions, and with a broad spectrum of applications (iv) the classical trajectory Monte Carlo method [23], which simulates the electronic system in terms of a statistical ensemble of classical point charges.

A little different in its philosophy is the *basis generator method* (BGM) [24], which allows optimization of the finite solution space in the sense that the individual basis functions dynamically adapt themselves to the momentary structure of the KS orbitals. In its present implementation [25] the BGM proves to be successful for collisions between ions and atoms [26], or ions and small molecules [27], and, very recently, for laser–assisted atomic collisions [28].

12.4 Extraction of Observables

The solution of the KS equations (12.9) is the one-particle density that describes the time propagation of the dynamical system to its final state. As the many-particle wave function is a functional of n any observable of the system must be a functional of the one-particle density as well. This is, however, a critical point in the application of time-dependent DFT as the functional dependence on n is only known for a few observables.

12.4.1 Exact Functionals

One observable that is accessible in all kinds of atomic–scattering systems irrespective of the complex individual collision processes involved is the energy-loss of the system

$$\mathcal{E}(t) - \mathcal{E}_0 = \int_{t_0}^{t} \dot{\mathcal{E}}(t')\, \mathrm{d}t' \,. \tag{12.17}$$

If one calculates the time derivative of the total energy taking into account that only the external potential is explicitly time dependent [4,29]

$$\dot{\mathcal{E}} = \frac{\mathrm{d}}{\mathrm{d}t}\langle \Psi | \hat{H}(t) | \Psi \rangle = \langle \Psi | \frac{\partial \hat{H}}{\partial t} - \mathrm{i}[\hat{H}, \hat{H}] | \Psi \rangle$$
$$= \langle \Psi | \frac{\partial}{\partial t} \hat{V}(t) | \Psi \rangle = \int n(\boldsymbol{r}, t)\dot{v}(\boldsymbol{r}, t)\, \mathrm{d}^3 r \,, \tag{12.18}$$

one obtains the energy loss as an explicit functional of the density

$$\mathcal{E}(t) - \mathcal{E}_0 = \int_{t_0}^{t} \int n(\boldsymbol{r}, t')\dot{v}(\boldsymbol{r}, t)\, \mathrm{d}^3 r\, \mathrm{d}t' \,. \tag{12.19}$$

For $t \to \infty$ the energy loss depends on the control parameters of the scattering system: (i) (b, E) impact parameter and energy of the impinging ion or (ii) (I, ω) intensity and color of the laser.

The expectation value of the total momentum of a dynamical system can either be written as a functional of the KS current

$$\mathcal{P}(t) = \langle \Psi | \hat{P} | \Psi \rangle = \frac{1}{i} \sum_{j=1}^{N} \langle \Psi | \nabla_j | \Psi \rangle$$

$$= \frac{1}{2i} \int (\nabla - \nabla') \, n(\boldsymbol{r}, \boldsymbol{r}', t)_{|\boldsymbol{r}=\boldsymbol{r}'} \, \mathrm{d}^3 r = \int \boldsymbol{j}(\boldsymbol{r}, t) \, \mathrm{d}^3 r \,, \qquad (12.20)$$

with

$$\boldsymbol{j}(\boldsymbol{r}, t) = \frac{1}{2i} \sum_{j=1}^{N} \{ \phi_j^*(\boldsymbol{r}, t) \nabla \phi_j(\boldsymbol{r}, t) - \phi_j(\boldsymbol{r}, t) \nabla \phi_j^*(\boldsymbol{r}, t) \} \qquad (12.21)$$

or with the aid of the continuity equation

$$\partial_t n(\boldsymbol{r}, t) = -\nabla \boldsymbol{j}(\boldsymbol{r}, t) \qquad (12.22)$$

and Green's theorem as a functional of the time derivative of the KS density [4]

$$\mathcal{P}(t) = \int \boldsymbol{r} \dot{n}(\boldsymbol{r}, t) \, \mathrm{d}^3 r = -\int \boldsymbol{r} (\nabla \boldsymbol{j}(\boldsymbol{r}, t)) \, \mathrm{d}^3 r = \int \boldsymbol{j}(\boldsymbol{r}, t) \, \mathrm{d}^3 r \,. \qquad (12.23)$$

In the case of a heavy-particle collision the longitudinal and transversal components of the total momentum are experimentally observable

$$\mathcal{P}_{\|}(t) = \int z \dot{n}(\boldsymbol{r}, t) \, \mathrm{d}^3 r$$

$$\mathcal{P}_{\perp}(t) = \int \sqrt{x^2 + y^2} \dot{n}(\boldsymbol{r}, t) \, \mathrm{d}^3 r \,, \qquad (12.24)$$

allowing the determination of the scattering angle (k_i denotes the momentum of the incoming projectile)

$$\tan \theta \approx \theta = \frac{\mathcal{P}_{\perp}}{\mathcal{P}_{\|} + k_i} \,, \qquad (12.25)$$

if the Coulomb repulsion between the nuclei is negligible.

Another set of global observables of the scattering system that can be expressed as functionals of the density under certain conditions are *net-probabilities* corresponding to the average number of electrons in a final state. If, for a heavy-ion collision, the density

$$n(\boldsymbol{r}, t) = n_{\mathrm{T}}(\boldsymbol{r}, t) + n_{\mathrm{P}}(\boldsymbol{r}, t) + n_{\mathrm{I}}(\boldsymbol{r}, t) \qquad (12.26)$$

can be split into finite regions around the target (T), the projectile (P), and an infinite complement (I) in such a way that the total volume (V=T+P+I) of the one-particle configuration space does not contain interference terms between these parts the particle number can be written as

$$N = \int_V n(\boldsymbol{r},t)\,\mathrm{d}^3r = \int_T n(\boldsymbol{r},t)\,\mathrm{d}^3r + \int_P n(\boldsymbol{r},t)\,\mathrm{d}^3r + \int_I n(\boldsymbol{r},t)\,\mathrm{d}^3r$$
$$= P_T^{\mathrm{net}} + P_P^{\mathrm{net}} + P_I^{\mathrm{net}}\,. \tag{12.27}$$

P_x^{net} corresponds to the average number of electrons that can be detected within the subvolume x. For the case of an initially neutral target and bare projectile these particle numbers can be interpreted as *net-ionization*, *net-capture*, and *net-loss* if one defines the electron–loss probability

$$P_{\mathrm{loss}}^{\mathrm{net}} = P_P^{\mathrm{net}} + P_I^{\mathrm{net}} = N - P_T^{\mathrm{net}}\,, \tag{12.28}$$

which in this situation corresponds to the average charge of the target.

12.4.2   Approximate Functionals

The situation becomes more involved if one is interested in less–global information about the scattering system. In (12.3) the q-particle density was introduced as a measure for the inclusive probability of finding q electrons at given positions in space while the remaining $N - q$ electrons are somewhere. Formally, these q-particle densities are related

$$\gamma^q(\boldsymbol{x}_1,\dots\boldsymbol{x}_q,t) = \frac{q+1}{N-q}\int \mathrm{d}^4x_{q+1}\,\gamma^{q+1}(\boldsymbol{x}_1,\dots\boldsymbol{x}_{q+1},t)\,, \tag{12.29}$$

which can be readily seen inserting the definition (12.3). The q-particle densities are normalized

$$\int_{V^q} \mathrm{d}^4x_1\dots\mathrm{d}^4x_q\,\gamma^q(\boldsymbol{x}_1,\dots\boldsymbol{x}_q,t) = \binom{N}{q}\,, \tag{12.30}$$

where V^q indicates that all q electron coordinates run over the entire volume V. If one splits the volume in which the many-particle state is analyzed into subvolumes – e.g., V=T+I for a two-electron system that can either be excited or ionized by a laser beam – the normalization integral yields [30]

$$1 = \int_{V^2} \gamma^2 = \int_{T^2} \gamma^2 + 2\int_{TI} \gamma^2 + \int_{I^2} \gamma^2$$
$$= P_{T^2} + P_{TI} + P_{I^2}\,. \tag{12.31}$$

The probabilities in this case correspond to the sum over elastic scattering, single and double excitation (P_{T^2}), single ionization with possible simultaneous excitation (P_{TI}), and double ionization (P_{P^2}), respectively. The extension to the more involved situation of a heavy-particle collision with an N-electron target is straightforward

$$1 = \int_{V^N} \gamma^N = \int_{(T+P+I)^N} \gamma^N = \sum_{\nu=0}^{N}\sum_{\mu=0}^{\nu}\binom{N}{\nu}\binom{\nu}{\mu}\int_{T^{N-\nu}P^{\nu-\mu}I^{\mu}} \gamma^N\,, \tag{12.32}$$

where the terms of the sum correspond to the probability of finding simultaneously $N - \nu$ electrons with the target, $\nu - \mu$ electrons with the projectile, and μ electrons ionized. It is interesting to note that the integration over the inifinite volume I can always be reduced to an integration over T or T and P. This is demonstrated for the simplest case (12.31), where the single–ionization probability can be expressed as

$$P_{\mathrm{TI}} = 2 \int_{\mathrm{TI}} \gamma^2 = 2 \int_{\mathrm{T}(V-\mathrm{T})} \gamma^2 = \int_{\mathrm{T}} \gamma - 2 \int_{\mathrm{T}^2} \gamma^2$$

$$= P_{\mathrm{T}}^{\mathrm{net}} - 2 \int_{\mathrm{T}^2} \gamma^2 \,, \tag{12.33}$$

using (12.29). Together with the knowledge of the initial state this type of probability allows analysis of the final charge state of the collision system. There is obviously a tremendous number of simultaneous processes that can be observed in a coincident experiment if N electrons become activated by an external time-dependent field (see the Appendix for a list of examples).

However, these inclusive probabilities depend on the q-particle density and there is not much known about the relation between the q-particle and one-particle density on a mathematically exact level. One, thus, has to evaluate these probabilities within the independent–particle picture [6], where the q-particle density is represented by a $q \times q$ determinant of the one-particle density matrix. Although this approach corresponds to neglecting the correlation in the final state it presents a way to calculate *any* kind of inclusive probability [31]. Nevertheless it is this part of the theory that has to be developed in order to make the power of time-dependent DFT fully available for the discussion of collisional systems.

12.5   Applications

12.5.1   Many-Electron Atoms in Strong Laser Fields

Atoms in strong laser fields have received increasing attention with the advent of strong femtosecond lasers [32]. New phenomena have been investigated in connection with multiphoton ionization: above threshold ionization, the stabilization of atoms with high laser intensity and frequency and the formation of harmonic spectra.

The external potential (12.7) for an atom in a linearly polarized laser pulse with shape function $f(t)$ is

$$v(\boldsymbol{r}, t) = -\frac{Q_t}{r} + E_0 f(t) \sin(\omega_0 t) z \,. \tag{12.34}$$

Solutions of the time-dependent KS equations (12.9) are compared within the ALDA and OPM approaches to the KS potential, where the numerical procedure relies on a finite difference method in cylindrical coordinates employing a Crank–Nicholson algorithm for the time integration [33]. Typical observables of the system are multiple ionization and the harmonic spectrum. The latter can be formulated as an *exact* functional of n by calculating the Fourier transform of the induced dipole moment

$$d(t) = \int z n(\mathbf{r}, t)\, \mathrm{d}^3 r\,. \tag{12.35}$$

Consequently comparison with experiment is very promising. Multiple ionization can, however, only be calculated within the x-only approach of the q-particle density which for an initial He ground state exposed to a laser pulse yields, according to (12.31)

$$P_1 = 2p(1 - p)$$
$$P_{1^2} = (1 - p)^2$$
$$p = \frac{1}{2} \int_T n(\mathbf{r}, t \to \infty)\, \mathrm{d}^3 r\,. \tag{12.36}$$

There is obviously a problem with the x-only approximation (12.36)

$$P_1 = 2\sqrt{P_{1^2}}\left(1 - \sqrt{P_{1^2}}\right), \tag{12.37}$$

as the relation between P_1 and P_{1^2} [30] always predicts a maximum of 0.5 for the single ionization while double ionization can become much larger.

As mentioned above this is so far the bottle-neck for applications based on time-dependent DFT: no matter how exact the density of the propagating system can be obtained [34] one depends upon approximate functionals for the evaluation of some of the observables.

12.5.2   Ion–Atom Collisions Involving Many Active Electrons

Modern experimental techniques like COLTRIMS (cold target recoil ion momentum spectroscopy) allow investigation of atomic collisions with high accuracy on a very detailed level. A considerable amount of theoretically unexplained experimental data have been collected over the past ten years [35]. In contrast to the experimental situation it was only very recently that a nonperturbative description of atomic collisions involving *many* active electrons became feasible (for a collection of references see [36]). With increasing numbers of electrons there is hardly an alternative to time-dependent DFT if one is tackling the many-particle problem in a systematic way. In this context it appears to be useful to investigate the influence of different approximations of the exact KS Hamiltonian on effects associated with the electronic interaction during the collision process.

For that purpose the KS potential (12.11) is rewritten

$$v_{\mathrm{KS}}[n](\mathbf{r}, t) = v(\mathbf{r}, t) + v_{\mathrm{ee}}[n](\mathbf{r}, t)\,, \tag{12.38}$$

where v_{ee} which includes the Hartree- and xc-potentials is decomposed into

$$v_{\mathrm{ee}}[n](\mathbf{r}, t) = v_{\mathrm{ee}}[n_0](\mathbf{r}) + \delta v_{\mathrm{ee}}[n](\mathbf{r}, t) \tag{12.39}$$

a stationary part $v_{ee}[n_0]$ including the potential of the undisturbed system and the response potential $\delta v_{ee}[n]$ depending on the time-dependent density.

- Within the no-response approach $\delta v_{ee} = 0$ one finds that the LDA approximation of the exchange potential notoriously overestimates the electron–loss process (net-capture and ionization) (12.28), whereas the OPM exchange yields accurate results [26,37]. This corresponds to the fact that a correct prediction of the first ionization potential depends on the correct treatment of the exchange potential. The incorrect asymptotic behavior of the LDA potential leads to additional artificial structures in the doubly differential cross section for inclusive single-electron emission [38].
- The inclusion of response effects becomes important with decreasing impact energy as the electrons have more time to adapt to the actual potential. In particular, q-fold ionization is considerably reduced by response effects as the binding energies of the residual electrons are increased during the collision [39]. q-fold capture might be reduced by the fact that the projectile charge decreases due to consecutive electron transfer [40].

These issues are discussed in closer detail in Chap. 24 of this book. So far it is difficult to judge the importance of correlation effects for the dynamical calculation. It certainly seems to be more important to include correlation in the evaluation of observables, which again requires the knowledge of the functional dependence of these observables on the one-particle density.

12.5.3 Fragmentation of Atomic Clusters in Collisions with Ions

The fragmentation of atomic clusters exposed to an external laser field or as a result of a heavy–particle collision are studied within the nonadiabatic molecular dynamics formalism. The theory combines the time-dependent LDA approach to the electronic motion with a molecular dynamics description for the classical paths of the cluster fragments. The relevant KS equations are either solved on a grid (for a review see [41]) or in a finite–basis expansion of the time-dependent KS orbitals [42]. The latter formalism has been successfully applied to collisions between ions and sodium clusters followed by electron transfer [43] and fragmentation processes [44].

12.6 Conclusion

For time-dependent systems with many active electrons DFT provides a realistic if not the only practicable approach to the quantum many-body problem. The basic theorems state that the many-particle TDSE can be mapped onto a set of single-particle equations from which the exact one-particle density can be calculated. Any observable of the system is, in principle, exactly related to the density in terms of density-functionals.

Approximations are, however, necessary due to the fact that (i) the single-particle (KS) potential including the many-electron effects is not known exactly and (ii) the functional dependence of the observables on the density can so far be formulated only for a few cases.

The structural simplicity of the time-dependent KS equations opens up the possibility for many applications in atomic and molecular physics that require a microscopic quantum theory but are far too complex for traditional methods.

Acknowledgements

The author gratefully acknowledges friendly and fruitful collaborations with Reiner Dreizler, Tom Kirchner, and Marko Horbatsch.

Appendix

A few examples of probabilities as functionals of the q-particle density are collected in this appendix.

- The q-fold ionization of a N-particle system exposed to a laser pulse is

$$
P_{\mathrm{I}q} = \binom{N}{q} \int_{\mathrm{T}^{N-q}\mathrm{I}^q} \gamma^N
$$
$$
= \sum_{\nu=0}^{q} (-1)^\nu \binom{N-q+\nu}{N-q} \int_{\mathrm{T}^{N-q+\nu}} \gamma^{N-q+\nu} . \tag{12.40}
$$

- Contrary to the q-fold ionization one defines the *inclusive* q-fold ionization, the probability of finding *at least* q electrons emitted to the continuum

$$
P_{\mathrm{I}q\Sigma} = \binom{N}{q} \int_{\mathrm{V}^{N-q}\mathrm{I}^q} \gamma^N = \sum_{\nu=0}^{q} (-1)^\nu \int_{\mathrm{T}^\nu} \gamma^\nu . \tag{12.41}
$$

- The transfer ionization in a collision between a bare ion and an initially neutral target can be calculated using (12.32): k-fold capture in coincidence with l-fold ionization is thus

$$
P_{\mathrm{P}k\mathrm{I}l} = \binom{N}{k+l}\binom{k+l}{l} \int_{\mathrm{T}^{N-k-l}\mathrm{P}k\mathrm{I}l} \gamma^N = \tag{12.42}
$$
$$
\sum_{\nu=0}^{l}\sum_{\mu=0}^{\nu} (-1)^\nu \frac{(N-l+\nu)!}{\mu!(\nu-\mu)!k!(N-k-l)!} \int_{\mathrm{T}^{N-k-l+\mu}\mathrm{P}k+\nu-\mu} \gamma^{N-l+\nu} .
$$

- Neutralization of the projectile in collisions between He$^+$ and a neutral target ((N+1)-electron system):

$$
P_{\mathrm{He}} = \int_{\mathrm{P}^2} \gamma^2 - 3\int_{\mathrm{P}^3} \gamma^3 + 6\int_{\mathrm{P}^4} \gamma^4 \mp \ldots \approx \int_{\mathrm{P}^2} \gamma^2 . \tag{12.43}
$$

The higher order terms correct the inclusive capture probability for the production of negative ions. These probabilities are, however small.

- Ionization of the projectile for the same scattering system:

$$P_{\mathrm{He}^{2+}} = 1 - \int_{\mathrm{P}} \gamma + \int_{\mathrm{P}^2} \gamma^2 - \int_{\mathrm{P}^3} \gamma^3 \pm \ldots \approx 1 - \int_{\mathrm{P}} \gamma + \int_{\mathrm{P}^2} \gamma^2 . \tag{12.44}$$

All these probabilities can only be evaluated explicitly within the independent particle picture, where the q-particle density is given in terms of the one-particle density matrix elements

$$\gamma^q(x_1, \ldots x_q) = \frac{1}{q!} \begin{vmatrix} \gamma(x_1, x_1) & \gamma(x_1, x_2) & \ldots & \gamma(x_1, x_q) \\ \vdots & \vdots & \vdots & \vdots \\ \gamma(x_q, x_1) & \gamma(x_q, x_2) & \ldots & \gamma(x_q, x_q) \end{vmatrix} . \tag{12.45}$$

References

1. J.S. Briggs and J.M. Rost: Eur. Phys. J. D **10**, 311 (2000)
2. E.K.U. Gross, J.F. Dobson, and M. Petersilka: Density Functional Theory of Time-Dependent Phenomena. In: *Topics in Current Chemistry 181*, ed. by R.F. Nalewajski (Springer, Berlin, Heidelberg, New York 1996)
3. K. Burke and E.K.U. Gross: A Guided Tour of Time-Dependent Density Functional Theory. In: *Density Functionals: Theory and Applications, Proceedings of the Tenth Chris Engelbrecht Summer School in Theoretical Physics* ed. by D. Joubert (Springer, Berlin, Heidelberg, New York 1997)
4. N.T. Maitra, K. Burke, H. Appel, E.K.U. Gross, and R. van Leeuwen: Ten Topical Questions in Time-Dependent Density Functional Theory. In: *Review in Modern Quantum Chemistry: A Celebration of the Contributions of R.G. Parr* ed. by K.D. Sen (World Scientific, Singapore 2001)
5. R.M. Dreizler and E.K.U. Gross: *Density Functional Theory* (Springer, Berlin, Heidelberg, New York 1990)
6. H.J. Lüdde and R.M. Dreizler: J. Phys. B **18**, 107 (1985)
7. P.-O. Löwdin: Phys. Rev. **97**, 1474 (1955)
8. E. Runge and E.K.U. Gross: Phys. Rev. Lett. **52**, 997 (1984)
9. M. Levy: Phys. Rev. A **26**, 1200 (1982); E.H. Lieb: Density Functionals for Coulomb Systems. In: *Density Functional Methods in Physics* ed. by R.M. Dreizler and J. da Providencia (Plenum, New York 1985)
10. R. van Leeuwen: Int. J. Mod. Phys. B **15**, 1969 (2001)
11. S.H. Vosko, L. Wilk, and M. Nusair: Can. J. Phys. **58**, 1200 (1980); J.P. Perdew and Y. Wang: Phys. Rev. B **45**, 13244 (1992)
12. R. Latter: Phys. Rev. **99**, 510 (1955); J.P. Perdew: Chem. Phys. Lett. **64**, 127 (1979); J.P. Perdew and A. Zunger: Phys. Rev. B **23**, 5048 (1981)
13. J.D. Talman and W.F. Shadwick: Phys. Rev. A **14**, 36 (1976)
14. E. Engel and R.M. Dreizler: J. Comp. Chem. **20**, 31 (1999)
15. A. Facco Bonetti, E. Engel, R.N. Schmid, and R.M. Dreizler: Phys. Rev. Lett. **86**, 2241 (2001)

16. K.C. Kulander (editor): *Thematic Issue on Time-Dependent Methods for Quantum Dynamics* Comp. Phys. Commmun. **63**, No: 1–3 (1991); T. Kirchner, H.J. Lüdde, O.J. Kroneisen, and R.M. Dreizler: Nucl. Instrum. Methods B **154**, 46 (1999)
17. D.R. Schultz, M.R. Strayer, and J.C. Wells: Phys. Rev. Lett. **82**, 3976 (1999)
18. D.C. Ionescu and A. Belkacem: Phys. Scr. **T80**, 128 (1999)
19. M. Chassid and M. Horbatsch: J. Phys. B **31**, 515 (1998); Phys. Rev. A **66**, 012714 (2002); A. Kolakowska, M. Pindzola, F. Robicheaux, D.R. Schultz, and J.C. Wells: Phys. Rev. A **58**, 2872 (1998); A. Kolakowska, M. Pindzola, and D.R. Schultz: Phys. Rev. A **59**, 3588 (1999)
20. B. Pons, Phys. Rev. A **63**, 012704 (2001); ibid. **64**, 019904 (2001); K. Sakimoto, J. Phys. B **33**, 5165 (2000); A. Igarashi, S. Nakazaki, and A. Ohsaki: Phys. Rev. A **61**, 062712 (2000); X.M. Tong, D. Kato, T. Watanabe, and S. Ohtani: Phys. Rev. A **62**, 052701 (2000)
21. J. Fu, M.J. Fitzpatrick, J.F. Reading, and R. Gayet: J. Phys. B **34**, 15 (2001); E.Y. Sidky, C. Illescas, and C.D. Lin: Phys. Rev. Lett. **85**, 1634 (2000); N. Toshima: Phys. Rev. A **59**, 1981 (1999)
22. E.A. Solov'ev: Sov. Phys. Usp. **32**, 228 (1989); M. Pieksma and S.Y. Ovchinnikov: J. Phys. B **27**, 4573 (1994)
23. C.L. Cocke and R.E. Olson: Phys. Rep. **205**, 163 (1991); M. Horbatsch: Phys. Rev. A **49**, 4556 (1994); C. Illescas and A. Riera: Phys. Rev. A **60**, 4546 (1999)
24. H.J. Lüdde, A. Henne, T. Kirchner, and R.M. Dreizler: J. Phys. B **29**, 4423 (1996)
25. O.J. Kroneisen, H.J. Lüdde, T. Kirchner, and R.M. Dreizler: J. Phys. A **32**, 2141 (1999)
26. T. Kirchner, L. Gulyás, H.J. Lüdde, A. Henne, E. Engel, R.M. Dreizler: Phys. Rev. Lett. **79**, 1658 (1997)
27. O.J. Kroneisen, A. Achenbach, H.J. Lüdde, and R.M. Dreizler: Model Potential Approach to Collisions Between Ions and Molecular Hydrogen. In: *XXII International Conference on Photonic, Electronic, and Atomic Collisions. Abstracts of Contributed Papers* ed. by S. Datz, M.E. Bannister, H.F. Krause, L.H. Saddiq, D. Schultz, and C.R. Vane (Rinton Press Princeton 2001)
28. T. Kirchner: Phys. Rev. Lett. **89**, 093203 (2002)
29. P. Hessler, J. Park, and K. Burke: Phys. Rev. Lett. **82**, 378 (1999)
30. H.J. Lüdde and R.M. Dreizler: J. Phys. B **16**, 3973 (1983)
31. P. Kürpick, H.J. Lüdde, W.D. Sepp, and B. Fricke: Z. Phys. D **25**, 17 (1992); H.J. Lüdde, A. Macias, F. Martin, A. Riera, and J.L. Sanz: J. Phys. B **28**, 4101 (1995)
32. W. Becker and M.V. Fedorov (editors): *Focus Issue on Laser-Induced Multiple Ionization* Opt. Express. **8**, No. 7 (2001)
33. M. Petersilka and E.K.U. Gross: Laser Phys. **9**, 105 (1999); M. Lein, E.K.U. Gross, and V. Engel: Phys. Rev. Lett. **85**, 4707 (2000); J. Phys. B **33**, 433 (2000); Phys. Rev. A **64**, 023406 (2001); M. Lein, V. Engel, and E.K.U. Gross: Opt. Express **8**, 411 (2001)
34. D.G. Lappas and R. van Leeuwen: J. Phys. B **31**, L249 (1998)
35. J. Ullrich, R. Moshammer, R. Dörner, O. Jagutzki, V. Mergel, H. Schmidt-Böcking, and L. Spielberger: J. Phys. B **30**, 2917 (1997); R. Dörner, V. Mergel, O. Jagutzki, L. Spielberger, J. Ullrich, R. Moshammer, and H. Schmidt-Böcking: Phys. Rep. **330**, 95 (2000)

36. H.J. Lüdde, T. Kirchner, and M. Horbatsch: In: Quantum Mechanical Treatment of Ion Collisions with Many-Electron Atoms. In: *XXII International Conference on Photonic, Electronic, and Atomic Collisions. Invited Papers* ed. by J. Burgdörfer, J. Cohen, S. Datz, and C.R. Vane (Rinton Press, Princeton 2002) p. 708

37. T. Kirchner, L. Gulyás, H.J. Lüdde, E. Engel, and R.M. Dreizler: Phys. Rev. A **58**, 2063 (1998)

38. L. Gulyás, T. Kirchner, T. Shirai, and M. Horbatsch: Phys. Rev. A **62**, 022702 (2000)

39. T. Kirchner, M. Horbatsch, H.J. Lüdde, and R.M. Dreizler: Phys. Rev. A **62**, 042704 (2000)

40. T. Kirchner, M. Horbatsch, and H.J. Lüdde: Phys. Rev. A **64**, 012711 (2001)

41. F. Calvayrac, P.G. Reinhard, E. Suraud, and C. Ullrich: Phys. Rep. A **337**, 493 (2000)

42. U. Saalmann and R. Schmidt: Z. Phys. D **38**, 153 (1996)

43. O. Knospe, J. Jellinek, U. Saalmann, and R. Schmidt: Eur. Phys. J. D **5**, 1 (1999); Phys. Rev. A **61**, 022715 (1999); Z. Roller-Lutz, Y. Wang, H.O. Lutz, U. Saalmann, and R. Schmidt: Phys. Rev. A **59**, R2555 (1999)

44. U. Saalmann and R. Schmidt: Phys. Rev. Lett. **80**, 3213 (1998); T. Kunert and R. Schmidt: Phys. Rev. Lett. **86**, 5258 (2001)

13 Electronic Collisions in Correlated Systems: From the Atomic to the Thermodynamic Limit

J. Berakdar

13.1 Introduction

This chapter gives a brief overview on recent advances in the treatment of nonrelativistic electronic collisions in finite and extended systems[1]. A proper description of electronic collisions [1–4] is a prerequisite for the understanding of a variety of material properties. Emphasis is put on analytical concepts that unravel common features and differences between scattering events in finite few-body (atomic) systems and large, extended systems (molecules, metal clusters, and solid surfaces). The properties of few-body Coulomb scattering states are discussed for two-, three-, four- and N- particle systems. For large, finite systems the concept of Green's function is utilized as a powerful tool for the description of electronic excitations as well as for the study of collective and thermodynamic properties. For the description of highly excited electronic states in solids and surfaces, the Green's function method, developed in field theory, is used. When available, the theoretical models are contrasted with experimental findings.

13.2 Two Charged-Particle Scattering

For an introduction, we consider the nonrelativistic scattering of two charged particles with charges z_1 and z_2. The Schrödinger equation for the wave function describing the relative motion of the particles is[2]:

$$\left[\Delta - \frac{2\mu z_1 z_2}{r} + k^2 \right] \Psi_k(r) = 0 , \qquad (13.1)$$

where r is the two-particle relative coordinate and k is the momentum conjugate to r. The energy of the relative motion is $E = k^2/2\mu$ and μ is the reduced mass of the particles. The effect of the Coulomb potential is exposed by making the ansatz $\Psi_k(r) = e^{ik \cdot r} \bar{\Psi}_k(r)$. The asymptotic behavior of (13.1) is unravelled by neglecting terms that fall off at large r faster than the Coulomb potential, which leads to

$$\left[ik \cdot \nabla - \frac{\mu z_1 z_2}{r} \right] \bar{\Psi}_k(r) = 0 . \qquad (13.2)$$

---

[1] This work is dedicated to John S. Briggs on the occasion of his sixtieth birthday.
[2] Atomic units are used unless otherwise stated.

This equation admits a solution of the form $\bar{\Psi} = \exp(i\phi)$, where $\phi_{\boldsymbol{k}}^{\pm}(\boldsymbol{r}) = \pm\frac{z_1 z_2 \mu}{k} \ln k(r \mp \hat{\boldsymbol{k}} \cdot \boldsymbol{r})$. The factor $z_1 z_2 \mu/k = z_1 z_2/v$ (v is the relative velocity) is called the *Sommerfeld parameter* and is a measure for the strength of the interaction potential. The key point is that the natural coordinate for Coulomb scattering is the so-called *parabolic* coordinate $\xi^{\pm} = r \mp \hat{\boldsymbol{k}} \cdot \boldsymbol{r}$, where the signs $+$ or $-$ corresponds to incoming- or outgoing-wave boundary conditions, respectively.

13.3   Three-Particle Coulomb Continuum States

The three-body Coulomb–scattering problem is still receiving much attention [5–15]. This is because, in contrast to the two-body problem, an exact derivation of the three-body quantum states is not possible. Only under certain (asymptotic) assumptions can analytical solutions be obtained that contain some general features of the two-body scattering, such as the characteristic asymptotic phases. As in the preceding section the center-of-mass motion of a three-body system can be factored out. The internal motion of the three charged particles with masses m_i and charges z_i; $i \in 1, 2, 3$ can be described by one set of the three Jacobi coordinates $(\boldsymbol{r}_{ij}, \boldsymbol{R}_k)$; $i, j, k \in \{1, 2, 3\}$; $\epsilon_{ijk} \neq 0$; $j > i$. Here, $\boldsymbol{r}_{ij}$ is the relative internal separation of the pair ij, and $\boldsymbol{R}_k$ is the position of the third particle (k) with respect to the center-of-mass of the pair ij. The scalar product

$$(\boldsymbol{r}_{ij}, \boldsymbol{R}_k) \cdot \begin{pmatrix} \boldsymbol{k}_{ij} \\ \boldsymbol{K}_k \end{pmatrix}$$

is invariant for all three sets of Jacobi coordinates. The kinetic energy operator H_0 is then diagonal and reads

$$H_0 = -\Delta_{\boldsymbol{r}_{ij}}/(2\mu_{ij}) - \Delta_{\boldsymbol{R}_k}/(2\mu_k), \quad \forall (\boldsymbol{r}_{ij}, \boldsymbol{R}_k),$$

where $\mu_k = m_k(m_i + m_j)/(m_1 + m_2 + m_3)$ and $\mu_{ij} = m_i m_j/(m_i + m_j)$; $i, j \in \{1, 2, 3\}$; $j > i$ are reduced masses. The eigenenergy of H_0 is $E_0 = k_{ij}^2/(2\mu_{ij}) + K_k^2/(2\mu_k)$, $\forall (\boldsymbol{k}_{ij}, \boldsymbol{K}_k)$. Defining $z_{ij} = z_i z_j$ the time-independent Schrödinger equation of the system reads

$$\left[H_0 + \sum_{\substack{i,j \\ j>i}}^{3} \frac{z_{ij}}{r_{ij}} - E \right] \langle \boldsymbol{r}_{kl}, \boldsymbol{R}_m | \Psi_{\boldsymbol{k}_{kl}, \boldsymbol{K}_m} \rangle = 0. \tag{13.3}$$

The relative coordinates r_{ij} occurring in the Coulomb potentials have to be expressed in terms of the appropriately chosen Jacobi-coordinate set $(\boldsymbol{r}_{kl}, \boldsymbol{R}_m)$.

13.3.1   Coulomb Three-Body Scattering in Parabolic Coordinates

Asymptotic scattering solutions of (13.3) for large interparticle distances r_{ij} have been considered in [5,6,8,9,16]. The derivation of general scattering solutions of the Schrödinger equation is a delicate task. One approach is to consider the three-body system as the subsum of three noninteracting two-body subsystems [11]. Since we know the appropriate coordinates for each of these two-body subsystems (the parabolic coordinates), we formulate the three-body problem in a similar coordinate frame with

$$\{\xi_k^{\mp} = r_{ij} \pm \hat{\boldsymbol{k}}_{ij} \cdot \boldsymbol{r}_{ij}\}, \quad \epsilon_{ijk} \neq 0; \quad j > i, \tag{13.4}$$

where $\hat{\boldsymbol{k}}_{ij}$ denote the directions of the momenta $\boldsymbol{k}_{ij}$. Since we are dealing with a six-dimensional problem, three other independent coordinates are needed in addition to (13.4). To make a reasonable choice for these remaining coordinates we remark that, usually, the momenta $\boldsymbol{k}_{ij}$ are determined experimentally, i.e. they can be considered as the laboratory-fixed coordinates. In fact it can be shown that the coordinates (13.4) are related to the Euler angles. Thus, it is advantageous to choose body-fixed coordinates. Those are conveniently chosen as

$$\{\xi_k = r_{ij}\}, \quad \epsilon_{ijk} \neq 0; \quad j > i. \tag{13.5}$$

Upon a mathematical analysis it can be shown that the coordinates (13.4), (13.5) are linearly independent [11] except for some singular points where the Jacobi determinant vanishes. The main task is now to rewrite the three-body Hamiltonian in the coordinates (13.4) and (13.5). To this end, it is useful to factor out the trivial plane-wave part by making the ansatz

$$\Psi_{\boldsymbol{k}_{ij},\boldsymbol{K}_k}(\boldsymbol{r}_{ij}, \boldsymbol{R}_k) = N \exp(\mathrm{i}\boldsymbol{r}_{ij} \cdot \boldsymbol{k}_{ij} + \mathrm{i}\boldsymbol{R}_k \cdot \boldsymbol{K}_k)\overline{\Psi}_{\boldsymbol{k}_{ij},\boldsymbol{K}_k}(\boldsymbol{r}_{ij}, \boldsymbol{R}_k). \tag{13.6}$$

Inserting the ansatz (13.6) into the Schrödinger equation (13.3) leads to

$$\left[\frac{1}{\mu_{ij}}\Delta_{\boldsymbol{r}_{ij}} + \frac{1}{\mu_k}\Delta_{\boldsymbol{R}_k} + 2\mathrm{i}\left(\frac{1}{\mu_{ij}}\boldsymbol{k}_{ij} \cdot \boldsymbol{\nabla}_{\boldsymbol{r}_{ij}} + \frac{1}{\mu_k}\boldsymbol{K}_k \cdot \boldsymbol{\nabla}_{\boldsymbol{R}_k} \right) - 2\sum_{\substack{m,n \\ n>m}}^{3} \frac{Z_{ij}}{r_{mn}} \right]$$
$$\times \overline{\Psi}(\boldsymbol{r}_{ij}, \boldsymbol{R}_k) = 0. \tag{13.7}$$

In terms of the coordinates (13.4) and (13.5), (13.7) casts

$$H\overline{\Psi}_{\boldsymbol{k}_{ij},\boldsymbol{K}_k}(\xi_1, \dots, \xi_6)$$
$$= [\,H_{\mathrm{par}} + H_{\mathrm{in}} + H_{\mathrm{mix}}]\,\overline{\Psi}_{\boldsymbol{k}_{ij},\boldsymbol{K}_k}(\xi_1, \dots, \xi_6) = 0. \tag{13.8}$$

The operator H_{par} is differential in the *parabolic* coordinates $\xi_{1,2,3}$ only, whereas H_{int} acts on the internal degrees of freedom $\xi_{4,5,6}$. The mixing term H_{mix} arises from the off-diagonal elements of the metric tensor and plays the role of a rotational coupling in a hyperspherical treatment. The essential point is that the differential operators H_{par} and H_{int} are exactly separable in the coordinates $\xi_{1\ldots3}$ and $\xi_{4\ldots6}$, respectively, for they can be written as [11]

$$H_{\text{par}} = \sum_{j=1}^{3} H_{\xi_j} \; ; \quad [H_{\xi_j}, H_{\xi_i}] = 0 \; ; \quad \forall i, j \in \{1, 2, 3\} \; , \quad \text{and} \tag{13.9}$$

$$H_{\text{int}} = \sum_{j=4}^{6} H_{\xi_j} \; ; \quad [H_{\xi_j}, H_{\xi_i}] = 0 \; ; \quad \forall i, j \in \{4, 5, 6\} \; , \quad \text{where} \tag{13.10}$$

$$H_{\xi_j} = \frac{2}{\mu_{lm} r_{lm}} \left[\partial_{\xi_j} \xi_j \partial_{\xi_j} + i k_{lm} \xi_j \partial_{\xi_j} - \mu_{lm} z_{lm} \right] \; ; \quad \epsilon_{jlm} \neq 0 \; ; \tag{13.11}$$

$$H_{\xi_4} = \frac{1}{\mu_{23}} \left[\frac{1}{\xi_4^2} \partial_{\xi_4} \xi_4^2 \partial_{\xi_4} + i 2 k_{23} \frac{\xi_1 - \xi_4}{\xi_4} \partial_{\xi_4} \right] \; , \tag{13.12}$$

$$H_{\xi_5} = \frac{1}{\mu_{13}} \left[\frac{1}{\xi_5^2} \partial_{\xi_5} \xi_5^2 \partial_{\xi_5} + i 2 k_{13} \frac{\xi_2 - \xi_5}{\xi_5} \partial_{\xi_5} \right] \; , \tag{13.13}$$

$$H_{\xi_6} = \frac{1}{\mu_{12}} \left[\frac{1}{\xi_6^2} \partial_{\xi_6} \xi_6^2 \partial_{\xi_6} + i 2 k_{12} \frac{\xi_3 - \xi_6}{\xi_6} \partial_{\xi_6} \right] \; . \tag{13.14}$$

The operator $H_{\text{mix}} = H - H_{\text{par}} - H_{\text{int}}$ derives from the expression

$$H_{\text{mix}} := \sum_{u \neq v = 1}^{6} \left\{ (\boldsymbol{\nabla}_{\boldsymbol{r}_{ij}} \xi_u) \cdot (\boldsymbol{\nabla}_{\boldsymbol{r}_{ij}} \xi_v) + (\boldsymbol{\nabla}_{\boldsymbol{R}_k} \xi_u) \cdot (\boldsymbol{\nabla}_{\boldsymbol{R}_k} \xi_v) \right\} \partial_{\xi_u} \partial_{\xi_v} \; .$$

$$\tag{13.15}$$

Noting that $H_{\xi_j}, j = 1, 2, 3$ is simply the Schrödinger operator for the two-body scattering rewritten in parabolic coordinates (after factoring out the plane-wave part), one arrives immediately, as a consequence of (13.9), at an expression for the three-body wave function as a product of three two-body continuum waves (the so-called 3C-model or the Ψ_{3C} wave function) with the correct boundary conditions at large interparticle separations. This result is valid if the contributions of H_{int} and H_{mix} are negligible as compared to H_{par}, which is, in fact, the case for large interparticle separations [11] *or* at high particles' energies. The above procedure can be performed within the Jacobi coordinate system, however, the operators H_{par}, H_{int}, and H_{mix} have a much more complex representation in the Jacobi coordinates (see [11]).

13.3.2 Remarks on the Structure of the Three-Body Hamiltonian

The structure of (13.9–13.15) and the mathematical properties of the operators H_{par}, H_{int}, and H_{mix} deserve several remarks.

1. The total potential is contained in the operator H_{par}, as can be seen from (13.11). Thus, the eigenstates of H_{par} treat the total potential in an exact manner. This means, on the other hand, that the operators H_{int} and H_{mix} are parts of the kinetic energy operator. This situation is to be

contrasted with other treatments of the three-body problem in regions of the space where the potential is smooth, e.g., near a saddle point. In this case one usually expands the potential around the fix point and accounts for the kinetic energy in an exact manner.

2. In (13.11) the total potential appears as a sum of three two-body potentials. It should be stressed that this splitting is arbitrary, since the dynamics is controlled by the total potential. Thus, any other splitting that leaves the total potential invariant is equally justified. We will use this factbelow for the construction of three-body states. For large interparticle separation the operators H_{int} and H_{mix} are negligible as compared to H_{par} and the splitting of the total potential as done in (13.11) becomes unique. For large $k_{ij}r_{ij}$, $\forall ij$, i.e., at high interparticle relative energies the three-body scattering dynamics is controlled by sequential two-body collisions. This is of particular importance for the interpretations of the outcome of experiments that test the three-body continuum problem (see the discussion of the theoretical and experimental results given below).

3. The momentum vectors $\boldsymbol{k}_{ij}$ enter the Schrödinger equation via the asymptotic boundary conditions. Thus, their physical meaning, as two-body relative momenta, is restricted to the asymptotic region of large interparticle distances. The consequence of this conclusion is that, in general, any combination or functional form of the momenta $\boldsymbol{k}_{ij}$ is legitimate, as long as the total energy is conserved and the boundary conditions are fulfilled (the energies and the wave vectors are linked via a parabolic dispersion relation). This fact has been employed in [9] to construct three-body wave functions with position-dependent momenta $\boldsymbol{k}_{ij}$ and in [17] to account for offshell transitions.

4. The separability of the operators (13.12)–(13.14) may be used to deduce representations of three-body states [18] that diagonalize simultaneously H_{par} and H_{int}. It should be noted, however, that generally the operator H_{mix}, which has to be neglected in this case, falls off with distance as fast as H_{int}.

5. As is well known, each separability of a system implies a related conserved quantity. In the present case we can only speak of an approximate separability and hence of approximate conserved quantum numbers. If we discard H_{int} and H_{mix} in favor of H_{par}, which is justified for large $k_{ij}\xi_k$, $\epsilon_{ijk} \neq 0, k \in [1,3]$ (i.e., for large ξ_k or for high two-particle momenta k_{ij}), the three-body good quantum numbers are related to those in a two-body system in parabolic coordinates. The latter are the two-body energy, the eigenvalue of the component of the Lenz–Runge operator along a quantization axis z, and the eigenvalue of the component of the angular momentum operator along z. In our case the quantization axis z is given by the linear momentum direction $\hat{\boldsymbol{k}}_{ij}$. In [11] the three-body problem has been formulated in hyperspherical-parabolic coordinates. In this case, the operator H_{int} takes on the form of the grand angular momentum opera-

tor. This observation is useful to expose the relevant angular momentum quantum numbers in case H_{mix} can be neglected.

6. In [8] the three-body system has been expressed in the coordinates $\eta_j = \xi_j^+$, $j = 1, 2, 3$ and $\bar{\eta}_j = \xi_j^-$, $j = 1, 2, 3$. From a physical point of view this choice is not quite suitable, for scattering states are sufficiently quantified by outgoing- or incoming-wave boundary conditions (in contrast to standing waves, such as bound states whose representation requires a combination of incoming and outgoing waves). Therefore, to account for the boundary conditions in scattering problems, either the coordinates η_j or $\bar{\eta}_j$ are needed. The appropriate choice of the remaining three coordinates should be made on the basis of the form of the forces governing the three-body system. In the present case where external fields are absent we have chosen $\xi_k = r_{ij}$, $k = 4, 5, 6$ as the natural coordinates adopted to the potential energy operator.

7. The approximate separability of the three-body Hamiltonian in the high energy regime (see (13.12)–(13.14)) results in the commutation relation of the two-body Hamiltonians $[H_{\xi_i}, H_{\xi_j}]$, $i, j \in \{1, 2, 3\}$. This fact can be expressed in terms of Green operators, which offers a connection to well-established methods of many-body theories [19,20] (see below for a brief summary of the Green's function method): let $G^{(3)}(z) = (z - H)^{-1}$ be the Green operator of the system. In a three-body system we consider, in a first step, two particles, say particle 1 and particle 2 to move independently *and* on the two-body energy-shell in the field of particle 3. In this case we find $G^{(3)} = g_1 g_2$, where $g_{1/2}$ are the two-body Green operators describing the independent motion of the respective particles. At high interparticle relative energies we can write $H = H_{\mathrm{par}} = \sum_{j=1}^{3} H_{\xi_j}$ (see (13.8)). In Green's function language this means

$$G^{(3)} \approx G_{3\mathrm{C}}^{(3)} = g_1 \left(I + g_{12}\, v_{12} \right) g_2 \,, \tag{13.16}$$

where v_{12} is the interaction potential between the particles 1 and 2 and $g_{12} = G_0 + G_0 v_{12} g_{12} = G_0 + G_0 v_{12} G_0 + G_0 v_{12} G_0 v_{12} G_0 + \cdots$. Here G_0 is the Green's operator in absence of interactions. Upon insertion of g_{12} in (13.16) and noting that $G_0(z) = (z - H_0)^{-1}$ we obtain

$$G^{(3)} \approx g_1 g_2 + g_1\, \kappa_{12}\, g_2 + g_1\, \kappa_{12}\, \kappa_{12}\, g_2 + g_1\, \kappa_{12}\, \kappa_{12}\, \kappa_{12}\, g_2 \cdots \,, \tag{13.17}$$

where κ_{12} is a dimensionless coupling parameter that measures the strength of the interaction potential (v_{12}) as compared to the kinetic energy (H_0). From (13.17) the following picture emerges: particle 1 and 2 can be considered as quasiparticles that interact successively an infinite number of times via κ_{12}. This is the exact counterpart of what is known as the *ladder approximation* in many-body theory [19,20]. This means that the approximation $G^{(3)} \approx G_{3\mathrm{C}}^{(3)}$ amounts to an exact sum of all the ladder diagrams. On the other hand, it is documented [19,20] that the ladder approximation results upon disregarding a number of (crossed) diagrams that are as well not accounted for by $G_{3C}^{(3)}$ for the three-body case. A way to incorporate these "higher-order" many-body effects is to consider them as a renormalization of the single-particle properties (mass, charge, etc.) and of the two-body interactions (v_{12}). The renormalized single-particle Green's functions and the mutual interaction are labeled, respectively, by $\tilde{g}_{1/2}$ and $\tilde{v}_{12}$. This procedure is outlined in the next section.

13.3.3   Dynamical Screening

In the realm of many-body theory it is well established that under certain conditions (small perturbations, low excitations) interaction effects can be accounted for by renormalizing the single-particle properties so that the one-particle picture remains viable. In the context of the three-body problem these ideas can be utilized and generalized as follows: correlations effects whose description goes beyond the ladder approximation should be incorporated as a redefinition of the two-particle properties so that we can still operate within the ladder approximation. In terms of wave functions this means we seek three-body wave functions that are eigenfunctions not only of H_{par}, but also of parts of H_{int} and H_{mix}. In addition, the structure of the total Hamiltonian, in the sense of (13.9) and (13.11), should be maintained.

As it turned out this can be achieved by introducing renormalized two-particle coupling strengths, namely, instead of z_{ij} for the bare two-body interaction, we define a variable $\bar{z}_{ij}$ and determine its functional dependence such that the structure of (13.9) and (13.11) is preserved. Using these $\bar{z}_{ij}$ one can then write down the three-body Green's operator in terms of the *dressed* one-particle Green's operators $\tilde{g}_{1/2}$, e.g., $\tilde{g}_1 = g_{01} + g_{01}(\bar{z}_{13}/r_{13})g_1$, where g_{01} is the free Green's operator of particle 1 and r_{13} is the relative position of particle 1 with respect to particle 3.

The actual derivation of $\bar{z}_{ij}$ is quite involved and is based on the following observations. (i) In a three-body system the form of the two-body potentials z_{ij}/r_{ij} are generally irrelevant, as long as the total potential is conserved. (ii) To keep the mathematical structure of the operators (13.9) and (13.11) unchanged and to introduce a splitting of the total potential, while maintaining the total potential's rotational invariance, one assumes the strength of the individual two-body interactions, characterized by z_{ij}, to be dependent on $\xi_{4,5,6}$. This means we introduce position-dependent product charges as $\bar{z}_{ij} = \bar{z}_{ij}(\xi_4, \xi_5, \xi_6)$, with $\sum_{j>i=1}^{3} \frac{\bar{z}_{ij}}{r_{ij}} = \sum_{j>i=1}^{3} \frac{z_{ij}}{r_{ij}}$. To obtain the "re-normalized" *many-body* potentials $\bar{V}_{ij} := \bar{z}_{ij}/r_{ij}$ we write $\bar{V}_{ij}$ as a linear combination of the isolated *two-body* interactions $V_{ij} := z_{ij}/r_{ij}$, i.e.,

$$\begin{pmatrix} \overline{V}_{23} \\ \overline{V}_{13} \\ \overline{V}_{12} \end{pmatrix} = \mathcal{A} \begin{pmatrix} V_{23} \\ V_{13} \\ V_{12} \end{pmatrix}, \tag{13.18}$$

where $\mathcal{A}(\xi_4, \xi_5, \xi_6)$ is a 3×3 matrix. The matrix elements are then determined according to (1) the properties of the total potential surface, (2) to reproduce the correct asymptotic of the three-body states and (3) in a way that minimizes H_{int} and H_{mix}. It should be stressed that the procedure until this stage is exact. It is merely a splitting of the total potential that leaves this potential and hence the three-body Schrödinger equation unchanged.

For an electron pair moving in the field of a positive ion the determination of $\bar{z}_{ij}$ has been accomplished in [10,11,21]. The resulting wave function has been termed the dynamically screened three-body continuum wave function (DS3C) Ψ_{DS3C}.

13.4   Theory of Excited N-Particle Finite Systems

Unfortunately, the curvilinear coordinate system (13.4), (13.5) used for the three-body problem does not have a straightforward generalization to the N–body case. The speciality of the three-body problem is that the number of interaction lines is equal to the number of particles. Therefore, the N-body Coulomb scattering problem has to be approached differently. For large N our system resembles that of the interacting electron gas (EG). In contrast to the present case, however, conventional treatments of EG [1,3,19] are focused on ground-state properties and (low-energy) excitations in the linear response regime.

13.5   Continuum States of N-Charged Particles

The motion of $N-1$ charged particles (with charges z_j, $j \in [1, N-1]$) in the field of a massive residual ion with charge z, at energies above the complete fragmentation threshold, is described by the Schrödinger equation

$$\left[H_0 + \sum_{j=1}^{N} \frac{z z_j}{r_j} + \sum_{\substack{i,j \\ j>i=1}}^{N} \frac{z_i z_j}{r_{ij}} - E \right] \Psi(\boldsymbol{r}_1, \cdots, \boldsymbol{r}_N) = 0, \tag{13.19}$$

where $\boldsymbol{r}_j$ is the position of particle j with respect to the residual charge z and $\boldsymbol{r}_{ij} = \boldsymbol{r}_i - \boldsymbol{r}_j$ denotes the relative coordinate between particles i and j. The kinetic energy operator H_0 has the form (in the limit $m/M \to 0$, where M is the mass of the charge z and m is the mass of the continuum particles, which are assumed to have equal masses) $H_0 = -\sum_{\ell=1}^{N} \Delta_\ell / 2m$, where Δ_ℓ is the Laplacian with respect to the coordinate $\boldsymbol{r}_\ell$. We note here that for a system of general masses the problem is complicated by an additional mass-polarization term that arises in (13.19). Upon introduction of N-body Jacobi coordinates, H_0 becomes diagonal, however, the potential terms acquire a much more complex form. The continuum states are characterized by the $N-1$ asymptotic momentum vectors $\boldsymbol{k}_j$. The Sommerfeld parameters α_j, α_{ij} are given by $\alpha_{ij} = \frac{z_i z_j}{v_{ij}}$, $\alpha_j = \frac{z z_j}{v_j}$.

Here, v_j is the velocity of the particle j and $v_{ij} := \mathbf{v_i} - \mathbf{v_j}$. The total energy of the system E is given by the asymptotic value of the kinetic energy, i.e., $E = \sum_{l=1}^{N} E_l$, where $E_l = \frac{k_l^2}{2m}$. Scattering eigenstates $\Psi(r_1, \cdots, r_N)$ of (13.19) have been derived using the ansatz [15]

$$\Psi(r_1, \cdots, r_N) = \mathcal{N}\Phi_I(r_1, \cdots, r_N)\Phi_{II}(r_1, \cdots, r_N)\chi(r_1, \cdots, r_N), \quad (13.20)$$

where Φ_I, Φ_{II} are appropriately chosen functions, $\mathcal{N}$ is a normalization constant and $\chi(r_1, \cdots, r_N)$ is a function of an arbitrary form. The function Φ_I is chosen to describe the motion of N-independent Coulomb particles moving in the field of the charge z at the total energy E, i.e., Φ_I is determined by the differential equation

$$\left(H_0 + \sum_{j=1}^{N} \frac{zz_j}{r_j} - E \right) \Phi_I(r_1, \cdots, r_N) = 0. \quad (13.21)$$

The regular solution Φ_I is $\Phi_I(r_1, \cdots, r_N) = \prod_{j=1}^{N} \bar{\xi}_j(r_j)\varphi_j(r_j)$, where $\bar{\xi}_j(r_j)$ is a plane wave dependent on the coordinate r_j and characterized by the wave vector k_j. Furthermore, $\varphi_j(r_j)$ is a confluent-hypergeometric function in the notation of [22] $\varphi_j(r_j) = {}_1F_1[\alpha_j, 1, -\mathrm{i}(k_j r_j + k_j \cdot r_j)]$. The function Φ_I describes the motion of the continuum particles in the extreme case of very strong coupling to the residual ion, i.e., $|zz_j| \gg |z_j z_i|$; $\forall i, j \in [1, N-1]$. In order to incorporate the other extreme case of strong correlations among the continuum particles ($|z_j z_i| \gg |zz_j|$; $\forall i, j \in [1, N-1]$) we choose Φ_{II} to possess the form

$$\Phi_{II}(r_1, \cdots, r_N) = \overline{\Phi}_{II}(r_1, \cdots, r_N) \prod_{j=1}^{N} \bar{\xi}_j(r_j),$$

$$\overline{\Phi}_{II}(r_1, \cdots, r_N) = \prod_{j>i=1}^{N} \varphi_{ij}(r_{ij}), \quad (13.22)$$

where $\varphi_{ij}(r_{ij}) := {}_1F_1[\alpha_{ij}, 1, -\mathrm{i}(k_{ij}r_{ij} + k_{ij} \cdot r_{ij})]$. It is straightforward to show that the expression $\varphi_{ij}(r_{ij}) \prod_{l=1}^{N} \bar{\xi}_l(r_l)$ solves for the Schrödinger equation (13.19) in the case of extreme correlations between particle i and particle j, i.e., $|zz_l| \ll |z_i z_j| \gg |z_m z_n|$, $\forall l, m, n \neq i, j$. In terms of differential equations this means

$$\left(H_0 + \frac{z_i z_j}{r_{ij}} - E \right) \varphi_{ij}(r_{ij}) \prod_{j=1}^{N} \bar{\xi}_j(r_j) = 0. \quad (13.23)$$

It should be stressed, however, that the function (13.22) does not solve for (13.19) in the case of weak coupling to the residual ion ($z \to 0$), but otherwise comparable strength of correlations between the continuum particles, because the two-body subsystems formed by the continuum particles are coupled to each other. This is the equivalent situation to the interacting electron gas.

The determination of the properties of the function χ that occurs in the ansatz (13.20) is a lengthly procedure and has been discussed in full detail in [15]. It turned out that, at higher energies or at large interparticle distances, χ has the form of a product of plane waves. In general, however, no closed analytical expression for χ has been found yet.

13.6 Green Function Theory of Finite Correlated Systems

With increasing number of particles the treatment of correlated systems becomes more complex and new phenomena appear whose description requires the knowledge of the collective behavior of the system. Thus, an approach is needed that is different from the wave function treatment. The method of choice for this purpose is the Green–function technique, which we will outline in this section.

A principle task in many-body systems is to deal appropriately with the correlations between the particles, for the independent-particle problem can be solved is a standard way. Therefore, for a canonical ensemble, a non-perturbative method has been developed [23] that allows dilution of the inter-particle interaction strength to a level where the problem can be solved by conventional methods (perturbation theory, mean-field approach, etc.). This can be achieved by an incremental procedure in which the N correlated particle system is mapped exactly onto a set of systems in which only $N - M$ particles are interacting ($M \in [1, N - 2]$), i.e., in which the strength of the potential energy part is damped.

The total potential is assumed to be of the class $U^{(N)} = \sum_{j>i=1}^{N} v_{ij}$ without any further specification of the individual potentials v_{ij}. The key point is that the potential $U^{(N)}$ satisfies the recurrence relations

$$U^{(N)} = \frac{1}{N-2} \sum_{j=1}^{N} u_j^{(N-1)}, u_j^{(N-1)} = \frac{1}{N-3} \sum_{k=1}^{N-1} u_{jk}^{(N-2)}, \quad j \neq k,$$

(13.24)

where $u_j^{(N-1)}$ is the total potential of a system of $(N-1)$ interacting particles in which the j particle is missing ($u_j^{(N-1)} = \sum_{m>n=1}^{N} v_{mn}, m \neq j \neq n$).

As shown in [23] the decomposition (13.24) can be used to derive similar recurrence relations for the transition $T^{(N)}$ and the Green's operator $G^{(N)}$, namely $T^{(N)} = \sum_{j=1}^{N} T_j^{(N-1)}, j \in [1, N]$ and

$$\begin{pmatrix} T_1^{(N-1)} \\ T_2^{(N-1)} \\ \vdots \\ T_{N-1}^{(N-1)} \\ T_N^{(N-1)} \end{pmatrix} = \begin{pmatrix} t_1^{(N-1)} \\ t_2^{(N-1)} \\ \vdots \\ t_{N-1}^{(N-1)} \\ t_N^{(N-1)} \end{pmatrix} + [K^{(N-1)}] \begin{pmatrix} T_1^{(N-1)} \\ T_2^{(N-1)} \\ \vdots \\ T_{N-1}^{(N-1)} \\ T_N^{(N-1)} \end{pmatrix}.$$

(13.25)

Here, $t_j^{(N-1)}$ is the transition operator of a system, in which only $N-1$ particles are interacting. The kernel $[K^{(N-1)}]$ is given in terms of $t_j^{(N-1)}$ only [23]. From (13.24) it is clear that $t_j^{(N-1)}$ can also be expressed in terms of the transition operators of the system where only $(N-2)$ particles are interacting leading to a recursive scheme. From the relation $G^{(N)} = G_0 + G_0 T^{(N)} G_0$ similar conclusions are made for the Green operator $G^{(N)} = G_0 + \sum_{j=1}^{N} G_j^{(N-1)}$. The operators $G_j^{(N-1)}$ are related to the Green operators $g_j^{(N-1)}$ of the systems in which only $(N-1)$ particles are correlated.

13.6.1   Application to Four-Body Systems

For the four-body system, $G^{(4)}$ can be expressed in terms of three-body Green's functions, namely, $G^{(4)} = \sum_{j=1}^{4} g_j^{(3)} - 3G_0$. Here, $g_j^{(3)}$ is the Green operator of the system where only three particles are interacting. In terms of state vectors the above procedure amounts to $|\Psi^{(4)}\rangle = |\psi_{234}^{(3)}\rangle + |\psi_{134}^{(3)}\rangle + |\psi_{124}^{(3)}\rangle + |\psi_{123}^{(3)}\rangle - 3|\phi_{\text{free}}^{(4)}\rangle$, where $|\psi_{ijk}^{(3)}\rangle$ is the state vector of the system in which the three particles i, j, and k are interacting, whereas $|\phi_{\text{free}}^{(4)}\rangle$ is the state vector of the noninteracting four-body system. Here, $|\psi_{ijk}^{(3)}\rangle$ is assumed known.

13.6.2   Thermodynamics and Phase Transitions in Finite Systems

Strictly speaking, finite systems do not exhibit phase transitions [24]. However, one expects to observe the onset of a critical behavior when the system approaches the thermodynamic limit. The traditional theory concerned with these questions is finite-size scaling theory [26]. For interacting systems the methods outlined above together with the ideas developed in [24,25], can be utilized for the study of critical phenomena in finite correlated systems: The canonical partition function is expressed in terms of the many-body Green function as $Z(\beta) = \int dE \, \Omega(E) \, e^{-\beta E}$. Here, $\Omega(E)$ is the density of states that is related to the imaginary part of the trace of $G^{(N)}$ via $\Omega(E) = -\frac{1}{\pi} Tr \Im G^{(N)}(E)$. The recurrence relation outlined above for the N-body Green function can be utilized to calculate $\Omega(E)$, leading to a recursion relation for the partition function

$$Z^{(N)} = \sum_{j=1}^{N} Z_j^{(N-1)} - (N-1)Z_0 \,. \tag{13.26}$$

Here, Z_0 is the partition function of the independent particle system (taken as a reference), while $Z_j^{(N-1)}$ is the canonical partition function of a system in which the interaction strength is diluted by cutting all interaction lines that connect to particle j. Equation (13.26) allows the thermodynamic properties of finite systems to be studied on a microscopic level as well as investigation of the inter-relation between the thermodynamics and the strength of correlations. Critical phenomena can be studied using the idea put forward by Yang and Lee [24,25]. For example, if one is interested in the onset of condensation in a quantum Bose gas, the ground-state occupation number $\eta_0(N, \beta)$ has to be considered

$$\eta_0(N, \beta) = -\frac{1}{\beta} \frac{\partial_{\epsilon_0} Z^{(N)}(\beta)}{Z^{(N)}(\beta)} = -\frac{1}{\beta} \frac{\sum_{j=1}^{N} \partial_{\epsilon_0} Z_j^{(N-1)} - (N-1)\partial_{\epsilon_0} Z_0}{Z^{(N)}}.$$

(13.27)

Here, ϵ_0 is the ground-state energy. By means of this equation one can study systematically the influence of the interaction on the onset of the critical regime or one may chose to find the roots of (13.26) in the complex β plane. Zero points of $Z^{(N)}(\beta)$ that approach systematically the real β axis signify the presence of transition points in the thermodynamic limit.

13.7   Collective Response Versus Short-Range Dynamics

In the thermodynamic limit (large volume V, large N, and finite number density $n = N/V$) the characteristic response of the system will be dominated by the cooperative behavior of all the electrons. For example, the fluctuations of the electronic density are determined by the polarization operator $\Pi(\boldsymbol{q}, \omega)$, which depends on the momentum transfer $\boldsymbol{q}$ and the frequency ω. On the other hand, the polarization of the medium modifies the properties of the particle–particle interaction $U(\boldsymbol{q}, \omega)$. The modified potential U_{eff} is related to U and $\Pi(\boldsymbol{q}, \omega)$ through the integral equation [2,3], i.e., $U_{\text{eff}} = U + U\Pi U_{\text{eff}}$; or $U_{\text{eff}} = U/(1 - U\Pi)$. Thus, the *screening* is quantified by $\kappa(\boldsymbol{q}, \omega) = 1/(1 - U\Pi)$, which is called the generalized dielectric function [3]. To determine U_{eff} and κ one needs the polarization function Π that describes the particle–hole excitations. The lowest–order approximation Π_0 is given by the random-phase approximation (RPA) as $i\Pi_0(\boldsymbol{q}, \omega) = \frac{2}{(2\pi)^4} \int d\boldsymbol{p} d\xi G_0(\boldsymbol{q} + \boldsymbol{p}, \omega + \xi) G_0(\boldsymbol{p}, \xi)$. Here G_0 is the free, single–particle Green function. The evaluation of Π_0 can be performed analytically for a homogeneous system [3]. In the long wavelength limit we have $\Pi_0 \approx -2N(\mu)$, where $N(\mu)$ is the density of states at the Fermi level μ. This means that, in the presence of the medium, the electron–electron interaction has the form $U_{\text{TF}} = 4\pi/[q^2 + 8\pi N(\mu)]$. In configuration space one obtains $U_{\text{TF}} = e^{-r/\lambda}/r$.

Hence, in contrast to atomic systems where collisions with small momentum transfer are predominant ($U \propto 1/q^2$), in a polarizable medium scattering events with small q are cut out due to the finite range of the renormalized scattering potential U_{eff}. The essential difference between the screening effects we are dealing with here and those introduced in the context of the

DS3C theory is that in the DS3C screening is incorporated into an explicit many-body theory to account for higher-order diagrams (beyond the ladder approximation) whereas here the screening accounts for certain correlations effects (linear response theory) within a mean-field theory.

Apart from the case of the extended, homogeneous electron gas the evaluations of U_{eff} is generally a challenging task (plane waves are, in general, not the single-particle eigenstates of the system). For an inhomogeneous electronic system, like solids and surfaces, the GW approximation [27] offers a direct extension of the RPA (G stands for the Green's function and $W \equiv U_{\text{eff}}$ for the screened interaction). In [28] this scheme has been discussed and results for the dielectric functions of copper and nickel surfaces have been presented.

13.7.1 Manifestations of Collective Response in Finite Systems

For finite systems the spectrum is generally discrete, which hinders fluctuations around the ground state. However, on increasing the size and number density n collective effects set in. The influence of the fluctuations is demonstrated nicely when considering the ionization channel of large molecules or metal clusters upon electron impact. Within the RPAE (RPA with exchange) the screened interaction U_{eff} between the electron and the target is

$$
\left\langle \bm{k}_1\bm{k}_2 \left| U_{\text{eff}} \right| \phi_\nu \bm{k}_0 \right\rangle = \left\langle \bm{k}_1\bm{k}_2 \left| U \right| \phi_\nu \bm{k}_0 \right\rangle \tag{13.28}
$$

$$
+ \sum_{\substack{\varepsilon_{\text{p}} \leq \mu \\ \varepsilon_{\text{h}} > \mu}} \left(\frac{\left\langle \varphi_{\text{p}}\bm{k}_2 \left| U_{\text{eff}} \right| \phi_\nu \varphi_{\text{h}} \right\rangle \left\langle \varphi_{\text{h}}\bm{k}_1 \left| U \right| \bm{k}_0 \varphi_{\text{p}} \right\rangle}{\epsilon_0 - (\varepsilon_{\text{p}} - \varepsilon_{\text{h}} - \mathrm{i}\delta)} \right.
$$

$$
\left. - \frac{\left\langle \varphi_{\text{h}}\bm{k}_2 \left| U_{\text{eff}} \right| \phi_\nu \varphi_{\text{p}} \right\rangle \left\langle \varphi_{\text{p}}\bm{k}_1 \left| U \right| \bm{k}_0 \varphi_{\text{h}} \right\rangle}{\epsilon_0 + (\varepsilon_{\text{p}} - \varepsilon_{\text{h}} - \mathrm{i}\delta)} \right).
$$

φ_{p} and φ_{h} are, respectively, the intermediate particle's and hole's states with the energies ε_{p}, ε_{h}, whereas δ is a small positive real number and μ is the chemical potential. The first term of (13.29) on the RHS amounts to neglecting the electron–hole (de)excitations, as done in [30]. In [29] (13.29) has been evaluated self-consistently for the C_{60} cluster and the ionization cross section has been calculated. The results are shown in Fig. 13.1, which clearly demonstrates the significance of screening in shrinking the effective size of the scattering region and thus leading to a suppression of the ionization cross section.

13.8   The Quantum Field Approach: Basic Concepts

For strongly correlated systems or multiple excitations in extended systems (such as one-electron or one-photon double–electron emission (γ, 2e) or (e,2e))

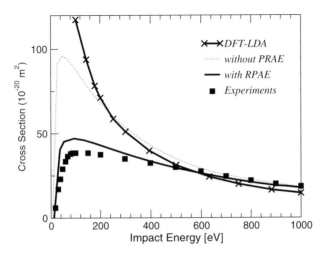

Fig. 13.1. Total electron-impact ionization cross section of C_{60} as a function of the impact energy. The *absolute* experimental data *(full squares)* for the production of stable C_{60}^{+} [31,32] are shown. The *solid line with crosses* is the DFT results [33], the *dotted line (solid)* is the result of the present calculation without RPAE (with RPAE)

methods that go beyond RPA are needed. There are a number of theories available, however most of them, like the hole-line expansion or the coupled-cluster methods [3,4,34], are restricted to the treatment of ground-state properties. For the treatment of correlated excited states the Green's function approach is well suited, however, the method as introduced in previous sections, becomes intractable with increasing N, since in this case one works within first quantization, i.e., the states have to be (anti)symmetrized.

Applying methods of field theory, Migdal and Galitskii as well as Martin and Schwinger [35,36] developed a theory that connects, by means of Feynman diagrams, higher-order propagators to the single particle (sp) propagator. The latter is then related to the free unperturbed propagator. The system symmetry enters through (anti)commutation relations of the operators [3,19,20]. This (perturbative) route has found extensive applications in various fields of physics. Here, we focus on the aspects that are of immediate relevance to $(\gamma, 2e)$ and (e,2e) reactions.

13.8.1   The Single–Particle Green's Function for Extended Systems

The sp Green's function $g(\alpha t, \beta t')$ can be considered as an expectation value for the time-ordered product of two operators evaluated using the correlated, exact (normalized) ground state $|\Psi_0\rangle$ of the N electron system, i.e.,

$$ig(\alpha t, \beta t') = \langle \Psi_0 | \mathcal{T}[a_{H\alpha}(t) a_{H\beta}^{\dagger}(t')] | \Psi_0 \rangle ,$$

where $\mathcal{T}$ is the time-ordering operator. $a^{\dagger}_{\mathrm{H}\beta}(t')$ and $a_{\mathrm{H}\alpha}(t)$ stand, respectively, for the fermionic creation and annihilation operators in the Heisenberg picture represented in an appropriate basis, the members of which are characterized by quantum numbers α and β. For a translationally invariant system, the appropriate basis states are the momentum eigenstates, labeled by $\boldsymbol{k}$. The effect of the chronological operator $\mathcal{T}$ can be described in terms of the step function $\Theta(t - t')$, in which case the Green's function is given by

$$
\begin{aligned}
ig(k, t - t') &= \Theta(t - t')\langle\Psi_0|a_{\mathrm{H}k}(t)a^{\dagger}_{\mathrm{H}k}(t')|\Psi_0\rangle \\
&\quad -\Theta(t' - t)\langle\Psi_0|a^{\dagger}_{\mathrm{H}k}(t')a_{\mathrm{H}k}(t)|\Psi_0\rangle \\
&= \Theta(t - t')\sum_{\gamma} e^{-i[E^{(N+1)}_{\gamma} - E^{(N)}_0](t-t')} \left|\langle\Psi^{(N+1)}_{\gamma}|a^{\dagger}_k|\Psi_0\rangle\right|^2 \\
&\quad -\Theta(t' - t)\sum_{\delta} e^{-i[E^{(N)}_0 - E^{(N-1)}_{\delta}](t-t')} \left|\langle\Psi^{(N-1)}_{\delta}|a_k|\Psi_0\rangle\right|^2 .
\end{aligned}
$$
(13.29)

$\Psi^{(N+1)}_{\gamma}$ and $\Psi^{(N-1)}_{\delta}$ stand for a complete set of eigenstates of the $(N+1)$- and the $(N-1)$ particle system, respectively. The energies $E^{(N)}_0$, $E^{(N+1)}_{\gamma}$, and $E^{(N-1)}_{\delta}$ refer to the exact energies for the correlated ground state of, respectively, the N, the $(N+1)$, and the $(N-1)$ particle system. The exponential with the energies in (13.30) is due to the Hamiltonians in the exponential functions in the definition of the Heisenberg operators. Noting that the step function has the integral representation

$$
\Theta(t) = -\lim_{\eta\to 0}\frac{1}{2\pi i}\int_{-\infty}^{\infty}d\omega\,\frac{e^{-i\omega t}}{\omega + i\eta},
$$

the Green's function in energy space can be obtained via Fourier transforming the time difference $t - t'$ to the energy variable ω. This yields the spectral or Lehmann representation of the sp Green's function [37],

$$
g(k, \omega) = \lim_{\eta\to 0}\left[\sum_{\gamma}\frac{\left|\langle\Psi^{(N+1)}_{\gamma}|a^{\dagger}_k|\Psi_0\rangle\right|^2}{\omega - [E^{(N+1)}_{\gamma} - E^{(N)}_0] + i\eta}\right.
$$
$$
\left. + \sum_{\delta}\frac{\left|\langle\Psi^{(N-1)}_{\delta}|a_k|\Psi_0\rangle\right|^2}{\omega - [E^{(N)}_0 - E^{(N-1)}_{\delta}] - i\eta}\right].
$$
(13.30)

This relation underlines that the sp Green's function is expressible in terms of measurable quantities: the poles of $g(k, \omega)$ correspond to the change in energy (with respect to $E^{(N)}_0$) if one particle is added ($E^{(N+1)}_{\gamma} - E^{(N)}_0$) or one particle is removed ($E^{(N)}_0 - E^{(N-1)}_{\delta}$) from the reference ground state with N interacting particle. The residua of these poles are given by the *spectroscopic factors*, i.e., the measurable probabilities of adding and removing one particle with wave vector $\boldsymbol{k}$ to produce the specific state γ (δ) of the residual system. Clearly, the latter probability is of direct relevance to the (e,2e) process. The infinitesimal quantity η shifts the poles below the Fermi energy (the states of the $(N-1)$ system) to slightly above the real axis and those above the Fermi energy [the states of the $(N+1)$ system] to slightly below the real axis.

It is useful to write the single-particle Green's function in terms of the hole and particle spectral functions, which are given for $\omega \leq \epsilon_{\mathrm{F}}$ by

$$S_{\mathrm{h}}(k,\omega) = \frac{1}{\pi}\Im g(k,\omega) = \sum_{\gamma}\left|\langle\Psi_{\gamma}^{(N-1)}|a_k|\Psi_0\rangle\right|^2\delta(\omega - (E_0^{(N)} - E_{\gamma}^{(N-1)})),$$

(13.31)

and for $\bar{\omega} > \epsilon_{\mathrm{F}}$

$$S_{\mathrm{p}}(k,\bar{\omega}) = \frac{1}{\pi}\Im g(k,\bar{\omega}) = \sum_{\gamma}\left|\langle\Psi_{\gamma}^{(N+1)}|a_k^{\dagger}|\Psi_0\rangle\right|^2\delta(\bar{\omega} - (E_{\gamma}^{(N+1)} - E_0^{(N)})).$$

(13.32)

The sp Green's function is then written as

$$g(k,\omega) = \lim_{\eta\to 0}\left(\int_{-\infty}^{\epsilon_{\mathrm{F}}}d\omega'\frac{S_{\mathrm{h}}(k,\omega')}{\omega - \omega' - i\eta} + \int_{\epsilon_{\mathrm{F}}}^{\infty}d\omega'\frac{S_{\mathrm{p}}(k,\omega')}{\omega - \omega' + i\eta}\right).$$

(13.33)

The single-particle Green's function is particularly important since it establishes a direct link to experimental processes that study the effect of a removal or an addition of a particle to the correlated system. As mentioned above, the (e, 2e) process is related to the hole spectral function. In addition, the Green's function allows the evaluation of the expectation value for *any* single-particle operator $\hat{O}$ (this is because $\langle\hat{O}\rangle = \sum_{\alpha\beta}\int_{-\infty}^{E_{\mathrm{F}}}d\omega S_{\mathrm{h}}(\alpha\beta,\omega)\langle\alpha|O|\beta\rangle$, where $\langle\alpha|O|\beta\rangle$ is the matrix representation of $\hat{O}$ in the basis $|\alpha\rangle$). This in turn highlights the importance of single-particle removal or addition spectroscopies, such as single photoemission [38,39] and (e,2e) processes, which allow insight into the respective part of the Green's function. A further advantage of the Green's function approach is that it offers a systematic way for approximations using the diagram technique [20]. In the diagrammatic expansion for g one introduces the concept of the self-energy Σ [27]. The knowledge of Σ allows the evaluation of g via the Dyson equation

$$g(\alpha\beta;\omega) = g_0(\alpha\beta;\omega) + \sum_{\gamma\delta}g_0(\alpha\gamma;\omega)\Sigma(\gamma\delta;\omega)g(\delta\beta;\omega),$$

(13.34)

where g_0 is the Green's function of a (noninteracting) reference system. The self-energy Σ accounts for all excitations due to the interaction of the particle with the surrounding medium and acts as a nonlocal, energy-dependent, and complex single-particle potential.

13.8.2   Particle-Particle and Hole-Hole Spectral Functions

As discussed in detail in [20], the Dyson equation (13.34) can be derived algebraically and the single-particle propagator $g(\alpha t, \alpha' t')$ can be related to the two-particle Green's function $g^{\mathrm{II}}(\beta t_1, \beta' t'_1, \gamma t_2, \gamma t'_2)$. This is a first cycle in a hierarchy that links the N-particle propagator to the $(N+1)$-particle propagator [36,35]. Of direct relevance to this work is the two-particle propagator $g^{\mathrm{II}}(\beta t_1, \beta' t'_1, \gamma t_2, \gamma t'_2)$.

Repeating the steps outlined above for the single-particle case, one arrives at the Lehmann representation of the two-particle Green's function in terms of energies and states of the systems with N and $N \pm 2$ particles (the $(N-2)$-particle state of the system is achieved upon a $(\gamma, 2e)$ reaction):

$$
g^{\mathrm{II}}(\alpha\beta, \gamma\delta; \Omega) = \sum_n \frac{\langle \Psi_0^{(N)} | a_\beta a_\alpha | \Psi_n^{(N+2)} \rangle \langle \Psi_n^{(N+2)} | a_\gamma^\dagger a_\delta^\dagger | \Psi_0^{(N)} \rangle}{\Omega - [E_n^{(N+2)} - E_0^{(N)}] + i\eta}
$$
$$
- \sum_m \frac{\langle \Psi_0^{(N)} | a_\gamma^\dagger a_\delta^\dagger | \Psi_m^{(N-2)} \rangle \langle \Psi_m^{(N-2)} | a_\beta a_\alpha | \Psi_0^{(N)} \rangle}{\Omega - [E_0^{(N)} - E_m^{(N-2)}] - i\eta} . \quad (13.35)
$$

Upon analogous considerations made for the one-particle case to arrive at the single-particle spectral functions, one can obtain from g^{II} the hole-hole spectral function as $S_{\mathrm{hh}}(\boldsymbol{k}_1, \boldsymbol{k}_1, \Omega) = \Im g^{\mathrm{II}}(\boldsymbol{k}_1, \boldsymbol{k}_1, \Omega), \Omega \leq 2\epsilon_\mathrm{F}/\pi$, which is intimately related to the $(\gamma, 2e)$ reaction.

The two-particle Green's function involves two kinds of diagrams: the first type includes two noninteracting single-particle propagators (see (13.34)) and is supplemented by similar diagrams that include all possible self-energy insertions [3]. The second defines the vertex function Γ. The latter involves all generalization of the lowest-order correction to the two-particle propagator in which two particles interact once. To visualize the role of Γ we write g^{II} in the form

$$
g^{\mathrm{II}}(\alpha t_1, \alpha' t'_1, \beta t_2, \beta' t'_2)
$$
$$
= \mathrm{i}\,[g(\alpha\beta, t_1 - t_2)\, g(\alpha'\beta', t'_1 - t'_2) - g(\alpha\beta', t_1 - t'_2)\, g(\alpha'\beta, t'_1 - t_2)]
$$
$$
\times \int \mathrm{d}t_a \mathrm{d}t_b \mathrm{d}t_c \mathrm{d}t_d \sum_{abcd} g(\alpha a, t_1 - t_a)\, g(\alpha' b, t'_1 - t_b)
$$
$$
\times \langle ab | \Gamma(t_a, t_b; t_c, t_d) | cd \rangle g(c\beta, t_c - t_2) g(d\beta', t_d - t'_2) . \quad (13.36)
$$

From this equation it is clear that Γ can be considered as the effective interaction between dressed particles. In addition, Γ plays a decisive role in the determination of its single-particle counterpart, the self-energy Σ [3].

In energy-space, the result for the noninteracting (free) product of dressed propagators including the exchange contribution, i.e., the zero-order term of (13.36) with respect to Γ, reads

$$g_{\mathrm{f}}^{\mathrm{II}}(\alpha\beta, \gamma\delta; \Omega) =$$

$$\frac{\mathrm{i}}{2\pi} \int \mathrm{d}\omega \left[g(\alpha, \gamma; \omega)\, g(\beta\,\delta; \Omega - \omega) - g(\alpha, \delta; \omega)\, g(\beta\,\gamma; \Omega - \omega) \right] =$$

$$\sum_{mm'} \frac{\langle \Psi_0^{(N)} | a_\alpha | \Psi_m^{(N+1)} \rangle \langle \Psi_m^{(N+1)} | a_\gamma^\dagger | \Psi_0^{(N)} \rangle \langle \Psi_0^{(N)} | a_\beta | \Psi_{m'}^{(N+1)} \rangle \langle \Psi_{m'}^{(N+1)} | a_\delta^\dagger | \Psi_0^{(N)} \rangle}{\Omega - \{ [E_m^{(N+1)} - E_0^{(N)}] + [E_{m'}^{(N+1)} - E_0^{(N)}] \} + \mathrm{i}\eta} -$$

$$\sum_{nn'} \frac{\langle \Psi_0^{(N)} | a_\gamma^\dagger | \Psi_n^{(N-1)} \rangle \langle \Psi_n^{(N-1)} | a_\alpha | \Psi_0^{(N)} \rangle \langle \Psi_0^{(N)} | a_\delta^\dagger | \Psi_{n'}^{(N-1)} \rangle \langle \Psi_{n'}^{(N-1)} | a_\beta | \Psi_0^{(N)} \rangle}{\Omega - \{ [E_0^{(N)} - E_n^{(N-1)}] + [E_0^{(N)} - E_{n'}^{(N-1)}] \} + \mathrm{i}\eta} -$$

$$(\gamma \longleftrightarrow \delta) . \tag{13.37}$$

The integration over ω has been carried out by utilizing the Lehmann representation for the single-particle Green's functions. The ladder approximation to the two-particle propagator is

$$g_{\mathrm{L}}^{\mathrm{II}}(\alpha\beta, \gamma\delta; \Omega) = g_{\mathrm{f}}^{\mathrm{II}}(\alpha\beta, \gamma\delta; \Omega)$$
$$+ \frac{1}{4} \sum_{\epsilon\eta\theta\zeta} g_{\mathrm{f}}^{\mathrm{II}}(\alpha\beta, \epsilon\eta; \Omega) \langle \epsilon\eta | V | \theta\zeta \rangle g_{\mathrm{L}}^{\mathrm{II}}(\theta\zeta, \gamma\delta; \Omega) , \tag{13.38}$$

where V stands for the naked two-body interaction. This integral relation can now be iterated to yield a set of ladder diagrams. The corresponding ladder sum for the effective interaction Γ, as it appears in (13.36) can be deduced from this result as

$$\langle \alpha_1, \beta_2 | \Gamma_{\mathrm{L}}(\Omega) | \alpha_1', \beta_2' \rangle = \langle \alpha_1 \beta_2 | V | \alpha_1' \beta_2' \rangle$$
$$+ \frac{1}{4} \sum_{\epsilon\eta\theta\zeta} \langle \alpha_1 \beta_2 | V | \epsilon\eta \rangle g_{\mathrm{f}}^{\mathrm{II}}(\epsilon\,\eta, \theta\,\zeta; \Omega)$$
$$\times \langle \theta, \zeta | \Gamma_{\mathrm{L}}(\Omega) | \alpha_1', \beta_2' \rangle . \tag{13.39}$$

The aforementioned RPA for the particle-hole (polarization) propagator Π means that only the term (13.37) is taken into account. The calculations of higher-order (vertex) corrections entail an evaluation of the sum in (13.38). It is also interesting to contrast (13.38) with the DS3C approach (13.16).

As noticed above, g^{II} is of direct relevance to the $(\gamma, 2e)$ reaction. It should be noted, however, that the ladder approximation (13.38) for the two-particle Green's function can be employed to define the self-energy Σ [20,35], which can then be used to obtain the single-particle Green's function via (13.34). On the other hand, this Green's function enters in the definition of the two-particle Green's function, as is clear, e.g., from (13.37) and (13.38). Thus, in principle, the Dyson (13.34) and (13.38) for the one-body and two-body Green's functions have to be solved in a self-consistent manner. As in the single-particle case where we established the relevance of the spectral representation to the (e,2e) experiments, one can relate g^{II} to the $(\gamma, 2e)$ measurements by means of (13.35): g^{II} shows poles at energies (relative to the ground state) corresponding to adding $[E_n^{(N+2)} - E_0^{(N)}]$ or removing $[E_0^{(N)} - E_n^{(N-2)}]$ two particles from the unperturbed ground state. The residua of these poles are related to the measurable spectroscopic factors for the addition or removal of the two particles, e.g., as done in the $(\gamma, 2e)$ experiment [40]. From the above discussion we conclude thus that $(\gamma, 2e)$ and (e,2e) provide quite different information. On the other hand they are related in as much as the single-particle and the two-particle spectral functions are related to each other.

13.8.3   The Two-Particle Photocurrent

The $(\gamma, 2e)$ experiments from surfaces have been conducted recently [40]. The two-photoelectron current $\mathcal{J}$ is characterized by the wave vectors $\mathbf{k}_1$ and $\mathbf{k}_2$ of the two photoelectrons. It has the form [41]

$$\mathcal{J} \propto \langle \mathbf{k}_1, \mathbf{k}_2 | g^{IIr} \Delta S_{hh}^{II}(\mathbf{k}_1', \mathbf{k}_2', E) \Delta^\dagger g^{IIa} | \mathbf{k}_1, \mathbf{k}_2 \rangle \,, \tag{13.40}$$

where Δ is the dipole operator, S_{hh}^{II} is the hole-hole spectral function, and g^{IIa} (g^{IIr}) is the advanced (retarded) two-particle Green's function. As for the single photoemission, the photocurrent can be represented by a two-particle Caroli diagram [41], which explicitly shows no signature of time ordering. This is due to the assumption that the experimental time resolution (typically 200 ns) is much longer than any other timescale in the system and hence a time integration has to be performed to arrive at (13.40). On the other hand, one can tune to initial- or final-state interactions (FSI) by a suitable choice of $\mathbf{k}_{1/2}$: if $\mathbf{k}_1$ and $\mathbf{k}_2$ are very large (compared to the Fermi wave vector), one can expect FSI to be limited to a small region in phase space where the two electrons escape with almost the same velocities. Apart from this regime, FSI become less important, which allows the effect of initial-state correlations to be highlighted. On the other hand, if the two electrons escape with very low (vacuum) velocities, FSI become the determining factor.

Numerical Realization for Metal Surfaces. As is clear from (13.40), the evaluation of g^{II} is the key ingredient for the numerical evaluation of the two-photoelectron current. On the other hand, we have seen in the preceding section that the single-particle Green's function g is needed to obtain g^{II}, and in turn g^{II} goes into the determination of Σ and hence g. Until now this self-consistent loop has been too complicated to be realized numerically within a realistic description of the surface, i.e., for an inhomogeneous electron gas or for a few-electron atomic system.

Incorporating the Single-Particle Band Structure. A possible approach to the evaluation of $\mathcal{J}$ is the following: assuming the two photoelectrons to be independent, then, according to (13.38), the two-particle Green's function g^{II} reduces to $g^{II} = g_f^{II}$. This means, as in the RPA case for Π, g^{II} simplifies to an antisymmetrized product of sp Green's function $g_j(\boldsymbol{k}_i, E_i)$, $i = 1, 2$ that can be used to generate the single-photoelectron states, e.g., by means of the layer Korringa–Kohn–Rostoker (LKKR) method [2]. This method utilizes a density-functional approach [2] combined with a semi-empirical function for the complex part of the self–energy. For the evaluation of $\mathcal{J}$ from metal surfaces one can utilize the interaction potential U_{TF}, with $N(\mu)$ being calculated by the (ab initio) LKKR method. Furthermore, we can write

$$U_{TF} = \frac{Z_1}{r_1} + \frac{Z_2}{r_2} \quad \text{with} \quad Z_j = a_j^{-1} \exp\left(-\frac{2a_j}{\lambda} r_j\right), \quad j = 1, 2, \quad (13.41)$$

where $a_j = r_{12}/(2r_j)$. Equation (13.41) indicates that the effect of the electron–electron interaction potential can be viewed as a modification Z_j/r_j to the single-particle potentials. The interelectronic correlation is subsumed into a dynamic nonlocal screening of the interaction w_j of the electron with the lattice. The behavior of this screening is dictated by the functions Z_j, and has the following features: when the two electrons are on top of each other ($r_{12} \to 0$) the potential w_j turns repulsive so as to simulate the strong, short-range electron–electron repulsion. If the two electrons are far away from each other ($r_i \gg r_j$, $i \neq j \in [1, 2]$), the screening strengths Z_1 and Z_2 become negligible and we end up with two independent particles. For the numerical evaluation of the two-photoelectron current Z_j have been approximated by $\bar{Z}_j$, where $\bar{Z}_j = \bar{a}_j^{-1} \exp\left(-2\bar{a}_j r_j/\lambda\right)$ and $\bar{a}_j = k_{12}/(2k_j)$. Here, $\boldsymbol{k}_{12} = \boldsymbol{k}_1 - \boldsymbol{k}_2$ is the interelectronic relative wave number. With this screening being included, the modification $\bar{Z}_j/r_j$ to the original single-particle potential w_j can be taken into account and the single-particle Green's function $\bar{g}_j$ is generated. However, in contrast to g, each $\bar{g}_j$ is dependent on the wave vectors of *both* electrons as well as on the mutual, relative wave vector of the escaping electrons. The two-particle Green's function is approximated by $\bar{g}^{II}$ which is the antisymmetrized, direct product of the modified single-particle Green's functions $\bar{g}_j$ (the first term in the ladder approximation, (13.38)). This model yielded useful results for the (e,2e) cross sections [42–44].

Angular Pair Correlation Functions for Cu(001). Here, the $(\gamma, 2e)$ calculation from Cu(001) are described. The ground-state potentials of the Cu(001) surface are calculated self-consistently with the scalar-relativistic LMTO method. Lifetime broadening of the spectra is simulated by employing a complex optical potential and the photoelectrons' current due to emission from the first 20 outermost layers is calculated. Convergence of the results with respect the maximum angular momentum, number of reciprocal lattice vectors, and accuracy of the energy integration is achieved.

Figure 13.2 shows the angular distributions of the current for single photoemission (labeled SPE) and the double photoemission (indicated by DPE) for an incoming s-polarized photon. The intensity variation is shown as a

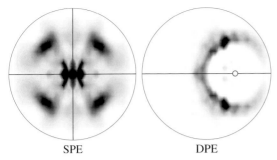

SPE   DPE

Fig. 13.2. The angular distribution of the photoemission intensity from Cu(001) in single-electron (labeled SPE) and double-electron photoemission (labeled DPE). The kinetic energy of the photoelectrons is 9 eV, the photon energy of the s-polarized light is 15.5 eV in the SPE case whereas in DPE the photon energy is increased to 31 eV as to compensate for the additional energy needed to emit two electrons instead of one. For the DPE case, the small circle indicates the emission direction of one of the electrons (polar angle with respect to the surface normal is $\vartheta = 40°$). Low (high) intensities correspond to *light (dark) gray* scale in the stereographic projection. *Horizontal* and *vertical lines* emphasize the symmetries of the angular distributions

function of the angular position of one photoelectron with 9 eV kinetic energy. The photon energy is 15.5 eV in the single photoemission case and 31 eV for double photoemission. We note that in the latter case both photoelectrons escape with the same kinetic energy of 9 eV. For SPE, the point group of the surface 4 mm is reduced to 2 mm in the angular distribution (as indicated by the horizontal and vertical lines in Fig. 13.2) because the electric-field vector of the incident photon (which lies in a mirror plane of the surface) is not invariant under the operations C_4 and C_4^{-1}. On the other hand in DPE, the group 2 mm is reduced further to m (horizontal line in Fig. 13.2), due to the presence of the second photoelectron. From Fig. 13.2 it is clear that the repulsion between the two escaping photoelectrons leads to a vanishing photoelectron current when the two electrons are close to each other. This is the origin of the *correlation hole* surrounding the fixed detector position. On the other hand, if the two electrons are far from each other, the electron–electron interaction diminishes in strength. Consequently, the two-electron current drops dramatically, for this current must vanish in the absence of correlation. The interplay between these two effects leads to a localization of the angular-intensity distribution of one of the photoelectrons around the position of the second one, as observed in Fig. 13.2. It should be stressed that the two detected photoelectrons are not only coupled to each other but also to the crystal potential. Therefore, the *correlation hole* is not isotropic in space and depends sensitively on the photoelectrons' energies.

13.9   Conclusion

This chapter gives a general overview on the foundations of many-body techniques for the treatment of single and multiple excitations in few and many-body systems. The author's goal has been to emphasis common features and differences when the system size and/or the number of interactions increase. While the wave function approach is well suited for the treatment of few interacting particles it becomes less valuable for large compounds. In the thermodynamic limit the well-established Green's function technique is a powerful tool for the description of correlated excitations in many-body systems.

References

1. D. Pines: Phys. Rev. **92**, 626 (1953); E. Abrahams: Phys. Rev. **95**, 839 (1954); D. Pines, P. Nozieres: *The Theory of Quantum Liquids* (Addison-Wesley, Reading 1966)
2. A. Gonis: *Theoretical Materials Science: Tracing the Electronic Origins of Materials Behavior* (Materials research society, Warrendale 2000)
3. A.L. Fetter, J.D. Walecka: *Quantum Theory of Many-particle Systems* (McGraw-Hill, New York 1971)
4. I. Lindgren, J. Morrison: *Atomic Many-Body Theory* (Springer, Berlin, Heidelberg, New York 1982)
5. L. Rosenberg: Phys. Rev. D **8**, 1833 (1973)
6. M. Brauner, J.S. Briggs, H. Klar: J. Phys. B **22**, 2265 (1989)
7. J.S. Briggs: Phys. Rev. A **41**, 539 (1990)
8. H. Klar: Z. Phys. D **16**, 231 (1990); J. Berakdar: (thesis) University of Freiburg i.Br. (1990) (unpublished)
9. E.O. Alt, A.M. Mukhamedzhanov: Phys. Rev. A **47**, 2004 (1993); E.O. Alt, M. Lieber: ibid. **54**, 3078 (1996)
10. J. Berakdar, J.S. Briggs: Phys. Rev. Lett. **72**, 3799 (1994); J. Phys. B **27**, 4271 (1994)
11. J. Berakdar: Phys. Rev. A **53**, 2314 (1996)
12. S.D. Kunikeev, V.S. Senashenko: Zh.E'ksp. Teo. Fiz. **109**, 1561 (1996); [Sov. Phys. JETP **82**, 839 (1996)]; Nucl. Instrum. Methods B **154**, 252 (1999)
13. D.S. Crothers: J. Phys. B **24**, L39 (1991)
14. G. Gasaneo, F.D. Colavecchia, C.R. Garibotti, J.E. Miraglia, P. Macri: Phys. Rev. A **55**, 2809 (1997); G. Gasaneo, F.D. Colavecchia, C.R. Garibotti: Nucl. Instrum. Methods B **154**, 32 (1999)
15. J. Berakdar: Phys. Lett. A **220**, 237 (1996); ibid. **277**, 35 (2000); Phys. Rev. A **55**, 1994 (1997)
16. R.K. Peterkop: *Theory of Ionisation of Atoms by the Electron Impact*, (Colorado Associated University Press, Boulder 1977)
17. J. Berakdar: Phys. Rev. Lett. **78**, 2712 (1997)
18. J. Berakdar: to be published
19. G.D. Mahan: *Many-Particle Physics*, second edn (Plenum Press, London 1993)
20. A.A. Abrikosov, L.P. Korkov, I.E. Dzyaloshinski: *Methods of Quantum Field Theory in Statistical Physics* (Dover, New York 1975)

21. J. Berakdar: Aust. J. Phys. **49**, 1095 (1996)
22. M. Abramowitz, I. Stegun: *Pocketbook of Mathematical Functions* (Verlag Harri Deutsch, Frankfurt 1984)
23. J. Berakdar: Phys. Rev. Lett. **85**, 4036 (2000)
24. C.N. Yang, T.D. Lee: Phys. Rev. **97**, 404 (1952); ibid. **87**, 410 (1952)
25. S. Grossmann, W. Rosenhauer: Z. Phys. **207**, 138 (1967); ibid. **218**, 437 (1969)
26. M.N. Barber: *Finite-size Scaling* (Phase Transitions and Critical Phenomena) eds. C. Domb, J.L. Lebowity (Academic Press, New York 1983) pp. 145–266
27. L. Hedin: J. Phys. C **11**, R489 (1999)
28. J. Berakdar, O. Kidun, A. Ernst: in *Correlations, Polarization, and Ionization in Atomic Systems*, eds. D.H. Madison, M. Schulz (AIP, Melville, New York 2002) pp. 64–69
29. O. Kidun, J. Berakdar: Phys. Rev. Lett. **87**, 263401 (2001)
30. S. Keller: Eur. Phys. J. D **13**, 51 (2001)
31. S. Matt, B. Dünser, M. Lezius, K. Becker, A. Stamatovic, P. Scheier, T.D. Märk: J. Chem. Phys. **105**, 1880 (1996)
32. V. Foltin, M. Foltin, S. Matt, P. Scheier, K. Becker, H. Deutsch, T.D. Märk: Chem. Phys. Lett. **289**, 181 (1998)
33. S. Keller, E. Engel: Chem. Phys. Lett. **299**, 165 (1999)
34. K.A. Brueckner: Phys. Rev. **97**, 1353 (1955); H.A. Bethe: Ann. Rev. Nucl. Sci. **21**, 93 (1971); J.P. Jeukenne, A. Legeunne, C. Mahaux: Phys. Rep. **25**, 83 (1976)
35. A.B. Migdal: *Theory of Finite Fermi Systems* (Interscience Pub., New York 1967)
36. P.C. Martin, J. Schwinger: Phys. Rev. **115**, 1342 (1959)
37. H. Lehmann: Nuovo Cimento A **11**, 342 (1954)
38. S.D. Kevan (Ed.): *Angle-Resolved Photoemission: Theory and Current Application*, (Studies in Surface Science and Catalysis) (Elsevier, Amsterdam 1992)
39. S. Hüfner: *Photoelectron Spectroscopy* (Springer Series in Solid-State Science, Vol. 82) (Spinger, Berlin, Heidelberg, New York 1995)
40. R. Herrmann, S. Samarin, H. Schwabe, J. Kirschner: Phys. Rev. Lett. **81**, 2148 (1998); J. Phys. (Paris) IV **9**, 127 (1999)
41. N. Fominykh, J. Henk, J. Berakdar, P. Bruno, H. Gollisch, R. Feder: Solid State Commun. **113**, 665 (2000)
42. J. Berakdar, H. Gollisch, R. Feder: Solid State Commun. **112**, 587 (1999)
43. N. Fominykh, J. Henk, J. Berakdar, P. Bruno, H. Gollisch, R. Feder: in *Many-particle Spectroscopy of Atoms, Molecules, Clusters and Surfaces*. Eds. J. Berakdar, J. Kirschner, (Kluwer Acad/Plenum Pub., New York 2001)
44. N. Fominykh, J. Berakdar, J. Henk, P. Bruno: Phys. Rev. Lett. **89**, 086402 (2002)

14 From Atoms to Molecules

R. Dörner, H. Schmidt-Böcking, V. Mergel, T. Weber, L. Spielberger,
O. Jagutzki, A. Knapp, and H.P. Bräuning

14.1 Introduction

In the present chapter, we discuss direct photo double ionization by single-photon absorption (Sect. 14.2) and Compton scattering (Sect. 14.3). We do not discuss the closely related phenomenon of multiple ionization by two-step processes such as photoionization followed by single- or multiple-Auger decay. We concentrate on the two most fundamental two-electron target systems: the helium atom (Sects. 14.2 and 14.3) and molecular hydrogen (deuterium) (Sect. 14.4). The subject of photo double ionization of helium is now a mature field in which an impressive experimental and theoretical breakthrough has been achieved in the previous 10 years. The theoretical progress is described in Part III of this book, we therefore restrict ourselves here to a phenomenological description and intuitive interpretation of the physical phenomena. For the problem of two-electron processes in molecules, in contrast, the major challenges for experimentalist and theoretician still lie ahead.

14.2 Double Ionization of Helium by Photoabsorption

14.2.1 Energy, Momentum, and Angular Momentum Considerations

Double ionization of helium by photoabsorption becomes possible if the energy of the photon E_γ is higher than the sum of the binding energies of both electrons $E_{\text{ion}}^{2+} = 24.6\,\text{eV} + 54.4\,\text{eV} = 79\,\text{eV}$. The excess energy $E_{\text{exc}} = E_\gamma - E_{\text{ion}}^{2+}$ can be shared among the two electrons in the continuum $E_1 + E_2 = E_{\text{exc}}$. The kinetic energy of the He^{2+} nucleus is negligible due to its heavy mass. In momentum space, however, the momenta of the electrons and the nucleus are of the same order of magnitude. From momentum conservation we obtain (assuming the atom to be at rest in the initial state):

$$\boldsymbol{k}_\gamma = \boldsymbol{k}_1 + \boldsymbol{k}_2 + \boldsymbol{k}_{\text{He}^{2+}}\,. \tag{14.1}$$

At nonrelativistic energies the photon momentum can be neglected against the electron and ion momenta ($k_\gamma \approx 0$). Hence, in the final state, the sum of the two-electron momenta is balanced by the ion momentum (see Chap. 1 for a more detailed discussion). At photon energies of below typically 1 keV the dipole approximation is expected to hold. Therefore, the absorption of the photon leads to a change of the angular momenta $\Delta L = 1$ and a change in parity between the initial and final state. Since the He ground state is an S state with gerade parity, the three-body final state is $^1P^\circ$. Angular momentum is not a good quantum number for the individual electron, but the two electrons have to couple to angular momentum $L = 1$ with odd parity. This $^1P^\circ$ character of the 3-body final state shapes the momentum and angular distributions, as will be discussed below in more detail.

The criterion for the validity of the dipole approximation is $k_\gamma r \ll 1$, where r is the typical size of the system (e.g., 1 a.u.). For single ionization there are detailed calculations including higher-order contributions [1], confirming the validity of the dipole approximation at $E_\gamma < 1$ keV. For double ionization, no experimental evidence of any deviation from the dipole approximation has been found so far. Kornberg and Miraglia [2] performed the only theoretical study of double ionization beyond the dipole approximation. They find no deviation for the ratio of double to single ionization cross section R_γ and only a small deviation in the angular distribution at 1 keV. The further discussion in this chapter will, therefore, be restricted to phenomena and arguments within the dipole approximation.

The three particles in the final state are determined by nine momentum components. Due to momentum and energy conservation, however, only five of them are linearly independent. The double-ionization process is therefore fully determined by a 5-fold differential cross section (FDCS). Sometimes this is also called a triply differential cross section. In this notion the linearly independent polar ϑ and azimuthal $\varPhi$ angle of the electrons are combined to a solid angle $\varOmega$, the fully differential cross section is then noted as $\mathrm{d}^3\sigma/\mathrm{d}E\mathrm{d}\varOmega_1\mathrm{d}\varOmega_2$. The dipole approximation results in a further symmetry axis in the final state (rotational symmetry around the polarization axis for linear light). This results in a further reduction to an only four-fold differential cross section. To measure such a cross section the experimentalist can freely choose which five out of the nine momentum components to measure. Using dispersive [3–9], time-of-flight [10,11] electron spectrometers or advanced imaging techniques [12], several groups succeeded in detecting the momenta of both electrons without detection of the ion. Alternatively COLTRIMS has been used to measure the momentum vector of the ion in coincidence with one of the electrons [13–19].

14.2.2 Probability and Mechanisms of Double Ionization

In most cases, the absorption of a photon will lead to single ionization of the helium atom with the He^{1+} ion in the ground state. The two-electron processes of ionization plus excitation and double ionization are of the order of a few per cent of the absorption cross section. They are solely a consequence of the electron–electron interaction. Today, the absolute value of R_γ is settled to

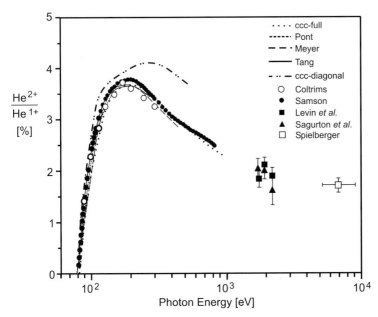

Fig. 14.1. Ratio of the total double to the total single ionization cross sections in helium by photoabsorption. *Open circles*: COLTRIMS data [20], *open square*: COLTRIMS data for photoionization only [23], *full dots* Samson et al. [21], *full triangles* [31], *full square* [32], *dotted and dash dotted line* [26], *short dashed line* [33], *long dashed line* [34], *full line* [35]

an accuracy of a few per cent experimentally and theoretically. R_γ rises almost linearly from the threshold, reaches a maximum of 3.7% at $E_\gamma = 200\,\text{eV}$ and slowly approaches the high-energy asymptotic value of 1.67% (Fig. 14.1). Below 1 keV the precision experiments by Dörner et al. [20] and Samson et al. [21] are in good agreement with each other and supersede older experiments, which were about 25% higher (see [20] for a comparison and discussion of these older experiments). In the high energy regime, the pioneering work of Levin and coworkers [22] reported an experimental value of $R_\gamma = 1.6 \pm 0.3\%$ at 2.8 keV. A measurement by Spielberger and coworkers [23] at 7 keV found $R = 1.72 \pm 0.12\%$ and thus confirmed that the high-energy limit has been reached. A collection of the data and some of the theoretical results are shown in Fig. 14.1.

What are the "mechanisms" leading to the ejection of both electrons? This seemingly clear-cut question does not necessarily have a quantum-mechanical answer. The word "mechanism" mostly refers to an intuitive mechanistical picture. It is not always clear how this intuition can be translated into theory and even if one finds such a translation the contributions from different mechanisms have to be added coherently to obtain the measurable final state of the reaction [24–26]. With these words of caution in mind, we list the most discussed mechanisms leading to double ionization:

1. *Shake-Off:* If one electron is removed rapidly (sudden approximation, nonadiabatic) from an atom or a molecule, the wave function of the remaining electron has to relax to the new eigenstates of the altered potential. Some of these states are in the continuum, so that a second electron can be "shaken-off" in this relaxation process. The overlap of the initial state $\Psi(k_1, k_2)$ with the continuum depends on the momentum k_1 of the primary electron [27], i.e., the shake-off probability for electron 2 is a function of k_1. Photoabsorption in the dipole approximation selects the fraction of the initial-state wave function, where the initial bound momentum is equal to the continuum momentum $k_1 = \sqrt{2E_{exc}}$. In the limit $k_1 \to \infty$ one obtains a shake-off ratio of $R_\gamma = 1.67\%$ for the best correlated initial-state wave function [28,29], in perfect agreement with experiment (see Fig. 14.1). In coordinate space this limit corresponds to picking electron 1 at the nucleus. Due to this dependence of the shake-off probability on k_1, Compton scattering leads to a much smaller ratio R (see discussion in Sect. 14.3).

2. *Two-Step-One (TS1):* A simplified picture of TS1 is that one electron absorbs a photon and knocks out the second electron via an electron–electron collision on its way through the atom [30]. A close connection between the electron-impact ionization cross section and R_γ as a function of the excess energy is seen experimentally [30] and theoretically [26], supporting this simple picture.

Thus, the high energy value of R_γ is given by the shake-off process, the shape of the curve from threshold to a few hundred eV can be understood in analogy to electron impact. The rise at threshold, like E_{exc}^α, with the Wannier coefficient $\alpha = 1.056$ is identical for double photoionization [36] and electron-impact ionization [30]. It is a consequence of the final-state phase space density and is described by the Wannier threshold law [37]. If one fragments a system of charged particles with very little excess energy, the evolution of the many-body-continuum wave function is governed by the saddle region of the potential energy surface in the continuum. The system does not carry any memory of the ionization process nor of the initial state it emerged from. The final state is, however, constraint by energy, parity, and angular momentum conservation laws. For two electrons and one positive particle, the Wannier configuration is simply given by the nucleus in the center between the two electrons screening their repulsion. This configuration is, however, prohibited by the odd parity of the final state. Extending the description of the saddle region to fourth order allows one a reasonable description of even the fully differential cross section up to about 20 eV [13,14,38,39].

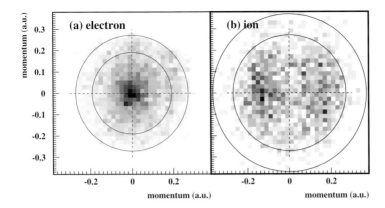

Fig. 14.2. Density plots of projections of the momentum distribution from double ionization of He by 80.1 eV linearly polarized photons (compare with Fig. 14.7 for the corresponding presentation for single ionization). The polarization vector of the photon is in the horizontal direction and the photon propagates in the vertical direction. (a) The distribution of single-electron momenta (k_1 or k_2). Only events with momentum components out of the plane of $-0.1 < k_{He2+} < 0.1$ are projected onto the plane. The *outer circle* locates the momentum of an electron that carries the full excess energy. (b) The recoil momentum distribution. The *outer circle* indicates the maximum calculated recoil momentum (from [13])

14.2.3   Electron and Ion Momentum Distributions

For single ionization, the ion- and electron-momentum distributions are identicall, since momentum conservation requires back-to-back emission. An example for 80–eV linearly polarized photons is shown in Fig. 14.7. The outer ring shows ions in the ground state, which exhibit a $\cos^2(\vartheta)$ dipolar distribution. The inner rings are excited ions, corresponding to the satellite lines in the electron energy spectrum. Figure 14.2 compares this to the momentum distributions of the He^{2+} ion and one of the electrons from double ionization. The nucleus clearly shows a dipolar emission pattern as a result of the absorption of the photon. This characteristic of the primary absorption process is completely washed out in the electronic-momentum distribution. This highlights the fact, that the nucleus as the center of positive charge in the system always participates in the absorption of the photon. It is the electron–electron interaction that is always required for double ionization, which smears this out, reminiscent of the photon angular momentum in the momentum distribution of a single electron. A more detailed discussion of this problem can be found in [13,15,39,40].

The energy distribution of the electrons is almost flat up to a photon energy of about 100 eV, i.e., all energy sharings are about equally likely [41–44]. Far from threshold, however, the electron energy distribution is extremely u-shaped. In this case, the fast electron has a β parameter of almost 2.[1] This indicates that at high photon energies, the photon energy and angular momentum is absorbed predominantly by one electron, which, in addition, can be experimentally distinguished.

14.2.4 Fully Differential Cross Sections

The internal structure of the square of the correlated two electron continuum wave function is shown in Fig. 14.4. Neglecting the small photon momentum the vector momenta of ion and both electrons have to be in one plane.

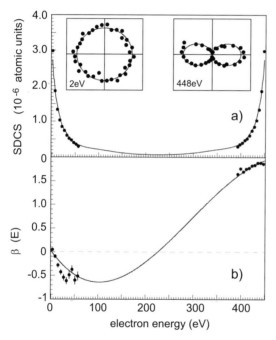

Fig. 14.3. Photo double ionization of He at $\hbar\omega = 529\,\text{eV}$. (**a**) SDCS $d\sigma/dE$. The line is the CCC calculation. The insets show the electron angular distribution $d\sigma^2/(d\Omega dE)$ at electron energies of $E = 2\,\text{eV}$ and $448\,\text{eV}$ (the vertical axis is the light propagation), *the line* is obtained using (14.2). The experimental data are normalized to the CCC calculation. (**b**) The asymmetry parameter β versus the electron energy. The *full line* is a polynomial fit through points calculated in CCC theory

---

[1] The angular distribution of electrons and ions is given by

$$\frac{d^2\sigma(\vartheta,\phi)}{d\Omega} = \frac{\sigma}{4\pi}\left(1 + \beta\left(\frac{3}{2}\cos\vartheta^2 - \frac{1}{2}\right)\right), \tag{14.2}$$

where $\beta = 2$ corresponds to a pure dipole distribution.

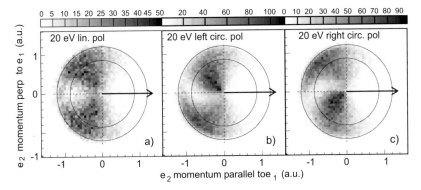

Fig. 14.4. Photo double ionization of He at 20 eV above threshold by linearly (**a**) polarized light, left (**b**) and right (**c**) circular-polarized light. Shown is the momentum distribution of electron 2 for fixed direction of electron 1 as indicated by the *arrow*. The plane of the figure is the momentum plane of the three particles. The data of (**a**) are integrated over all orientations of the polarization axis with respect to this plane. The figure samples the full cross sections, for all angular and energy distributions of the fragments. The *outer circle* corresponds to the maximum possible electron momentum, the *inner one* to the case of equal energy sharing. In (**b**) and (**c**) the light propagates into the plane of the figure, the electrons are confined to the plane perpendicular to the light propagation (from [54] and [18])

Figure 14.4a shows the electron-momentum distribution in this plane for linearly polarized light. The data are integrated over all orientations of the polarization axis with respect to this plane, the x-axis is chosen to be the direction of one electron. The structure of the observed momentum distribution is dominated by two physical effects. First the electron–electron repulsion leads to almost no intensity for both electrons in the same half-plane. Secondly, the $^1P^\circ$ symmetry leads to a node in the square of the wave function at the point $k_1 = -k_2$ [3,45–47]. The corresponding data for left- and right-circular-polarized light are shown in Fig. 14.4b,c. They show a strong circular dichroism, i.e., a dependence on the chirality of the light. This might be surprising since the helium atom is perfectly spherical symmetric. Berakdar and Klar [48] first pointed out that for circular dichroism to occur it is sufficient that the direction of light propagation and the momentum vectors of the electrons span a tripod of defined handedness. This is the case if the two electrons and the light direction are non coplanar and the two electrons have unequal energy (see [49–51] for a detailed discussion and experimental results [9,10,17,52,53]).

The fully differential cross section for linear polarized light is obtained from Fig. 14.4a by also fixing the direction of polarization and then plotting the countrate along a circle. Data from the pioneering work of Schwarzkopf et al. [3] for equal energy sharing (corresponding to the inner circle in Fig. 14.4a

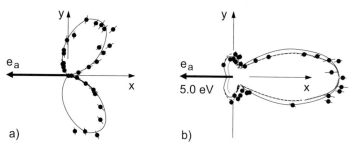

Fig. 14.5. FDCS of the He photo double ionization at (**a**) 20 eV and (**b**) 52.9 eV excess energy. Electron a is indicated by the *arrow*, polarization horizontal. (**a**) $E_a = E_b = 10$ eV, (**b**) $E_a = 5$ eV, $E_b = 47.9$ eV. *Full line* (**a**) fit with Gaussian correlation function (see text), (**b**) full line 3C from Maulbetsch and Briggs [55], *dotted* Pont and Shakeshaft [56], *chain* CCC [57] (adapted from [3,4,58])

are shown in Fig. 14.5a. Again the electron repulsion and the node for back-to-back emission is visible. The full curve shows a fit using a parametrization suggested by Huetz and others [45,47]. They have shown that within the dipole approximation the FDCS can be written as:

$$\frac{\mathrm{d}^4\sigma}{\mathrm{d}E_1 d\cos\vartheta_1 d\cos\vartheta_2 d\phi} \sim |(\cos\vartheta_1 + \cos\vartheta_2)a_g(E_1,\vartheta_{12})$$
$$+ \quad (\cos\vartheta_1 - \cos\vartheta_2)a_u(E_1,\vartheta_{12})|^2 \qquad (14.3)$$

with two arbitrary complex functions functions a_u and a_g of the angle between the two electrons ϑ_{12} and the energy sharing. The amplitude a_u is antisymmetric under exchange of the electrons so that $a_u = 0$ for $E_1 = E_2$. The advantage of this approach is that it splits the cross section into a trivial part which describes the symmetry of the $^1P^\circ$ state, and two functions of lower dimension, that describe the three-body dynamics. For a_g a Gaussian with FWHM of 91° has been used in Fig. 14.5a. For very unequal energy sharing a_u and a_g contribute and the selection rules allow for a much richer pattern (Fig. 14.5b shows an example).

From (14.3) one can also read the three main selection rules [46] that impose restriction on the cross section:

1. For equal energy sharing back-to-back emission is forbidden (see node in Fig. 14.4). This holds for linear and circular polarization.
2. For equal energy sharing and linearly polarized light $\vartheta_1 \neq 180° - \vartheta_2$, where ϑ denotes the polar angle of the electron to the polarization. This node is a cone around the polarization axis and the back-to-back emission is part of this cone.
3. For all energy sharings both electrons can not emerge perpendicular to the polarization ($\vartheta_1 = \vartheta_2 = 90°$ is forbidden).

While at low energies the long-range final-state interaction shapes the FDCS, at very high energies one electron leaves rapidly (see Fig. 14.3) and clear traces

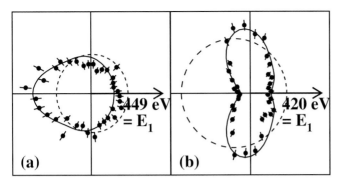

Fig. 14.6. FDCS of the He PDI at 529 eV photon energy. The primary photoelectron 1 indicated by the *arrow*, the polarization is horizontal, the angular distribution of the complementary electron 2 with energy E_2 given by the symbols. (**a**) $447 < E_1 < 450$ eV, $0 < E_2 < 3$ eV, (**b**) $410 < E_1 < 430$ eV, $20 < E_2 < 40$ eV. (**a**) shows the dominance of shake-off, the 90° emission in (**b**) indicates the importance of TS1 at this energy. The *solid line* shows the full CCC calculation, the *dashed line* is the shake-off only part of the CCC calculation (from [19])

of the shake-off and TS1 mechanism can be found in the angular distribution of the slow electron (see also [25,44]). The shake-off electron is expected to be isotropic or slightly backward directed with respect to the primary electron, while TS1 will yield 90° between the two electrons. At 529 eV photon energy the electron angular distributions show a dominance of the shake-off mechanism for secondary electrons that have very low energy (2 eV) and display clear evidence that an inelastic electron–electron scattering is necessary to produce secondary electrons of 30 eV [19] (see Fig. 14.6).

14.3   Double Ionization of Helium by Compton Scattering

At photon energies above 6 keV, the ionization cross section of helium by Compton scattering exceeds the photoabsorption cross section [59,60]. To experimentally determine the ratio of the total double to total single ionization cross section, it is, therefore, necessary to detect not only the charge state of the ions, but also determine wether they are created by absorption or Compton scattering. This can be done most easily by measuring the ion momentum. Ions from photoabsorption compensate the electron momenta and, hence, have comparably high momenta. In a Compton scattering event, however, the electron momentum is provided by the scattered photon, while the ion core is only a spectator. As a consequence cold ions are produced (see Chap. 1). This has been used by Samson and coworkers to measure the single-ionization Compton scattering cross section [61]. Spielberger and coworkers pioneered COLTRIMS to measure R for photoabsorption (R_γ) and Compton

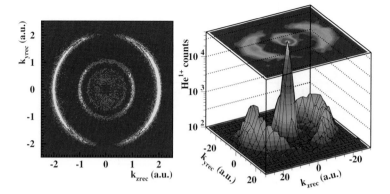

Fig. 14.7. Momentum distributions of singly charged ions created by 80 eV photons (*left*: photoabsorption only) and 7 keV photons (*right*: outer rim – absorption, *narrow peak* – Compton scattering). The polarization is horizontal (see text)

scattering (R_C) [23] separately (see also [62–64]). Figure 14.7 shows the measured He^{1+} momentum distribution created by about 9 keV photon impact. The circular rim results from highly energetic ions from photoabsorption, while the narrow peak at the origin are ions from Compton scattering. Since Compton scattering produces a continuum of electron energies, very high photon energies are necessary to approach the asymptotic shake-off limit [63]. The shake-off probability for Compton scattering is predicted to be 0.86% [63] and is not fully reached at 100 keV photon energy. $R_C^{\infty} = 0.86\%$ differs significantly from $R_{\gamma}^{\infty} = 1.67\%$ because high-energy photoabsorption removes electrons with a high momentum component in the initial state, while Compton scattering samples the full initial momentum space. Up to now, there are no differential experimental data on double ionization by Compton scattering available. In principle, eight degrees of freedom would have to be determined for a fully differential cross section. Such data are very desirable for the future since they complement (e,3e) and ion-impact double ionization studies, but avoid some of the problems since there are only three charged particles in the final state.

14.4   Double Ionization of H_2

Double ionization of H_2 is, from the experimental as well as from the theoretical side, much more challenging than atomic double ionization, since it is a 4-body problem. In most cases, however, the electronic and nuclear motion can be decoupled (Born–Oppenheimer approximation) (see [65,66] for a theoretical discussion and [67] for the relationship of ion and electron energies). Within this approximation, the four-body problem is reduced to the problem of two electrons moving and scattering in a two-center potential. Long after the two electron have left the molecule, the two protons will Coulomb–explode. Since the molecular rotation is slow compared to the fragmentation one obtains the alignment of the internuclear axis at the instant of photoabsorption from the measured direction of the protons (axial recoil approximation) and the internuclear distance can be inferred from the proton energy (reflection approximation). This technique for measuring electron angular distribution with respect to the molecular axis for one electron is discussed in detail in two other chapters of this book (see, e.g., [68–74]).

The main difference between atomic and molecular photoionization is that in the atomic case the angular momentum is a good quantum number of the continuum electron wave function and within the dipole approximation only transition with $\Delta L = 1$ and a change in parity are allowed. For linear molecules angular momentum conservation also requires $\Delta L = 1$, but this is the angular momentum of the total wave function, including the nuclei. Total angular momentum is no longer a good quantum number of the electronic part of the wave function alone. Angular momentum can be exchanged between the electronic and nuclear wave function. The electron(s) escaping from the molecule can leave a rotating molecule behind (this rotation can be seen experimentally [75]). Hence, for a linear molecule only the projection of the electronic angular momentum onto the molecular axis is a good quantum number of the electronic wave function. One may think of the angular momentum transfer between electron and nuclei as a scattering of the electron wave at the nuclei. At an electron energy of 1 a.u. and a typical distance of 0.7 a.u. between the molecule center-of-mass and the nucleus one can expect angular momentum exchange of up to a few a.u. These higher angular momentum components in the electronic wave function allow for a rich structure already in the angular distribution of one electron in the molecule-fixed frame [76–78]. For double ionization, such higher momentum components are also predicted [79].

For H_2 double ionization in pioneering experiments Kossmann et al. [80] measured the angular distribution of the protons, without detecting the electrons. They found a strong dependence of the double-ionization cross section on the molecular orientation, i.e., a strong anisotropy of the heavy fragments. No physical explanation for this observation has been reported so far. Reddish and coworkers studied the angular correlation between the two electrons without detection of the protons. They found a surprising similarity between He and H_2 (see Fig. 14.8 and references [81–83]). From these experiments it seems that the molecular effects on the two-electron wave function are small, at least for equal energy sharing. One reason for this similarity between He and H_2 is that back-to-back emission of equal energy electrons is forbidden also for H_2 (see [79,84,85]). This is because, in this case, the sum momentum of the two electrons vanishes. There is no momentum coupling between electronic and nuclear wave functions, the nuclei must have opposite and equal momenta and hence defined gerade parity. The additional selection rule pro-

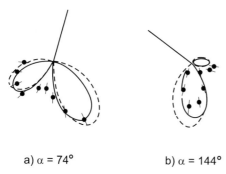

a) α = 74°   b) α = 144°

Fig. 14.8. FDCS for double ionization of D_2. Both electrons have 10 eV, the direction of the first electron is indicated by the arrow, both electrons are coplanar. The *dashed line* is a fit to the equivalent data for He with a Gaussian correlation function of FWHM 91°, the *full line* is FWHM=78°. From [81]

hibiting $\vartheta_1 = 180° - \vartheta_2$, which is valid for He, does not hold for D_2 [84,85]. Due to a limited experimental angular resolution of the data in Fig. 14.8, this gives rise to an apparent filling of the node [83,85].

The first kinematically complete experiment in which both electrons and both protons are detected in coincidence have been performed only very recently by Weber and coworkers [86,87]. Some of the physical effects that have been predicted by theory to become visible in such experiments are:

1. Contributions from higher angular momentum components to the two-electron wave function [79,84] as discussed above.
2. Interference from the two centers in the H_2 molecule [88,79].
3. Different electron-emission patterns for parallel and perpendicular orientation of the molecule to the polarization [79].
4. Dependence of the electron-emission pattern on the internuclear distance, i.e. on the kinetic energy release [65,66]. This can be expected because the initial state electron wave function and, in addition, the scattering of the electron wave at the nuclei in the exit channel depend on the internuclear distance.

To explore this fascinating physics in the seemingly simple process of fragmentation of H_2 remains a major challenge to experiment and theory.

14.5   Conclusions and Open Questions

Decisive experimental and theoretical progress has been made concerning double ionization of helium by photoabsorption. For ground-state helium, a very good agreement between theory and experiment has been reached for experiment from $E_{exc} = 100$ meV to $E_{exc} = 450$ eV. It is, however, unclear at which energies the dipole approximation will break down and which new features in the differential cross section can be expected beyond this. Also, the promise that photo double ionization can be used as a tool for correlation spectroscopy of the ground state [89,90] has not been kept. For further work in this direction, a comparison of the ground-state double ionization to double ionization from metastable excited states of helium or even a spin-polarized triplet state seems promising. Also, the relation of double ionization by photons and charged-particle impact (double ionization and transfer ionization) is still presently heavily discussed (see, e.g., a contribution by Schmidt-Böcking given in Chap. 20).

For double ionization by Compton scattering only total cross sections are available. Much work is required here on the experimental as well as on the theoretical side. Experiments are particulary difficult since the cross section is only $10^{-26}\,\mathrm{cm}^2$ and due to the photon in the final state there are three more degrees of freedom compared to photoabsorption. Again, the close relationship between Compton scattering and ionization by a binary charged particle collision further fuel the interest in this subject.

Due to the rapid progress in multiparticle imaging techniques, the step from the kinematically complete double-ionization experiment in atoms to molecu-les is feasible experimentally. The first theoretical results in this field are available already. The exciting prospects include the search for a breakdown of the Born–Oppenheimer approximation, for a coupling between electronic and nuclear wave functions and for a dependence of the double-ionization process on the internuclear distance.

References

1. J.W. Compton: Phys. Rev. A **47**, 1841 (1993)
2. M.A. Kornberg and J.E. Miraglia: Phys. Rev. A **52**, 2915 (1995)
3. O. Schwarzkopf, B. Krässig, J. Elmiger, and V. Schmidt: Phys. Rev. Lett. **70**, 3008 (1993)
4. O. Schwarzkopf, B. Krässig, V. Schmidt, F. Maulbetsch, and J. Briggs: J. Phys. **B27**, L347–50 (1994)
5. O. Schwarzkopf and V. Schmidt: J. Phys. B **28**, 2847 (1995)
6. P. Lablanquie, J. Mazeau, L. Andric, P. Selles, and A. Huetz: Phys. Rev. Lett. **74**, 2192 (1995)
7. C. Dawson, S. Cvejanovic, D.P. Seccombe, T.J. Reddish, F. Maulbetsch, A. Huetz, J. Mazeau, and A.S. Kheifets: J. Phys. B **34**, L525 (2001)
8. S. Cvejanovic, J.P. Wightman, T.J. Reddish, F. Maulbetsch, M.A. MacDonald, A.S. Kheifets, and I. Bray: J. Phys. **B33**, 265 (2000)
9. K. Soejima, A. Danjo, K. Okuno, and A. Yagishita: Phys. Rev. Lett. **83**, 1546 (1999)
10. J. Viefhaus, L. Avaldi, G. Snell, M. Wiedenhöft, R. Hentges, A. Rüdel, F. Schäfer, D. Menke, U. Heinzmann, A. Engelns, J. Berakdar, H. Klar, and U. Becker: Phys. Rev. Lett. **77**, 3975 (1996)
11. J. Viefhaus, L. Avaldi, F. Heiser, R. Hentges, O. Gessner, A. Rüdel, M. Wiedenhöft, K. Wielczek, and U. Becker: J. Phys. **B29**, L729 (1996)

258   R. Dörner et al.

12. A. Huetz and J. Mazeau: Phys. Rev. Lett. **85**, 530 (2000)
13. R. Dörner, J. Feagin, C.L. Cocke, H. Bräuning, O. Jagutzki, M. Jung, E.P. Kanter, H. Khemliche, S. Kravis, V. Mergel, M.H. Prior, H. Schmidt-Böcking, L. Spielberger, J. Ullrich, M. Unverzagt, and T. Vogt: Phys. Rev. Lett. **77**, 1024 (1996); see also erratum in Phys. Rev. Lett. **78**, 2031 (1997)
14. R. Dörner, H. Bräuning, J.M. Feagin, V. Mergel, O. Jagutzki, L. Spielberger, T. Vogt, H. Khemliche, M.H. Prior, J. Ullrich, C.L. Cocke, and H. Schmidt-Böcking: Phys. Rev. **A57**, 1074 (1998)
15. H.P Bräuning, R. Dörner, C.L. Cocke, M.H. Prior, B. Krässig, A. Bräuning-Demian, K. Carnes, S. Dreuil, V. Mergel, P. Richard, J. Ullrich, and H. Schmidt-Böcking: J. Phys. **B30**, L649 (1997)
16. H.P Bräuning, R. Dörner, C.L. Cocke, M.H. Prior, B. Krässig, A. Kheifets, I. Bray, A. Bräuning-Demian, K. Carnes, S. Dreuil, V. Mergel, P. Richard, J. Ullrich, and H. Schmidt-Böcking: J. Phys. **B31**, 5149 (1998)
17. V. Mergel, M. Achler, R. Dörner, K. Khayyat, T. Kambara, Y. Awaya, V. Zoran, B. Nyström, L. Spielberger, J.H. McGuire, J. Feagin, J. Berakdar, Y.Azuma, and H. Schmidt-Böcking: Phys. Rev. Lett. **80**, 5301 (1998)
18. M. Achler, V. Mergel, L. Spielberger, Y. Azuma R. Dörner, and H. Schmidt-Böcking: J. Phys. **B34**, L965 (2001)
19. A. Knapp, A. Kheifets, I. Bray, T. Weber, A.L. Landers, S. Schössler, T. Jahnke, J. Nickles, S. Kammer, O. Jagutzki, L.P. Schmidt, T. Osipov, J. Rösch, M.H. Prior, H. Schmidt-Böcking, C.L. Cocke, and R. Dörner: Phys. Rev. Lett. **89**, 033004 (2002)
20. R. Dörner, T. Vogt, V. Mergel, H. Khemliche, S. Kravis, C.L. Cocke, J. Ullrich, M. Unverzagt, L. Spielberger, M. Damrau, O. Jagutzki, I. Ali, B. Weaver, K. Ullmann, C.C. Hsu, M. Jung, E.P. Kanter, B. Sonntag, M.H. Prior, E. Rotenberg, J. Denlinger, T. Warwick, S.T. Manson, and H. Schmidt-Böcking: Phys. Rev. Lett. **76**, 2654 (1996)
21. J.A.R. Samson, W.C. Stolte, Z.X. He, J.N. Cutler, Y. Lu, and R.J. Bartlett: Phys. Rev. **A57**, 1906 (1998)
22. J.C. Levin, D.W. Lindle, N. Keller, R.D. Miller, Y. Azuma, N. Berrah Mansour, H.G. Berry, and I.A. Sellin: Phys. Rev. Lett. **67**, 968 (1991)
23. L. Spielberger, O. Jagutzki, R. Dörner, J. Ullrich, U. Meyer, V. Mergel, M. Unverzagt, M. Damrau, T. Vogt, I. Ali, K. Khayyat, D. Bahr, H.G. Schmidt, R. Frahm, and H. Schmidt-Böcking: Phys. Rev. Lett. **74**, 4615 (1995)
24. K. Hino, T. Ishihara, F. Shimizu, N. Toshima, and J.H. McGuire: Phys. Rev. **A48**, 1271 (1993)
25. S. Keller: J. Phys. **B33**, L513 (2000)
26. A. Kheifets: J. Phys. **B34**, L247 (2001)
27. T.Y. Shi and C.D. Lin: Phys. Rev. Lett., submitted for publication
28. F.W. Byron and C.J. Joachain: Phys. Rev. **164**, 1 (1967)
29. L.R. Andersson and J. Burgdörfer: Phys. Rev. Lett. **71**, 50 (1993)
30. J.A.R. Samson: Phys. Rev. Lett. **65**, 2861 (1990)
31. M. Sagurton, R.J. Bartlett, J.A.R. Samson, Z.X. He, and D. Morgan: Phys. Rev. **A52** 2829 (1995)
32. J.C. Levin, G.B. Armen, and I.A. Sellin: Phys. Rev. Lett. **76**, 1220 (1996)
33. M. Pont and R. Shakeshaft: Phys. Rev. **A51**, 494 (1995)
34. K.W. Meyer, J.L. Bohn, C.H. Green, and B.D. Esry.: J. Phys. **B30**, L641 (1997)

35. J.Z. Tang and I. Shimamura: Phys. Rev. **A52**, 1 (1995)
36. H. Kossmann, V. Schmidt, and T. Andersen: Phys. Rev. Lett. **A60**, 1266 (1988)
37. G.H. Wannier: Phys. Rev. **90**, 817 (1953)
38. J.M. Feagin: J. Phys. **B28**, 1495 (1995)
39. J.M. Feagin: J. Phys. **B29**, l551 (1996)
40. M. Pont and R. Shakeshaft: Phys. Rev. **A54**, 1448 (1996)
41. R. Wehlitz, F. Heiser, O. Hemmers, B. Langer, A. Menzel, and U. Becker: Phys. Rev. Lett. **67**, 3764 (1991)
42. D. Proulx and R. Shakeshaft: Phys. Rev. **A48**, R875 (1993)
43. M.A. Kornberg and J.E. Miraglia: Phys. Rev. **A48**, 3714 (1993)
44. Z.J. Teng and R. Shakeshaft: Phys. Rev. **A49**, 3597 (1994)
45. A. Huetz, P. Selles, D. Waymel, and J. Mazeau: J. Phys. **B24**, 1917 (1991)
46. F. Maulbetsch and J.S. Briggs: J. Phys. **B28**, 551 (1995)
47. L. Malegat, P. Selles, and A. Huetz: J. Phys. **B30**, 251 (1997)
48. J. Berakdar and H. Klar: Phys. Rev. Lett. **69**, 1175 (1992)
49. J. Berakdar, H. Klar, A. Huetz, and P. Selles: J. Phys. **B26**, 1463 (1993)
50. J. Berakdar: J. Phys. **B31**, 3167 (1998)
51. J. Berakdar: J. Phys. **B32**, L25 (1999)
52. A. Kheifets, I. Bray, K. Soejima, A. Danjo, K. Okuno, and A. Yagishita: J. Phys. **B32**, L501 (1999)
53. A. Kheifets and I. Bray: Phys. Rev. Lett. **81**, 4588 (1998)
54. R. Dörner, V. Mergel, H. Bräuning, M. Achler, T. Weber, K. Khayyat, O. Jagutzki, L. Spielberger, J. Ullrich, R. Moshammer, Y. Azuma, M.H. Prior, C.L. Cocke, and H. Schmidt-Böcking: Atomic Processes in Plasmas AIP Conference Proceedings Eds.: E. Oks, M. Pindzola, 443, (1998)
55. F. Maulbetsch and J.S. Briggs: J. Phys. **B26**, L647 (1994)
56. M. Pont and R. Shakeshaft: Phys. Rev. **A51**, R2676 (1995)
57. A. Kheifets and I. Bray: J. Phys. **B31**, L447 (1998)
58. J. Briggs and V. Schmidt: J. Phys. **33**, R1 (2000)
59. P.M. Bergstrom, Jr., K. Hino, and J. Macek: Phys. Rev. **A51**, 3044 (1995)
60. J.A.R. Samson, C.H. Green, and R.J. Bartlett: Phys. Rev. Lett. **71**, 201 (1993)
61. J.A.R. Samson, Z.X. He, R.J. Bartlett, and M. Sagurton: Phys. Rev. Lett. **72**, 3329 (1994)
62. L. Spielberger, O. Jagutzki, B. Krässig, U. Meyer, K. Khayyat, V. Mergel, T. Tschentscher, T. Buslaps, H. Bräuning, R. Dörner, T. Vogt, M. Achler, J. Ullrich, D.S. Gemmel, and H. Schmidt-Böcking: Phys. Rev. Lett. **76**, 4685 (1996)
63. L. Spielberger, H. Bräuning, A. Muthig, J.Z. Tang, J. Wang, Y. Qui, R. Dörner, O. Jagutzki, T. Tschentscher, V. Honkimäki, V. Mergel, M. Achler, T. Weber, K. Khayyat, J. Burgdörfer, J. McGuire, and H. Schmidt-Böcking: Phys. Rev. **59**, 371 (1999)
64. B. Krässig, R.W. Dunford, D.S. Gemmell, S. Hasegawa, E.P. Kanter, H. Schmidt-Böcking, W. Schmitt, S.H. Southworth, T. Weber, and L. Young: Phys. Rev. Lett. **83**, 53 (1999)
65. H. LeRouzo: J. Phys. **B19**, L677 (1986)
66. H. LeRouzo: Phys. Rev. **A37**, 1512 (1988)
67. R. Dörner, H. Bräuning, O. Jagutzki, V. Mergel, M. Achler, R. Moshammer, J. Feagin, A. Bräuning-Demian, L. Spielberger, J.H. McGuire, M.H. Prior, N. Berrah, J. Bozek, C.L. Cocke, and H. Schmidt-Böcking: Phys. Rev. Lett. **81**, 5776 (1998)

68. E. Shigemasa, J. Adachi, M. Oura, and A. Yagishita: Phys. Rev. Lett. **74**, 359 (1995)
69. E. Shigemasa, J. Adachi, K. Soejima, N. Watanabe, A. Yagishita, and N.A. Cherepkov: Phys. Rev. Lett. **80**, 1622 (1998)
70. F. Heiser, O. Geßner, J. Viefhaus, K. Wieliczec, R. Hentges, and U. Becker: Phys. Rev. Lett. **79**, 2435 (1997)
71. S. Motoki, J. Adachi, Y. Hikosaka, K. Ito, M. Sano, K. Soejima, G. Raseev, A. Yagishita, and N.A. Cherepkov: J. Phys. **B33**, 4193 (2000)
72. A. Landers, T. Weber, I. Ali, A. Cassimi, M. Hattass, O. Jagutzki, A. Nauert, T. Osipov, A. Staudte, M.H. Prior, H. Schmidt-Böcking, C.L. Cocke, and R. Dörner: Phys. Rev. Lett. **87**, 013002 (2001)
73. T. Weber, O. Jagutzki, M. Hattass, A. Staudte, A. Nauert, L. Schmidt, M.H. Prior, A.L. Landers, A. Bräuning-Demian, H. Bräuning, C.L. Cocke, T. Osipov, I. Ali, R. Diez Muino, D. Rolles, F.J. Garcia de Abajo, C.S. Fadley, M.A. Van Hove, A. Cassimi, H. Schmidt-Böcking, and R. Dörner: J. Phys. **B34**, 3669 (2001)
74. T. Jahnke, T. Weber, A.L. Landers, A. Knapp, S. Schössler, J. Nickles, S. Kammer, O. Jagutzki, L. Schmidt, A. Czasch, T. Osipov, E. Arenholz, A.T. Young, R. Diez Muino, D. Rolles, F.J. Garcia de Abajo, C.S. Fadley, M.A. Van Hove, S.K. Semenov, N.A. Cherepkov, J. Rösch, M.H. Prior, H. Schmidt-Böcking, C.L. Cocke, and R. Dörner: Phys. Rev. Lett. **88**, 073002 (2002)
75. H.C. Choi, R.M. Rao, A.G. Mihill, S. Kakar, E.D. Poliakoff, K. Wang, and V. McKoy: Phys. Rev. Lett. **72**, 44 (1994)
76. J.L. Dehmer and D. Dill: Phys. Rev. Lett. **35**, 213 (1975)
77. J.L. Dehmer and D. Dill: J. Chem. Phys. **65**, 5327 (1976)
78. J.L. Dehmer and D. Dill: Phys. Rev. **A18**, 164 (1978)
79. M. Walter and J.S. Briggs: J. Phys. **B32**, 2487 (1999)
80. H. Kossmann, O. Schwarzkopf, B. Kämmerling, and V. Schmidt: Phys. Rev. Lett. **63**, 2040 (1989)
81. T.J. Reddish, J.P. Wightman, M.A. MacDonald, and S. Cvejanovic: Phys. Rev. Lett. **79**, 2438 (1997)
82. J.P. Wightman, S. Cveejanovic, and T.J. Reddish: J. Phys. **B31**, 1753 (1998)
83. T.J. Reddish and J. Feagin: J. Phys. **B32**, 2473 (1998)
84. M. Walter and J.S. Briggs: Phys. Rev. Lett. **85**, 1630 (2000)
85. J.M. Feagin: J. Phys. **B31**, L729 (1998)
86. T. Weber et al.: to be published
87. T. Weber: Dissertation, Frankfurt (2002)
88. I.G. Kaplan and A.P. Markin: Sov. Phys. Dokl. **14**, 36 (1969)
89. V.G. Levin, V.G. Neudatchin, A.V. Pavlitchankov, and Y.F. Smirnov: J. Phys. **B17**, 1525 (1984)
90. Y.F. Smirnov, A.V. Pavlitchenkov, V.G. Levin, and V.G. Neudatschin: J. Phys. **B11**, 3587 (1978)

15 Vector Correlations
in Dissociative Photoionization
of Simple Molecules Induced
by Polarized Light

D. Dowek

15.1 Introduction

The field of dissociative photoionization covers a large class of processes that feature interesting examples of reaction dynamics of isolated molecules. The detailed study of such reactions at the level achieved by the vector-correlation approach as discussed in this chapter is a source of new information about the properties of excited neutral and ionic molecular states, their electronic and geometrical structure, and the dynamics of ionization and dissociation involving, e.g, electron scattering by the nonspherical molecular potential, electronic correlation, or coupling between electronic and nuclei motions. The advantage of vector properties for the characterization of molecular photoejection dynamics was first demonstrated in the study of photodissociation [1–4].

   Dissociative photoionization (DPI) of a molecule yields, in general, few electrons, ionic and neutral fragments. Here we discuss the information that is gained (i) using the correlated momentum spectroscopy, i.e., the determination of the three-dimensional velocity vectors V_n for the maximum number of particles produced in each DPI event and detected in coincidence in a 4π (or a large fraction) collection solid angle (ii) using linearly or circularly polarized weak radiation field light sources ($\hat{e}$) ($\hat{e}$ represents P, the linear polarization of the light, or k, the light propagation axis, respectively). Velocity-mapping studies, where the combination of resonance-enhanced multiphoton ionization (REMPI) spectroscopy and imaging techniques using charge-coupled devices for the analysis of one of the reaction products, photoelectron or photoion, also gives access to correlated vector properties of the photofragmentation process, reviewed recently [5,6], are not discussed here. Multiple ionization in strong laser fields is discussed in Chap. 18.

   The main part of this chapter applies to the study of single photoionization reactions induced by valence-shell excitation of diatomic or simple polyatomic molecules yielding a free electron (e) and two heavy particles ($A^+ + B$) for which a kinematically complete experiment is carried out (Sect. 15.2). Selected examples for diatomic molecules are presented to illustrate the output of the two types of observables derived from the (V_e, V_{A^+}, $\hat{e}$) vector correlation [7–9], (i) the energetic correlations, and (ii) the angular correlations. We focus in particular on the derivation of the complete $I_{AB}(\theta_e, \phi_e)$ molecular-frame photoelectron angular distributions (MFPADs) for AB space-fixed molecules, where the polar and azimuthal angles (θ_e, ϕ_e) characterize the electron emission direction in the molecular frame, the orientation AB being inferred from the recoil velocity $\boldsymbol{V}_{A^+}$. These observables represent the electronic wave functions in the continuum, and provide the most complete information about the photoionization dynamics [10–12]. Other recent studies, which fit into the scope of vector correlations in valence-shell photoionization in conditions leading to the polar angle $I_{AB}(\theta_e)$ MFPADs (see, e.g., [13–15]), are included in the discussion of Sect. 15.2.3.

An alternative access to $I_{AB}(\theta_e, \phi_e)$ MFPADs is to measure photoelectron angular distributions (PAD) in the laboratory frame for an ensemble of optically aligned molecules in, e.g., REMPI or intense-laser field experiments (see, e.g., [16–20]), or for an ensemble of molecules adsorbed on a surface [21]. A link can also be made with binary (e,2e) spectroscopy, where electron momentum spectroscopy gives access to, currently pherically averaged, molecular wave functions [22].

In inner-shell photoionization, photoelectron emission is, in most cases, followed by Auger-electron ejection, and subsequent multiple-ionization and fragmentation of the molecule. The photoelectron molecular-frame angular distributions provide a similar level of information as for the outer-shell excitation case, as illustrated below in the example of K-shell photoionisation of the CO molecule [23]. MFPADs corresponding to the Auger-electron emission for the same process [24] are analyzed in Chap. 16. This distinction between the two electrons is not relevant for the double photoionization of H_2 discussed in the preceding chapter, which features a prototype example of four-body fragmentation [25]. Two other examples are briefly described below.

Combining the electron–ion vector correlation analysis with pump-probe ultrafast laser techniques, it is shown in the example of multiphoton ionization of NO_2 that time-resolved MFPADs can be achieved, and used to probe photodissociation of neutral species [26]. Finally, Sect. 15.3 addresses other topics related to the dynamics of molecular ions that can be investigated by the analysis of the $(\boldsymbol{V}_{A^+}, \boldsymbol{V}_{B^+}, \boldsymbol{V}_{C^+} \ldots, \hat{\boldsymbol{e}})$ correlation of the velocity vectors of the ionic fragments with the polarization axis, illustrated by the example of nonsymmetric nuclear motion in CO_2 core excited Renner–Teller states [27]. Through the examples presented, we emphasize how the observables derived from the vector-correlation approach in dissociative photoionization enable theoretical predictions to be probed at a very sensitive level.

15.2 $(V_e, V_{A^+}, \hat{e})$ Photoelectron–Photoion Vector Correlations in Dissociative Photoionization of Small Molecules

Dissociative photoionization of a molecule is induced by single-photon absorption when the photon energy exceeds the threshold of ionization and dissociation, and it may occur along the two following reaction path schemes:

$$AB(X) + h\nu(\hat{e}) \rightarrow AB^{**} \rightarrow AB^+ + e \rightarrow A^+ + B^* + e, \qquad (15.1)$$

$$AB(X) + h\nu(\hat{e}) \rightarrow AB^{**} \rightarrow A + B^* \rightarrow A^+ + B^* + e. \qquad (15.2)$$

The AB^{**} transient neutral state belongs to the ionization continua of the underlying ionic molecular states, which are structured by shape resonances and autoionizing Rydberg states [28]. Reaction (15.1) describes direct photoionization of the AB molecule, or fast autoionization of a resonant state in the Franck–Condon (FC) region: the electron is emitted first, while the nuclei stay space-fixed, then the molecular ion dissociates in a slower process. Reaction (15.2) corresponds to the excitation of an autoionizing state whose lifetime is comparable to, or longer than, the dissociation time. There, ionization occurs during or after dissociation has taken place, possibly leading to atomic autoionization. The vector-correlation approach consists in measuring for each DPI event the V_e and V_{A^+} velocity vectors of the nascent electron and ionic fragment, ensuring a kinematically complete analysis of the (e, A^+, B) three-body fragmentation.

15.2.1 Experimental Approaches

The main experimental requirements for the achievement of the vector-correlation studies are described in the second chapter of this book. The experiments illustrated in this section [7–9,23,26] combine in general (i) a cold target (supersonic molecular expansion), (ii) a 4π solid-angle collection (or a large fraction) of electrons and ions extracted from the interaction region and driven to two time- and position-sensitive detectors by means of uniform and/or nonuniform [29] electric fields, eventually associated with the use of a magnetic field for the control of the electron trajectories [30], (iii) time-of-flight and impact-position measurements for both particles, providing the three components of the nascent V_e and V_{A^+} velocity vectors for each (e, A^+) coincident event. We refer to the specific papers for a detailed description of the setups used. Tunable, polarized, and pulsed synchrotron radiation constitute very appropriate light sources for such photoionization studies. Energy and angular correlations in DPI are also currently investigated using the velocity imaging photoionization coincidence method (VIPCO) developed by Takahashi et al. [14], where the two-dimensional electron/ion imaging technique [31] is combined with the coincidence detection of both particles: the V_e and

$\boldsymbol{V}_{A^+}$ velocity distributions are then obtained, under specific symmetry assumptions, by means of a back-projection of the electron and ion position distributions [32]. The most recent developments of the VIPCO technique also report time focus for the ion fragment [33].

In the following, we first describe the $(\boldsymbol{V}_e, \boldsymbol{V}_{A^+}, \boldsymbol{P})$ vector-correlation studies in the conditions developed by Lafosse et al. [7–9]. After a proper filtering of the (e,A^+) coincident events based on the A^+ ion TOF determination, the $(\boldsymbol{V}_e, \boldsymbol{V}_{A^+}, \boldsymbol{P})$ vector correlation is analyzed in two steps: (i) the electron-ion kinetic energy correlation is derived from the magnitudes of the $(\boldsymbol{V}_e, \boldsymbol{V}_{A^+})$ velocities, as illustrated below for DPI of the O_2 molecule (ii) the angular distributions are derived from the spatial analysis of the $(\boldsymbol{V}_e, \boldsymbol{V}_{A^+}, \boldsymbol{P})$ triplet. The reported experiments were performed at the synchrotron radiation facility SuperACO (LURE) in Orsay.

15.2.2   Electron–Ion Kinetic Energy Correlation

The example of DPI of O_2 [8] is chosen to illustrate the output of the energy-correlation analysis for (i) the identification of the DPI processes in terms of their reaction pathway, including the intermediate ionic (or neutral) state formed and the dissociation limit (ii) the determination of the branching ratios between the distinct processes. Apart from the fundamental aspects in molecular spectroscopy and in the study of reaction mechanisms, this type of data constitutes valuable information for several applications, e.g., in plasma and atmospherical studies, where cross sections, internal and kinetic energy distributions of electrons, ionic and neutral fragments arising from photofragmentation reactions are needed as inputs for modeling chemical reactions in the relevant medias. The chosen photon excitation energy of 23.15 eV allows one to illustrate the excitation of the two types of processes described in (15.1) and (15.2).

The (E_e, E_{O^+}) kinetic energy-correlation diagram (KECD) shown in Fig. 15.1a represents a bidimensional histogram of the (e,O^+) coincident events distributed as function of the E_e and E_{O^+} kinetic energies, derived from the magnitudes of the $(\boldsymbol{V}_e, \boldsymbol{V}_{O^+})$ velocities. In this diagram, a process appears as an accumulation of data points, characterized by the set of kinetic energies of the electron and ion (E_e, E_{O^+}). The KECD shows three distinct structures and illustrates the resolving power of the energy correlation. The dominant peak I corresponds to a process of type (15.1): the E_e electron kinetic energy scales the potential energy difference between the O_2^{**} transient state in the continuum and the O_2^+ ionic state formed in the FC region, here identified as the $O_2^+(B^2\Sigma_g^-, v \leq 4)$ state. The kinetic energy release of the fragments (KER=$E_{O^+} + E_O$) is equal to the potential energy difference between the O_2^+ ionic state and the dissociation limit L_D, hereby identified as $L_1[O^+(^4S) + O(^3P)]$:

$$O_2(X^3\Sigma_g^-) + h\nu \rightarrow O_2^+(B^2\Sigma_g^-, \nu \leq 4) + e \rightarrow O^+\left(^4S\right) + O(^3P) + e . (15.3)$$

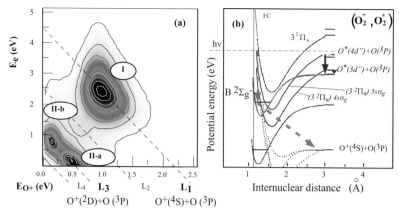

Fig. 15.1. (a) Kinetic energy correlation diagram for DPI of the O_2 molecule at $h\nu = 23.15\,\text{eV}$, from [8]; the *dashed straight lines* of slope -2 correspond to the accessible dissociation limits $L_1,\dots,L_4$. (b) Scheme of the relevant potential energy curves for O_2 and O_2^+ molecular states: identification of the reaction pathways for processes I and II of **a** using *dashed* (I) and *full* (II) *arrows*

On the other hand, peaks II (a,b), which remain at fixed E_e electron energy when the photon energy is varied nearby, are attributed to the excitation of discrete neutral Rydberg states, here $O_2^*(3\,^2\Pi_u, 4s\sigma_g)$ and $O_2^*(3\,^2\Pi_u, 5s\sigma_g)$, which dissociate into the $O(^3P) + O^*(^2P, 3d'')$ [34] and $O(^3P) + O^*(^2P, 4d'')$ neutral limits, and then encounter atomic autoionization to the ionic limit L_3. All the (e, O^+) DPI processes associated with the same dissociation limit L_D (potential energy E_D) lie on a single straight line of slope -2 for a homonuclear molecule, since the $[E_e + 2E_{O^+}] = h\nu - E_D$ linear relationship is imposed by energy and momentum conservation. Excitation of an extended vibrational distribution of the ionic state, or that of a repulsive part of its potential energy curve, elongates the peak along such a diagonal. The (O_2^{*+}, L_D) or (O_2^*, L_D) reaction pathways associated with these processes are schematized on Fig. 15.1b. The $(\Delta E_e, \Delta E_{O^+})$ energy resolutions are governed by the uncertainties on the three components of each velocity vector: they depend on the process studied, i.e., the actual TOF and position distributions [8,29].

Since the extraction field magnitude is chosen such that a 4π collection of electrons and ions is achieved for the processes of interest, the integration of the number of events in each structure directly enables a safe evaluation of the branching ratios for the different processes (70%, 10%, and 8% for processes I, II(a), and (b) respectively in the example above). Finally, the KECD enables a proper selection of each process in order to perform their complete angular analysis [41], as described in the next section.

DPI of a polyatomic molecule breaking into two heavy fragments leads to additional structures in the KECD since, for a given electronic transition pathway, a molecular fragment may be produced in an internally excited state: this excitation is now characterized by an elongation of the peak along the E_{A+} axis, which reveals the vibrational and rotational energy distribution of the molecular fragment. Such a distribution is illustrated, e.g., in the recent studies of photoionization of N_2O into the $N_2O^+(C^2\Sigma^+)$ ionic state [33,35], where the KECD for DPI leading to $N^+(^3P)+NO(X^2\Pi,\nu)$ allows resolution of the vibrational distribution of the NO ground state.

Similar energetic correlations between photoelectrons and photofragments have been examined in the investigation of dissociative photodetachment of negative ions produced in fast beams [36].

15.2.3 $(V_e, V_{A+}, \hat{e})$ Angular Correlations

Analyzing the directions of the $(V_e, V_{A+}, \hat{e})$ vectors requires three angles, as shown in Fig. 15.2. The direction of V_{A+} with respect to $\hat{e}$ is defined by the polar angle χ_{A+}, whereas two angles are necessary to define the direction of V_e since the polarization axis breaks the cylindrical symmetry of the electron emission in the molecular frame: we introduce the polar angle θ_e with respect to V_{A+}, and the azimuthal angle ϕ_e with respect to the plane that contains V_{A+} and the polarization axis $\hat{e}$.

The interpretation of the angular distributions is at its simplest when the axial recoil approximation is valid [37], i.e., when dissociation is fast with respect to molecular rotation: then the direction of V_{A+} is the signature of the molecular- axis orientation at the instant of photoabsorption, and θ_e and ϕ_e characterize the electron-emission pattern in the molecular frame as illustrated in Fig. 15.2. The examples below are discussed in the frame of the axial-recoil approximation. However, when the dissociation time amounts to a significant fraction of the rotational period, the photofragment angular distributions are influenced by molecular rotation [38–40]: the analysis presented in this section has been generalized recently to include the case where the influence of molecular rotation is significant [41].

The first level of information is the angular distribution of the ionic fragments with respect to the polarization axis (V_{A+}, P), which reflects the

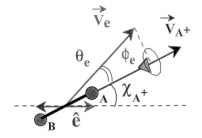

Fig. 15.2. Relevant angles in the spatial analysis of the $(V_e, V_{A+}, \hat{e})$ vector correlation, see text. $\hat{e}$ represents P, the polarization axis for linearly polarized light, or k, the light propagation axis for circularly polarized light

spatial distribution of the molecular axes for the photoexcited molecules. In the dipole approximation, for linear molecules and in the Hund's cases a and b, the photoabsorption transition induced by linearly polarized light obeys the selection rule $\Delta \Lambda = 0$, or ± 1, where Λ is the projection of the electronic angular momentum of the molecule onto the internuclear axis [42]. Except when a Rydberg autoionizing state of well-defined Λ symmetry is resonantly excited, the photon absorption populates two states of different Λ symmetry in the continuum, which are energy degenerated. For a gas of randomly oriented molecules, the spatial distribution of the molecular axes after photoabsorption has the form:

$$I(\chi_{A+}) = \frac{\sigma}{4\pi}\left[1 + \beta_{A+}\, P_2(\cos(\chi_{A+}))\right] , \tag{15.4}$$

where the β_{A+} asymmetry parameter ranges between $\beta_{A+} = -1$ for a perpendicular transition ($\Delta \Lambda = \pm 1$) and $\beta_{A+} = +2$ for a parallel transition ($\Delta \Lambda = 0$). Its actual value in the $[-1, 2]$ interval characterizes the symmetry of the state in the continuum [43]. The angular distribution of the photoelectrons with respect to the polarization axis has the same form [44].

The general expression of the molecular frame photoelectron angular distributions for space-fixed molecules of definite orientation, on which we focus the discussion in this section, was first derived by Dill in a reference paper [10]. It is of the general form:

$$I\left(\theta_e, \phi_e\right) \propto \sum_{K=0}^{2l_{max}} \sum_M A_{KM} Y_{KM}\left(\theta_e, \phi_e\right) , \tag{15.5}$$

where l_{max} is the largest orbital momentum component of the photoelectron amplitude. For two very specific geometries only involving polarized light, i.e., when the molecular axis is parallel to the polarization axis $\boldsymbol{P}$ of linearly polarized light, or when the molecular axis is parallel to the propagation axis $\boldsymbol{k}$ of circularly polarized light, the azimuthal dependence vanishes and the general form results in the simpler expansion:

$$I\left(\theta_e\right) \propto \sum_{K=0}^{2l_{max}} A_K P_K\left(\cos \theta_e\right) . \tag{15.6}$$

This expression naturally holds when the molecular axis is parallel to the propagation axis of unpolarized light. The determination of the $A_{KM}(A_K)$ coefficients of the $I\left(\theta_e, \phi_e\right)$ ($I(\theta_e)$) expansions gives access to the molecular-frame transition amplitudes, i.e., the dipole matrix elements, magnitudes, and relative phases [10–12]. The MFPAD observables therefore most completely characterize the dynamics of the photoionization process.

A number of experiments have been designed in order to probe electron emission in the molecular frame, for space-fixed molecules. First, evidences of electron emission anisotropies with respect to the molecular axis were inferred from the shape of the ion fragment time-of-flight spectra by photoelectron-photoion coincidence (PEPICO) spectroscopy [45]. The $I\,(\theta_e)$ polar angle MFPADs for molecules excited with weakly polarized light, or aligned parallel or perpendicular to the polarization axis of linearly polarized light P, have been determined in a series of experiments for selected inner-shell (e.g., [46–51]) and outer-shell photoionization of diatomic [13,52–57] or polyatomic molecules [15,58,59], using different types of angular-resolved photoelectron–photoion coincidence setups (AR-PEPICO) or the VIPCO technique. In these devices, where the ϕ_e azimuthal dependence was not determined, the $I\,(\theta_e)$ MFPAD for a molecule aligned perpendicular to the polarization axis corresponds either to the detection of photoelectrons with a fixed azimuthal angle ϕ_e (for example, electrons in the plane defined by the polarization and the molecular axes, see, e.g., [50]) or to that of all photoelectrons, averaging de facto over ϕ_e.

These studies enable one to analyze distinctively the electron angular distribution for the two degenerated continua corresponding to the parallel and the perpendicular transitions. Probing, in particular, the l partial wave composition of the σ shape resonances excited in core-level or valence-shell ionization, which was predicted to have predominantly a $l = 3$ character [60], they allowed for a critical test of calculations based on models like the one-electron multiple-scattering method [61,62], and the random-phase approximation, including intershell electronic correlation [63,64]. The richly structured angular distributions in electron emission from spatially oriented molecules, as illustrated by the examples referred to above, can also be discussed in terms of diffraction patterns of the emitted electron probing the two-center molecular potential [65]. When a long-lived autoionizing state is excited, electron emission takes place on the same timescale as molecular dissociation: a backward-forward asymmetry along the molecular axis of a homonuclear molecule, like the one observed in inner-valence shell ionization of O_2 around 22–23 eV (process II (a) above) can also be interpreted as the signature of intramolecular scattering of the outgoing electron, autoionized by one O fragment and scattered on the second nucleus [53,41]. In all these examples, the richness of the angular structure of the $I\,(\theta_e)$ MFPADs is magnified by the existence of resonance phenomena: the actual influence of the molecular potential depends then on the photon energy, and on the excited orbital. This is supported, e.g., by the angular analysis of the C–Cl or C–F bond fragmentation induced by valence-shell photoionization of the CH_3Cl and CH_3F molecules [15,58]. The $I\,(\theta_e)$ distributions probe the degree of coupling of the photoelectron with the molecular potential and the light electric field, respectively. A significant influence of the molecular potential gives rise to a rich angular structure, whereas the dominant influence of the light field imposes electron emission along the polarization axis [83].

$I\left(\theta_{\mathrm{e}}, \phi_{\mathrm{e}}, \chi_{A+}\right)$ **MFPADs for Dissociative Photoionization induced by Linearly Polarized Light.** Here we emphasize and illustrate in recent examples the advantage for the investigation of these subjects of measuring the complete $I\left(\theta_{\mathrm{e}}, \phi_{\mathrm{e}}, \chi_{A+}\right)$ angular distribution, including the polar and azimuthal dependence of the MFPADs, and we present the corresponding analysis formalism. The azimuthal dependence of the MFPADs has been investigated previously in a few specific conditions, e.g., experimentally in the study of rotationally resolved REMPI of NO [16,17], or theoretically for photoionization of molecules adsorbed on surfaces [21] and Auger emission from fixed-in-space molecules [66].

For linear molecules ionized with linearly polarized light the general expression of the MFPADs as expressed by Lucchese [9,41] takes the remarkably simple functional form:

$$
\begin{aligned}
I\left(\theta_{\mathrm{e}}, \phi_{\mathrm{e}}, \chi_{A+}\right) &= F_{00}\left(\theta_{\mathrm{e}}\right) + F_{20}\left(\theta_{\mathrm{e}}\right) P_2^0\left(\cos \chi_{A+}\right) \\
&\quad + F_{21}\left(\theta_{\mathrm{e}}\right) P_2^1\left(\cos \chi_{A+}\right) \cos \left(\phi_{\mathrm{e}}\right) \\
&\quad + F_{22}\left(\theta_{\mathrm{e}}\right) P_2^2\left(\cos \chi_{A+}\right) \cos \left(2 \phi_{\mathrm{e}}\right),
\end{aligned} \tag{15.7}
$$

where the azimuthal dependence is limited to the second harmonic in ϕ_{e}, and the dependence upon the orientation of the molecule appears explicitly in terms of simple Legendre polynomials. The maximum information that can be derived from such an experiment is then contained in the four F_{LN} functions. These functions are developed in Legendre polynomial expansion $F_{LN}\left(\theta_{\mathrm{e}}\right) \propto \sum_{L'} C_{L'LN} P_{L'}^N\left(\cos \theta_{\mathrm{e}}\right)$, where the $C_{L'LN}$ coefficients (with $L' \leq 2l_{\max}$) are functions of the dipole matrix elements.

There are several possible presentations of the complete spatial analysis of the $\left(\boldsymbol{V}_{\mathrm{e}}, \boldsymbol{V}_{A+}, \boldsymbol{P}\right)$ vector correlations. We illustrate three of them here (i,ii,iii), in the example of DPI of NO leading to the $c^3 \varPi$ state of NO^+ [7,9]:

$$
NO(^2\varPi) + h\nu \rightarrow NO^+(c^3\varPi) + \mathrm{e} \rightarrow N^+\left(^3P\right) + O(^3P) + \mathrm{e}. \tag{15.8}
$$

(i) Presentation in terms of the four F_{LN} functions: The F_{LN} functions are extracted from the $I\left(\theta_{\mathrm{e}}, \phi_{\mathrm{e}}, \chi_{A+}\right)$ distribution by a three-angle fitting according to (15.7) [9], using all the registered events. This presentation of the data enables a detailed comparison between experiment and theory, as shown in Fig. 15.3, where the measured F_{LN} functions for reaction (15.8) are compared with the results of multichannel Schwinger configuration interaction method [67–69] (MCSCI) calculations [9], as well as with the calculations convoluted with the apparatus function. The main features of the data are very well predicted by the calculations, although some discrepancies remain. These may be due to the high sensitivity of the calculated F_{LN} functions to the electron energy and to the inclusion of correlation terms as discussed below, or may also include an imperfect description of the apparatus function.

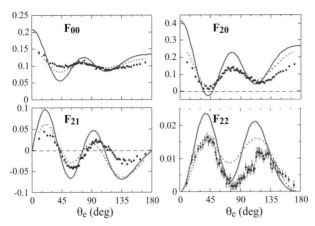

Fig. 15.3. Measured F_{LN} functions in Mb (*dots*) for reaction (15.8) at $h\nu =$ 23.65 eV, compared with the results of the MCSCI calculations before (*full line*) and after (*dashed line*) convolution with the apparatus function [9]. The normalization is performed by adjusting the total photoionization cross section for the studied process. θ_e is referred to the N end of the NO molecule

(ii) Presentation in terms of the $I_{AB}(\theta_e, \phi_e)$ MFPADs for three independent χ_{A^+} orientations of the molecule: Once the F_{LN} functions have been determined, the $I_{AB}(\theta_e, \phi_e)$ MFPADs for an orientation of the molecule parallel, at the magic angle, and perpendicular to the polarization axis are obtained from (15.7) according to:

$$I_{\chi=0°}(\theta_e) = F_{00}(\theta_e) + F_{20}(\theta_e) \tag{15.9}$$

$$I_{\chi=54.7°}(\theta_e, \phi_e) = F_{00}(\theta_e) + \sqrt{2}F_{21}(\theta_e)\cos(\phi_e)$$
$$+ 2F_{22}(\theta_e)\cos(2\phi_e) \tag{15.10}$$

$$I_{\chi=90°}(\theta_e, \phi_e) = F_{00}(\theta_e) - \frac{1}{2}F_{20}(\theta_e) + 3F_{22}(\theta_e)\cos(2\phi_e). \tag{15.11}$$

These are plotted in Fig. 15.4 for reaction (15.8). Equations (15.9)–(15.11) explicitly show how the F_{21} and F_{22} functions characterize the azimuthal dependence of the MFPADs for any orientation of the molecule, except that parallel to the polarization axis $\boldsymbol{P}$. In particular, the azimuthal dependence of the MFPAD for a molecule oriented perpendicular to $\boldsymbol{P}$ provides direct insight into the symmetry of the initial (neutral) and final (ionic) molecular states involved: e.g., if these states both have Σ^+ or Σ^- symmetry the MFPAD reduces to the form $I_\perp(\theta_e, \phi_e) \propto \cos^2(\phi_e)$ ($F_{22} > 0$), although it reduces to the form $I_\perp(\theta_e, \phi_e) \propto \sin^2(\phi_e)$ ($F_{22} < 0$) if the initial and final states have a Σ symmetry of opposite reflection symmetry [9]. On the other hand, the F_{21} function is the only one giving access to the relative phase between the dipole moments for the parallel and the perpendicular transition; its determination requires the measurement of the MFPAD for a molecule,

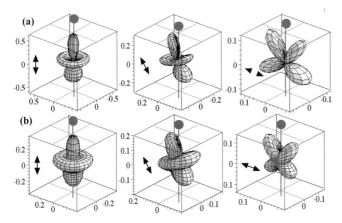

Fig. 15.4. Measured (**a**) and calculated (**b**) $I(\theta_e, \phi_e)$ MFPADs for reaction (15.8) at $h\nu = 23.65\,\text{eV}$, when the molecule is oriented parallel, at the magic angle and perpendicular to $\boldsymbol{P}$ [9]. The *arrows* indicate the direction of the polarization axis, and the *top sphere* represents the N end of the NO molecular axis

e.g., aligned at the magic angle with respect to the polarization axis, which is neither parallel nor perpendicular to $\boldsymbol{P}$.

The parallel orientation shows an angular distribution close to that of a $d\sigma$ wave, whereas for the perpendicular orientation the dominant contribution to the shape of the angular distribution is that of a $d\pi$ wave where only the lobes in the plane defined by the molecular and the polarization axes are populated. The $I(\theta_e)$ profiles for these two orientations measured using the VIPCO method at 40.8–eV photon energy close to threshold [14] show a clear evolution since a 40.8–eV electron emission is strongly favored in the direction of the O fragment. The strong azimuthal anisotropy of the MFPAD for the perpendicular transition in Fig. 15.4 is close to a $\cos^2(\phi_e)$ distribution, demonstrating that reaction (15.8) identifies with a transition between Σ states of same reflection symmetry. The opposite azimuthal dependence, where electron emission is favored in a plane perpendicular to the polarization axis ($F_{22} < 0$), has been found so far for two photoionization transitions $- \text{O}_2(X^3\Sigma_g^-) \to \text{O}_2^+(3\,^2\Pi_u)$ [41] and $\text{N}_2\text{O}(X^1\Sigma^+) \to \text{N}_2\text{O}^+(B^2\Pi_u)$ close to threshold [70].

The studied reaction (15.8) shows that the MFPAD observables are very sensitive to the influence of electronic correlation: we stress that the good agreement between the measured and calculated angular distributions at $h\nu = 23.65\,\text{eV}$ was only achieved with a 17–channel calculation [9,71]. This is due, in particular, to the open-shell electronic structure of the initial state. The MCSCI method, which accounts for static (intrachannel) and dynamic (interchannel) electronic correlation [72,73], provides a very accurate description at this level.

(iii) Presentation in terms of the dipole matrix elements: The conditions for the extraction of the dipole matrix elements and relative phase shifts, thereby approaching a "quantum-mechanically complete" experiment, from the $I(\theta_e)$ MFPADs measured with linearly polarized light and related A_K expansion coefficients given by (15.6), are discussed in detail in the literature (e.g. [46,48–51,54,55]). For the reaction (15.8), the analysis of the four $F_{LN}(\theta_e)$ functions and related $C_{L'LN}$ coefficients, with a maximum angular momentum $l_{\max} = 3$ and $l_{\max} = 4$, for $N = 0$ and $N > 0$, respectively, provides a complete determination of the magnitudes and relative phases of the dipole matrix elements, except for the sign of the relative phase between the transition amplitudes for the parallel and the perpendicular transition. The "completeness" is achieved because the azimuthal behavior allows one to reduce the $\Pi \to \Pi$ reaction (15.8) to a transition between Σ states of the same reflection symmetry [9]. Determination of the missing phase sign requires an experiment involving circularly polarized light. This was achieved for reaction (15.8) at 40.8 eV by combining linearly polarized light experiments using the VIPCO method [14] and circularly polarized light experiments using a multianalyzer AR-PEPICO setup [74]. The 23.6 eV and 40.8 eV energies lie at threshold and above the shape resonance of the $NO^+(c^3\Pi)$ state [69,75–77], respectively. The results obtained in this energy range constitute the basis for studying the dynamical evolution of this photoionization process, which is clearly far beyond a one-electron description. We refer to the papers for the detailed presentation of the dipole matrix elements.

Among the other studied diatomic molecules, H_2 attracts special interest as the simplest molecular system and prototype for photoionization. Since the pioneering experiments of Dehmer and Dill on photoion angular distributions at 40.8 eV photon excitation energy [43], results have been obtained related in particular with the proton kinetic energy distributions [78–80] and the $I(\theta_e)$ MFPADs characterizing direct photoionization to the four lowest $H_2^+(nl\lambda_{g/u})$ excited states of different symmetries [13,56,57] for a molecule aligned parallel or perpendicular to the linear polarization of the light. In the 26–40 eV photon energy range autoionization of the Q_1 and Q_2 doubly excited states of H_2 constitutes an interesting example where the electronic and nuclei motions cannot be disentangled, and where interference occurs between resonant (15.2) and nonresonant (15.1) de-excitation pathways [81]. The $I(\theta_e, \phi_e)$ MFPADs for DPI following the resonant excitation of Q_1 and Q_2 excited states obtained recently [82] show specific characteristics with respect to those obtained for direct ionization, and should bring new information about this interference mechanism.

Extending the vector-correlation approach to valence-shell photoionization of a polyatomic molecule, breaking into two heavy fragments, opens up the possibility to study the fragmentation of a specific molecular bond in a complex molecular system [15]. DPI of N_2O into the $N_2O^+(C^2\Sigma^+)$ ionic state, leading mainly to the dissociation limits $NO^+(X^1\Sigma^+)+N(^2P)$

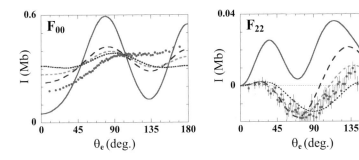

Fig. 15.5. F_{00} and F_{22} functions for PI into the N_2O^+ ($C\,^2\Sigma^+$) state computed using the MCSCI method at 21.5 eV for different lifetimes of the C state: $t = 0$ (—), 1 ps (- - -), 2 ps (- - - -), and infinite (...), compared with the measured F_{00} and F_{22} (*dots*) [35]. θ_e is referred to the O end of N_2O

and $N^+(^3P)+NO(X^2\Pi,\nu)$, constitutes an example where the analysis of the $I(\theta_e,\phi_e)$ MFPADs in terms of F_{LN} functions is extended to photofragmentation of a linear triatomic molecule [35]. It is also a showcase where the predissociation time amounts to a significant fraction of the rotational period, restricting the validity of the axial recoil approximation, a situation that is more likely to occur when complex systems are considered. For that purpose, the rotational motion has been introduced in the quantal derivation of the MFPADs by Lafosse et al. [41]. The obtention of meaningful angular information requires then a reliable estimation of the rotational temperature of the molecular beam, as well as the knowledge of the predissociation time of the molecular ion formed after photoionization. This benefits from the results of high-resolution spectroscopy of molecular ions like two-photon absorption [84] or pulsed-field ionization-photoelectron [85] techniques. When the predissociation time is not known from complementary experiments or calculations, the comparison of the measured and calculated F_{LN} functions taking into account molecular rotation allows estimation of its value [35]. This is illustrated in Fig. 15.5 for the above DPI process of N_2O, where a predissociation time of the order of 2 ps is inferred from the comparison between the measured and computed F_{LN} functions.

$I(\theta_e,\phi_e,\chi_{A^+})$ **MFPADs for Dissociative Photoionization induced by Circularly Polarized Light.** Already present in the general formulation of the MFPADs by Dill [10], circular dichroism in photoelectron angular distributions (CDAD) from oriented linear molecules in the frame of the electric-dipole approximation, i.e., the influence of left- or right-handed circularly polarized light (LHC/RHC) on the MFPADs, was specifically pointed out several years ago theoretically [11,86,87]. It was subsequently probed in a two color REMPI investigation of the NO molecule via the excitation of the $NO(A^2\Sigma^+)$ state [88]. A complete description of photoionization of optically

274   D. Dowek

O
|
C

a)   b)

Fig. 15.6. MFPADs for $C(1s)$ photoelectrons emitted at the maximum of the shape resonance from a CO molecule oriented perpendicular to the light propagation axis, by absorption of LHC (**a**) and RHC (**b**) polarized light [23]. The sense of rotation of the polarization vector is indicated by the *spiral* drawn around the light propagation axis, consistent with the convention of classical optics [93]

aligned $NO(A^2\Sigma^+)$ molecules ionized by circularly polarized light was later demonstrated by Reid et al. [89,90] by measuring rotationally resolved PADs.

In photoionization of ground-state molecules, CDAD for fixed-in-space molecules was identified for $CO(O(1s)$ ionization) [91], N_2 ($2\sigma_g$ ionization) [92] and $NO(4\sigma$ ionization) [74]. The first $I(\theta_e, \phi_e)$ MFPADs combining the vector correlation approach with circularly polarized light were measured by Jahnke et al. [23] for DPI following K-shell ionization of CO and N_2. Figure 15.6 illustrates the circular dichroism observed at the peak of the $C(1s\sigma)$ shape resonance for a CO molecule oriented perpendicular to the light propagation axis: the strong asymmetry of the MFPAD in the plane perpendicular to the propagation axis under excitation with a LHC-polarized photon (a), is reversed when a RHC-polarized photon is absorbed (b).

The detailed comparison of the experimental results with recent one-electron-model multiple scattering calculations [23], and many-electron-correlation random–phase approximation calculations [63,64], show that CDAD is a sensitive probe of the theoretical description of this shape resonance process, in particular at low electron energies.

Considering the general expression of the MFPADs at the same level as in (15.7) for linear molecules randomly oriented in the ground state and ionized with LHC/RHC circularly polarized light, characterized by an helicity $h = \pm 1$ along the $\boldsymbol{k}$ propagation axis (Stokes parameter $S_3 = \mp 1$, respectively [93]), the functional form derived by Lafosse et al. is as follows [41]:

$$
\begin{aligned}
I_{h=\pm 1}(\theta_e, \phi_e, \chi_{A^+}) = {}& F_{00}(\theta_e) - \frac{1}{2}F_{20}(\theta_e)\,P_2^0(\cos\chi_{A^+}) \\
& - \frac{1}{2}F_{21}(\theta_e)\,P_2^1(\cos\chi_{A^+})\cos(\phi_e) \\
& - \frac{1}{2}F_{22}(\theta_e)\,P_2^2(\cos\chi_{A^+})\cos(2\phi_e) \\
& \pm F_{11}(\theta_e)\,P_1^1(\cos\chi_{A^+})\sin(\phi_e)\,.
\end{aligned}
\tag{15.12}
$$

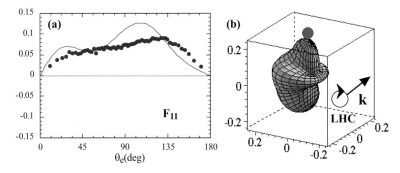

Fig. 15.7. (a) Measured F_{11} function in Mb (*dots*) at $h\nu = 23.65$ eV compared with the MCSCI calculations (*full line*); (b) $I(\theta_e, \phi_e)$ MFPAD for a molecule oriented perpendicular to the k vector, and ionized with LHC-polarized light [71]

The orientation of the molecule χ_{A^+} is now referred to the k propagation vector of the light, and the azimuthal angle ϕ_e with respect to the plane that contains the molecular axis and the k vector (see Fig. 15.2). This expression demonstrates that measuring the complete $I(\theta_e, \phi_e, \chi_{A^+})$ distribution with a single circular polarization of the light provides a complete experiment, since the four F_{LN} functions defined above for linearly polarized light, plus the additional F_{11} term that characterizes the circular dichroism, are determined. Such a complete experiment has been recently performed for reaction (15.8) at the 23.65 eV photon energy; the measured F_{11} function displayed in Fig. 15.7a compares very well with the MCSCI calculations [71]. Its magnitude reveals a strong circular dichroism, consistently reflected by the $I(\theta_e, \phi_e)$ MFPAD presented for a NO molecule aligned perpendicular to the propagation axis and photoionized by LHC polarized light (Fig. 15.7b).

The corresponding CDAD parameter defined as $(I_{\text{LHC}} - I_{\text{RHC}})/(I_{\text{LHC}} + I_{\text{RHC}})$ can be evaluated from (15.12) as a function of the F_{LN} for any orientation of the molecule [71]. For a molecular orientation perpendicular to the propagation axis it reaches a maximum value of 1 for reaction (15.8) [71,74]. Vector correlations then provide an efficient route to measure CDAD for oriented molecules, giving access to complete experiments with a single circular polarization of the light. This opens up an interesting field for the study of the different types of electronic interactions taking place in a large variety of species, including chiral molecules [94–96].

Time–Resolved MFPADs. Probing electron–ion vector correlation during the dissociation of a parent molecule by using femtosecond time-resolved experiments enables one to study the time-dependent separation and reorientation of the dissociating photofragments. Various stages of the dissociation process

$$NO_2 \rightarrow NO(C\,{}^2\Pi) + O({}^3P), \tag{15.13}$$

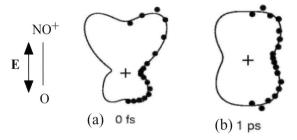

Fig. 15.8. $I(\theta_e)$ MFPADs measured in photoionization of NO_2 for pump-probe time delays of 0 fs (**a**) and 1 ps (**b**)

where probed using ionization of the NO fragment to $NO^+(X)$, at different time delays with respect to the pulse initiating the dissociation [26]. The $I\,(\theta_e)$ MFPADs measured for a dissociating molecular axis parallel to the linear polarization of both pump and probe lasers constituted the first demonstration of this type of experiment. The strong forward-backward asymmetry observed in the MFPAD for a 0–fs pump-probe time delay, as shown in Fig. 15.8, is interpreted as the effect of a nonspherical scattering of the ejected electron by the close recoiling O-atom fragment. For delays equal to or larger than 1 ps the shape of the distribution is modified and the asymmetry vanishes, indicating ionization of free NO(C) after complete dissociation of NO_2.

Diverse applications in the study of photofragmentation of molecules and clusters, involving time-resolved coincidence spectroscopies combined with velocity-vector correlations, have been reviewed recently [36].

15.3 $(V_{A+}, V_{B+}, \ldots \hat{e})$ Vector Correlations in Multiple Ionization of Small Polyatomic Molecules

Dissociative multiple photoIonization (DMPI) of small polyatomic molecules leads to the production of several ions and electrons, and the extension of the vector-correlation approach requires use of the multihit capacities of the position-sensitive detectors (see [97,98] and Chap. 2.1). We refer again to Chap. 14 for DMPI of H_2, and to Chap. 16 for the study of relaxation dynamics probed by Auger electron–ion coincidences. The topics adressed in DPMI studies concern, in particular, the multibody fragmentation dynamics of multiply charged ions, i.e., the identification of the different types of sequential or concerted decay reactions ([99,100] and refs. therein); the momentum correlation of the fragments also provides insight into the geometrical structure of the molecular ions and their dissociation lifetimes. Related problems are investigated in the study of photoion-pair formation [101]. The multi-particle fragmentation of neutral systems into neutrals is discussed in Chaps. 4 and 17.

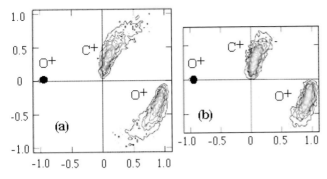

Fig. 15.9. Newton diagrams for the 3-body breakup of CO_2^{3+} for the excitation of the A_1 (**a**) and B_1 (**b**) Renner–Teller states of the $CO_2(1s^{-1}2\pi_u)$ $^1\Pi_u$ molecule, recorded at 535.4 eV energy. The amplitude of the linear momentum of the first O^+ is normalized to unity and placed along the negative x-axis

In the example presented, the vector correlation involves the fragment-ion momenta and the linear polarization of the exciting light ($\boldsymbol{P}$), and a triple-ion-coincidence momentum imaging technique is used. This enables the detailed spectroscopy of the dynamics of the nuclear motion in core excited states of the CO_2 molecule [27]. The bent and linear geometries of the $CO_2(1s^{-1}2\pi_u)$ $^1\Pi_u$ Renner–Teller pair states, A_1 and B_1, respectively, are probed for the first time using the property that the bending motion proceeds in the direction parallel to $\boldsymbol{P}$ for the A_1 state, but perpendicular to $\boldsymbol{P}$ for the B_1 state. The Newton diagrams for the three-body breakup of CO_2^{3+}, where the fragmentation events corresponding to the excitation of the A_1 or the B_1 states are selected, are displayed in Fig. 15.9a and b, respectively.

The shape of the C^+ and O^+ emission lobes, in particular the longer tail in (a) than in (b), is interpreted as the signature of the significant bent character of the A_1 state.

15.4 Conclusion and Perspectives

The vector-correlation approach for the investigation of dissociative photoionization of small molecules induced by polarized light has encountered important developments in recent years. The ability to collect most of the particles produced in the same fragmentation event, without restriction on their emission angles, and to measure their velocity vectors, allows one to perform kinematically complete experiments in a very efficient way. Molecular-frame photoelectron angular distributions, obtained for any orientation of the molecular axis with respect to physical vectorial quantities like the polarization axis of linearly polarized light or the propagation vector of circularly polarized light, give access to the most complete dynamical description of the photoionization processes. The field of one-photon single photoionization of small molecules induced by linearly polarized light is now at a mature level, and the recent theoretical developments achieve quite precise predictions of the $I(\theta_e, \phi_e, \chi_{A^+})$ distributions. Due to the significant improvements in the instrumentation and subsequent energy and angular resolutions, most of the problems involving multi-photofragmentation should benefit from the vector correlation approach. Several directions are promising in the future, some of them being already explored and productive. Implementing the vector-correlation approach for the study of the photofragmention of more complex systems constitutes a first challenge, in two directions. One consists in the investigation of the breaking of a specific molecular bond at the level of the MFPADs, the other involves the study of the multi-fragmentation mechanisms. Breaking complex systems produces an increasing number of neutral fragments; in order to detect these efficiently one must initiate photoionization in a fast molecular or cluster beam and communicate a significant initial translational velocity to the particles: the state-of-the-art in this approach is discussed, e.g., in other chapters of this book by Zajfman, Schwalm, Wolf, and Müller and Helm. The access to new light sources of increasing performance, like ultrashort and ultraintense lasers, high-order harmonic generation, new-generation synchrotron radiation including preparation of exotic polarizations, free-electron lasers, and others that can be efficiently combined with extended vector-correlation methods, opens up new perspectives in the study of the molecular dynamics enlightened in photofragmentation processes.

Acknowledgements

All the coworkers involved in the course of this work are gratefully acknowledged. I am very grateful to J.C. Houver, A. Lafosse, and M. Lebech for their help in the preparation of this chapter, to R.R. Lucchese for many helpful discussions and to A. Lahmann-Bennani for a careful reading of the manuscript.

References

1. C.H. Greene and R.N. Zare: Annu. Rev. Phys. Chem. **33**, 119 (1982)
2. G.E. Hall and P.L. Houston: Annu. Rev. Phys. Chem. **40**, 375 (1989)
3. R.N. Dixon: J. Chem. Phys. **85**, 1866 (1986)
4. L.D.A. Siebbeles, M. Glass-Maujean, O.S. Vasyutinskii, J.A. Beswick, and O. Roncero: J. Chem. Phys. **100**, 3610 (1994) and references therein
5. D.H. Parker: Photoionization and Photodetachment **10A**, ed. by C.Y. Ng (World Scientific, Singapore 2000) 3
6. A.G. Suits and R.E. Continetti: Imaging in Chemical Dynamics (ed. A.G. Suits and R.E. Continetti, ACS Symp. Ser. 770 (Am. Chem. Soc., Washington DC 2001)
7. A. Lafosse, M. Lebech, J.C. Brenot, P.M. Guyon, O. Jagutzki, L. Spielberger, M. Vervloet, J.C. Houver, and D. Dowek: Phys. Rev. Lett. **84**, 5987 (2000)

8. A. Lafosse, J.C. Brenot, A.V. Golovin, P.M. Guyon, K. Hoejrup, J.C. Houver, M. Lebech, and D. Dowek: J. Chem. Phys. **114**, 6605 (2001)
9. R.R. Lucchese, A. Lafosse, J.C. Brenot, P.M. Guyon, J.C. Houver, M. Lebech, G. Raseev, and D. Dowek: Phys. Rev. A **65**, 020702 (2002)
10. D. Dill: J. Chem. Phys. **65**, 1130 (1976)
11. N.A. Cherepkov and V.V. Kuznetsov: Z. Phys. D **7**, 271 (1987)
12. N.A. Cherepkov and G. Raseev: J. Chem. Phys. **103**, 8283 (1995)
13. J.H.D. Eland, M. Takahashi, and Y. Hikosaka: Faraday Discuss. **115**, 119 (2000)
14. M. Takahashi, J.P. Cave, and J.H.D. Eland: Rev. Sci. Instrum. **71**, 1337 (2000)
15. P. Downie and I. Powis: Phys. Rev. Lett. **82**, 2864 (1999); J. Chem. Phys. **111**, 4535 (1999)
16. K.L. Reid, D.J. Leahy and R.N. Zare: J. Chem. Phys. **95**, 1746 (1991)
17. D.J. Leahy, K.L. Reid and R.N. Zare: J. Chem. Phys. **95**, 1757 (1991)
18. K.L. Reid and T.A. Field, M. Towrie, and P. Matousez: J. Chem. Phys. **111**, 1438 (1999)
19. M. Tsubouchi, B.J. Whitaker, L. Wang, H. Kohguchi, and T. Suzuki: Phys. Rev. Lett. **86**, 4500 (2001)
20. F. Rosca-Pruna and M.J.J. Vrakking: Phys. Rev. Lett. **87**, 153902 (2001)
21. V.V. Kuznetsov, N.A. Cherepkov, and G. Raseev: J. Phys. Condens. Matter. **8**, 10327 (1996)
22. C.E. Brion: Int. J. Quantum Chem. *XXIX*, 1397 (1986)
23. T. Jahnke, T. Weber, A.L. Landers, A. Knapp, S. Schössler, J. Nickles, S. Kammer, O. Jagutzki, L. Schmidt, A. Czasch, T. Osipov, E. Arenholz, A.T. Young, R. Díez Muiño, D. Rolles, F.J. García de Abajo, C.S. Fadley, M.A. Van Hove, S.K. Semenov, N.A. Cherepkov, J. Rösch, M.H. Prior, H. Schmidt-Böcking, C.L. Cocke, and R. Dörner: Phys. Rev. Lett. **88**, 073002 (2002)
24. R. Guillemin, E. Shigemasa, K. Le Guen, D. Ceolin, C. Miron, N. Leclercq. P. Morin and M. Simon: Phys. Rev. Lett. **87**, 203001 (2001)
25. R. Dörner, H. Bräuning, O. Jagutzki, V. Mergel, M. Achler, R. Moshammer, J.M. Feagin, T. Osipov, A. Bräuning-Demian, L. Spielberger, J.H. McGuire, M.H. Prior, N. Berrah, J.D. Bozek, C.L. Cocke, and H. Schmidt-Böcking: Phys. Rev. Lett. **81**, 5776 (1998)
26. J.A. Davies, J.E. LeClaire, R.E. Continetti, and C.C. Hayden: J. Chem. Phys. **111**, 1 (1999); J.A. Davies, R.E. Continetti, D.W. Chandler, and C.C. Hayden: Phys. Rev. Lett. **84**, 5983 (2000)
27. Y. Muramatsu, K. Ueda, N. Saito, H. Chiba, M. Lavollée, A. Czasch, T. Weber, O. Jagutzki, H. Schmidt-Böcking, R. Moshammer, U. Becker, K. Kubozuka, and I. Koyano: Phys. Rev. Lett. **88**, 133002 (2002)
28. J.L Dehmer: "Resonances in molecular photoionization" in Handbook on Synchrotron Radiation, edited by G.V. Marr, Vol. 2 (North-Holland, Amsterdam 1987) p. 241
29. M. Lebech, J.C. Houver and D. Dowek: Rev. Sci. Instrum. **73**, 1866 (2002)
30. R. Moshammer, M. Unverzagt, W. Schmitt, J. Ullrich and H. Schmidt-Böcking: Nucl. Instrum. Methods B **108**, 425 (1996)
31. A.T.J.B. Eppink and D.H. Parker: Rev. Sci. Instrum. **68**, 3477 (1997)
32. C. Bordas, F. Paulig, H. Helm, and D.L. Huestis: Rev. Sci. Instrum. **67**, 2257 (1996)

33. Y. Hikosaka and J.H.D. Eland: Chem. Phys. **281**, 91 (2002)
34. P.M. Guyon, A.V. Golovin, C.J.K. Quayle, M. Vervloet, and M. Richard-Viard: Phys. Rev. Lett. **76**, 600 (1996)
35. M. Lebech, J.C. Houver, R.R. Lucchese, and D. Dowek: J. Chem. Phys. **117**, 9248 (2002)
36. R.E. Continetti: Annu. Rev. Phys. Chem. **52**, 165 (2001) and references therein
37. R.N. Zare: J. Chem. Phys. **47**, 204 (1967); Mol. Photochem. **4**, 1 (1972)
38. C. Jonah: J. Chem. Phys. **55**, 1915 (1971)
39. S. Yang and R. Bersohn: J. Chem. Phys. **61**, 4400 (1974)
40. S. Mukamel and J. Jortner: J. Chem. Phys. **61**, 5348 (1974)
41. A. Lafosse, J.C. Brenot, A.V. Golovin, P.M. Guyon, J.C. Houver, M. Lebech, D. Dowek, P. Lin, and R.R. Lucchese: J. Chem. Phys. **117**, 8368 (2002)
42. G. Herzberg: Molecular Spectra and Molecular Structure I. Spectra of Diatomic Molecules (Krieger, Malabar, Florida, 1989)
43. J.L. Dehmer and D. Dill: Phys. Rev. A **18**, 164 (1978)
44. A.D. Buckingham, B.J. Orr, and J.M. Sighel: Philos. Trans. Roy. Soc. Lond. A **268**, 147 (1970)
45. J.H.D. Eland: J. Chem. Phys. **70**, 2426 (1979)
46. E. Shigemasa, J. Adachi, M. Oura, and A. Yagishita: Phys. Rev. Lett. **74**, 359 (1995)
47. F. Heiser, O. Gessner, J. Viefhaus, K. Wieliczek, R. Hentges, and U. Becker: Phys. Rev. Lett. **79**, 2435 (1997)
48. N. Watanabe, J. Adachi, K. Soejima, E. Shigemasa, A. Yagishita, N.G. Fominykh, and A.A. Pavlychev: Phys. Rev. Lett. **78**, 26 (1997)
49. K. Ito, J. Adachi, Y. Hikosaka, S. Motoki, K. Soejima, A. Yagishita, G. Raseev, and N.A. Cherepkov: Phys. Rev. Lett. **85**, 46 (2000)
50. S. Motoki, J. Adachi, Y. Hikosaka, K. Ito, M. Sano, K. Soejima, A. Yagishita, G. Raseev, and N.A. Cherepkov: J. Phys. B **33**, 4193 (2000)
51. N.A. Cherepkov, G. Raseev, J. Adachi, Y. Hikosaka, K. Ito, S. Motoki, M. Sano, K. Soejima, and A. Yagishita: J. Phys. B **33**, 4213 (2000)
52. A.V. Golovin, N.A. Cherepkov and V.V. Kuznetsov: Z. Phys. D **24**, 371 (1992)
53. A.V. Golovin, F. Heiser, C.J.K. Quayle, P. Morin, M. Simon, O. Gessner, P.M. Guyon, and U. Becker: Phys. Rev. Lett. **79**, 4554 (1997)
54. Y. Hikosaka and J.H.D. Eland: J. Phys. B **33**, 3137 (2000)
55. Y. Hikosaka and J.H.D. Eland: Phys. Chem. Chem. Phys. **2**, 4663 (2000)
56. K. Ito, J.I. Adachi, R. Hall, S. Motoki, E. Shigemasa, K. Soejima, and A. Yagishita: J. Phys.B **33**, 527 (2000)
57. Y. Hikosaka and J.H.D. Eland: Chem. Phys. **277**, 53 (2002)
58. Y. Hikosaka, J.H.D. Eland, T.M. Watson, and I. Powis: J. Chem. Phys. **115**, 4593 (2001)
59. T. Kinugawa, Y. Hikosaka, A.M. Hodgekins and J.H.D. Eland: J. Mass Spectrom. **37**, 854 (2002)
60. J.L. Dehmer and D. Dill: Phys. Rev. Lett. **35**, 213 (1975)
61. D. Dill and J.L. Dehmer: J. Chem. Phys. **61**, 692 (1974)
62. D. Dill, J. Siegel, and J.L. Dehmer: J. Chem. Phys. **65**, 3158 (1976)
63. S. Semenov, N.A. Cherepkov, G.H. Fecher and G. Schöhense: Phys. Rev. A **61**, 0327704 (2000)
64. N.A. Cherepkov, S.K. Semenov, Y. Hikosaka, K. Ito, S. Motoki and A. Yagishita: Phys. Rev. Lett. **84**, 250 (2000)

65. A. Landers, Th. Weber, I. Ali, A. Cassimi, M. Hattass, O. Jagutzki, A. Nauert, T. Osipov, A. Staudte, M.H. Prior, H. Schmidt-Böcking, C.L. Cocke, and R. Dörner: Phys. Rev. Lett. **87**, 013002 (2001)
66. S. Bonhoff, K. Bonhoff, and K. Blum: J. Phys. B **32**, 1139 (1999)
67. R.R. Lucchese, K. Takatsuka, and V. McKoy: Phys. Rep. **131**, 148 (1986)
68. R.E. Stratmann, and R.R. Lucchese: J. Chem. Phys. **102**, 8493 (1995)
69. R.E. Stratmann, R.W. Zurales, and R.R. Lucchese, J. Chem. Phys. **104**, 8989 (1996)
70. M. Lebech, J.C. Houver, R.R. Lucchese, and D. Dowek: J. Chem. Phys. (in preparation, to be published in 2003)
71. M. Lebech, J.C. Houver, A. Lafosse, D. Dowek, C. Alcaraz, L. Nahon, and R.R. Lucchese: J. Chem. Phys. **118**, 9653 (2003)
72. B. Basden, and R.R. Lucchese: Phys. Rev. A **34**, 5158 (1986)
73. P. Lin and R.R. Lucchese: J. Chem. Phys. **113**, 1843 (2000)
74. O. Gessner, Y. Hikosaka, B. Zimmermann, A. Hempelmann, R.R. Lucchese, J.H.D. Eland, P.M. Guyon, and U. Becker: Phys. Rev. Lett. **88**, 193002 (2002)
75. T. Gustafsson, and H.J. Levinson: Chem. Phys. Lett. **78**, 28 (1981)
76. S.H. Southworth, C.M. Truesdale, P.H. Kobrin, D.W. Lindle, W.D. Brewer, and D.A. Shirley: J. Chem. Phys. **76**, 143 (1982)
77. Y. Lu, W.C. Stolte, and J.A.R. Samson: J. Electron Spectrosc. Relat. Phenom. **87**, 109 (1997)
78. S. Strathdee and R. Browning: J. Phys. B **12**, 1789 (1979)
79. C.J. Latimer, K.F. Dunn, F.P. O'Neill, M.A. MacDonald, and N. Kouchi: J. Chem. Phys. **102**, 722 (1995)
80. K. Ito, R. Hall and M. Ukai: J. Chem. Phys. **104**, 8449 (1996)
81. I. Sanchez and F. Martin: Phys. Rev. Lett. **79**, 1654 (1997) and **82**, 3775 (1999)
82. A. Lafosse, M. Lebech, J.C. Brenot, P.M. Guyon, L. Spielberger, O. Jagutzki, J.C. Houver, and D. Dowek: J. Phys. B (in preparation, to be published in 2003)
83. D. Dowek, I. Powis: Faraday Discuss. **115**, 181 (2000), pp. 175–204
84. P.O. Danis, T. Wyttenbach, and J.P. Maier: J. Chem. Phys. **88**, 3451 (1988)
85. C.Y. Ng: Annu. Rev. Phys. Chem. **53**, 101 (2002)
86. N.A. Cherepkov: J. Phys. B **14**, L623 (1981)
87. R.L. Dubs, S.N. Dixit and V. McKoy: Phys. Rev. Lett. **54**, 1249 (1985)
88. J.R. Appling, M.G. White, R.L. Dubs, S.N. Dixit, and V. McKoy: J. Chem. Phys. **87**, 6927 (1987)
89. K.L. Reid, D.J. Leahy and R.N. Zare: Phys. Rev. Lett. **68**, 3527 (1992)
90. D.J. Leahy, K.L. Reid, H. Park, and R.N. Zare: J. Chem. Phys. **97**, 4948 (1992)
91. F. Heiser, O. Gessner, U. Hergenhahn, J. Viehaus, K. Wieliczek, N. Saito and U. Becker: J. Electron Spectrosc. Relat. Phenom. **79**, 415 (1996)
92. S. Motoki, J. Adachi, K. Ito, K. Ishii, K. Soejima, A. Yagishita, S.K. Semenov, and N.A. Cherepkov: Phys. Rev. Lett. **88**, 063003 (2002)
93. M. Born and E. Wolf: Principles of Optics (Pergamon Press, New York 1970)
94. G. Schönhense and J. Hormes: VUV and Soft X-Ray Photoionization, edited by U. Becker and D.A. Shirley (Plenum, New York 1996) p. 607
95. I. Powis: J. Chem. Phys. **112**, 301 (2000)
96. N. Böwering, T. Lischke, B. Schmidtke, N. Müller, T. Khalil, and U. Heinzmann: Phys. Rev. Lett. **86**, 1187 (2001)

97. M. Lavollée: Rev. Sci. Instrum. **70**, 2968 (1999)
98. O. Jagutzki, V. Mergel, K. Ullmann-Pfleger, L. Spielberger, U. Spillmann, R. Dörner, and H. Schmidt-Böcking: Nucl. Instrum. Methods A **477**, 244 (2002)
99. S. Hsieh and J.H.D. Eland: J. Phys. B **30**, 4515 (1997)
100. M. Lavollée and V. Brems: J. Chem. Phys. **110**, 918 (1999)
101. Y. Hikosaka and J.H.D. Eland: Rapid Commun. Mass Spectrom. **14**, 2305 (2000) and references therein

16 Relaxation Dynamics
of Core Excited Molecules
Probed by Auger-Electron–Ion Coincidences

Marc Simon, Catalin Miron, and Paul Morin

16.1 Introduction

Core ionization or excitation of molecules using synchrotron radiation has been revealed as a unique tool to store a large amount of energy in a system, while keeping open several possibilities: a choice of the excited site, a choice of the symmetry of the intermediate resonant state, and the amount of vibrational energy one can store in the system.

The general interest in these studies relies on the understanding of energy dissipation and degradation of matter after photon irradiation. At the present time, for instance, a phenomenological description of the role of core ionization in the primary process of biological materials destruction after irradiation is still lacking. From an application point of view, it is also important to understand these processes in order to optimize technologies involving photon-induced methods like, for instance, chemical vapor deposition (CVD). In addition, gas-phase studies are ideally suited to develop spectroscopical methods because a large number of parameters can be accurately measured at the same time.

After photon absorption, the molecules follow different relaxation pathways through which numerous particles can be ejected: emission of a photoelectron and/or emission of an Auger electron induced by the inner-vacancy filling; fragmentation of the species in the neutral or ionized state. In addition, a strong change in the geometry of the molecule may appear during these steps leading to the probe of the system in totally "unusual" conformations. Experimentally, because so many particles are ejected with different kinetic energies and angular distributions, it is essential to measure all of these quantities simultaneously in a single photon-absorption event (coincidence method) to distinguish various competing decay channels.

During the last decade, we have tried to develop an experimental approach combining simultaneously a high spectral resolution both at the excitation (a photon beam) and analysis (electrons, ions) stage. In the present chapter, we will review both experimental approaches with the actual limitations of coincidental measurements, and recent results related to nuclear dynamics of CO_2, BF_3 molecules, selective fragmentation of $BrCH_2Cl$, $Fe(CO)_2(NO)_2$, and continuum characteristics like dynamical angular correlation observed in

CO. We have restricted our studies to the soft X-ray regime for three main reasons:

- This energy range covers the K-shell edges of the most abundant atoms (C, N, O).
- In this domain, the inner-vacancy lifetimes are long enough to allow nuclear motion to play a role in the relaxation processes.
- The kinetic energies released (electron, ions) are not too high so that particles can be efficiently detected and accurately characterized.

Core ionization of free molecules is an efficient way to produce highly excited and/or multiply charged ions with well-defined internal energies. Indeed, such a process is very weak in the threshold energy range because of dipole selection rules, but is the main decay process when core levels are excited and resonant Auger and normal Auger mechanisms take place on a very short timescale.

It is usually assumed that in a first step, the soft X-ray photon is absorbed with the concomittent promotion of the inner-shell electron into the continuum when ionization occurs, or to an empty molecular orbital when excitation into discrete states take place. In a second step, the Auger (or resonant Auger) process occurs: one valence electron fills the hole, another valence electron is ejected into the continuum. The lifetime value of the inner-shell vacancy is between 1 and 10 fs in the soft X-ray regime. In a third step, the doubly charged ion (singly charged for resonant excitation) dissociates on a much shorter timescale.

Due to the short lifetime, the Auger electrons leave the molecule extremely rapidly, providing a fast picture of the different processes occurring after formation of the intermediate state (nuclear motion, Franck–Condon factors). On the other hand, dissociation is a much slower process so that the detection of ions gives a picture at much longer time. Measuring Auger electrons and ions in coincidence is of interest not only because of the correlation between a final state and its subsequent dissociation, but also because, in specific cases, these measurements allow one to reconstruct the dynamics between two extreme times giving access to the topology of the potential energy surfaces.

16.2   Experimental Approaches

For many years, scientists have developed different techniques for the electron kinetic energy and angular-distribution measurements. However, one has to choose a compromise between a high energy resolution and a high angular acceptance for fast electrons. Different techniques have been used, mainly electrostatic or time-of-flight dispersion ones. A time-of-flight technique coupled with a position-sensitive detector is of particular interest for relatively slow electrons ($< 30\,\mathrm{eV}$) because of the 4π detection angle that this technique can offer. Another main advantage of the time-of-flight technique is the possibility to perform a multichannel analysis. This explains the beautiful results obtained with them, widely described elsewhere in this book. However, one must note that because the time-of-flight is inversely proportional to $\sqrt{E_k}$ (E_k being the electron kinetic energy, the resolution is limited by the temporal resolution of the electronics and is not suited for Auger electrons that have high kinetic energies because the energy resolution depends on the electron kinetic energy. On the contrary, electrostatic deflection plate analyzers have the advantage of a constant energy resolution over a wide energy range, because they can be equipped with efficient retarding lens systems. They are commonly used in research fields like photoemission, ion spectrometry, and others (see Leckey [1] for a review).

Most of the electrostatic deflection plates analyzers may be considered as a borderline case of the toroidal deflection plate analyzer. The toroidal deflection plate analyzer consists of two concentric toroidal deflection plates, and possesses the main qualities of the commonly used electron energy analyzers such as the hemispherical deflection-plate analyzer (SDA) and the cylindrical-mirror analyzer (CMA).

The SDA electron analyzer is commonly used because of its very high resolution, a dispersion over a wide energy range, and the possibility to easily perform electron angular-distribution studies. Moreover, it can be efficiently coupled to more or less complex retarding lens systems to perform high kinetic energy electron analysis. A very high performance SDA electron spectrometer, with a resolution better than 2 meV at 2 eV analyzer pass energy, has recently been demonstrated by Wannberg's group at Uppsala University in Sweden [2]. But, because of their relatively low angular acceptance, they are not used for Auger-ion coincidence experiments.

The CMA electron analyzer is also widely used, mainly because of its high angular acceptance (several per cent of 4π) and its quite high energy resolution. In contrast with the SDA electron analyzer, the CMA usually operates in a single-channel energy mode, which reduces its efficiency. To perform angular-distribution studies, several electron detectors have been used [3]. The main disadvantage of the CMA is the difficulty to design an efficient retarding lens system.

Toroidal analyzers are particularly suited to measure simultaneously the kinetic energy of fast electrons and their angular distribution with a high luminosity. This luminosity is high mainly because of the cylindrical symmetry allowing one to detect simultaneously all the ejection angles without any rotation of the analyzer. Several groups have developed such analyzers. Leckey and coworkers [4] conceived and realized the first toroidal electron analyzer. They are able to detect all the electrons emitted in a plane that contains the photon propagation vector and the polarization vector. This analyzer has been succesfully used for surface-science photoemission studies. Huetz and coworkers [5] have developed the same kind of electron analyzer dedicated to double-photoionization studies. Taking advantage of the large acceptance angle of the toroidal analyzer, they have combined a hemispherical analyzer with the former analyzer: the detection of two angularly resolved electrons is of fundamental interest in order to understand electron–electron correlations. In the design of Reddish et al. [6], of a dual-toroidal electron spectrometer, the electrons are detected in a plane perpendicular to the photon propagation vector allowing them to obtain in coincidence, two angularly resolved electrons in order to study the double-photoionization dynamics. Lahmam-Bennani and coworkers [7] have developed a dual double-toroidal analyzer allowing one to study the double ionization by electron impact of neutral target by the so-called (e,3e) coincidence experiments technique. The high luminosity of their experimental setup is a key in the success of such challenging experiments, for which up to three particles have to be simultaneously angularly and energetically analyzed.

16.2.1   Description of the DTA (Double–Toroidal Analyzer)

An original double-toroidal ion spectrometer, dedicated to ion scattering spectroscopy studies, has been built by Brongersma's group [8] in Eindhoven, and served as a starting point for our design. The full details of the design of our double-toroidal electron spectrometer have been already described [9] and we summarize them briefly in this section.

We have tried to combine the advantages of all types of electron analyzers previously mentioned namely:

- a high acceptance angle (CMA);
- a multichannel energy analysis (SDA);
- a high energy resolution independent of electron kinetic energy (SDA);
- a retarding lens system (SDA).

Hellings et al. [10] suggested the use of appropriate noncircular numerically calculated shapes for the deflection plates, providing perfect circular equipotential curves in a meridian plane containing the axis of symmetry. The design of charged particle analyzers with a noncircular deflection plates shape improves the particle focusing and decreases the aberrations. Nevertheless, it leads to a not perfectly linear energy dispersion function that is determined by known electron kinetic-energy lines.

As is shown in Fig. 16.1, we detect electrons emitted from the source volume with respect to the cylindrical symmetry axis of the analyzer around the angle of 54.7°, called the magic angle, which cancels any angular anisotropy for photoionization studies for linearly polarized light, allowing nondipole effects to be emphasized [11]. We choose the planar acceptance angle of the collimator of $\pm 3°$ (corresponding to a 5% of 4π angular aperture of the DTA), a compromise between the countrate and the resolving power.

The operation mode of the DTA is with a fixed pass energy and variable retarding potential, and the lens is adjusted at the appropriate kinetic

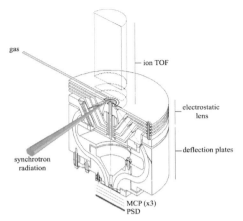

Fig. 16.1. Three-dimensional view of the double-toroidal analyzer, the time-of-flight mass spectrometer and the interaction region

energy. This mode has the main advantage of giving a constant energy resolution proportional to the pass energy. We recently designed a four-element conical electron lens [12] that improves the focusing on the entrance slit of the analyzer and allows us to work with a decelerating factor up to 30 (ratio between the initial kinetic energy and the pass energy). By using the small spot size ($300\,\mu$m$\times300\,\mu$m H$\times$V) of the third-generation synchrotron source MAX lab in Sweden, we obtained a resolution of 150 meV for C1s Auger electrons of the CO molecule that has a kinetic energy of about 250 eV at a pass energy of 20 eV. Note that this 150 meV resolution is low enough to observe the effect of stretching modes for small molecules in dissociation dynamics studies of doubly charged ions.

The double-toroidal analyzer is conceived in such a way that the electron trajectories in the analyzer after dispersion are focused onto the plane of the position-sensitive detector. We can then simultaneously detect electrons over a wide energy range (15% of the pass energy). When angular distributions have to be measured, the angular response of the apparatus is precisely calibrated by tabulated distributions [13].

16.2.2   Mass Spectrometer and Coincidence Regime

The ions created in the same ionization event that the Auger electrons are analyzed by a time-of-flight Wiley–MacLaren mass spectrometer. In order to keep a good electron resolution, we use a pulsed extraction field in the ionization region triggered by the electron detection. Because the extraction field is high (1 kV/cm), the rise time is low (15 ns) and the drift tube is short (12 cm), we minimize the angular discrimination against fast ions emitted in a direction perpendicular to the mass spectrometer axis.

We call electron–ion coincidence the *simultaneous* detection of an electron and an ion (within a time window). The coincidence measurements provide additional information compared to the separate measurements, as long as their origin is the same ionization event. When the electron and the ion do not come from the same event, this coincidence is called *random* coincidence. While true coincidence rates are proportional to the counting rate, random coincidences are proportional to the square of the counting rate. In order to keep a relatively low random coincidences rate, one has to keep a moderate counting rate. In our case, we keep the ion counting rate at 10^4 detected ions per second. The electron counting rate, depending on the pass energy, is of the order of 100 detected electrons per second, allowing detection of about 30 coincidences per second. The true and random coincidences appear at the same time-of-flight. The random coincidences are subtracted from the collected data with a random trigger.

16.3   Nuclear Motion in Competition with Resonant Auger Relaxation

An important question, which has provided a lot of interest in the last two decades, concerns the interplay between the electronic decay and the nuclear motion of core excited species. Indeed, it has long been considered that the lifetime of core-excited states is so short (a few tens of femtoseconds) that no nuclear motion has enough time to occur. However, the resonant excitation of a bound core electron into an empty molecular orbital results in a rearrangement of the whole electronic cloud, and the combined effect of an additional electron in the valence shell and of an additional positive charge in the core may induce a strong conformational change in the molecule depending on several parameters: the bonding/antibonding character of the resonant state, the core-hole lifetime of the inner vacancy and the mass of the involved atoms. The first example, revealed by Morin and Nenner almost twenty years ago [14], concerns the relaxation of a core excited HBr molecule – the strongly antibonding character of the σ^* intermediate state – and is shown to lead to a situation where the neutral dissociation of the core excited molecule may take place before the Auger decay. On the other hand, in the case when a transition is made into a bound intermediate state, it is possible to observe a specific vibrational distribution in the final ionic state as a fingerprint of the geometrical changes between ground/intermediate and final states.

In order to be able to accurately investigate such situations, one must combine several experimental techniques allowing the complete decay process starting from the photon–molecule collision and up to the ejection of one or several ionic fragment to be probed. For that purpose, the high-resolution electron spectrocopy provides a unique tool for probing the first femtoseconds, while the coincidence measurements between energy-selected Auger electrons and fragment ions allow probing of the longer timescale processes, i.e., dissociation, and to correlate the various dissociation patterns to the exact electronic/vibrational states of the dissociating ions. We will try to illustrate in the following the use of these techniques by two examples.

16.3.1   Mapping Potential Energy Surfaces by Core Electron Excitation: BF_3

The nuclear motion taking place in the B1s→$2a_2''$ core excited state in gas-phase BF_3 has been intensively studied, e.g., by Ueda et al. [15,16] and Simon et al. [17]. Strong evidence of a conformational change of the molecule from plane (D_{3h} point group) in the ground state to pyramidal (C_{3v} point group) in the $2a_2''$ state upon boron 1s excitation was given. In a more recent study by Simon and collaborators [18], a significantly improved resolution (about 0.2 eV) for both the excitation and the electron analysis channels, together with support from *ab initio* quantum-chemistry calculations [19], allowed one to emphasize a new effect, the so-called "dynamical Auger emission". This effect was observed as being a continuous extension, or "tail" towards the lower kinetic energy side of the resonant Auger lines and related to the nuclear motion in the core excited state. These early studies stimulated general interest on the scientific case of the fast nuclear motion induced by core excitation in polyatomic systems, and therefore further investigations have been performed [20–23].

Within a simple picture, the dynamical Auger emission reflects the motion of the nuclei in the core excited state, and it is observed as the result of a superposition of contributions associated with the decay of the propagating wavepacket for a wide range of internuclear distances. We will show how the intensity and the energy distribution of the dynamical Auger emission can be used to characterize the relevant potential energy surfaces of the system under consideration.

The experiment [24] has been carried out at the undulator beamline I411 [25] of the third-generation synchrotron radiation facility MAX II in Lund, Sweden. This beamline is equipped with a modified Zeiss SX 700 plane-grating monochromator and with a rotatable hemispherical Scienta SES 200 electron spectrometer. During the experiments, the monochromator band-width was chosen to be better than 30 meV at 200 eV photon energy and the electron spectrometer has been operated at the magic angle of 54.7° with a pass energy of 10 eV providing a kinetic-energy resolution of about 35 meV. The overall experimental resolution of the resonant Auger spectra was thus better than 45 meV.

Figure 16.2 shows the direct photoemission spectrum excited by 185 eV photons and the resonant Auger spectrum measured on top of the B1s→ $2a_2''$ resonance at 195.52 eV. The spectra are normalized with respect to the photon flux and gas pressure and plotted on a binding-energy scale, $E_b = h\nu - E_k$, where $h\nu$ is the photon energy and E_k is the Auger electron kinetic energy. The assignment of the various lines, taken from the theoretical work of Haller

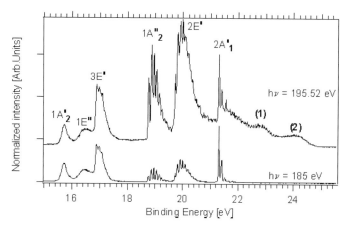

Fig. 16.2. Comparison of the direct photoemission spectrum and the resonant Auger spectrum measured on top of the B1s→ $2a_2''$ resonance (195.52 eV) for the BF_3 molecule on a binding-energy scale

et al. [26], is indicated in the figure. A striking feature to note is the exceptionally wide vibrational extension (about 3.6 eV) above the onset of the $2A_1'$ state for the resonant spectrum as compared to the off-resonance spectrum for which the intensity already vanishes 0.6 eV above its ionization energy. Moreover, both the $2E'$ and $2A_1'$ states show a pronounced asymmetric tail towards higher binding energies that exhibits two broad structures, labeled (1) and (2) in Fig. 16.2.

The v_2 out-of-plane bending mode is not observed in the direct photoemission spectrum. Indeed, this mode is symmetry forbidden according to the dipole-selection rules. However, the vibrational extension seen in the resonant Auger decay above 21.6 eV has a characteristic spacing of about 80 meV. This spacing is very close to the v_2 out-of-plane mode in the neutral ground state [27], so we assign this progression to the v_2 bending mode based on the ground-state geometry. This forbidden vibrational excitation is allowed here due to the change of geometry in the intermediate state where the out-of-plane vibration is strongly excited.

The intensity distribution within the progression is governed by the wave-packet dynamics in the resonant state and the decay to the final state. According to calculations by Tanaka et al. [19] within the framework of the dynamical Auger emission model, several broad maxima are expected within this wide vibrational progression. In addition to those related to the inner and outer classical turning points of the relevant potential-energy curves [28], the main contribution is predicted to arise from a region, situated in the vicinity of the minimum energy of the core excited state, where the intermediate and final-state potential-energy curves are locally parallel. We assign structure (1) to $2E'$ and structure (2) to $2A_1'$ based on the difference between the ionization

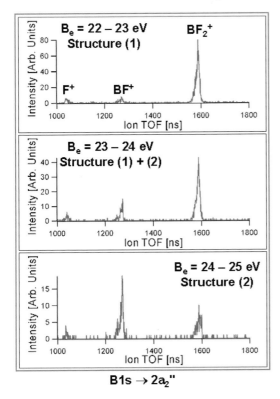

Fig. 16.3. Time-of-flight mass spectrum in coincidences with resonant Auger electron measured on the top of the B1s→ $2a_2''$ resonance at 195.52 eV

potentials. In order to definitely elucidate the origin of these features, their photon-energy dependence has been studied, assigned to the outer "classical turning point" and allowed one to qualitatively determine the relative slopes of the intermediate and final states potential-energy curves [24].

We have performed resonant Auger electron–ion coincidences on the top of the B1s→ $2a_2''$ resonance at 195.52 eV, which are presented in Fig. 16.3. We emphasize in the figure the fragmentation of the ion along the tail with its two structures. On structure (1), the dominating ion in the mass spectrum is the BF_2^+ ion. On structure (2), the BF^+ channel dominates. The aperture of this dissociation channel explains why the v_2 out-of-plane bending mode is less and less well resolved at higher binding energies in the resonant Auger spectrum: the dissociation is faster than the vibration period. Our results are in complete agreement with already published thermodynamical data [29].

16.3.2   Molecular Dissociation Mediated by Bending Motion in the Core–Excited CO_2

The carbon dioxide molecule is linear in its electronic ground state, but it is known to bend upon $C1s \longrightarrow \pi^*$ excitation [30]. Within a more refined picture, one must consider that the $1s \rightarrow \pi^*$ core excited states are doubly degenerate in a linear conformation, but this degeneracy is lifted through the Renner–Teller effect by the bending of the molecule, so that the core excited state exhibits two nondegenerate components $\pi^*_{in-plane}$ and $\pi^*_{out-of-plane}$, which correspond, respectively [31], to bent and linear conformations. Nice illustrations of the effects induced by this physical situation on the decay processes of the $C1s \rightarrow ^2 \Pi_u$ core-excited state in CO_2 were given by the recent studies of Morin et al. [32], Kukk et al. [33] and Muramatsu et al. [34] and for the $N_t1s \rightarrow \pi^*$ core excited state in N_2O by Miron et al. [20]. Namely, it was shown by the energy-selected resonant Auger electron–ion coincidence measurements (ES-RAEPICO) of Morin and coworkers that completely different dissociation channels became accessible from the $\widetilde{A}^2\Pi_u$ electronic state of CO_2^+, following resonant Auger decay of either the $C1s \rightarrow a_1\pi^*_{in-plane}$ or of the $C1s \rightarrow b_1\pi^*_{out-of-plane}$ RT split components of the π^* core-excited state in a C_{2v} symmetry [32].

We recently measured, at high resolution for the electron analysis channel, the coincidence spectra between energy-selected resonant Auger electrons and ions after $C1s \rightarrow ^2 \Pi_u$ excitation in CO_2. The measurements were done at the undulator beamline I411 at MaxLab in Lund using our "double-toroidal" electron energy analyzer [9] and a conventional time-of-flight spectrometer triggered by a high-voltage extraction field. The overall resolution achieved in the coincidence measurements shown here is 300 meV. This is mainly due to the DTA instrumental resolution, the resolution in the excitation channel being about an order of magnitude higher (in order to improve the true/false coincidence rate, the monochromator was operated with a 1–μm slit).

Figure 16.4 shows the ion mass spectra corresponding (through the formula: $E_b = h\nu - E_k$, where $h\nu$ is the photon energy and E_k is the Auger electron kinetic energy) to the highest populated vibrational levels of the $\widetilde{A}^2\Pi_u$ electronic state ($E_b = 19.6 \pm 0.15$ eV) of CO_2^+ created through Auger decay after resonant excitation at three different excitation energies corresponding to the top ($h\nu = 291$ eV), left side ($h\nu = 290.6$ eV), and right side ($h\nu = 291.4$ eV) of the $C1s \rightarrow \pi^*$ resonance. The contributions of the direct photoionization, as well as that due to the fortuitous coincidences, have been subtracted from the spectra.

Two interesting observations can be made from the spectra in Fig. 16.4: (i) For the same internal energy stored in the CO_2^+ ion (19.6 eV), at least two distinct dissociation patterns can be "preferentially induced" after Auger decay as a function of the initial energy transfer to the neutral molecule (excitation energy): $CO_2^+ \longrightarrow O^+ + CO$ and $CO_2^+ \longrightarrow O + CO^+$. (ii) Contrary to the normal behavior observed in photodissociation measurements, here the

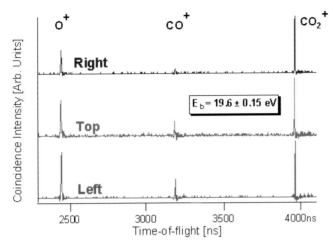

Fig. 16.4. Ion mass spectra in coincidence with highest vibrational states of the $\tilde{A}^2\Pi_u$ electronic state (E_b=19.6±0.15 eV) of CO_2^+ at three different photon energies: top ($h\nu = 291$ eV), left side ($h\nu = 290.6$ eV) and right side ($h\nu = 291.4$ eV) of the $C1s \rightarrow \pi^*$ resonance

fragmentation rate becomes weaker as the excitation energy increases from left to right! Indeed, it is clear that the O^+ ion production decreases as photon energy increases. In order to explain such an unusual behavior, one must take into account that in the considered energy range the allowed dissociation channels of CO_2^+ may only lead to CO^+ in a linear conformation. However, dissociation into O^++CO may become possible through predissociation by a $^4\Sigma^-$ electronic state [35], which needs a bent conformation (C_{2v}). One is thus able to control the dissociation of the system (namely the O^+ ion production) by controling the molecular conformation (linear vs. bent in this specific case) of the core excited state via the excitation energy. This kind of experiment may be used as an accurate probe of molecular dynamics in many other cases, as soon as a conformational change is expected to occur between the ground and the excited state.

These two examples illustrate that the use of high-resolution spectroscopy with a high-quality third-generation synchrotron radiation source may allow investigation of the details of molecular dynamics as well as the topology of complex potential energy surfaces in core excited systems.

16.4   Selective Photofragmentation

A simple and attractive idea that has emerged in the last few years is to take advantage of the hole localization to induce a specific bond breaking in a complex molecular system. The localized character of the hole appears clearly in the Auger spectrum due to the predominance of intra- vs. inter-molecular processes: only orbitals with a significant overlap with the created hole participate in the Auger relaxation, at least for nonionic compounds. But this condition is not sufficient to ensure selective dissociation. Indeed, there is a severe competition between dissociation and redistribution of the internal energy among the various vibration modes of the ion. We can imagine two extreme cases:

(i) The dissociation is direct and very fast, like in ultrafast dissociation. The electronic relaxation (Auger effect) is achieved after fragmentation.

(ii) The dissociation is fairly slow and is driven by statistical laws.

Generally, when the molecule is photoexcited in one electronic state, there is a fast internal conversion process from electronic energy into vibrational energy of the ground electronic state. This vibrational energy is redistributed statistically over the various degrees of freedom and the molecule dissociates. The fragmentation rate and branching ratio depend upon the internal energy and the number of rovibrational states above the activation barrier along the reaction coordinate. Such branching ratios are always in favor of the rupture of the weakest bond that occurs only after conversion of electronic internal energy into vibrational excitation of all the degrees of freedom of the system. These two borderline cases also feature two extreme cases of site selectivity: The first case corresponds to a pure site selective effect, because it is governed by the core excited electronic state, whereas the second case corresponds, in contrast, to the total loss of memory of any initial localisation of the excitation process. The final result of fragmentation after core photoionization is thus the net result between these two extreme behaviors. Investigation on site-selective fragmentation has been reported for a large variety of molecules including halogeno-hydrocarbons [36–40], organometallic compounds [41–46], silicon compounds [47–51], and other species [52,53] We wish to focus here on two selected examples corresponding to the extreme cases discussed above: the $Br(CH_2)Cl$ and $Fe(CO)_2(NO)_2$ molecules.

In a recent investigation, Miron et al. [54] reported a clear example of selective bond breaking, comparing the fragmentation after Br 3d or Cl2p hole formation on the bromochloromethane molecule. The experiment consists of measuring coincidence between the emitted Auger electron and the corresponding fragments. At a photon energy located above both Br3d and Cl2p ionization thresholds, it is possible to distinguish between two types of Auger relaxation because they give rise to very distinct kinetic energies. A strong selectivity is observed as shown in Fig. 16.5: after Br3d hole relaxation, Br^+ clearly dominates the fragmentation pattern, whereas the $BrCH_x^+$ fragment is almost negligible. On the contrary Cl^+ is now dominating, whereas $BrCH_2^+$ fragment is as intense as Br^+ after Cl2p ionization. Thus, after excitation of the Br3d edge the bond with Br and the rest of the molecule is preferentially broken whereas at the Cl2p edge, the bond with Cl is preferentially broken despite the fact that the C–Br bond strength is lower than the C–Cl one. A

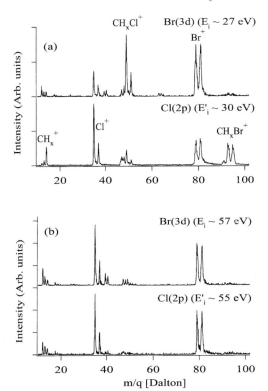

Fig. 16.5. $Br(CH_2)Cl$ Auger-ion coincidence spectra for low (**a**) and the high (**b**) binding energy states of $Br(CH_2)Cl^{++}$. At low binding energy one can note a strong site-selective fragmentation of the doubly charged ion (*top panel*), whereas at high binding energy atomization processes are dominant (*bottom panel*)

very interesting feature of the experiment is that the coincidence measurement allows one to point out the origin of the observed selectivity. Indeed, it can be noted that at high binding energy (lower spectra) the coincidence mass spectra are almost identical: the initial localization of the energy deposit is lost. On the contrary, the selectivity is maximum at low binding energy (upper spectra): the localized character of these selected final states (two holes on lone pairs of Br and Cl respectively) allows selective fragmentation to be observed. In this example, the memory of the initial core-hole localization, combined with the localization of the valence vacancies in the doubly charged final ion allows selective effects to be observed.

Another system of interest with respect to the site selectivity is the $Fe(CO)_2(NO)_2$ molecule. Indeed, in such a system one can selectively excite the carbon or the nitrogen 1s shells that are well separated in energy. Moreover, both CO and NO ligands exhibit strong and sharp π^* resonances and broad σ^* shape resonances that can be excited from the 1s shell. The fragmentation of this molecule has been investigated [45] using the TOF multicoincidence technique. Mass spectra along the C1s→ π^* (287.8 eV) and N1s→ π^* (400.3 eV) resonances have been recorded. The most striking observation is that the spectra are very similar: no memory effects due to the initial localization of the hole (selected site) and excitation (π^* resonances) can be seen. Simon et al. [45] pointed out, however, some slight differences when comparing (at the same edge) π^* and σ^* excitation, which were interpreted as a change in the single/double ionization branching ratio along the two resonances: in the continuum, one expects only double ionization to be observed due to the dominant normal Auger effect, whereas on the π^* resonance, resonant Auger decay favors single-ionization channels (spectator model). The loss of memory effects was interpreted as being due to the statistical fragmentation of the residual ground state multiply charged ion after the Auger decay. This is because the ground state of the ion is the same (localization of the charges, equilibrium geometry) independent of the location of the initial vacancy. Furthermore, the fragmentation pattern that occurs after the redistribution of the electronic into vibration energy is governed by the internal energy of the ion and the electronic structure of the molecular ion populated by the core-hole decay.

In conclusion, these two prototype examples show that core ionization can be used to produce site-selective fragmentation of complex molecules. However, whereas core-hole localization and valence-hole localization after Auger decay are necessary conditions to be fulfilled, it is also necessary that internal energy-conversion processes are slow enough. Coincidence measurements are essential to disentangle the origin of selective effects.

16.5   Dynamical Angular Correlation of the Photoelectron and the Auger Electron

Angular distribution (AD) of photo-electrons and Auger electrons allows one to obtain a refined characterization of the continuum wave functions as it accounts for both amplitudes and phase differences of the various matrix elements (dipole, quadrupole, etc. and Coulomb) involved in photoionization and subsequent decay. Due to the high luminosity of our electron analyzer (DTA), we are able to extract angular distributions for fixed-in-space molecules giving much more information than from randomly oriented molecules. Indeed, in cases where the ion dissociation is fast enough, the initial orientation of the molecule can be deduced from the fragmentation axis (axial recoil approximation) which is determined by coincidence measurements.

Chapter 15 clearly shows the richness of the information one can obtain by vectorially correlated photoelectrons and ions. In the case of initial core ionization, the processes are more complex because fast Auger electrons are emitted. For a better understanding of dynamical processes occurring after a soft X-ray photon absorption by a molecule, it is necessary to characterize the continuum electronic wave functions obtained by measuring their angular distributions. It is generally admitted to consider the processes ionization/Auger decay as occurring in two distinct steps: after the fast (few attosecond) inner-shell electron emission, the Auger effect occurs after a few femtoseconds. Several theoretical models [55–57] predicted, using the sequential model, that the Auger angular distribution in the molecular frame is independent of the photon energy as well as molecular-axis orientation with respect to the polarization vector of the incoming light.

We have tested this prediction by measuring the Auger electron angular distribution after C1s ionization of CO molecule, in coincidence with initially oriented molecules parallely or perpendicularly to the polarization vector. We took advantage of the rapidity of Auger relaxation and fragmentation processes compared to the rotation of the molecule by measuring angular distribution of electrons in coincidence with fragment ions emitted at $0°$ and $90°$ that select the internuclear axis orientation for a diatomic molecule.

For the $\widetilde{B}$ electronic state of the CO^{++} ion, the Auger angular distribution has been measured for the internuclear axis parallel or perpendicular to the polarization vector at three different photon energies (299 eV, 305 eV, and 400 eV) and are shown in Fig. 16.6 [58]. At 400 eV, we obtain the same angular distribution for both molecular orientations as theoretically predicted. At lower photon energies, these distributions become very different with rich structures. At 305 eV, the wave function looks like a d wave, whereas at 299 eV, it looks like an f wave, similarly to the 1s photoelectron measured for fixed-in-space molecules [59]. This suggests a dynamical angular correlation of the slow photoelectron with the Auger electron.

The Auger process is basically an intra-atomic process. When emitted from a particular atom, the Auger electron is scattered by the surrounding atoms. This scattering process (well known for the photoelectron where it leads to the EXAFS modulations) should be strongly reflected in the angular distribution of these electrons when established in the frame of the molecule. Zähringer et al. [60] reported theoretical predictions of such scattering. Such Auger electrons scattered by the surrounding nucleus have been measured and reported very recently by Dörner et al. [61] on high electronic states of a doubly charged CO molecule. This kind of experiment is only beginning and there is no doubt that a promising field of research is now open.

16.6 Conclusions and Perspectives

We have shown that coincidence experiments are ideally suited to bring a deep insight into nuclear dynamics following core ionization or excitation. In this energy domain, where high-energy electrons are emitted together with energetic fragment ions, electrostatic electron analyzers coupled with TOF mass spectrometers appear the best compromise to achieve good resolution and reasonable counting rate. Challenging in the near future is to still improve

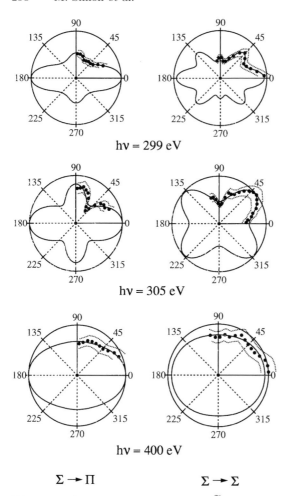

$\Sigma \rightarrow \Pi$ $\Sigma \rightarrow \Sigma$

Fig. 16.6. Angular distribution of the $\tilde{B}$ electronic state of the doubly charged CO dication at three different photon energies for $\Sigma \rightarrow \Sigma$ (*right*) and $\Sigma \rightarrow \Pi$ (*left*) ionization channels. The internuclear axis is given by the 0°–180° axis

the electron-energy resolution to access vibrational resolution of triatomic molecules. The use of third-generation sources is important in this respect. A systematic analysis of full momentum characterization (amplitude and angle) of the emitted particles will also help to understand these dynamical effects.

So far, most of the experiments have been carried out for simple isolated molecules. It is important both from the fundamental and applied viewpoints to explore more complex systems, like weakly bonded complexes (van der Waals, mixed organic clusters) or large molecules, like biomolecules. Certainly, of special interest is the study of molecules in a solvent environment, like biomolecules in water, in order to study the behavior of irradiated systems in realistic conditions.

References

1. R.C.G. Leckey: J. Electron. Spectrosc. **43**, 183 (1987)
2. N. Mårtensson, P. Baltzer, P.A. Brühwiler, J.-O. Forsell, A. Nilsson, A. Stenborg and B. Wannberg: J. Electron. Spectrosc. **70**, 117 (1994)
3. J.M. Bizau, D. Cubaynes, P. Gérard and F.J. Wuilleumier: Phys. Rev. A **40**, 3002 (1989)
4. F. Toffoletto, R.C.G. Leckey and J.D. Riley: Nucl. Instrum. Methods B 12, 282 (1985)
5. J. Mazeau, P. Lablanquie, P. Selles, L. Malegat and A. Huetz: J. Phys. B **30**, L293 (1997)
6. T.J. Reddish, G. Richmond, G.W. Bagley, J.P. Wightman and S. Cvejanovic: Rev. Sci. Instrum. **68**, 2692 (1997)
7. A. Duguet, A. Lahmam-Bennani, M. Lecas and B. El Marji: Rev. Sci. Instrum. **69**, 3524 (1998)
8. G.J.A. Hellings, H. Ottevanger, S.W. Boelens, C.L.C.M. Knibbeler and H.H. Brongersma: Surf. Sci. **162**, 913 (1985)
9. C. Miron, M. Simon, N. Leclercq and P. Morin: Rev. Sci. Instrum. **68**, 3728 (1997)
10. G.J.A. Hellings, H. Ottevanger, C.L.C.M. Knibbeler, J. Van Engelshoven and H.H. Brongersma: J. Electron. Spectrosc. **49**, 359 (1989)
11. R. Guillemin, O. Hemmers, D.W. Lindle, E. Shigemasa, K. Le Guen, D. Ceolin, C. Miron, N. Leclercq, P. Morin, M. Simon and P.W. Langhoff: Phys. Rev. Lett. **89**, 033002 (2002)
12. K. Le Guen, D. Céolin, R. Guillemin, C. Miron, N. Leclercq, M. Bougeard, M. Simon, P. Morin, A. Mocellin, F. Burmeister, A. Naves de Brito and S.L. Sorensen: Rev. Sci. Instrum. **73**, 3885 (2002)
13. R. Guillemin, E. Shigemasa, K. Le Guen, D. Céolin, C. Miron, N. Leclercq, K. Ueda, P. Morin and M. Simon: Rev. Sci. Instrum. **71**, 4387 (2000)
14. P. Morin and I. Nenner: Phys. Rev. Lett. **56**, 1913 (1986)
15. K. Ueda, K. Ohmori, M. Okunishi, H. Chiba, Y. Shimizu, Y. Sato, T. Hayaishi, E. Shigemasa and A. Yagishita: Phys. Rev. A **52**, R1815 (1995)
16. K. Ueda, K. Ohmori, M. Okunishi, H. Chiba, Y. Shimizu, Y. Sato, T. Hayaishi, E. Shigemasa and A. Yagishita: J. Electron Spectrosc. Relat. Phenom. **79**, 411 (1996)
17. M. Simon, P. Morin, P. Lablanquie, M. Lavollée, K. Ueda and N. Kosugi: Chem. Phys. Lett. **238**, 42 (1995)
18. M. Simon, C. Miron, N. Leclercq, P. Morin, K. Ueda, Y. Sato, S. Tanaka and Y. Kayanuma: Phys. Rev. Lett. **79**, 3857 (1997)
19. S. Tanaka, Y. Kayanuma and K. Ueda: Phys. Rev. A **57**, 3437 (1998)
20. C. Miron, M. Simon, P. Morin, S. Nanbu, N. Kosugi, S.L. Sorensen, A. Naves de Brito, M.N. Piancastelli, O. Björneholm, R. Feifel, M. Bässler and S. Svensson: J. Chem. Phys. **115**, 864 (2001)
21. C. Miron, R. Guillemin, N. Leclercq, P. Morin and M. Simon: J. El. Spectrosc. Relat. Phenom. **93**, 95 (1998)

22. K. Ueda, M. Simon, C. Miron, N. Leclercq, P. Morin and S. Tanaka: Phys. Rev. Lett. **83**, 3800 (1999)
23. K. Ueda, S. Tanaka, Y. Shimizu, Y. Muramatsu, H. Chiba, T. Hayaishi, M. Kitajima and H. Tanaka: Phys. Rev. Lett. **85**, 3129 (2000)
24. C. Miron, R. Feifel, O. Björneholm, S. Svensson, A. Naves de Brito, S.L. Sorensen, M.N. Piancastelli, M. Simon and P. Morin: Chem. Phys. Lett. **359**, 48 (2002)
25. M. Bässler, A. Ausmees, M. Jurvansuu, R. Feifel, J.-O. Forsell, P. de Tarso Fonseca, A. Kivimäki, S. Sundin, S.L. Sorensen, R. Nyholm and S. Svensson: Nucl. Instrum. Methods A **469**, 382 (2001)
26. E. Haller, H. Köppel, L.S. Cederbaum, W. von Nissen and G. Bieri: J. Chem. Phys. **78**, 3 (1983)
27. E. Haller, H. Köppel, L.S. Cederbaum, G. Bieri and W. von Nissen: Chem. Phys. Lett. **85**, 12 (1982)
28. The calculations in [19] were made within a simplified model where only the relevant internuclear coordinate is considered, i.e., $Q_{a_2''}$ describing the distance between the boron atom with respect to the plane defined by the three fluorine atoms. For this reason, we use the term of "potential energy curve" instead of "potential energy surface".
29. M. Farber and R.D. Srivastava: J. Chem. Phys. **81**, 241 (1984)
30. P. Morin, M. Lavollée, M. Meyer and M. Simon: ICPEAC Conf. Proc. AIP **295**, p. 139, Editors: T. Andersen et al., NY 1993
31. J. Adachi N. Kosugi, E. Shigemasa and A. Yagishita: J. Chem. Phys. **107**, 4919 (1997)
32. P. Morin, M. Simon, C. Miron, N. Leclercq, E. Kukk, J.D. Bozek and N. Berrah: Phys. Rev. A **61**, 050701 (2000)
33. E. Kukk, J.D. Bozek and N. Berrah: Phys. Rev. A **62**, 032608 (2000)
34. Y. Muramatsu, K. Ueda, N. Saito, H. Chiba, M. Lavollée, A. Czasch, T. Weber, O. Jagutzki, H. Schmidt-Böcking, R. Moshammer, U. Becker, K. Kubozuka and I. Koyano: Phys. Rev. Lett. **88**, 133002 (2002)
35. M.T. Praet, J.C. Lorquet and G. Raseev: J. Chem. Phys. **77**, 4611 (1982)
36. P. Morin, T. LeBrun and P. Lablanquie: J. Chimie Phys. (France) **86**, 1833 (1989)
37. P. Morin, T. LeBrun and P. Lablanquie: Bull. Sté Royale Sci. Liège **58**, 135 (1989)
38. H.C. Schmelz, C. Reynaud, M. Simon and I. Nenner: J. Chem. Phys. **101**, 3742 (1994)
39. K. Müller-Dethlefs, M. Sander, L.A. Chewter and E.W. Schlag: J. Chem. Phys. **85**, 5755 (1986)
40. I. Nenner, P. Morin, M. Simon, P. Lablanquie and G.G.B. de Souza: Desorption Induced by Electronic Transitions, DIET III, Springer Series in Science vol. 13, Eds. Stulen R.H. and Knotek, M.L. (Springer, Berlin Heidelberg New York 1988) pp. 10–31
41. S. Nagaoka, I. Koyano, K. Ueda, E. Shigemasa, Y. Sato, A. Yagishita, T. Nagata and T. Hayaishi: Chem. Phys. Lett. **154**, 363 (1989)
42. K. Ueda, E. Shigemasa, Y. Sato, S. Nagaoka and I. Koyano: Chem. Phys. Lett. **154**, 18 (1989)
43. K. Ueda, Y. Sato, S. Nagaoka, I. Koyano, A. Yagishita and T. Hayaishi: Chem. Phys. Lett. **170**, 389 (1990)

17.2   Signatures of Many-Body Interactions in Predissociation

When a transient molecular state such as H_3^* is described by a purely repulsive potential energy surface, the fragmentation process is usually termed a *direct dissociation*. However, when this state is chemically bound, but couples to a continuum state, the process of dissociation can be delayed, taking place after many vibrational or rotational periods. This indirect fragmentation is frequently termed a *predissociation*. Using Fermi's Golden Rule, the rate for photodissociation can be written as

$$dP_D/dt \propto |\langle \Psi^i|T|\Psi^f\rangle|^2, \qquad (17.4)$$

where $\Psi^{i,f}$ are wave functions of the initial and final states, and T denotes a coupling operator connecting the wave functions in their respective bases. Note that the transient state does not explicitly appear in this most general approach to photodissociation. As will be shown below, under certain conditions it is possible to separate the excitation process from the dissociation process. If this is the case, a similar probability expression can be defined for the predissociation rate of the laser-prepared excited state Ψ^p

$$dP_P/dt \propto |\langle \Psi^p|T_p|\Psi^f\rangle|^2. \qquad (17.5)$$

The nature of the predissociation process is frequently named in terms of the nature of T_p, e.g., *vibrational, rotational,* or *electronic* predissociation. The arrows in (17.2) and (17.3) represent the transformation of rovibronic molecular energy into translational (W), rotational (j), and vibrational (v) energy of the product states. For an isolated molecule this transformation is solely mediated by the internal dynamics of the molecule. In other words, the coupling term between the prepared state and final states vectors

$$\langle \Psi^p|T_P|\Psi^f\rangle \qquad (17.6)$$

is nonzero only when the symmetries (the irreducible representations) of the two states are equal, $\Gamma_p = \Gamma_f$. As a result, the selection rules are that parity, total angular momentum, and energy be preserved in the transformation from Ψ^p to Ψ^f.

Only in the case where a molecular motion can be separated into vibrational, electronic, and rotational parts, can the additional selection rules be invoked for each degree of freedom. This simplification is increasingly difficult to justify in polyatomic systems at high excitation energy. As a consequence, the proper evaluation of the coupling term (17.6) requires full account of the rovibronic function Ψ^p and its projection into the final-state continuum Ψ^f through the transition operator T_P. Nevertheless, in order to illustrate some of the underlying physics it is often useful to separate these parts. Current theories are restricted to make such a separation even in the simple case of triatomic hydrogen. A general discussion of the rules for transition operators as they have been used in wavepacket simulations of predissociation of H_3 [28] can be found in [29].

17.2.1   Scalar Observables

In the limit of low light intensity, the absorption cross section $\sigma(\omega)$ is a measure of the opacity of the molecule for photoabsorption

$$\sigma(\omega) \propto \omega_{\mathrm{fi}} |\langle \Psi^{\mathrm{i}} | e \cdot \mu | \Psi^{\mathrm{f}} \rangle|^2, \tag{17.7}$$

where e is the unit vector of polarization of the light field and μ is the electric dipole operator of the molecule. The absorption cross section does not merely reflect the transition to the transient state, but entails the evolution through the transition states to the exit channel. Frequently, a predissociation is slow, hence the energy dependence of the absorption cross section is highly structured with Lorentzian, but more generally Fano-type, lineshapes. The width of the lines (another scalar observable) is frequently viewed as the lifetime of the transient complex but, in principle, it depends on the specific initial state from which excitation occurred. The width and shape of the excitation profile reflect the energy-dependent view of the exit continuum from a given initial state. The dependence on the initial state is particularly striking when more than one bound and/or continuum state appears in the exit channel.

   Partial cross sections are refined scalar observables that are a measure of the cross section that leads to a specific product of quantum state α:

$$\sigma(\omega, \alpha) \propto \omega_{\mathrm{fi}} |\langle \Psi^{\mathrm{i}} | e \cdot \mu | \Psi_{\alpha}^{\mathrm{f}} \rangle|^2. \tag{17.8}$$

Obviously, $\sigma(\omega) = \Sigma_{\alpha}\, \sigma(\omega, \alpha)$ and the product distribution $P(\omega, \alpha)$ is given as $P(\omega, \alpha) = \sigma(\omega, \alpha)/\sigma(\omega)$. A specific example in photodissociation of H_3 is the rovibrational product distribution $P(v, j)$ following excitation to a specific excited intermediate state at a specific wavelength (see Fig. 17.3).

17.2.2   Observation of Vector Correlations

By its nature, photodissociation is an anisotropic process since μ in (17.8) defines a specific axis in the molecule-fixed frame. As e is defined in the laboratory frame, a correlation between the body-fixed and space-fixed frames is, to some extent, preserved by the angular distribution of photofragments.

   For breakup into two heavy fragments, three vectors are sufficient to fully describe the dynamics of the photodissociation process in the laboratory frame: the relative recoil velocity of the fragments, u, the rotational angular momentum of the fragments j, and the direction of the light polarization e. A correlation exists between the three vectors, mediated by the nature of μ and the details of the transient states. When one photon is used on an isotropic initial state, the angular distribution of the framents $I(\theta)$ is given by the relation

$$I(\theta) \propto \frac{1}{4\pi} \left[1 + \beta\, P_2(\cos \theta) \right], \tag{17.9}$$

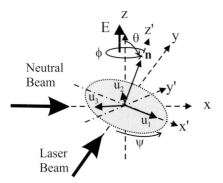

Fig. 17.2. Parameterization of the fragment momentum vectors in a three-body decay process

where θ defines the angle between $\boldsymbol{u}$ and $\boldsymbol{e}$. The second-order Legendre polynomial P_2 is weighted by the anisotropy parameter $-1 \leq \beta \leq 2$. Extremes of the anisotropy parameter are commonly labeled by designations deduced from the picture of direct, instantaneous dissociation where parallel ($\beta = 2$) refers to an alignment of $\boldsymbol{\mu} \| \boldsymbol{u}$, perpendicular ($\beta = -1$) refers to $\boldsymbol{\mu} \perp \boldsymbol{u}$, and $\beta = 0$ to the isotropic case.

To breakup a molecule into three heavy fragments, (17.9) may still be defined, but now the preferred axis is the normal to the plane in which the three nuclei recoil (see the angle θ in Fig. 17.2). For three-body breakup an additional vector correlation appears in terms of the relative orientation of the three individual momentum vectors $m_i\boldsymbol{u}_i$ ($i = 1, 2, 3$) in this plane. The coordinate system in this plane requires specification of the two angles ϕ and ψ in Fig. 17.2. As a result, six individual and independent parameters can be determined for each dissociating molecule. If linearly polarized light is used and the neutral target molecules are isotropically distributed in space, the angles ϕ and ψ carry no significant information on the underlying molecular dynamics. The angle θ and the three momentum vectors $m_i\boldsymbol{u}_i$ carry all center-of-mass specific information of the molecular process. Owing to momentum conservation, it is sufficient to specify three components of the three vectors. Thus, in a state-selected experiment, energy conservation can be further invoked to depict the correlation in a Dalitz plot, as discussed in Chap. 4.

17.3   Imaging Molecular Dynamics in Triatomics

Observing neutral fragments that slowly recede from the center-of-mass, as a consequence of photopreparation of an unstable molecular state, allows one to study correlations between fragment momentum vectors as they arise in the transition from the molecular into the separated atom frame. This correlation reflects the decay dynamics on the repulsive potential energy landscape as well as the nuclear and electronic geometry in which the molecular system enters the dissociative continuum. In contrast to breakup processes into

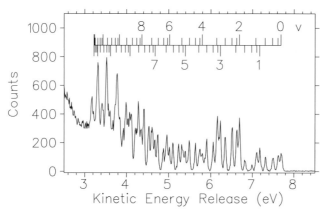

Fig. 17.3. Two-body decay spectrum of the kinetic energy release W of H_3 $3s^2$ A_1' $\{\nu_1 = \nu_2 = 0\}$ ($N = 1$, $G = 0$)

charged particles, the momentum correlations between neutral fragments are not affected by long-range post-collisional interaction which in part mask the correlation in the molecular frame.

The experimental approach to illucidate (17.6) is different from that of theory, in that the scalar and vector observables are determined for the transformation of specific prepared states Ψ^{p} into the final states Ψ^{f}. The roadmap that governs the projection in the coupling matrix element (17.6) is mapped out in initial- and final-state observables. The interpretation of these maps is a matter of active current discussion but lacks a rigorous theoretical approach.

To review this status we illustrate selected experimental results, compare them to wavepacket simulations, where available, and give a general overview of the meaning of the observed vector correlations.

17.3.1   Two-Body Decay

A spectrum of W from two-body decay of the $3s$ $^2A_1'$ state is shown in Fig. 17.3. Pronounced discrete peaks appear in the W-spectrum. They correspond to the rovibrational states of the $H_2(v,j)$ fragment as indicated by the stick spectrum in Fig. 17.3. The energy resolution is 50 meV (FWHM) at $W=5$ eV. The spectrum can be deconvoluted using basis functions determined in a Monte Carlo simulation that takes into account the finite geometric collection efficiency of the photofragment detector. The final-state rovibrational population of the $H_2(v,j)$ fragments is shown in Fig. 17.4(a) by the size of the squares. The corresponding distribution for the $3d$ $^2E''$ initial state is shown in Fig. 17.4(b). The highly structured distributions have no similarity with thermal rotational or vibrational distributions. Note that these distributions each arise from dissociation of a well-characterized

(a)

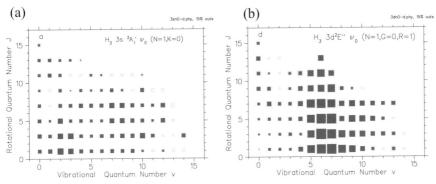

(b)

Fig. 17.4. Rovibrational distributions of the $H_2(v,J)$ fragment produced by two-body decay of the H_3 molecule in the (**a**) $3s\ ^2A_1'\{\nu_1 = 0, \nu_2 = 0\}$ (N = 1, G = 0) and (**b**) $3d\ ^2E''\{\nu_1 = 0, \nu_2 = 0\}$ (N = 1, G = 0) states

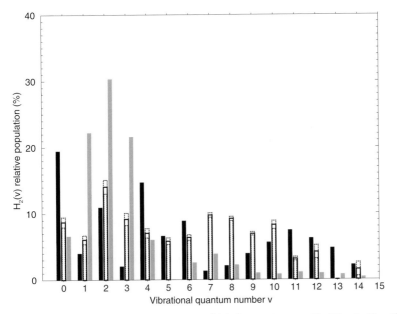

Fig. 17.5. Two-body decay of H_3 $3s\ ^2A_1'$ $\{\nu_1 = 0, \nu_2 = 0\}$ (N = 1, G = 0): Vibrational population of the H_2 fragment. Legend: *black bars*: theory by Schneider and Orel [29], *gray bars*: theory by Orel and Kulander [28], *open bars*: experiment with error bars indicated by dashed lines [21]

single rovibrational initial state. The $H_2(v,j)$ final-state populations strongly depend on the electronic initial state and reflect the mechanism of predissociation as well as the subsequent decay dynamics on the repulsive potential energy surface. The fragmentation process can only be understood within a quantum-dynamic treatment.

Figure 17.5 gives a comparison of the measured vibrational distributions [18] with the results of two-dimensional wavepacket calculations by Orel and Kulander [28], and by Schneider and Orel [29]. The newer calculations [29] are in remarkably good agreement with the experiment considering that the rotational degree of freedom cannot yet be tackled by theory. To achieve this high quality, the calculations had to explicitly include the nonadiabatic couplings between the potential energy surface of the laser-prepared initial state and the repulsive ground state. Only recently have such comprehensive calculations become possible due to the progress in computer power. Agreement between experiment and theory has also been achieved for the 2s $^2A_1'$ and the vibrationally excited 3s $^2A_1'$ states of H_3.

17.3.2 Three-Body Decay

When triple coincidence signals are collected, the momenta of the three hydrogen atoms $m_i u_i$ in the center-of-mass frame (see Fig. 17.2) are determined individually. Momentum conservation requires that $m_2 u_1 + m_2 u_2 + m_3 u_3 = 0$, which implies that the three momentum vectors have only six independent components. With the Freiburg spectrometer, the 6-fold differential cross section is determined, completely characterizing the final state of the three-body decay process. Note that such a measurement is carried out, one molecule at a time, at a rate of 10–100 three-body events per second. The most suitable projection of this high-dimensional set of information is into six parameters that uniquely describe the three fragment momentum vectors (see Fig. 17.2). The spatial orientation of the (x', y', z')-coordinate frame in Fig. 17.2 is determined by the spatial anisotropy of the photoexcitation process. For the remaining three parameters describing the arrangement of the three momenta in the (x', y')-plane, we may choose the absolute values of the momenta $p_i = m|u_i|$ or the individual fragment energies $\epsilon_i = p_i^2/(2m)$. We use the total kinetic energy W and two parameters showing the correlation among the fragment momenta. The experimental distribution of the total center-of-mass kinetic energy release $W = m \left(u_1^2 + u_2^2 + u_3^2 \right)/2$ represents a stringent test on the quality and precision of the data-acquisition and -reduction procedure. Since the energy of the laser-excited initial state above the three-body limit is laser selected, W has to appear as a discrete observable.

Figure 17.6 shows the kinetic energy release spectrum from triple coincidence data of the H_3 3s $^2A_1'$ and 3d $^2E''$ initial states binned at 10 meV resolution. In the spectrum of the 3s $^2A_1'$ state, a single narrow peak appears at 3.17 eV, close to the expected energy of the initial state [17] above the three-body limit of 3.155 eV. The excellent agreement between measured *absolute* values of W and the previously known state energies confirms the quality of the absolute energy calibration of the detector. The kinetic energy spectrum of the 3d $^2E''$ state also shows a strong peak near the expected position of 3.230 eV. Additionally, a small peak appears at lower energy at 2.71 eV revealing three-body breakup of a lower-lying state that is populated in a

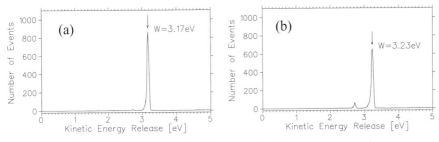

Fig. 17.6. Three-body decay of the H_3 molecule. Kinetic energy release spectra of the (**a**) 3s $^2A'_1\{\nu_1 = 0, \nu_2 = 0\}$ (N = 1, G = 0) and (**b**) 3d $^2E''\{\nu_1 = 0, \nu_2 = 0\}$ (N = 1, G = 0) states

competing radiative decay channel ($\lambda = 2.38\,\mu m$) of the predissociated state 3d $^2E'' \rightarrow$ 3p $^2E'$ [30] (see Fig. 17.1). No continuous background is observed in the kinetic energy release spectra of the three-body decay. This observation reveals that radiative transitions from the 3s $^2A'_1$ and 3d $^2E''$ states to the repulsive ground state lead to $H+H_2$ fragment pairs as the only exit channel [18]. It also indicates that the previously observed continuous photoemission spectra [31,32] are accompanied by two-body decay only.

17.3.3   Interpretation of Experimental Maps of Nonadiabatic Coupling

To show the correlation among the fragment momenta within the (x', y') plane, we use a Dalitz plot [33]. For each event, the value of $(\epsilon_3/W - 1/3)$ is plotted vs. $(\epsilon_2 - \epsilon_1)/(W \times \sqrt{3})$. Energy and momentum conservation require that the data points in this Dalitz plot lie inside a circle with radius $1/3$, centered at the origin. In a Dalitz plot, the phase-space density is conserved. As a result, a fragmentation process with a matrix element independent of the configuration leads to a homogeneous distribution in the kinematically allowed region. Preferred fragmention pathways can immediately be recognized from the event density in such a plot.

   In Fig. 17.7a and b, the triple-coincident events following three-body breakup of the laser-excited H_3 3s $^2A'_1$ and 3d $^2E''$ states are shown in Dalitz plots. Since the three hydrogen atoms are indistinguishable, points are drawn in Fig. 17.7 for six permutations of the fragment energies ϵ_i measured in each event. In Fig. 17.7c, the correspondence between the configuration of the fragment momenta and the location in the plot is visualized. The three-fold rotation symmetry around the origin and the mirror symmetry with respect to the dashed lines in Fig. 17.7c result from the equal masses of the fragments. In order to understand the meaning of the very pronounced islands of correlation appearing in the experimental data in Fig. 17.7a and b, the limited collection efficiency of the detector has to be considered.

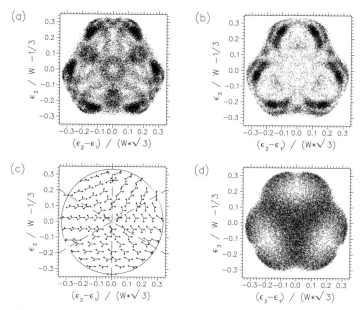

Fig. 17.7. Dalitz plot of the raw data from three-body decay of the 3s ^{2}A$_1'$ (**a**) and 3d ^{2}E″ (**b**) initial states of H$_3$. In (**c**), the correspondence between the location in the plot and the fragmentation configuration is indicated. In (**d**) a Monte Carlo simulation for a random distribution of fragmentation configuration in the phase space is shown, demonstrating the effect of the geometric detector collection efficiency

Currently, detection of fragment triples where one of the momenta is close to zero (linear configuration) and therefore hits the space between the two detectors, is excluded from detection as a 3-body event. Also, fragment hits that are too close in time as well as in spatial coordinates (H+ H–H configuration) are suppressed due to the finite pulse-pair resolution of 10 ns. The geometric and electronic detector collection efficiencies were determined by a Monte Carlo simulation generating a uniform distribution of fragmentation configurations and calculating the fragment propagation to the detector. A Monte Carlo simulation of a typical detector response is shown in Fig. 17.7d for the 3d state. The collection efficiency vanishes only for the linear and the H+H$_2$ configurations on the circle boundary. The remainder of the Dalitz plot area shows a smooth variation of the detection efficiency. As a consequence of the Monte Carlo results, we are able to attribute the islands of high point density in Fig. 17.7a and b to the correlation of the fragment momenta produced by the dissociation process itself. We can definitely exclude experimental artefacts as contributing to this patterning. Nevertheless, we correct the measured distributions by weighting the data points with the geometric collection efficiency. In Fig. 17.8, the resulting three-body final-state distributions of the 3s ^{2}A$_1'$ and 3d ^{2}E″ states are shown in false-color images.

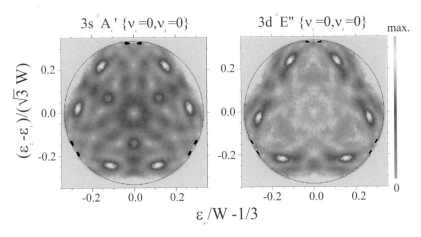

Fig. 17.8. Dalitz plots of the three-body decay of the 3s ^{2}A$'_1$ (**a**) and 3d ^{2}E$''$ (**b**) initial states of H$_3$. The geometric collection efficiency was calculated by a Monte Carlo simulation and the measured data were corrected.

Despite the high symmetry (D_{3h}) of the initial molecular states, asymmetric fragmentation configurations are very much preferred in finding a path into the 3-particle continuum. Neither the totally symmetric configuration (center of the plot) nor isosceles configurations (dashed lines in Fig. 17.7c show a preferred population. It may appear surprising that the preferred fragmentation configurations sensitively depend on the initial state, although the absolute energies of the states investigated here differ by only 75 meV, and the nuclear equilibrium configuration for each initial state is extremely close to that of the vibrationless H$_3^+$ ion in its ground electronic state and its lowest rotational level. The striking difference of the final-state distributions reflects the different coupling mechanism active between the initial state and the two sheets [34] of the repulsive ground-state potential energy surface. The same argument should hold for the marked differences in the final-state distributions of the two-body decay shown in Fig. 17.4. We observe a general tendency, that the two-body as well as the three-body distributions of the 3s ^{2}A$'_1$ state show extremely rich structures with numerous minima and maxima. The distributions of the 3d ^{2}E$''$ state are more concentrated on a few features. The suggestion is that in the two cases discussed here the zero point vibration, while exploring practically identical regions of internuclear configuration in the two cases considered here, finds individual regions of coupling to the continuum that are specific to each electronic state.

From a symmetry point of view, the failure of the Born–Oppenheimer approximation for the 3s ^{2}A$'_1$ state is mediated by the zero-point motion in the degenerate vibration. In the case of the 3d ^{2}E$''$ state, the coupling is induced by the rotational tumbling motion [19] that is much slower than the vibrational motion. In a semiclassical picture, the internal motion of the protons explores, on a femtosecond timescale, the phase space associated with the rovibronic molecular wave function Ψ^p, thereby probing the geometries under which transitions between the initial state and the ground-state surface may occur. Trajectories starting from geometries with favorable access into the ground-state surface will interfere, producing the rich patterns observed in the final state distributions.

While the coupling between the bound states and the continuum mediates the first entry of the quasibound system into the continuum, a series of avoided crossings between the upper sheet of the repulsive ground-state surface and the s- and d-Rydberg states of $^2\Sigma_g^+$ symmetry in linear geometry (Petsalakis et al. [34]) will govern the further evolution of the continuum state.

17.4 Outlook

Fast-beam translational spectroscopy is a powerful method to investigate laser-photofragmentation of triatomic molecules into neutral products. Recent progress in detector performance has even opened up the additional avenue of studying breakup processes into three atoms and a photon. The detection of the fragments in coincidence and the accurate measurement of their impact positions and arrival-time differences allows determination of the vectorial fragment momenta in the center-of-mass frame. Studies that combine quantum mechanically complete preparation of the initial state and kinematically complete final-state analysis in a photodissociation experiment have become feasible.

The performance of the method has so far been demonstrated for dissociation of laser-excited H_3 molecules. Two-body decay into $H+H_2(v,J)$ fragment pairs and three-body decay into neutral hydrogen atoms is observed. In the case of two-body decay, the partial cross sections for specific $H_2(v,j)$ fragments carry the signature of the complex proton–electron rearrangement when evolving from the bound state into the continuum. In the case of three-body decay, the kinematically accessible phase space is filled with a rich structure of momentum correlation, which likely represents the most detailed view an experiment may provide for (17.6).

While two-dimensional wavepacket calculations are able to account for the rotationally integrated partial cross section for formation of $H_2(v,j)$ in two-body decay, theorists have not yet attempted to predict the three-body decay, except for estimates of the two-body/three-body branching ratios [28]. The detailed maps of momentum partition among the three fragments obtained in the experiment now pose a significant challenge to quantum-structure and quantum-dynamics calculations.

References

1. U. Müller, T. Eckert, M. Braun, and H. Helm: Phys. Rev. Lett. **83**, 2718 (1999)
2. C. Maul, T. Haas, and K.H. Gericke: J. Phys. Chem. **101**, 6619 (1997); C. Maul and K.H. Gericke; Int. Rev. Phys. Chem. **16**, 1 (1997)
3. Y. Tanaka, M. Kawaski, Y. Matsumi, H. Futsiwara, T. Ishiwata, L.J. Rogers, R.N. Dixon, and M.N.R. Ashfold: J. Chem. Phys. **109**, 1315 (1998)
4. J.J. Lin, D.W. Hwang, Y.T. Lee, and X. Yang: J. Chem. Phys. **108**, 10061 (1998)
5. M. Lange, O. Pfaff, U. Müller, and R. Brenn: Chem. Phys. **230**, 117 (1998)
6. K. Wilson: Disc. Faraday Soc. **44**, 234 (1967)
7. J.T. Moseley et. al.: J. Chem. Phys. **70**, 1474 (1979)
8. H. Helm: Ion Interactions Probed by Photofragment Spectroscopy in *Electronic and Atomic Collisions*, Inv. Papers of the XII ICPEAC, Berlin (North Holland, Amsterdam 1984) p. 275–293
9. H. Helm, D. DeBruijn, and J. Los: Phys. Rev. Lett. **53**, 1642 (1984)
10. A.B. van der Kamp, P.B. Athmer, R.S. Hiemstra, J.R. Peterson, and W.J. van der Zande: Chem. Phys. **193**, 181 (1995); L.D.A. Siebbeles, E.R. Wouters, and W.J. van der Zande: Phys. Rev A **54**, 531 (1996); E.A. Wouters, L.D.A. Siebbeles, P.C. Schuddeboom, B.R. Chalamala, and W.J. van der Zande: Phys. Rev. A **54**, 522 (1996); B. Buijsse, E.R. Wouters, and W.J. van der Zande: Phys. Rev. Lett. **77**, 243 (1996); E.R. Wouters, B. Buijsse, J. Los, and W.J. van der Zande: J. Chem. Phys. **106**, 3974(1997)
11. L.D. Gardner, M.M. Graff, and J.L. Kohl: Rev. Sci. Instrum. **57**, 177 (1986)
12. R.E. Continetti, D.R. Cyr, D.L. Osborn, D.J. Leahy, and D.M. Neumark: J. Chem. Phys. **99**, 2616 (1993)
13. K.A. Hanold, A.K. Luong, and R.E. Continetti: J. Chem. Phys. **109**, 9215 (1998)
14. K.A. Hanold, A.K. Luong, T.G. Clements, and R.E. Continetti: Rev. Sci. Instrum. **70**, 2268 (1999)
15. H. Helm and P.C. Cosby: J. Chem. Phys. **90**, 4208 (1989)
16. C.W. Walter, P.C. Cosby, and H. Helm: J. Chem. Phys. **99**, 3553 (1993)
17. P.C. Cosby and H. Helm: Phys. Rev. Lett **61**, 298 (1988)
18. U. Müller and P.C. Cosby: J. Chem. Phys. **105**, 353, (1996)
19. U. Müller and P.C. Cosby: Phys. Rev. A **59**, 3632 (1999)
20. M. Braun, M. Beckert, and U. Müller: Rev. Sci. Instrum. **71**, 4535 (2000)
21. M. Beckert and U. Müller: Eur. Phys. J. D, **12**, 303 (2000)
22. D.P. de Bruin and J. Los: Rev. Sci. Instrum. **53**, 1020 (1982)
23. H. Helm and P.C. Cosby: J. Chem. Phys. **86**, 6813 (1987)
24. S.E. Sobottka and M.B. Williams: IEEE Trans. Nucl. Sci. **35**, 348 (1988)
25. K. Beckord, J. Becker, U. Werner, and H.O. Lutz: J. Phys. B **27**, L585 (1994)
26. Z. Amitay and D. Zajfman: Rev. Sci. Instrum. **68**, 1387 (1997)
27. O. Jagutzki, V. Mergel, K. Ullmann-Pfleger, L. Spielberger, U. Meyer, and H. Schmidt-Böcking: Proc. SPIE **3764**, 61 (1999) (in print)
28. A.E. Orel and K.C. Kulander: J. Chem. Phys. **91**, 6086 (1989); J.L. Krause, K.C. Kulander, J.C. Light, and A.E. Orel: J. Chem. Phys. **96**, 4283 (1992); J.L. Krause, A.E. Orel, B.H. Lengsfield, and K.C. Kulander: in *Time-Dependent Quantum Molecular Dynamics*, Ed. J. Broeckhove and L. Lathouwers (Plenum, New York 1992) p. 131

29. I.F. Schneider and A.E. Orel: J. Chem. Phys. **111**, 5873 (1999)
30. G. Herzberg, J.T. Hougen, and J.K.G. Watson: Can. J. Phys. **60**, 1261 (1982)
31. R. Bruckmeier, C. Wunderlich, and H. Figger: Phys. Rev. Lett. **72**, 2550 (1994)
32. A.B. Raksit, R.F. Porter, W.P. Garver, and J.J. Leventhal: Phys. Rev. Lett. **55**, 378 (1985)
33. R.H. Dalitz: Philos. Mag. **44**, 1068 (1953); Ann. Rev. Nucl. Sci. **13**, 339 (1963)
34. I.D. Petsalakis, G. Theodorakopoulos, and J.S. Wright: J. Chem. Phys. **89**, 6850 (1988)

18 Nonsequential Multiple Ionization in Strong Laser Fields

H. Rottke

18.1 Introduction

In 1965, it was experimentally observed that photoionization of an atom through absorption of many photons with an energy $\hbar\omega$ smaller than the ionization threshold is possible if only the light intensity is chosen high enough [1]. The formation of doubly charged ions in multiphoton ionization was first observed for alkaline-earth atoms with two electrons outside of closed shells [2] and later also for rare-gas atoms irradiated with Nd:YAG laser light [3]. The light intensity necessary to observe multiple ionization is in the 10^{13} W/cm^2 range and higher. This intensity corresponds to an electric field strength of at least 10^8 V/cm in the wave. At these field strengths, the light–atom interaction can no longer be characterized by perturbation theoretical methods since the external electric field approaches the Coulomb field the valence electrons are exposed to in an atom. Soon after its observation the question arose whether double ionization is a stepwise process [4]:

$$A + n_1\hbar\omega \to A^+ \tag{18.1a}$$
$$A^+ + n_2\hbar\omega \to A^{++}, \tag{18.1b}$$

or a direct one, where both electrons are removed simultaneously from the atom:

$$A + n\hbar\omega \to A^{++}. \tag{18.2}$$

Here, $n1$, $n2$, and n are the numbers of absorbed photons. An answer to the question was already given in the same publication: "In the lowest part of the intensity range, doubly charged ions are formed by direct multiphoton absorption from the atom (18.2). At higher intensity, they are formed by stepwise multiphoton absorption through the singly charged ion" (18.1a, 18.1b) [4]. A mechanism behind nonsequential ionization that needs the absorption of a large number of photons was not yet identified. The only integral measurements of the total yield of ions in the different charge states did not allow an identification. For a long time the experiments dealing with double or multiple ionization of atoms in a high-intensity laser pulse were restricted to the measurement of total ion yields. Electron spectroscopy was impossible since the ratio of the yield of doubly charged ions to that of singly charged ions is always very small (see, for example, [4]). Therefore, electron spectroscopy needed at least a coincident detection of the ion of interest together with the electron.

With the growing sophistication of laser technology, which allowed the generation of high-power, stable, ultrashort pulses (pulsewidth down to a few fs) with a high repetition rate it became possible to measure total ion yields for all rare-gas atoms with high precision. The light intensity range covered reached up into the PW/cm^2 regime, where it was possible to at least doubly ionize all of these atoms. The first experiments concentrated on double ionization of the fundamental 2-electron helium atom [5,6]. At that time, several theoretical models were available to calculate the total ionization rate of an atom or ion subjected to high-intensity light, provided only one electron is active at a time. This allowed to test precisely whether double or even multiple ionization is a sequential process (18.1a, 18.1b). For helium it was found that at "high" light intensity, where the first ionization step is saturated (18.1a, with no atoms left in the light beam), double ionization is sequential. First, a singly charged ion is formed in the ground state, which is subsequently further ionized by the light. At "low" light intensity the total yield of He^{++} observed in the experiments is several orders of magnitude higher than that expected from sequential ionization [5,6]. The deviation increases with decreasing light intensity. The experiments thus revealed unambiguously that double ionization, in this case of He, has to be nonsequential (18.2).

Several nonsequential ionization mechanisms have been proposed meanwhile. A "shake-off" mechanism [5] as is well known from double ionization with single high-energy photons [7]. Absorption of photons from the light beam in this case transfers one electron to the ionization continuum. This electron leaves the atom so fast that the other electrons cannot adapt adiabatically to the new interaction potential within the singly charged ion. Therefore an electron either may be shaken up into a bound excited state of the ion or directly shaken off into the ionization continuum. Collective tunnel ionization of two or more electrons was proposed [8]. This necessitates ionization of the atom to proceed in the quasistatic regime. A third mechanism [9], which is based on scattering of an electron, photoionized first, on its parent ion core will be outlined below in detail. It turned out to be the most relevant mechanism.

A first hint to the nonsequential ionization mechanism came from the dependence of the total yield of doubly charged ions on the state of polarization of the light beam. Nonsequential ionization is suppressed if the light becomes elliptically polarized (at a small deviation from linear polarization) and disappears for circular polarization [45]. This pointed to electron scattering, mentioned above, as the ionization mechanism.

Depending on the nonsequential ionization mechanism a specific correlation among the photoelectrons leaving the atom is expected. To reveal this correlation it is necessary to perform differential measurements, for example photoelectron spectroscopy or e–e correlation measurements concerning energy sharing or angular correlation. A first experiment along this line was based on an e–e and e–ion coincidence measurement [10]. Its outcome was not sufficient to allow an identification of the double-ionization mechanism.

A breakthrough concerning differential measurements on multiple photoionization in high-intensity laser pulses and the identification of mechanisms came with the application of cold target recoil ion momentum spectroscopy (COLTRIMS) to this problem [11,12]. These experiments soon were extended to allow a kinematically complete analysis of the final state of the electrons after strong-field ionization [13,14]. Following these experiments photoelectron spectroscopy in coincidence with doubly charged ions was again used by several groups to gain further insight into the ionization mechanism [15–17].

In this contribution I will present what was learned on strong-field non-sequential ionization of atoms by applying COLTRIMS as the experimental technique to arrive at a complete kinematical analysis of the final photoelectron state. I will start with the basic concepts concerning strong-field single ionization and the "simple-man" classical view of the motion of a free electron in a light wave. These prerequisites are necessary to understand nonsequential multiple ionization. COLTRIM spectroscopy relies on momentum conservation in the ionization process. I will therefore analyze the limitations momentum conservation of the atomic system alone, which is necessary to be fulfilled, puts on the light beam. Then I will describe the COLTRIMS setup we used, the experimental results and what they reveal about the nonsequential multiple ionization mechanism. The theoretical background of the experiments is presented in detail by Becker and Faisal in this book.

18.2   Strong–Field Single Ionization

Single ionization of atoms in an intense light pulse is successfully described by assuming that only one electron of the atom is active. This simplifies an analysis of the light–atom interaction significantly. Also, it was found that it is possible to subdivide the range of light intensities applied in experiments into two regimes [18]. At "low" intensity photoionization is most easily understood by assuming that it proceeds via simultaneous multiple absorption of photons (Fig. 18.1a). In this case the quantum nature of light becomes obvious in the multiphoton transition to the ionization continuum. The transition of the electron happens in the vicinity of the ion core, determined by the extension of the wave function in the initial state. Bound excited states of the atom may play a significant role in the ionizing transition even if they are not resonant with multiples $n\omega$ of the frequency of the light wave without the external perturbation. The AC-Stark effect shifts them transiently into n-photon resonance during the light pulse. This gives rise to an enhancement of the ionization rate that is observable in electron kinetic energy distributions

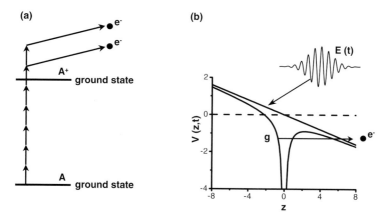

Fig. 18.1. Strong-field ionization in the multiphoton regime (**a**) and in the quasistatic regime (**b**). The *insert* in (**b**) shows the dependence of the electric field on time in a light pulse. $V(z,t)$ is the instantaneous "potential" energy the electron is exposed to. It is a superposition of the internal Coulomb and the external "potential" due to the electric field of the light wave

[19,20]. Absorption of photons does not stop after the electron reaches the ionization continuum (see Fig. 18.1a). Many more photons may be absorbed leading to above threshold ionization (ATI) [21].

At "high" light intensity a quasistatic picture is most adequate to describe strong-field ionization. It proceeds as electric-field ionization induced by the oscillating electric field $E(t) = E_0(t)\cos\omega t$ of the light wave. This is illustrated in Fig. 18.1b. The potential $V(z,t)$ seen instantaneously by the active electron mainly consists of the Coulomb potential (interaction with the ion core) and an external "potential" $zE_0(t)\cos\omega t$ (Fig. 18.1b) due to the electric field of the light wave. It is assumed here that the light is linearly polarized along the z-axis of the frame of reference. The resultant potential develops a barrier that is shown in Fig. 18.1b along the z-axis where it is lowest. Depending on the electric field strength two situations may occur. The electron in the ground state (g in Fig. 18.1b) is ionized by tunneling through the barrier. This is shown in Fig. 18.1b. Or, the electron escapes over the barrier. This will happen if its binding energy I_p becomes smaller than the height of the barrier $I_p \leq 2\sqrt{|E_0(t)\cos\omega t|}$. When the electron appears at the exit of the tunnel it is mainly influenced further on by the external field alone. The Coulomb interaction with the ion core becomes negligible.

The quasistatic view of strong-field ionization becomes applicable if the "time" it takes the electron to tunnel is small compared to the oscillation period of the light wave. The transition between the regimes is governed by the Keldysh adiabaticity parameter $\gamma = \sqrt{I_p/2U_p}$ [18], where $U_p = E_0^2/4\omega^2$ is the ponderomotive ("quiver") energy of a free electron in a light wave with frequency ω. E_0 is the electric field strength amplitude of the wave.

$\gamma > 1$ characterizes the multiphoton and $\gamma < 1$ the quasistatic regime. The transition between the regimes is smooth.

Based on the work of [22,23] the instantaneous tunneling rate $R(t)$ for the electron in the quasistatic regime is given by [24]:

$$R\left(t\right) = a_{n^*lm}\ \left|E\left(t\right)\right|^{1+\left|m\right|-2n^*}\ \exp\left(-\frac{2\left(2I_{\mathrm{p}}\right)^{3/2}}{3\left|E\left(t\right)\right|}\right). \tag{18.3}$$

In this expression the parameter a_{n^*lm} depends on the effective principle quantum number $n^* = Z/\sqrt{2I_{\mathrm{p}}}$ (Z is the charge of the ion), the angular momentum l and its projection onto the light beam polarization axis m of the initial bound state of the electron. The instantaneous electric field in (18.3) is given by $E(t) = E_0(t)\cos\omega t$. Equation (18.3) shows that tunnel ionization is most effective near maxima of the electric field strength that oscillates with the frequency of the light wave. The electrons therefore are set free during short time intervals around these maxima, i.e., at a particular phase. This phase-relatedness distinguishes the quasistatic regime from the multiphoton regime. There, the probability for ionization shows no dependence on the phase of the carrier wave.

Experiments on nonsequential strong-field ionization were always done in the transition region between multiphoton and quasistatic ionization with γ values between 0.3 and 1.5. Nevertheless, I will always use the quasistatic approximation for the analysis of experimental results, which is still reliable. Deviations concerning the sublaser-cycle ionization rate (18.3) are discussed in [25].

18.3   Electron Rescattering on the Ion Core

After the electron appears at the exit of the tunnel the light wave determines its further motion (see Fig. 18.1b). To analyze its motion and that of the ion I will restrict myself to a classical description and assume that the light wave is linearly polarized. In the nonrelativistic limit, which is applicable here, the Lorentz force on the electron due to the magnetic field of the light wave may first be neglected. In a first approximation, a force on the electron is only exerted by the electric field along the axis of polarization (here chosen to be the z-axis). In atomic units, used throughout this contribution, the equations of motion of a charged particle with charge q and mass m then may be written as ($m = 1$, $q = -1$ for an electron):

$$m\,\ddot{x} = 0; \quad m\,\ddot{y} = 0; \quad m\,\ddot{z} = q\,E_0\left(t\right)\cos\omega t\,. \tag{18.4}$$

Only the z-motion, which is influenced by the light, will be analyzed. With the reasonable assumption that after electric-field ionization the electron appears at the exit of the tunnel (chosen to be at $z_0 = 0$) at time t_0 with negligible momentum the evolution of the electron and ion momentum p and position z with time is given by:

$$p_z(t) = \frac{q\left[E_0(t)\sin\omega t - E_0(t_0)\sin\omega t_0\right]}{\omega}, \tag{18.5}$$

$$z(t) = \frac{q\left[-E_0(t)\cos\omega t + E_0(t_0)\cos\omega t_0 - \omega(t-t_0)E_0(t_0)\sin\omega t_0\right]}{m\omega^2}. \tag{18.6}$$

In integrating (18.4) it was assumed that the amplitude $E_0(t)$ varies slowly compared to the carrier $\cos\omega t$ of the wave. Expression 18.5 shows that after the light pulse passed by a charged particle, which starts motion at t_0 has a final momentum $p_z(t\to\infty) = -2q\sqrt{U_p(t_0)}\sin\omega t_0$ with $U_p(t) = E_0(t)^2/4\omega^2$ the ponderomotive (quiver) energy a free electron has at time t in the pulse. The charged particle (electron or ion) therefore gains the highest final momentum after the light pulse is gone, if it starts motion at times t_0 where the electric field crosses a zero ($\cos\omega t = 0$). It has zero final momentum if motion starts at the maxima of the electric field ($\cos\omega t \approx 1$). In the case of tunnel ionization this consideration applies to electrons and the corresponding ion as long as there is no further interaction between the two particles for $t \geq t_0$.

The analysis of $z(t)$ (18.6) shows that the light wave may drive the charged particle back to the position ($z = 0$) where the motion started. This happens only for certain start times t_0. For the $\cos\omega t$ carrier wave and a slowly varying amplitude $E_0(t)$ t_0 has to be in the intervals $[n\pi, (n+1/2)\pi]$ (n an integer) to allow at least one return. More than one return is possible, dependent on t_0. This is shown in Fig. 18.2a for three different start times t_0. An important parameter of the charged particle, when it returns to $z = 0$ at time t_r is the kinetic energy $E_{kin}(t_r)$ present in the motion along the z-axis. $E_{kin}(t_r)$ spans energies between zero and $E_{kin,max} = 3.17\, U_p(t_r)/m$. For an electron $E_{kin}(t_r)$ is shown in Fig. 18.2b in multiples of the ponderomotive energy $U_p(t_r)$ at the return time ($m = 1$). $E_{kin,max}$ may reach substantial values. As an example, at a typical light intensity of $1 \times 10^{14}\,\mathrm{W/cm^2}$ and a wavelength of $800\,\mathrm{nm}$ $E_{kin,max} = 19\,\mathrm{eV}$ for the electron.

An electron born by tunnel ionization of an atom at t_0 and returning to its parent ion core at t_r may interact strongly with the ion. Elastic scattering was found to contribute to above-threshold ionization of atoms. It gives rise to a high-energy plateau in the photoelectron spectrum [9,26–28] and to side lobes in the angular distribution [27–29]. Recombination of the rescattering electron into the atomic ground state with the emission of a high-energy photon gives rise to the generation of high-order harmonics of the driving laser radiation [9,30–32]. As will be outlined below, inelastic scattering of the returning electron on the parent ion core is responsible for strong-field nonsequential multiple ionization of atoms, too.

"Stepwise" strong-field ionization of an atom, consisting of electric-field ionization and acceleration of the free electron by the light wave, with a possible re-encounter and scattering (elastic, recombination, inelastic) on the par-

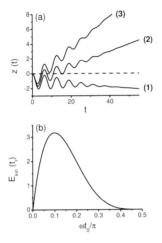

Fig. 18.2. (a) Electron motion along the polarization vector of the light beam (see (18.6)). Free-electron motion starts at $z = 0$. Three start times t_0 are shown: (1) The electron does not return to $z = 0$, (2) the electron returns several times, (3) the electron returns only once. (b) shows the kinetic energy of the returning electron $E_{\mathrm{kin}}(t_r)$ versus the start time t_0 in multiples of the ponderomotive energy U_{p}

ent ion core, as described here, may be cast into a strict quantum-mechanical formulation. The corresponding formulation is given in the chapter by Becker in this book.

18.4   Is Momentum Conserved in a Strong Laser Pulse?

The experimental techniques applied to analyze strong-field atomic ionization partly rely on conservation of momentum of an initially neutral atomic system during interaction with the light wave. These are experiments based on cold target "recoil" ion momentum spectroscopy (COLTRIMS). Momentum measurements on the ion formed after ionization are used to get information on the photoelectron momentum. The momentum of all photoelectrons together with the ion is certainly not a conserved quantity. This is most easily recognized if the interaction with the light wave is viewed quantum mechanically. The atom absorbs a certain number of photons before it is ionized. Each absorbed photon transfers a momentum $\hbar k$ to the atom (k the wave vector). This leads to a net transfer of momentum from the wave to the atom, breaking momentum conservation. Typically, the momentum of one photon is about 4.15×10^{-4} a.u. ($\lambda = 800\,\mathrm{nm}$). This is small compared to the typical experimental resolution. But one has to keep in mind that, depending on the final charge state of the ion, up to several hundred or thousand photons are absorbed. Then the momentum transfer may finally become non-negligible.

In the quasistatic limit of strong-field ionization and using classical mechanics to analyze the motion of free electrons and ions the momentum transfer is found to be induced by the magnetic field of the light wave through the Lorentz force term in the equation of motion of the charged particles after tunnel ionization. It is a relativistic effect that gains importance with increasing light intensity. In a plane-wave light pulse the momentum transfer may be calculated analytically from the fully covariant Lorentz equation of motion [33]:

$$\frac{dp^\mu}{d\tau} = \frac{q}{mc} F^{\mu\nu} p_\nu .$$ (18.7)

Here, p^μ is the 4-momentum of the particle, τ its proper time, m the rest mass, and $F^{\mu\nu}$ the field-strength tensor of the pulsed-plane electromagnetic wave. From this equation the final spatial components p_f of the 4-momentum after the light pulse is gone are derived to be:

$$\boldsymbol{p}_f = \boldsymbol{p}_i + \frac{q}{c} \boldsymbol{A}(\tau_0) + \frac{q}{c} \frac{\boldsymbol{p}_i \cdot \boldsymbol{A}(\tau_0) + \frac{q}{2c} \boldsymbol{A}^2(\tau_0)}{p_i^0 - \hat{\boldsymbol{e}}_k \cdot \boldsymbol{p}_i} \hat{\boldsymbol{e}}_k .$$ (18.8)

The final "time like" component p_f^0 of the 4-momentum is given by:

$$p_f^0 = p_i^0 + \frac{q}{c} \frac{\boldsymbol{p}_i \cdot \boldsymbol{A}(\tau_0) + \frac{q}{2c} \boldsymbol{A}^2(\tau_0)}{p_i^0 - \hat{\boldsymbol{e}}_k \cdot \boldsymbol{p}_i} .$$ (18.9)

In these equations p_i^0 and $\boldsymbol{p}_i$ are the initial momentum components at proper initial time τ_0, $\boldsymbol{A}(\tau_0)$ is the vector potential of the light wave at time τ_0 at the position where the motion starts, and $\hat{\boldsymbol{e}}_k$ is the unit vector along the propagation direction of the light pulse. It was assumed here that the vector potential vanishes at the end of the pulse.

Equation (18.8) shows that total momentum is not conserved. If one assumes, for example, an atom at rest gets singly ionized in the light wave at τ_0 by electric-field ionization (the quasistatic limiting case) the initial momenta of the free electron $\boldsymbol{p}_{i,e}$ and ion $\boldsymbol{p}_{i,I}$ add up to zero, the total momentum of the atom. According to (18.8) the final momentum sum $\boldsymbol{p}_f = \boldsymbol{p}_{f,e} + \boldsymbol{p}_{f,I}$, after the light pulse is gone, then becomes:

$$\boldsymbol{p}_f = \frac{1}{c} \left\{ \frac{\boldsymbol{p}_{i,I} \cdot \boldsymbol{A}(\tau_0) + \frac{1}{2c} \boldsymbol{A}^2(\tau_0)}{p_{i,I}^0 - \hat{\boldsymbol{e}}_k \cdot \boldsymbol{p}_{i,I}} - \frac{\boldsymbol{p}_{i,e} \cdot \boldsymbol{A}(\tau_0) - \frac{1}{2c} \boldsymbol{A}^2(\tau_0)}{p_{i,e}^0 - \hat{\boldsymbol{e}}_k \cdot \boldsymbol{p}_{i,e}} \right\} \hat{\boldsymbol{e}}_k .$$ (18.10)

As expected from the starting considerations, only the momentum component along the propagation direction of the wave is not conserved. The two components perpendicular to this direction are strictly conserved, even in the relativistic limit. If the initial momenta of the particles are zero, (18.10) becomes especially simple, and the most significant contribution to momentum non-conservation becomes visible:

$$\boldsymbol{p}_f = \frac{\boldsymbol{A}^2(\tau_0)}{2c^3} \left(1 + \frac{1}{m_I}\right) \hat{\boldsymbol{e}}_k = 2\frac{U_p(t_0)}{c} \left(1 + \frac{1}{m_I}\right) \sin^2 \omega t_0 \hat{\boldsymbol{e}}_k .$$ (18.11)

It was assumed here that the electric field of the wave is given by $\boldsymbol{E}(t) = \boldsymbol{E}_0(t) \cos \omega t$. $U_p(t_0)$ is the ponderomotive energy of the electron at t_0. Equation (18.11) shows that the contribution of the ion to $\boldsymbol{p}_f$ is smaller than that of the electron by the electron-to-ion mass ratio. Most of the momentum is transferred to the electron. For a typical light intensity of $10^{15}\,\mathrm{W/cm^2}$ and a light wavelength of $800\,\mathrm{nm}$ U_p reaches $2.2\,\mathrm{a.u.}$ The maximum momentum transfer to the atom is then approximately $|p_f| = 0.03\,\mathrm{a.u.}$ This transfer is still smaller than the momentum resolution along the light beam propagation direction in a typical COLTRIMS experiment. The example shows that momentum transfer becomes significant only at light intensities well beyond $10^{15}\,\mathrm{W/cm^2}$. In the expression for the momentum transfer (18.10) one has to keep in mind that it was derived with the assumption that the electron leaves the ion without any further interaction (rescattering). This limits its applicability to calculate the exact momentum transfer to the atom.

All considerations up to now were based on a plane light wave. In the real world this is an idealization since the light intensities of interest can only be reached in a focused beam. Depending on the size of the focal spot a steep intensity gradient transverse to the beam propagation axis may exist. This gradient also has a profound effect on momentum conservation. An analysis may start from the (nonrelativistic) Lorentz equation of motion of a charged particle in time-dependent and inhomogeneous electric $\boldsymbol{E}(\boldsymbol{x},t)$ and magnetic fields $\boldsymbol{B}(\boldsymbol{x},t)$ [34–37]:

$$m\ddot{\boldsymbol{x}}(t) = q\boldsymbol{E}(\boldsymbol{x},t) + \frac{q}{c}\dot{\boldsymbol{x}} \times \boldsymbol{B}(\boldsymbol{x},t)\,. \tag{18.12}$$

For an electromagnetic wave, the fields may be expressed by the associated vector potential $\boldsymbol{A}(\boldsymbol{x},t)$. In terms of $\boldsymbol{A}$ the equation of motion reads:

$$m\ddot{\boldsymbol{x}}(t) = -\frac{q}{c}\left[\frac{\partial \boldsymbol{A}}{\partial t} + (\dot{\boldsymbol{x}}\nabla)\,\boldsymbol{A} - \nabla\,(\dot{\boldsymbol{x}}\boldsymbol{A})\right]\,. \tag{18.13}$$

In this equation the vector potential is now assumed to oscillate in time with the center frequency ω: $\boldsymbol{A}(\boldsymbol{x},t) = \boldsymbol{a}_1(\boldsymbol{x},t)\cos\omega t + \boldsymbol{a}_2(\boldsymbol{x},t)\sin\omega t$. $\boldsymbol{a}_i$, $i = 1,2$ are amplitude functions varying slowly in time. An approximate solution of (18.13) may then be found in the form of a truncated "Fourier" expansion: $\boldsymbol{x}(t) = \boldsymbol{d}(t) + \boldsymbol{u}(t)\cos\omega t + \boldsymbol{v}(t)\sin\omega t$ with $\boldsymbol{d}$, $\boldsymbol{u}$, $\boldsymbol{v}$ slowly varying in time.

The quantities $\boldsymbol{u}(t)$ and $\boldsymbol{v}(t)$ are the amplitudes of a "quiver" motion. It was already found above in a plane electromagnetic wave (18.6). $\boldsymbol{d}(t)$ represents a slow drift motion which, in a focused light beam, no longer has a constant velocity as in a plane wave (see (18.6)). With this "Ansatz" approximate equations of motion for drift $\boldsymbol{d}(t)$ and "quiver" $\boldsymbol{u}(t)$, $\boldsymbol{v}(t)$ are found to be:

$$\boldsymbol{u}(t) = \frac{q}{mc\omega}\boldsymbol{a}_2[\boldsymbol{d}(t),t] \tag{18.14a}$$

$$\boldsymbol{v}(t) = -\frac{q}{mc\omega}\boldsymbol{a}_1[\boldsymbol{d}(t),t] \tag{18.14b}$$

$$m\ddot{\boldsymbol{d}}(t) = -\frac{q^2}{m}\nabla U_{\mathrm{p}}(\boldsymbol{d},t)\,. \tag{18.14c}$$

In (18.14c), the quantity

$$U_\mathrm{p}(\boldsymbol{d}, t) = \frac{\boldsymbol{a}^2(\boldsymbol{d}, t)}{4c^2} \tag{18.15}$$

is the ponderomotive potential in the light beam along the drift trajectory $\boldsymbol{d}(t)$ of the particle expressed as a function of the amplitude $\boldsymbol{a}(\boldsymbol{d}, t)$ of the vector potential. It was assumed that the amplitudes $\boldsymbol{a}_i$, $i = 1, 2$ may be written as $\boldsymbol{a}_1 = \boldsymbol{a}(\boldsymbol{x}, t) \cos \boldsymbol{kx}$ and $\boldsymbol{a}_2 = \boldsymbol{a}(\boldsymbol{x}, t) \sin \boldsymbol{kx}$ with $\boldsymbol{k}$ the wave vector. In the equation of motion for the drift (18.14c) U_p appears in the form of a potential energy. Equations (18.14a, 18.14b) show that the ponderomotive potential is just the cycle-averaged kinetic energy present in the "quiver" motion of the charged particle.

According to (18.14c) the acceleration of the drift motion depends on q^2, the charge squared. It does not depend on its sign. This directly shows that in a nonplane-wave momentum conservation after photoionization of an atom fails. In order to retain momentum conservation at least approximately it is necessary to use ultrashort light pulses to investigate strong-field ionization with COLTRIMS. The pulse width has to be chosen in such a way that the photoelectrons (they are fastest) do not move over a distance in the focal spot where the ponderomotive potential changes significantly while the light pulse passes by. Under these circumstances an integration of (18.14c) shows that the electron momentum does not change significantly.

The analysis of experimental data on strong-field double ionization given below is based on momentum conservation for the atomic system. It is assumed that the final momentum sum of the ionization products is equal to the initial momentum of the atom before arrival of the light pulse. This is allowed since in all discussed experiments relativistic momentum nonconservation and also ponderomotive acceleration in the focal spot have been smaller than the apparatus resolution. Conservation of momentum is an extremely powerful tool since it allows information on the momentum of the photoelectrons to be obtained from the ion momentum.

18.5   An Experimental Setup

An experimental setup for the investigation of strong field-multiple ionization should be able to allow the measurement of highly differential data of the electrons in the final state. A kinematically complete experiment, where the momenta of all charged particles, electrons and the ion, are measured, may be based on the COLTRIMS. It was very successfully applied to analyze strong-field multiple ionization. This experimental technique, which we used in our experiments, and the detection techniques employed are described in detail by Moshammer in this book. I will restrict myself to the description of the special spectrometer setup we used.

In order to reveal any correlation among the momenta of the electrons and the ion it is necessary to have at most one ionization event per laser pulse.

The accumulation of statistically significant data within a reasonable time therefore necessitates use of a laser with a pulse repetition rate as high as possible for the light intensity range to be investigated. Meanwhile, lasers with a repetition rate in a range between 1 kHz and 100 kHz have been employed in experiments. One such laser system typically used in investigations will be described here.

18.5.1   The Momentum Spectrometer

The momentum spectrometer setup is schematically shown in Fig. 18.3 [12,38]. It allows the simultaneous determination of the complete momentum vector for each electron and ion formed during the interaction of a focused light beam with an atomic beam. The atomic target is supplied as a well-collimated supersonic beam [39]. In this way, the initial momentum of the atoms, before they are hit by the laser pulse, is well defined. The beam is formed by expanding the target gas through an orifice 20 μm in diameter in a thin metal foil (thickness 13 μm). The backing pressure is adjustable in a range up to about 10 bar. Dependent on the gas, cooling to LN_2-temperature prior to expansion is possible. The expansion further cools the translational degrees of freedom far below the thermal distribution in the gas reservoir [39]. The axial reduction of the momentum distribution along the atomic beam in the expansion is important for the experiment. The transverse momentum distribution is reduced by beam collimation.

After expansion into a first differential-pumping stage the beam passes through a skimmer with a 0.5–mm orifice diameter. About 50 cm downstream of the skimmer a slit for final beam collimation is mounted. It is possible to set the slit to several widths between 25 μm and 0.5 mm. This controls the width of the atomic beam along the propagation direction of the light beam. After passage through a further orifice with a 4–mm diameter the atomic beam enters an ultrahigh vacuum (UHV) chamber where it is intersected by the light beam at right angles. The focal spot of the light beam is placed in the center of the atomic beam where the target gas density is adjusted to $\approx 10^8\,\mathrm{cm}^{-3}$. The base pressure in the UHV chamber has to be kept below 3×10^{-10} mbar to reduce background gas ionization to a tolerable level.

Ions and electrons created in the focal spot of the light beam are extracted by an electric field (1–7 V/cm) in opposite directions. After acceleration they pass field-free drift tubes. At the end of the respective drift tubes the particles hit multi-channel plates (MCP, diameter 80 mm) that are both equipped with a position-sensitive anode. Either delay-line or wedge-and-strip anodes are used for position encoding (see the contribution of Moshammer). The separation of the electron and ion detectors from the focal spot is 20 cm and 30 cm, respectively. This setup allows the detection of all ions created irrespective of the initial momentum they received in the ionization process. To allow a similar range of solid angle for the detection of electrons, a homogeneous magnetic field ($\Delta B/B < 10^{-3}$) is applied over the electron drift

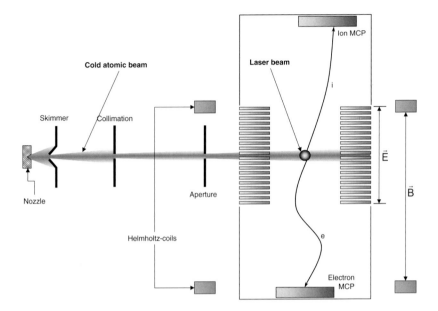

Fig. 18.3. Schematical diagram of the experimental setup used for the investigation of strong-field ionization of atoms

tube parallel to the spectrometer axis. The field strength is adjustable in a range up to about 2.5×10^{-3} Tesla. It forces electrons with an initial momentum perpendicular to the spectrometer axis into cyclotron orbits and thus directs them to the electron detector. Up to a transverse electron momentum of $|\boldsymbol{p}_\perp| = d\omega_c/4$ (d: diameter of the electron detector MCP in a.u., ω_c: the cyclotron frequency in a.u.) the solid angle of detection is larger than 2π. The exact value is determined by the total extraction voltage applied to the acceleration stage of the spectrometer.

For each ion and electron hitting the respective detector, the time-of-flight from the laser focal spot to the MCP and the position where the particle hits the MCP are measured. From this information it is possible to reconstruct the full initial momentum vector each particle started with after ionization (see Moshammer in this book). The momentum resolution of the spectrometer depends on the momentum component and the particle, electron or ion. Along the atomic beam the ion-momentum resolution is worst, limited by the residual axial thermal motion of the atoms. For Ne as target, cooled prior to the expansion to approximately $90\,\mathrm{K}$, the resolution achieved was about $0.2\,\mathrm{a.u.}$ [38]. It worsens if heavier atoms are used. An upper limit for the ion and electron momentum resolution along the spectrometer axis (the vertical direction in Fig. 18.3) can be derived from the sum momentum distribution of the photoelectron and ion from single ionization. Ideally $p_{e,\parallel} + p_{I,\parallel}$ should be zero. The corresponding measured distribution function for Ne had a width of 0.11 a.u. [38]. Perpendicular to the spectrometer axis an average electron momentum resolution of $\Delta p_{e,\perp} = 0.1$û. is estimated [38].

The electron detector, and in an improved setup also the ion detector, is capable of accepting several electrons/ions per laser shot. This, for example, opens up the possibility to measure the momenta of all three charged particles created in double ionization of an atom.

18.5.2   The Laser System

In our experiments, we used a Ti:sapphire laser system to excite the atoms. It was based on a Kerr-lens mode-locked Ti:sapphire oscillator with a pulse repetition rate of 76 MHz. The low-energy pulses of the oscillator were amplified at a pulse repetition rate of 1 kHz. The amplifier medium was also a Ti:sapphire laser rod pumped by the pulses of the second harmonic of a Nd:YLF laser at 532 nm. After amplification, a pulse energy adjustable up to ≈ 0.6 mJ and a final pulse width of 25 fs were achieved. The center wavelength of the laser radiation was 790 nm. The laser beam was focused into the atomic beam with a gold-coated spherical mirror in back reflection. The mirror focal length was 100 mm. The focal spot diameter was approximately 8 μm full width at half maximum. With this setup it was, in principle, possible to reach a maximum light intensity of $\approx 4 \times 10^{16}$ W/cm². At the pulse width of 25 fs ponderomotive acceleration of the photoelectrons in the spatial intensity gradient of the light beam is below the resolution limit of the momentum spectrometer. Thus, momentum conservation may be used to analyze multiple ionization.

In order to ensure that detected electron–ion pairs are from the same atom it is necessary to have less than one ionization event per laser pulse. This was achieved by adjusting the gas density in the focal spot to a value where the ion countrate was 10–100 Hz. Since the probability of nonsequential double ionization is smaller by a factor of 100–1000 than for single ionization this ion count rate gave rise to a countrate for true double-ionization events of about one per minute. The collection of a statistically significant amount of double-ionization events thus needed a beam-time of several days, putting high demands on the long-term stability of the laser system. To get around these long-term measurements the only possibility is an increase of the repetition rate of the laser. In recent experiments we therefore employed a Ti:sapphire laser system capable of a pulse repetition rate of 100 kHz (pulse width 30 ps) [40,41]. At a reduced pulse energy (up to 6 μJ) and focused light intensities up to 2×10^{14} W/cm² it was possible to raise the true coincidence rate for double-ionization events to several Hz. This setup heavily reduces the time necessary for a measurement and opens up the possibility for systematic studies of atomic or molecular multiple ionization and dissociation.

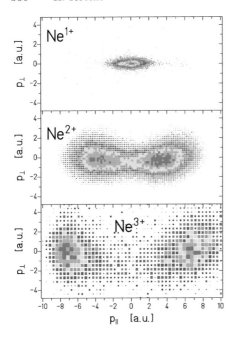

Fig. 18.4. Ne^{n+} (n = 1, 2, 3) ion momentum distributions measured at light intensities 1.3 PW/cm^2 (Ne^{1+}, Ne^{2+}) and 1.5 PW/cm^2 (Ne^{3+}). The 2-dimensional distributions were determined by projecting the full distribution onto a plane ($p_{\parallel}, p_{\perp}$) containing the light beam linear polarization direction (along the $p_{\parallel}$ momentum coordinate axis). The light pulse width was 30 fs. The spectra are taken from [12]

18.6   Neon Multiple Ionization

Neon was found to show in a very clean way, one route to nonsequential double ionization that also contributes in Ar [13,14,41–44] and He [11] double ionization. I will therefore restrict myself first to a presentation of results for neon. Figure 18.4 shows the momentum distributions found for Ne$^+$, Ne^{++}, and Ne^{3+} ions after multiple strong-field ionization at light intensities of 1.3 PW/cm^2 (Ne$^+$, Ne^{++}) and 1.5 PW/cm^2 (Ne^{3p}) [12]. The light beam was linearly polarized along the momentum component $p_{\parallel}$ in the figure. A projection of the full rotationally symmetric (about the polarization vector) 3d-distribution onto a plane is shown. The light pulsewidth in the experiment was 30 fs and the wavelength 795 nm. The light intensities were chosen to be smaller than the intensity where single ionization of neon saturates. At these intensities it is known from total ion yield measurements that double ionization of neon proceeds nonsequentially [45–47]. The ion momentum distributions shown are fully equivalent to sum-momentum distributions of the photoelectrons leaving the neon atom. Conservation of total momentum parallel to the polarization vector and approximate conservation of momentum along the second momentum direction $p_{\perp}$ (the light beam propagation direction) give rise to this equivalence.

A significant structural difference between the Ne$^+$ and Ne^{++}/Ne^{3+} ion momentum distributions is obvious. While for Ne$^+$ it develops one maximum along $p_{\parallel}$ at $p_{\parallel} = 0$, two are found for Ne^{++} at $p_{\parallel} \approx \pm 4$ a.u., and for Ne^{3p} at

$p_\parallel \approx \pm 7.5$ a.u. with a minimum in the distribution at $p_\parallel = 0$. This difference clearly indicates that the ionization mechanisms leading to single and multiple ionization, respectively, are different.

Concerning single ionization, the quasistatic ionization model is adequate to understand the momentum distribution in Fig. 18.4. The Keldysh parameter γ is 0.35 at $1.3\,\mathrm{PW/cm^2}$. The measured momentum distribution follows the theoretical distribution function (see the review [48] and [22,49]):

$$w\left(p_\parallel, \boldsymbol{p}_\perp\right) = w(0)\exp\left[-\frac{p_\parallel^2 \omega^2 \left(2I_\mathrm{p}\right)^{3/2}}{3E_0^3} - \frac{\boldsymbol{p}_\perp^2 \left(2I_\mathrm{p}\right)^{1/2}}{E_0}\right], \qquad (18.16)$$

derived for quasistatic ionization. $w(p_\parallel, \boldsymbol{p}_\perp)\mathrm{d}^3\boldsymbol{p}$ is the rate of formation of electrons/ions in $\mathrm{d}^3\boldsymbol{p}$ and E_0 is the amplitude of the electric field strength of the light wave. This is a Gaussian distribution with a maximum at $\boldsymbol{p} = 0$ and significantly different widths along and perpendicular to the light polarization vector as observed in the experiment. A maximum at $p_\parallel = 0$ is expected since in the quasistatic limit ionization happens with the highest rate near the maxima of the oscillating electric field strength. The final momentum $p_\parallel$ of these electrons/ions is zero (see above).

The observed Ne^{++} and Ne^{3+} momentum distributions with preferentially large $p_\parallel$ immediately ruled out two of the mechanisms proposed for strong-field nonsequential multiple ionization, "collective tunneling" [8] and "shake-off" [5]. Both mechanisms release electrons simultaneously near the maxima of the oscillating electric field strength. Therefore, similar to single ionization, one expects to find electrons and therefore also ions, preferentially with small final momentum parallel to the light polarization vector. The opposite is observed.

"Stepwise" ionization via inelastic rescattering of the first electron released, in the quasistatic limit, by tunnel ionization leads to significantly different ion-momentum distributions. Via this mechanism the second "photo"-electron or further ones are created by electron-impact ionization of the singly charged ion core after acceleration of the first electron in the light wave as introduced above. Impact ionization near zero crossings of the electric field where the kinetic energy of the impinging electron is close to its maximum gives rise to large final ion and therefore electron sum momenta through acceleration of the free charged particles by the light pulse. A further contribution to the final ion momentum originates in the excess energy after impact ionization $E_\mathrm{exc} = E_\mathrm{kin}(t_\mathrm{r}) - I_\mathrm{p}^+$. At the light intensity used in the experiment this contribution is significant. A detailed kinematical analysis of this two-step ionization mechanism showed that momentum distributions with a double-hump structure along $p_\parallel$ are expected where they are found in the experiment [12].

Further confirmation of the inelastic rescattering route to nonsequential double ionization came from model calculations of electron sum-momentum distributions based on strong-field approximation and electron rescattering

[50,51]. Up to now, the best agreement with measured momentum distributions was found with an electron–electron contact interaction for the scattering event [50]. The different theoretical approaches and comparisons with experimental results are treated more thoroughly by Becker et al. and Faisal in this book.

Perpendicular to the light beam polarization vector the Ne^{++} and Ne^{3+} ion momentum distributions show a single-hump structure with a maximum at $p_\perp = 0$. Along this direction the free ions and electrons are not accelerated by the electric field of the light wave after they are formed. The ion momentum distribution along $p_\perp$ therefore is determined by two facts: the momentum distribution of the rescattering electron just before impact on the ion core and the dynamics of the inelastic scattering event. The width of the momentum distribution $f(p_\perp)$ was found to depend on the light intensity and therefore on the kinetic energy of the rescattering electron (see, for example, the momentum distributions for He [11], Ne [12,41], and Ar [42,41]). $f(p_\perp)$ may thus be expected to give a deeper insight into the dynamics of the inelastic scattering event, whereas $f(p_\parallel)$, the momentum distribution along the polarization vector, is mainly determined by postcollision acceleration of the charged particles.

If the light intensity is increased beyond the intensity where single ionization saturates, double ionization becomes dominated by sequential electric-field (tunnel) ionization (in the quasistatic limit). In the first step one electron is removed from the atom and a singly charged ion is formed in the ground state. Then this ion is again electric-field ionized. The transition is impressively visible in a collapse of the double-hump structure of the Ne^{++} momentum distribution in Fig. 18.5. Figure 18.5a repeats the Ne^{++} momentum distribution of Fig. 18.4. The spectrum in Fig. 18.5b was taken at a light intensity of $4\,PW/cm^2$. Otherwise, the light beam characteristics are the same as above. As expected, at the high light intensity the Ne^{++} momentum distribution looks like the Ne^+ distribution in Fig. 18.4.

Ion, and therefore electron, sum-momentum distributions do not give any insight into momentum sharing among the electrons. To get access to this information for double ionization in addition to the ion momentum the final momentum of at least one of the photoelectrons has to be measured. From this information and overall momentum conservation it is possible to calculate the momentum of the second electron. The result of such a kinematically complete experiment for Ne as target gas is shown in Fig. 18.6. In the 2d-plot the e–e momentum correlation for the components parallel to the polarization vector $f(p_{1,\parallel}, p_{2,\parallel})$ is shown. The figure shows that both electrons are preferentially emitted with similar momentum components $(p_{1,\parallel}, p_{2,\parallel})$ into the same half-space. Only a few events are found where the electrons are emitted back-to-back. Since after impact ionization both electrons are accelerated by the light wave in the same way this result means that impact ionization preferentially gives rise to equal momentum sharing among the electrons along

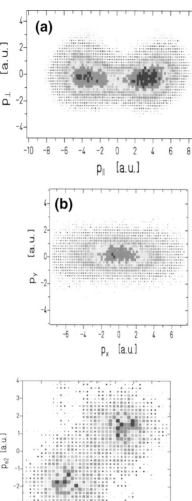

Fig. 18.5. Ne^{2+} ion momentum distribution in the regime of nonsequential (**a**) and in the regime of sequential ionization (**b**). The light intensities were 1.3 PW/cm^2 (**a**) and 4 PW/cm^2 (**b**), respectively. The light polarization vector (linear polarization) points along the horizontal axes. The light pulse width was 30 fs

Fig. 18.6. Correlation diagram of the electron momentum components parallel to the light beam polarization vector $f(p_{e1}, p_{e2})$ for double ionization of Ne. A light intensity in the nonsequential regime was chosen ($I = 1$ PW/cm^2). The light pulse width was 25 fs

the initial momentum vector of the impinging electron (the light polarization vector).

At the light intensity used, the maximum kinetic energy of the electron impinging on the singly charged ion is ≈ 190 eV. It is large compared to the ionization threshold of Ne$^+$ ($I_p = 41.07$ eV). One would therefore expect that the Coulomb–Born approximation applied to the inelastic scattering event should make sense. But model calculations showed that it gives rise to final electron momentum distributions that do not agree well with the experimental results [51,52]. With increasing light intensity an unequal momentum sharing is found. This is due to the Coulomb–Born approximation applied to scattering. With increasing kinetic energy it favors minimum momentum transfer to the electron kicked out of the ion and therefore unequal momentum sharing. Better qualitative agreement with the experimental result in Fig. 18.6 is achieved by using a contact e–e interaction in the strong-field model calculation [51]. This reproduces the electron momentum sharing found for Ne (see also Becker in this book).

18.7   Argon Double Ionization: The Differences Compared to Neon

For Ne, electron-impact ionization is the main nonsequential double-ionization mechanism over the whole intensity range investigated up to now, starting at $150\,\mathrm{TW/cm^2}$ up to $1.5\,\mathrm{PW/cm^2}$ [12,41]. Ar and also the fundamental two-electron system He seem to behave in a different way [11,13,14,41–44]. Over the whole investigated light-intensity range at least one further ionization mechanism seems to contribute strongly to nonsequential double ionization of these atoms. A first hint came from Ar^{++} ion-momentum distributions along the light-beam polarization axis. Contrary to Ne they show many events near $p_{\parallel} = 0$. Independent of the light intensity no pronounced double-hump structure develops, as was observed for Ne (see [41,42]). A similar result was found for He [11]. In the case of Ar a further characterization of the events with ion momentum close to zero was possible via the electron momentum correlation $f(p_{1,\parallel}, p_{2,\parallel})$ [13,14]. Figure 18.7 shows $f(p_{1,\parallel}, p_{2,\parallel})$ for Ar at a light intensity of $250\,\mathrm{TW/cm^2}$ a wavelength of $790\,\mathrm{nm}$ and a pulse width of $25\,\mathrm{fs}$. The light beam was linearly polarized. In the first and third quadrant of the plot the momentum distribution looks similar to that of Ne (Fig. 18.6) with the momenta scaled to take into account the smaller light intensity. In contrast to Fig. 18.6, a large number of events is seen in the second and fourth quadrants of Fig. 18.7. They correspond to electron pairs emitted into opposite half-spaces. It is these events that give rise to small Ar^{++} (electron sum-) momenta.

In [14] symmetry arguments were used to reveal the ionization mechanism behind these events. Kinematical constraints restrict events from Ar^+ impact ionization to the first and third quadrants of Fig. 18.7 [14]. A possible double-ionization mechanism of Ar that kinematically allows electrons to be found with opposite momentum components along the light-beam polarization vector was shown to be based also on electron rescattering on the Ar^+ ion core. The impinging electron now does not ionize Ar^+ but excites the ion to bound states. The ionization potential of these states is so small that they are "immediately" ionized during the next cycles of the electric field of the light wave (in the quasistatic limit by tunnel ionization). Kinematic constraints that apply to this double-ionization mechanism restrict the final

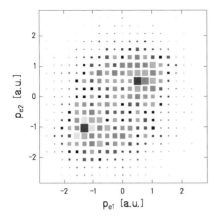

Fig. 18.7. Correlation diagram of the electron-momentum components parallel to the light-beam polarization vector $f(p_{e1}, p_{e2})$ for nonsequential double ionization of Ar. The light intensity was $250\,\mathrm{TW/cm^2}$ and the pulse width 25 fs. The figure is taken from [14]

electron momenta classically to just the regime where they were found in the experiment [14].

18.8   Conclusions and Perspectives

Besides "conventional" photoelectron spectroscopy in coincidence with doubly charged ions, especially kinematically complete final-state measurements based on COLTRIMS made a wealth of highly differential information accessible on strong-field nonsequential ionization of atoms. All results obtained so far for the noble gases Ar, Ne and the true 2-electron system He have shown that a rescattering mechanism is at the base of nonsequential ionization. Direct electron-impact ionization and impact excitation with subsequent electric-field ionization of the excited singly charged ion was found to contribute to double and multiple ionization. Scattering, especially at low kinetic energy of the returning electron, is expected to be influenced by the usually strong electric field of the light wave present at the instant of impact [41].

Among the measured distribution functions electron-difference momentum distributions and the distribution functions for the individual electron-momentum components perpendicular to the light beam polarization vector are not influenced by postionization acceleration of the charged particles through the electric field of the light wave. They therefore contain the most detailed information on the electron–ion scattering event.

There are several unexplored fields of nonsequential strong-field ionization and fields where detailed measurements are still missing. Up to now, no experiments have been done at low light intensities where the first ionization step is well within the multiphoton ionization regime ($\gamma \gg 1$). There, the phase relatedness of the first ionization step may disappear. The kinetic energy of the returning electron is small compared to the ionization threshold of the singly charged ion. In this regime double ionization may proceed by absorption of photons during recollision or it may be altogether a direct double multiphoton ionization process involving, for example, doubly excited bound states of the atom. The other extreme, high light intensity, where relativistic effects start to influence rescattering via the magnetic field of the light wave, also lacks experimental data up to now. The state of polarization of the radiation field and the pulsewidth open up the possibility of manipulating rescattering and its outcome. Few-cycle laser pulses ($t_p \approx 5\,\mathrm{fs}$ at $800\,\mathrm{nm}$ wavelength) have not yet been applied. They should give rise to new phenomena since the usually applied slowly varying amplitude approximation to analyze the phenomena breaks down. During one cycle the "amplitude" of the pulse changes heavily, giving rise to possibly new rescattring scenarios. New short-wavelength sources such as the TESLA free-electron laser open the possibility to study few (2-)photon nonsequential double ionization in the perturbative light intensity regime where precise theoretical studies concerning the electron–electron correlation should be possible. Besides atoms molecules also show nonsequential double ionization. This was found in total ion yield measurements [53–55]. The mechanism behind molecular nonsequential ionization has not yet been explored experimentally. It may be influenced by the new degrees of freedom, namely nuclear motion. Particularly for light molecules it may have an influence on the rescattering of the electron. Further influence may arise from the two or more scattering centers.

Acknowledgement

The experimental work presented in this contribution has been done in collaboration with the group of J. Ullrich, Max-Planck-Institut für Kernphysik in Heidelberg.

References

1. G.S. Voronov, N.B. Delone: Zh. Eksp. Teor. Fiz. **1**, 42 (1965) [Sov. Phys. JETP **23**, 54 (1966)]
2. I. Aleksakhin, N. Delone, I. Zapesochnyi, V. Suran: Zh. Eksp. Teor. Fiz. **76**, 887 (1979) [Sov. Phys. JETP **49**, 447 (1979)]
3. A. l'Huillier, L.A. Lompre, G. Mainfray, C. Manus: Phys. Rev. Lett. **48**, 1814 (1982)
4. A. l'Huillier, L.A. Lompre, G. Mainfray, C. Manus: Phys. Rev. A **27**, 2503 (1983)
5. D.N. Fittinghoff, P.R. Bolton, B. Chang, K.C. Kulander: Phys. Rev. Lett. **69**, 2642 (1992)
6. B. Walker, B. Sheehy, L.F. DiMauro, P. Agostini, K.J. Schafer, K.C. Kulander: Phys. Rev. Lett. **73**, 1227 (1994)
7. V. Schmidt: *Electron Spectrometry of Atoms using Synchrotron Radiation* (Cambridge University Press, Cambridge 1997)
8. U. Eichmann, M. Dörr, H. Maeda, W. Becker, W. Sandner: Phys. Rev. Lett. **84**, 3550 (2000)
9. P.B. Corkum: Phys. Rev. Lett. **71**, 1994 (1993)

10. B. Walker, E. Mevel, B. Yang, P. Breger, J.P. Chambaret, A. Antonetti, L.F. DiMauro, P. Agostini: Phys. Rev. A **48**, R894 (1993)
11. T. Weber, M. Weckenbrock, A. Staudte, O. Jagutzki, V. Mergel, F. Afaneh, G. Urbasch, M. Vollmer, H. Giessen, R. Dörner: Phys. Rev. Lett. **84**, 443 (2000)
12. R. Moshammer, B. Feuerstein, W. Schmitt, A. Dorn, C.D. Schröter, J. Ullrich, H. Rottke, C. Trump, M. Wittmann, G. Korn, K. Hoffmann, W. Sandner: Phys. Rev. Lett. **84**, 447 (2000)
13. T. Weber, H. Giessen, M. Weckenbrock, G. Urbasch, A. Staudte, L. Spielberger, O. Jagutzki, V. Mergel, M. Vollmer, R. Dörner: Nature **405**, 658 (2000)
14. B. Feuerstein, R. Moshammer, D. Fischer, A. Dorn, C.D. Schröter, J. Deipenwisch, J.R. Crespo Lopez-Urrutia, C. Höhr, P. Neumayer, J. Ullrich, H. Rottke, C. Trump, M. Wittmann, G. Korn, W. Sandner: Phys. Rev. Lett. **87**, 043003 (2001)
15. B. Witzel, N.A. Papadogiannis, D. Charalambidis: Phys. Rev. Lett. **85**, 2268 (2000)
16. R. Lafon, J.L. Chaloupka, P.M. Paul, P. Agostini, K.C. Kulander, L.F. DiMauro: Phys. Rev. Lett. **86**, 2762 (2001)
17. E.R. Peterson, P.H. Bucksbaum: Phys. Rev. A **64**, 053405 (2001)
18. L.V. Keldysh: Zh. Éksp. Teor. Fiz. **47**, 1945 (1964) [Sov. Phys. JETP **20**, 1307 (1965)]
19. R.R. Freeman, P.H. Buchsbaum, H. Milchberg, S. Darack, D. Schumacher, M.E. Geusic: Phys. Rev. Lett. **59**, 1092 (1987)
20. H. Rottke, B. Wolff, M. Brickwedde, D. Feldmann, K.H. Welge: Phys. Rev. Lett. **64**, 404 (1990)
21. P. Agostini, F. Fabre, G. Mainfray, G. Petite, N.K. Rahman: Phys. Rev. Lett. **42**, 1127 (1979)
22. A.M. Perelomov, V.S. Popov, M.V. Terent'ev: Zh. Éksp. Teor. Fiz. **50**, 1393 (1966) [Sov. Phys. JETP **23**, 924 (1966)]
23. A.M. Perelomov, V.S. Popov, M.V. Terent'ev: Zh. Éksp. Teor. Fiz. **51**, 309 (1966) [Sov. Phys. JETP **24**, 207 (1967)]; A.M. Perelomov, V.S. Popov: ibid. **52**, 514 (1967) [ibid. **25**, 336 (1967)]; V.S. Popov, V.P. Kuznetsov, A.M. Perelomov: ibid. **53**, 331 (1967) [ibid. **26**, 222 (1967)]
24. G.L. Yudin, M.Y. Ivanov: Phys. Rev. A **63**, 033404 (2001); M.V. Ammosov, N.B. Delone, V.P. Krainov: Zh. Éksp. Teor. Fiz. **91**, 2008 (1986) [Sov. Phys. JETP **64**, 1191 (1986)]; N.B. Delone, V.P. Krainov: J. Opt. Soc. Am. B **8**, 1207 (1991)
25. G.L. Yudin, M.Y. Ivanov: Phys. Rev. A **64**, 013409 (2001)
26. G.G. Paulus, W. Nicklich, H. Xu, P. Lambropoulos, H. Walther: Phys. Rev. Lett. **72**, 2851 (1994)
27. G.G. Paulus, W. Becker, W. Nicklich, H. Walther: J. Phys. B: At. Mol. Opt. Phys. **27**, L703 (1994)
28. W. Becker, A. Lohr, M. Kleber: J. Phys. B: At. Mol. Opt. Phys. **27**, L325 (1994)
29. B. Yang, K.J. Schafer, B. Walker, K.C. Kulander, P. Agostini, L.F. DiMauro: Phys. Rev. Lett. **71**, 3770 (1993)
30. A. McPherson, G. Gibson, H. Jara, U. Johann, T.S. Luk, I. McIntyre, K. Boyer, C.K. Rhodes: J. Opt. Soc. Am. B **4**, 595 (1987)
31. M. Ferray, A. L'Huillier, X.F. Li, L.A. Lompré, G. Mainfray, C. Manus: J. Phys. B: At. Mol. Opt. Phys. **21**, L31 (1988)

338   H. Rottke

32. M. Lewenstein, P. Balcou, M.Y. Ivanov, A. L'Huillier, P.B. Corkum: Phys. Rev. A **49**, 2117 (1994)
33. J.D. Jackson: *Classical Electrodynamics*, 3rd edn. (John Wiley & Sons, Inc., New York 1999) p. 580
34. T.W.B. Kibble: Phys. Rev. Lett. **16**, 1054 (1966)
35. P. Mulser: J. Opt. Soc. Am. B **2**, 1814 (1985)
36. W. Becker, R.R. Schlicher, M.O. Scully, K. Wódkiewicz: J. Opt. Soc. Am. B **4**, 743 (1987)
37. P.H. Bucksbaum, R.R. Freeman, M. Bashkansky, T.J. McIlrath: J. Opt. Soc. Am. B **4**, 760 (1987)
38. R. Moshammer, B. Feuerstein, D. Fischer, A. Dorn, C.D. Schröter, J. Deipenwisch, J.R. Crespo Lopez-Urrutia, C. Höhr, P. Neumayer, J. Ullrich, H. Rottke, C. Trump, M. Wittmann, G. Korn, W. Sandner: Opt. Express **8**, 358 (2001)
39. D.R. Miller: Free Jet Sources. In: *Atomic and Molecular Beam Methods, Vol. 1*, ed. by G. Scoles (Oxford University Press, Oxford 1988)
40. F. Lindner et al.: submitted
41. E. Eremina, H. Rottke, W. Sandner, A. Dreischuh, F. Lindner, F. Grasbon, G.G. Paulus, H. Walther, R. Moshammer, B. Feuerstein, J. Ullrich: submitted for publication
42. T. Weber, M. Weckenbrock, A. Staudte, L. Spielberger O. Jagutzki, V. Mergel, F. Afaneh, G. Urbasch, M. Vollmer, H. Giessen, R. Dörner: J. Phys. B: Atomic, Molec. Opt.Phys. **33**, L127 (2000)
43. M. Weckenbrock, M. Hattass, A. Czasch, O. Jagutzki, L. Schmidt, T. Weber, H. Roskos, T. Löffler, M. Thomson, R. Dörner: J. Phys. B: Atomic, Molec. Opt. Phys. **34**, L449 (2001)
44. R. Moshammer, B. Feuerstein, J. Crespo Lopez-Urrutia, J. Deipenwisch, A. Dorn, D. Fischer, C. Höhr, P. Neumayer, C. D. Schröter, J. Ullrich, H. Rottke, C. Trump, M. Wittmann, G. Korn, W. Sandner: Phys. Rev. A **65**, 035401 (2992)
45. D.N. Fittinghoff, P.R. Bolton, B. Chang, K.C. Kulander: Phys. Rev. A **49**, 2174 (1994)
46. P. Dietrich, N.H. Burnett, M. Ivanov, P.B. Corkum: Phys. Rev. A **50**, R3585 (1994)
47. V.R. Bhardwaj, S.A. Aseyev, G.L. Yudin, D.M. Villeneuve, D.M. Rayner, M.Yu. Ivanov, P.B. Corkum: Phys. Rev. Lett. **86**, 3522 (2001)
48. N.B. Delone, V.P. Krainov: Uspekhi Fiz. Nauk **168**, 531 (1998) [Physics–Uspekhi **41**, 469 (1998)]
49. A.I. Nikishov, V.I. Ritus: Zh. Eksp. Teor. Fiz. **50**, 255 (1966) [Sov. Phys. JETP **23**, 162 (1966)]
50. R. Kopold, W. Becker, H. Rottke, W. Sandner: Phys. Rev. Lett. **85**, 3781 (2000)
51. S.P. Goreslavskii S.V. Popruzhenko, R. Kopold, W. Becker: Phys. Rev. A **64**, 053402 (2001)
52. S.P. Goreslavskii, S.V. Popruzhenko: Opt. Express **8**, 395 (2001)
53. A. Talebpour, S. Larochelle, S.L. Chin: J. Phys. B: Atomic, Molec. Opt. Phys. **30**, L245 (1997)
54. C. Cornaggia, P. Hering: J. Phys. B: Atomic, Molec. Opt. Phys. **31**, L503 (1998)
55. C. Cornaggia, P. Hering: Phys. Rev. A **62**, 023403 (2000)

19 Helium Double Ionization in Collisions with Electrons

A. Dorn

19.1 Introduction

The investigation of electron-impact ionization of atoms contributed considerably to our understanding of the correlated fragmentation dynamics of atomic systems. This is mainly due to the early realization of kinematically complete experiments in the late 1960s [1], in which the momentum vectors of all participating continuum particles, in the initial and final states, were under control. These (e,2e) experiments allowed theoretical models to be tested critically and gave detailed insight into basic atomic-collision processes. Nevertheless, the fundamental three-body Coulomb continuum problem persisted from being solved satisfactory until in the last ten years when a number of very successful theoretical treatments were carried out that not only gave a precise description of the cross sections but also were able to explain the observed cross section pattern in terms of dominating interactions and contributing reaction mechanisms [2–4].

As the next logical step, experimental and theoretical efforts presently focus more and more on processes involving two active target electrons like electron-impact double ionization or simultaneous ionization and excitation. In these systems, correlation plays a dominant role in the initial target ground state, during the collision and in the final continuum state making the theoretical treatment highly demanding. Depending on the projectile velocity, different aspects of these collision systems can be investigated. For very fast electron impact, first-order projectile–target collisions dominate, where the projectile interacts only with one bound electron.

In this high-velocity regime, double ionization occurs only via the interaction of the target electrons and, therefore, it is sufficient to take the correlations between the three constituents of the helium target into account. Thus, theoretical techniques originally developed to describe the dynamics of three-body Coulomb systems like (e,2e) collisions or photo double ionization, (γ,2e), can be applied. For decreasing projectile velocity, higher-order collisions are also expected to contribute and in the vicinity of the ionization threshold the interactions in all two-body subsystems are equally important. It is this regime where the full complexity of the four-body Coulomb system is expected to appear.

The first kinematically complete experiments for electron-impact double ionization have been realized by extending a conventional (e,2e) apparatus with a third spectrometer in order to detect the second ejected electron in the (e,3e) process [5]. But due to the small combined solid angle of three-electron analyzers, which is of the order of 10^{-6} of 4π or below, this conventional technique reaches its limit for the investigation of double ionization. In particular, for the most fundamental and, therefore, theoretically most interesting helium-target experiments are very time consuming due to the small total cross sections which are of the order of 10^{-20} cm^2. Typical signal rates are a few coincidences per hour. This is despite major efforts to improve the efficiency of conventional electron spectrometry by introducing multiangle and multienergy detection techniques aiming to increase the fraction of the final-state momentum space that is covered by the experiment [6].

In order to overcome these experimental limitations, we have applied the combined multielectron–recoil-ion imaging technique ("Reaction Microscope") to investigate electron-impact processes and, in particular, to perform systematic studies for double ionization of helium by electron impact [7]. Similarly to ion-impact double-ionization experiments, performed with this technique, both ejected target electron momenta k_c and k_d are detected in coincidence with the momentum k_R of the recoiling helium ion. The momentum q transferred by the scattered projectile that is not detected is obtained by momentum conservation $q = (k_a - k_b) = k_R + k_c + k_d$. Here k_a and k_b are the momentum vectors of the incoming and the scattered projectile, respectively. The essential advantage of this concept with respect to the conventional technique is the large acceptance in solid angle and in momentum of all detected particles allowing comparatively high coincidence rates of about one per second.

In the following, after a brief presentation of the experimental set-up, we will discuss the structure and main features of the (e,3e) cross section for 2 keV projectile energy corresponding to 12 atomic units of velocity and for ejected electron energies between 5 eV and 20 eV. In order to understand the relation to the extensively studied photo double ionization process, first we will compare low momentum transfer collisions for which dipole transitions should dominate with the corresponding photo double ionization process. Particular cuts of the fully differential data set are identified that allow quantification of deviations from the dipole limit and contributions beyond the first order in the projectile-target interaction. Subsequently, these results are contrasted to collisions far off the dipole limit with a large momentum transfer. Here, the emission pattern is free to evolve without restrictions due to dipole selection rules, and detailed insight into underlying reaction mechanisms can be gained.

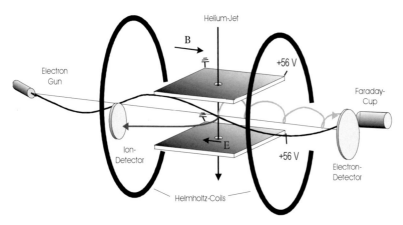

Fig. 19.1. Schematic view of the experimental setup.

19.2   Experimental Setup

A detailed discussion of the working principle of the Reaction Microscope is given by Moshammer in Chap. 2.1 of this book. Therefore, in the following we describe only the particularities of the present setup that has been designed to fit the requirements of electron-collision experiments. A scheme of the apparatus is given in Fig. 19.1. A conventional electron gun is used to produce a pulsed primary beam with a repetition rate of 500 kHz and a pulse length of $\tau = 1$–2 ns. The helium target was prepared in a triple-stage supersonic jet. Ions and low-energy electrons produced in (e,3e) collisions were extracted in opposite directions by a uniform 2.7 V/cm electric field applied along the apparatus axis and are detected by two-dimensional position-sensitive multichannel plates. A solenoidal magnetic field of 12 G produced by a pair of Helmholtz coils forces the slow electrons with nonzero transverse momenta into spiral trajectories. In this way, electrons with energies below 30 eV and essentially all ions are detected with the full solid angle of 4π. From the times-of-flight (TOF) and the measured positions on the detectors the trajectories of the particles can be reconstructed and their initial longitudinal and transverse momentum components are obtained. For electrons, the calculation of the initial transverse momentum is not unambiguous if their TOF is an integer number of cyclotron revolutions (for details see [8]).

   In the present experiment, the cross section for the corresponding longitudinal momenta was obtained by a second experimental run applying a slightly different electric extraction field and therefore changing the TOF of the electrons. The 80–mm active diameter electron detector is equipped with a fast delay-line readout and a multihit time-to-digital converter. Thus, positions as well as arrival times of both electrons emitted in a double-ionization event are determined if their flight-time difference exceeds the detectors deadtime of about 15 ns. This results in a small loss of momentum space in the final state for electrons having similar momenta in the longitudinal direction towards the electron detector.

One peculiarity with using an electron projectile beam with moderate energy in conjunction with the Reaction Microscope is the strong deflection of the primary beam by the applied extraction fields. Due to their large charge-to-mass ratio electrons are strongly influenced, in particular, by the applied magnetic extraction field. As indicated in Fig. 19.1, this can be used to focus the projectile beam into the target. The electron gun as well as the collision volume and the electron detector are aligned on the axis of the apparatus that coincides with the axis of the magnetic field. The TOF of the primary electrons ($E_a = 2\,\text{keV}$) from the gun to the target interaction point is equal the electron cyclotron revolution time in the magnetic field ($t_c = 26\,\text{ns}$). Therefore, the magnetic-lens effect images an electron beam focus at the exit of the electron gun into the helium jet where the beam diameter is below 0.5 mm. A superimposed initial transverse-momentum component results in an offset of 7 cm of the electron beam from the apparatus axis at the position of the ion detector (TOF of $t_c/2$) and at the position of the electron detector (TOF of $3t_c/2$). In this way, the projectile beam passes both detectors and is dumped in a Faraday cup next to the electron detector.

19.3   Low Momentum Transfer Collisions

Experimentally, the square of the four-particle wave function in the final state continuum was obtained from which fully differential cross sections can be extracted, in principle, for arbitrary coordinates that seem to be appropriate to study the process. Here we present five-fold differential cross sections (FDCS) differential in the energies $E_{c,d}$ and solid angles $\Omega_{c,d}$ of two slowly ejected electrons and in the solid angle Ω_b of a fast-emitted electron:

$$\text{FDCS} = \frac{\text{d}^5\sigma}{\text{d}\Omega_b\text{d}\Omega_c\text{d}\Omega_d dE_c dE_d}. \tag{19.1}$$

For the present conditions with a fast projectile ($E_a = 2\,\text{keV}$) and two slow final-state electrons ($E_{c,d} \leq 30\,\text{eV}$), the electron-exchange effects can be neglected. Therefore, the fast outgoing electron can be identified with the scattered projectile and by fixing the scattering angle the amount of momentum $|q|$ transferred to the target is determined.

In the following, the cross sections are presented in the angle-scanning mode: for a given momentum transfer and for fixed energies of the ejected electrons the cross section is plotted as a function of the ejected-electron emission angles. Thus, a direct comparison with data obtained by conventional techniques applying electrostatic electron spectrometers is possible. We chose a coplanar scattering geometry illustrated in Fig. 19.2 where the target electrons are ejected in the plane defined by the incoming and scattered projectile and low momentum transfer with $|q| = 0.5\,\text{a.u.}$ corresponding to a projectile

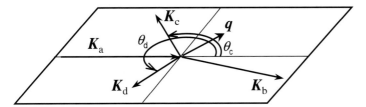

Fig. 19.2. Coplanar scattering geometry.

scattering angle of 2°. The energies of the ionized electrons are 5 eV each. In Fig. 19.3a the resulting FDCS is given as a function of the emission angles θ_c and θ_d of both electrons [9]. The density-plot representation allows one to obtain an overview of the important emission configurations for given values of $|q|$ and $E_{c,d}$. The solid circular lines in the diagrams indicate the angular range that is affected by the electron detector dead-time. For angular configurations outside the circles, the flight-time difference of the ejected electrons is less than 15 ns and, therefore, the experimental detection efficiency is reduced. In Fig. 19.3c the photo double ionization (PDI), or $(\gamma,2e)$, cross section is presented, which is obtained by using the generally accepted parametrization of the PDI cross section by Huetz et al. [10,11]. In order to enable a comparison, the light polarization vector E, which is the quantization axis for PDI, is oriented along the momentum transfer q, the corresponding quantization axis for a first-order projectile-target collision. Since for equal energy sharing both electrons are interchangeable, the diagrams are symmetric with respect to the diagonal line $\theta_c = \theta_d$. Both, the (e,3e) and the $(\gamma,2e)$ cross sections are governed by four peaks at about the same position. From the corresponding angles θ_c and θ_d it can be concluded that the electrons are emitted most probably with a relative angle of approximately 130° as indicated in Fig. 19.3b and d.

For the $(\gamma,2e)$ process, the electron sum momentum is directed along the electric-field axis. The nodal lines in between the cross section peaks can be attributed to Coulomb repulsion and to dipole-selection rules acting: Emission into small relative angles $\theta_c \approx \theta_d$, is suppressed due to electron–electron repulsion since both emitted electrons have equal energy. As can be seen in Fig. 19.3a, the corresponding angular range is not accessible experimentally but it can be confirmed experimentally that the cross section approaches small values already far off this diagonal line. Additionally, it is straightforward to show that for $k_c = -k_d$ the two-body final continuum state has an even parity. Therefore, dipole transitions from the even parity ^1S_e ground state that end up in an odd-parity state are prohibited, resulting in a cross section minimum along the dashed lines (i) in Fig. 19.3. Finally, for equal energies of the emitted electrons dipole transitions vanish if the vector sum momentum of the ejected electrons is perpendicular to q [12,13] (lines (ii) in Fig. 19.3). Obviously for $|q| = 0.5$ a.u. and $E_{b,c} = 5$ eV the observed (e,3e)

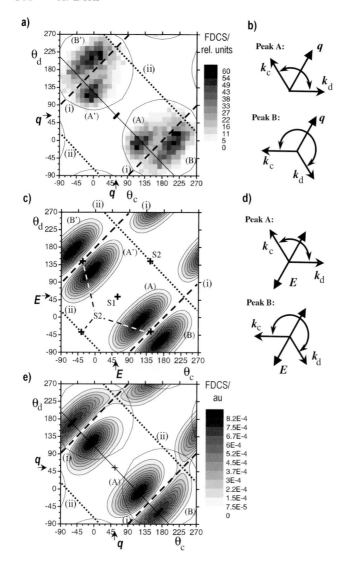

Fig. 19.3. Five-fold differential cross section (FDCS) in coplanar scattering geometry as a function of the ejected-electron emission angles θ_c and θ_d relative to the primary beam forward direction. The ejected electron energies are $E_{c,d} = 5 \pm 2$ eV. (a) Experimental cross section for $E_a = 2$ keV, $|q| = 0.5 \pm 0.2$ a.u. (b) The electron-emission configurations for the cross section maxima (A) and (B) in (a). (c) Photo double ionization cross section for $E_{c,d} = 5$ eV. (d) The electron emission configurations for the cross section maxima (A) and (B) in (c). (e) CCC calculation for $|q| = 0.5$ a.u. The direction of the momentum transfer q is marked by *arrows*. In both diagrams the angular range that is not affected by the experimental detector dead-time is encircled by *solid lines* (From [9]).

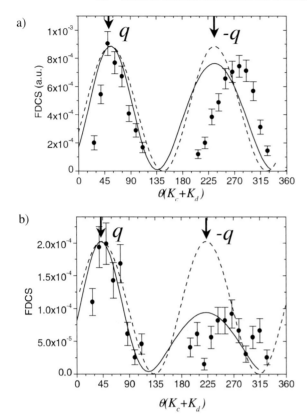

Fig. 19.4. Five-fold differential cross section for fixed relative angle $\theta_{cd} = |\theta_c - \theta_d| = 125°$ as function of $\theta_S = (\theta_b + \theta_c)/2$, the angle of the sum momentum $\boldsymbol{k}_c + \boldsymbol{k}_d$ of the ejected electrons. $E_c = E_d = 5\,\text{eV}$. *Dashed line*: Photo double ionization cross section obtained by the parametrization of Huetz et al. [10,11]. *Solid line*: CCC calculation. The experimental values and the PDI cross section have been normalized to fit the binary peak height of the CCC calculation close to $\theta_S = 60°$ (from [15]).

cross section pattern is governed by these selection rules and, therefore, dipole transitions dominate.

On the other hand, deviations from the strict symmetries present in the PDI case are also identified. First, since the light polarization denotes an axis in space and not a distinct direction there must be reflection symmetry of the $(\gamma,2e)$ cross section with respect to a plane perpendicular to the electric field axis E. As result, the emission configurations of peak A and peak B which are indicated in Fig. 19.3d, are equivalent and the PDI cross section pattern is symmetric to the points $S2$ in Fig. 19.3c. In contrast, for the (e,3e) process this reflection symmetry is broken, since for a finite modulus of the momentum transfer the $\boldsymbol{q}$ vector constitutes a direction in space and the peaks

(A) and (B) in Fig. 19.3a and the corresponding emission configurations in Fig. 19.3b, in general, may be different in shape and in their absolute height. Indeed, the peaks (A) and (B) in Fig. 19.3a are different at least where the shape is concerned. Therefore, we have clear signatures that higher multipole transitions are also of importance for the kinematic conditions discussed.

Secondly, PDI is a first-order process and, thus, the cross section is axially symmetric with respect to the light polarization vector. This forces the cross section pattern in Fig. 19.3c to be symmetric with respect to the simultaneous inversion of both electron angles at the direction of E, or equivalently at the points $S1$. For the (e,3e) process in Fig. 19.3a the peaks (B) and (B') break this symmetry since they are shifted from the position expected from the PDI case to larger values for both angles θ_c and θ_d. This observation reveals the influence of collision processes beyond first order.

In Fig. 19.3e, theoretical results obtained with the convergent close coupling (CCC) approach are shown [9]. The method, which treats the interaction of the slow ejected electrons nonperturbatively, is known to yield very reliable results for the PDI process [14]. In order to apply this model to the present (e,3e) process, the projectile–target interaction is treated in first order by applying the first Born approximation. Comparing with the experimental cross section, the agreement is reasonably good in shape and in the relative peak heights. Clearly, the finite momentum transfer gives rise to higher multipole contributions, which can be identified for angular combinations where the dipole selection rules hold, i.e., at the intersection points of the dashed and dotted lines. Disagreement between experiment and theory is presented for the node for back-to-back emission (dashed lines), which, in contrast to theory, is partially filled in the experimental data. A further disagreement is the shift of the peaks marked (B) to larger angles for both emitted electrons. As discussed above, this can be explained by higher-order projectile-target interactions that are not within the scope of a first Born treatment.

In the following, we discuss a distinct cut of the complete cross section pattern given in Fig. 19.3 that allows one to make a quantitative comparison of nondipole and higher-order contributions with the (γ,2e) case and with the theoretical results [15]. Therefore, we fix the relative angle of the two ejected electrons to $|\theta_c - \theta_d| = 125°$, a value close to the cross section maximum and scan both electron momentum vectors simultaneously. Thus, the influence of the Coulomb repulsion between the ejected electrons is constant and both electrons are treated as a quasiparticle with momentum $k_S = k_c + k_d$. The cross section obtained is presented in Fig. 19.4a for $E_c = E_d = 5\,\text{eV}$ and in Fig. 19.4b for $E_c = E_d = 20\,\text{eV}$. For these cuts through the final-state momentum space, the cross sections show characteristics that are typical for (e,2e) processes with a fast projectile and low momentum transfer. A first cross section maximum is obtained for k_S being directed along q corresponding to the peaks (A) in Fig. 19.3, and a second maximum of almost the same magnitude if k_S is directed opposite q corresponding to the peaks (B) in

Fig. 19.3. According to the (e,2e) nomenclature, we call these maxima *binary* peak and *recoil* peak, respectively. On the other hand, for a $(\gamma, 2e)$ process with equal energies $E_c = E_d$ we expect a pure $\cos^2\theta_S$ dependence of the cross section (dashed curves in Fig. 19.4a and b). In this representation, the differences of experimental data, the photoionization process and the CCC calculation (solid curve in Fig. 19.4), which have been discussed above, become very obvious.

The broken mutual symmetry of the binary and the recoil peak for the (e,3e) process that can be attributed to higher multipole transitions is reproduced fairly well by the CCC calculation. First, the binary peak, is slightly stronger compared to the recoil peak. This relative height of both peaks is reproduced nicely by the CCC theory. Secondly, the binary peak width is lower and recoil peak width larger compared to the dipole limit. This behavior is also reproduced by the calculation, although the theoretical peak widths are larger than observed in the experiment. On the other hand, the strong angular shift of the recoil peak is not described theoretically due to the first-order treatment of the projectile-target interaction. It is interesting to note that for (e,2e) collisions the recoil peak is generally shifted towards the backward direction. This often is attributed to the repulsion between the scattered projectile and the ejected electron. In the present (e,3e) case, however, the recoil peak is shifted to the forward direction. This counterintuitive behavior can be a critical test for more elaborate future calculations.

In Fig. 19.4b the cross section is displayed for higher energies of the ejected electrons $E_c = E_d = 20\,\mathrm{eV}$. Here, the asymmetry between binary and recoil peaks is very pronounced since the recoil peak is much weaker now. Obviously the relative importance of higher multipole transitions is not only determined by the amount of momentum transferred to the target but the ejected electron energies also play an important role. A possible explanation is that for lower energies $E_{c,d}$ outgoing partial waves with low orbital angular momenta are emphasized. Therefore, for low energies $E_{c,d}$ dipole transitions are more strongly weighted even if $|q|$ is still considerably higher than the optical limit.

In conclusion, this example displays how the fully resolved (e,3e) cross section allows us to disentangle signatures of dipole and multipole transitions, as well as of first- and higher-order projectile-target interaction. The data for higher energies of the ejected electrons reveal that a momentum transfer of $|q| = 0.5\,\mathrm{a.u.}$ is still far off the optical limit. On the other hand, we can restrict the process to dipole transitions by choosing low energies for the ejected electrons.

19.4   Impulsive Collisions with Large Momentum Transfer

Now we will discuss the cross section for kinematical conditions where the projectile transfers a large amount of momentum to the target. In Fig. 19.5

348   A. Dorn

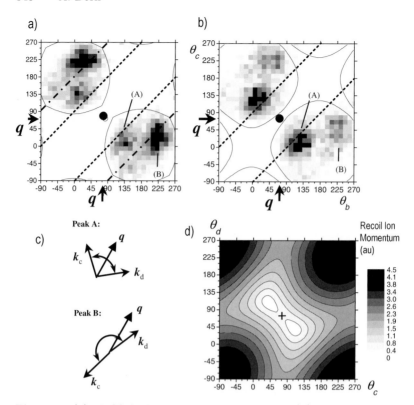

Fig. 19.5. (a) FDCS for large momentum transfer of $|q| = 2$ a.u. and $E_{c,d} = 5$ eV.
(b) FDCS for $|q| = 2$ a.u. and $E_{c,d} = 20$ eV. (c) The electron-emission configurations for the cross section maxima (A) and (B) marked in (a) and (b). (c) The recoil-ion momentum as a function of the ejected-electron emission angles.

the experimental FDCS is plotted for $|q| = 2$ a.u. and for two energies of the ejected electrons: $E_{c,d} = 5$ eV and 20 eV [16]. For low energies $E_{c,d} = 5$ eV (see Fig. 19.5a), a strong cross section maximum marked (B) is observed for back-to-back emission of both electrons with one electron going roughly in the direction of the momentum transfer. Since back-to-back emission is forbidden for dipole transitions from the helium ground state it is obvious that optical transitions play no role here, even for low energies of the ejected electrons. An additional lower peak (A) is observed for smaller relative electron angles $|\theta_c - \theta_d| \approx 110° - 120°$ with both electrons going into the half-plane of the momentum transfer.

The emission configurations of the cross section peaks (A) and (B) are indicated in Fig. 19.5c. If the energies of the ejected electrons are increased the relative magnitudes of the maxima change. For $E_c = E_d = 20$ eV (Fig. 19.5b) the peaks (B) become relatively unimportant compared to the maxima (A), which now dominate the cross section.

In order to identify possible mechanisms for double ionization, it is useful to consider the magnitude of the recoil-ion momentum. The kinematics with low electron energies $E_{c,d} = 5\,\text{eV}$ is not favored for clean binary knock-out collisions where the full momentum transfer is carried by the ionized electrons and the recoiling ion has very little momentum (the Bethe ridge condition). The energies of the ejected electrons correspond to rather small momenta of $|\mathbf{k}_{c,d}| = 0.6\,\text{a.u.}$ and thus the Bethe ridge condition $|\mathbf{k}_{\text{ion}}| \approx 0\,\text{a.u.}$ or equivalently $\mathbf{k}_c + \mathbf{k}_d \approx \mathbf{q}$, cannot be fulfilled even if both electrons are going in the momentum transfer direction. Therefore, double ionization can occur only if initial states with large ionic momentum in the tail of the Compton profile are involved or if the ion gains momentum in the collision. From (e,2e) processes it is known that for low energies of the ejected electron the interaction with the residual ion is strong. This results in backscattering with large residual ion momenta and a correspondingly strong recoil peak in the electron emission pattern. For larger energies $E_{c,d} = 20\,\text{eV}$ of the outgoing electrons the situation is different. Now, Bethe kinematics with vanishing recoil-ion momentum can be reached for a relative electron angle of about 90°. This can be seen in Fig. 19.5d where the recoil-ion momentum is plotted as a function of the ejected electron angles for the kinematical conditions of Fig. 19.5b. Roughly, for the angular configuration of peak A the recoil-ion momentum is very small indicating a binary knock-out collision. The value of the relative emission angle close to 90° is a signature of the two-step-1 (TS-1) process in which the target electron, hit by the projectile, ionizes in turn the second electron in a binary collision while leaving the atom.

19.5   Conclusion and Outlook

We have shown results from a kinematically complete experiment for double ionization of helium by 2-keV electron impact. From the large final-state momentum space mapped experimentally, fully differential cross sections for $|\mathbf{q}| = 0.5\,\text{a.u.}$ and for $|\mathbf{q}| = 2\,\text{a.u.}$ momentum transfer and for coplanar scattering geometry have been extracted. For low momentum transfer the cross section pattern is governed by nodal lines that can be attributed to dipole selection rules revealing the dominance of dipole transitions. Additionally, signatures of higher-order processes in the projectile–target interaction are found. In particular the position of the recoil peak, for which the ejected electron sum momentum is directed opposite to the momentum transfer direction, shows a strong symmetry break with respect to $\mathbf{q}$. The experimental data are compared to theoretical results obtained by the convergent close-coupling approach. The general agreement of the shape and the relative peak heights with experiment is rather good. On the other hand, the calculation that treats the projectile–target interaction in first order is not capable to reproduce the observed peak shifts and the filling of the node for back-to-back emission of the ejected electrons. For large momentum transfer, the dipole selection rules become unimportant and the cross section is free to evolve according to the underlying collision dynamics. We have seen that for kinematics, where the Bethe ridge condition can be fulfilled, maxima in cross section are observed. For low energies of the ejected electrons, their mutual repulsion is dominant and a relative emission angle of $180°$ is observed.

The results presented in this chapter demonstrate that the combined recoil-ion–multielectron imaging technique is well suited to perform kinematically complete studies of multiple ionization induced by electron impact. Three decades ago, (e,2e) pioneering studies for the kinematically complete investigation of the three-body Coulomb problem were made. Now, (e,3e) experiments play a similar role for the investigation of particle-induced double ionization. This is despite a number of studies of ion-induced double ionization that were kinematically complete but the statistics of the data did not allow fully differential cross sections to be extracted.

In future, experiments for lower projectile velocities will be performed where the full complexity of the four-body Coulomb problem is expected to show up. Additionally, experiments for simultaneous ionization and excitation of helium will be performed. The description of this process is less demanding for theory since the final state contains only two electrons. In a triple coincidence the (e,2e) process is detected in a kinematically complete way and, in addition, information on the state of the excited residual ion is obtained by detecting the fluorescence radiation emitted.

Acknowledgements

This work is supported by the Deutsche Forschungsgemeinschaft within the Sonderforschungsbereich 276, TP B7 and the Leibniz programm.

References

1. H. Ehrhardt, M. Schulz, T. Tekaat, and K. Willmann: Phys. Rev. Lett. **22**, 89 (1969)
2. T.N. Rescigno, M. Baertschy, W.A. Isaacs, and C.W. McCurdy, Science **286**, 2474 (1999)
3. C.T. Whelan, Science **286**, 2457 (1999)
4. J.S. Briggs, Comment. At. Mol. Phys. **23**, 155 (1989)
5. A. Lahmam-Bennani, C. Dupre and A. Duguet: Phys. Rev. Lett. **63**, 1582 (1989)
6. I. Taouil, A. Lahmam-Bennani, A. Duguet, and L. Avaldi: Phys. Rev. Lett. **81**, 4600 (1998)
7. A. Dorn, R. Moshammer, C.D. Schröter, T.J.M. Zouros, W. Schmitt, H. Kollmus, R. Mann and J. Ullrich: Phys. Rev. Lett. **82**, 2496 (1999)
8. R. Moshammer, M. Unverzagt, W. Schmitt, J. Ullrich and H. Schmidt-Böcking: Nucl. Instum. Methods Phys. B **108**, 425 (1996)
9. A. Dorn, A. Kheifets, C.D. Schröter, B. Najjari, C. Höhr, R. Moshammer, and J. Ullrich: Phys. Rev. Lett. **86**, 3755 (2001)

10. A. Huetz, P. Selles, D. Waymel and J. Mazeau: J. Phys. B **24**, 1917 (1991)
11. J.S. Briggs and V. Schmidt: J. Phys. B **33**, R1 (2000)
12. J. Berakdar and H. Klar: J. Phys. B **26**, 4219 (1993)
13. F. Maulbetsch and J.S. Briggs: J. Phys. B **26**, 1679 (1993)
14. A.S. Kheifets and I. Bray: J. Phys. B **31**, 5149 (1998)
15. A. Dorn, B. Najjari, G. Sakhelashvili, C. Höhr, C.D. Schröter, R. Moshammer, J. Ullrich A. Kheifets, and R.D. DuBois: Photonic and Atomic Collisions, 22. international conference, Santa Fe, July 2001, edited by J.Burgdörfer, J. S. Cohen, S. Daz and C. R. Vane, Rinton press, Princeton 2002, p.423
16. A. Dorn, A. Kheifets, C.D. Schröter, B. Najjari, C.Höhr, R. Moshammer, and J. Ullrich: Phys. Rev. A, **65**, 2709 (2002)

20 Fast p–He Transfer Ionization Processes: A Window to Reveal the Non-s² Contributions in the Momentum Wave Function of Ground–State He

H. Schmidt-Böking, V. Mergel, R. Dörner, H.J. Lüdde, L. Schmidt,
T. Weber, E. Weigold, and A.S. Kheifets

20.1 Introduction

It is commonly believed that the He ground-state wave function is perfectly understood, since the theoretically determined He ground-state binding energy [1,2] is in excellent agreement with high-precision experimental data [3]. In investigating the ground-state binding energy by high-resolution spectroscopy one probes, however, the wave function at the region of the maximum density at a distance close to the Bohr radius. The theoretical binding energies are obtained on the basis of a many-body approximation, such as the multi-configuration approach (MCA). Using variational methods, a wave function is generated that requires a huge basis of diagonal and off-diagonal matrix elements, or as in nuclear physics on-shell and off-shell states. These off-diagonal matrix elements represent highly correlated virtually excited contributions to the He ground state, which cannot be described by He independent-particle shell-model states, i.e., the lowest virtually excited p contributions for this He ground state are not the 2p states of He but are the so-called pseudostates [4–6] in the field of a nucleus with a nuclear charge larger than two. The MCA He ground-state wave function is then represented by a very long list of numbers, which account for the strength of all diagonal and off-diagonal matrix elements. Since the He ground state is a $^1S_0$ state, the three-particle ground state can contain only strongly correlated s^2, p^2, etc. $^1S_0$ contributions, and the MCA wave function can be separated into such angular-momentum contributions. In MCA ground-state energy calculations the s^2 states contribute about 99% of the energy, the virtually excited p^2 about 1%, etc.

Since the early days of atomic physics the correlated momentum wave function of the He ground state remained as one of the unsolved fundamental puzzles in modern physics [7]. Until recently, there was no experimental way to directly access the correlated momentum wave function of both electrons in the He ground state. Electron-momentum spectroscopy measurements to final He^+ ion states with symmetries, such as np, nd, etc., not contained in the Hartree–Fock ground state, provide such a method. However, in He^+ the $n\ell$ ($\ell \neq 0$, $n \geq 2$) states for a given n cannot be resolved, and since the ns states ($n \neq 0$) are not completely orthogonal to the He ground-state 1s wave function excitation of these states is possible, even in the Hartree–Fock (s^2) approximation. Here n and ℓ denote the principal and orbital quantum numbers.

Cook et al. [8] used the technique of electron-momentum spectroscopy (e,2e measurements) to investigate the non-s components and found that the recoil-ion momentum distributions to the $n = 2$ states of He^+ were dominated by the correlated part of the He ground-state wave function. However, since they could not separate the 2p from the 2s states in their measurements, there was still a small s contribution to the cross section. Thus, although these highly correlated *virtually* excited states are usually believed to be merely mathematical constructs, which cannot be observed in experiments, suitable experiments can project them out. The direct observation of off-diagonal or non-s^2 contributions would reveal a scarcely explored, but very fundamental, part of the correlated momentum wave function of the He ground state. Other possible experiments that select the correlated part of the wave functions are photo double ionization (γ,2e) at high photon energies, when both electrons are emitted with similar high energy and can be treated as plane wave (see [9,10] and also [11] for a recent review).

Several groups have recently performed the so-called (e,3e) [12–17] experiments with He targets and have measured fully differential cross sections in momentum space. In (e,3e)-like experiments, where post-collision interaction is relatively small, indications of discrete structures in the final-state momentum pattern were found [12,14]. However, these experiments have been performed in the regime where one or more of the outgoing electrons has a very low energy, and therefore the reaction dynamics has been dominated by second-order and post-collision effects. In addition, they have no direct access to the initial-state momentum vector of one of the bound electrons. The relative importance of the non-s^2 contributions should increase towards larger distance from the nucleus and in the regime of high-momentum components of the wave function.

From the Schrödinger equation we learn that the ground-state wave function at large distance from the nucleus is the tunneling part. It is commonly believed that all bound electrons at large distance from the nucleus have to move slowly. This is true for Rydberg state electrons, but not for tunneling ground-state electrons. Here, electrons can still have high kinetic energy, since tunneling means no deceleration by the Coulomb potential $\propto 1/r$. The tunneling electron to absorb a virtual photon has for a very short time (see, e.g., self-energy process), to enable it to move fast to large distances. These non-s^2 contributions at large distance from the nucleus are important for the long-range properties of the He atom at low temperatures, in particular, to understand the weak van-der-Waals forces in the He ground state [15,16].

In the study presented here a particular channel of the transfer ionization (TI) process, the so-called correlated tunneling transfer ionization process (TuTI) in $p+He \rightarrow H^0 + He^{2+} + e^-$ collisions is chosen. This capture process is a Brinkman–Kramers-like process when electron and proton velocities match. The advantage of this experimental technique is that TuTI in the case of electron capture by a fast proton at very small deflection angles (regime of distant collisions), allows one a determination of the initial-state momentum vector of the captured electron and gives detailed information on the correlated momentum wave function of the three-particle He ground state. This projection technique is based on a sudden but gentle reaction with a small perturbation in momentum space, where both He electrons undergo correlated transitions from the almost unperturbed initial to the well-resolved final momentum states. In this reaction channel, the fast proton captures, by tunneling through the two-center Coulomb barrier, one electron of the He atom (named number 1) almost exclusively into its ground state, while the remaining He^+ ion is left in a virtually exited state from which it instantaneously fragments due to a shake-off process leaving electron 2 in a continuum state.

In fast collisions, the He atom is dissected essentially instantaneously (reaction time < 1 a.u. $\approx 2.4 \times 10^{-17}$ s and proton impact velocity $v_p \approx 2\text{–}8$ a.u., 1 a.u. $\approx 2.2 \times 10^6$ m/s). Since the force between the fast departing neutral H^0 and the remaining He^{1+} decreases rapidly, postcollision interaction between the projectile and the fragments is negligibly small. Thus, electron 2 is transferred from a virtually exited state to the real energy continuum in the final state with only a minor change of its initial momentum. Since the initial-state momentum of electron 1 can be determined from the final-state H^0 deflection angle (as shown below), the initial-state momentum correlation between electron 1 and 2 can be directly revealed from the final-state momentum distributions obtained for the TuTI reaction channel [18].

The cold target recoil ion momentum spectroscopy (COLTRIMS) technique, applied here, combines a very high momentum resolution with very high coincidence efficiency as is described in the reviews [19,20]. We will show that it provides the possibility of making quasi-snapshots of the correlated momentum wave function similar as using an ultrafast stroboscopic camera with a resolution of $< 10^{-17}$ s. With the COLTRIMS Reaction Microscope one can project the total wave function, but also tiny fractions (here less than one part in 10^8) of the total momentum wave function onto a special kinematical final state. Thus, one can probe details of the ground-state wave function not seen before.

The electron transfer to the projectile can proceed via different reaction channels:

- electron-electron-Thomas TI (eeTTI) [21–27],
- nucleus-electron-Thomas TI (neTTI) [28–30] and
- uncorrelated ionization of electron 2 and tunneling capture of electron 1 (n/eTuTI) (n/e indicates that the H^0 transverse momentum is due to the scattering at the nucleus resp. at the emitted electron).

While eeTTI always leads to a transfer ionization where the second electron is ejected, the capture of the n/eTuTI- and neTTI-process is accompanied by ionization of the second electron either by shake-off (SO) from mainly s^2 contributions or by an independent binary collision (S2) of the proton with the second electron. The shake-off process is highly dependent on the correlated part of the wave function, which is virtually excited to the continuum. The momentum distribution of the shake-off electron 2 and its shake-off probability depend strongly on the momentum state of electron 1. This effect on the shake-off probability is well established: if electron 1 is removed via Compton scattering, which averages over all momenta of electron 1, the shake-off probability is 0.8%. If, however, electron 1 is picked from the high momentum component of the initial state by high-energy photo absorption, the shake-off probability rises to 1.67% [31–37].

One would expect that if the initial momentum state of electron 1 is not defined (as, say, on the total cross sections) the momenta of shake-off electron 2, which is mainly the s-electron, should peak at zero momentum in the laboratory frame. If the momentum of electron 1 is large and well defined, the momentum of shake-off electron 2 may also be large and may even be well defined. Again, a related effect has been theoretically predicted for double ionization at very high photon energies, where a contribution from two fast electrons is expected [38]. If the fully differential cross sections for the TuTI process are measured, such transitions can indeed be interpreted as a shake-off process, where electrons 1 and 2 initially occupy a well defined entangled and virtually excited off-diagonal state. All TI processes such as eeTTI, neTTI, and n/eTuTI, followed by double scattering with the He nucleus, will lead to characteristic locations in the final-state momentum phase space. The maxima of the distributions of the longitudinal H^0 final-state momentum and, in particular, of the recoil momentum, provide a unique signature for the different TI channels, as will be detailed below.

In order to distinguish the different channels experimentally the projectile momentum transfer (the transverse and longitudinal component) has to be measured with an extremely high resolution of about 0.3 a.u. that corresponds to $\approx 10^{-5}$ of the projectile momentum, which can never be achieved with standard techniques, but can easily be achieved by the COLTRIMS technique, where in inverse kinematics the recoil momentum is detected. In Fig. 20.1 the expected TI peak locations of the recoil-momentum distribution projected onto the H^0 scattering plane are shown schematically for fast-proton impact. The separation of the recoil energy between the channels is only of the order of a few meV.

20.2 Experimental Technique

Using the high momentum resolution and high multi-coincidence efficiency of COLTRIMS, the complete final-state momentum distributions for fast, i.e.,

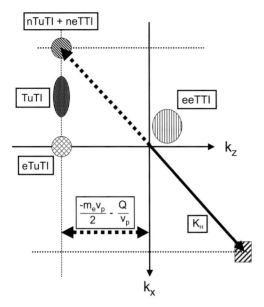

Fig. 20.1. Expected recoil-momentum locations for the different TI channels (neTTI, n/eTuTI, and eeTTI, for definitions see text) projected on the H^0 scattering plane. k_z defines the momentum components in the beam direction, k_x is the transverse momentum component in the direction of the scattered H^0. The *solid lines* represent $k_z = 0$, $k_x = 0$. k_H is the momentum change between initial and final state of the H particle. The *dashed lines* represent the positive and negative H transverse momenta and $-m_e v_p/2 - Q/v_p$ the He^{1+} recoil momentum in the lab frame for the pure electron capture into the H^0 ground state

with energy of 150 to 1400 keV collisions, transfer ionization processes (TI)

$$p + He \rightarrow H^0 + He^{2+} + e^-$$

have been systematically measured by Mergel [39] at the 2.5-MeV van-de-Graaf accelerator of the Institut für Kernphysik of the Universität Frankfurt.

The experimental setup is shown in Fig. 20.2 and is described in detail in [39]. The projectile beam was collimated to a diameter of < 0.5 mm and a divergence of < 0.25 mrad. The beam was charge-state selected in front and behind the target region by different sets of electrostatic deflector plates. A supersonic helium gas-jet is used as the target, as it combines the two most important features necessary for high-resolution recoil-ion spectroscopy: low internal temperature and localization of the target (diameter 5 mm). The helium gas is cooled to 14 K before it expands through a 30–μm nozzle into the source chamber. During the expansion the gas cools down to an internal temperature of < 50 mK. The gas-jet is formed by passing through a 0.7–mm skimmer, located 6 mm from the nozzle, resulting in a jet diameter of 5 mm at the intersection with the ion beam. A residual gas pressure without the

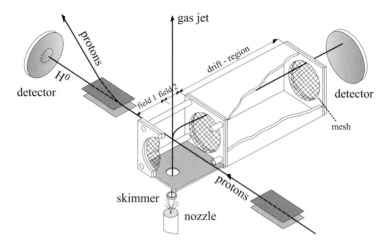

Fig. 20.2. Experimental setup of the COLTRIMS system with the H^0-recoil coincidence detection system

gas-jet of 1×10^{-8} mbar and a target density of 1.5×10^{12} were obtained. The recoil ions are extracted by a weak electric field of 9 V/cm transversely to the ion beam (see [19]). For this experiment we operated the spectrometer with a momentum resolution of 0.15 a.u. in favor of a higher target density.

The kinematics of the capture and TI reactions is described in detail in [19]. Here we repeat only the specific kinematical aspects of the two processes. In the TI process, the three free particles, diverging in the exit channel, have nine degrees of freedom. However, only five of them are independent, due to energy and momentum conservation. We have determined the polar and azimuthal scattering angle of the projectile in coincidence with the charge state and the 3-dimensional momentum vector of the recoiling target ion. Using the conservation laws, one obtains the 4-momentum components that have not been directly measured:[1]

$$k_{x,e2} = -k_{x,H} - k_{x,rec} \,, \tag{20.1}$$

$$k_{y,e2} = -k_{y,H} - k_{y,rec} \,, \tag{20.2}$$

$$k_{z,e2} = v_P \pm \sqrt{2(v_P^2 + Q + v_P k_{z,rec}) - k_{x,e2}^2 - k_{y,e2}^2} \,, \tag{20.3}$$

$$k_{z,H} = -k_{z,e2} - k_{z,rec} \,, \tag{20.4}$$

where $k_{x,rec}$, $k_{y,rec}$ and $k_{z,rec}$ are the momentum components of the He^{2+}-recoil ion, $k_{x,e2}$, $k_{y,e2}$, and $k_{z,e2}$ of the emitted electron and $k_{x,H}$, $k_{y,H}$ and $k_{z,H}$

---

[1] A discussion of the kinematics of recoil ion production can also be found in [19]. The laboratory coordinate system and atomic units (e= $\hbar = m_e = 1$) are assumed here, where e and m_e denote the electron charge and mass, respectively, and $\hbar$ the Planck constant.

the momentum transfer of the projectile including the transferred electron with respect to the laboratory frame. x, y denote the components perpendicular and z the component parallel to the incident projectile, v_p the proton velocity. The Q-value is equal to $Q = E_{He} - E_H$, where E_H denotes the binding energy of the hydrogen atom and $E_{He} = -2.9$ a.u. that of the helium ground state. Exploiting the rotational symmetry we rotate the coordinate frame around the z-axis from the laboratory frame into the scattering plane, defined by the incident and scattered projectile for each event. x and y denote the components parallel and perpendicular to the projectile scattering plane, where the projectile is scattered in the positive x-direction. This is equivalent to the definition of $k_{y,H} = 0$ for each event.

Equation (20.3) contains two unknown quantities:

- the final binding energy of the hydrogen bound state E_H that is included in the Q-value, and
- the sign of the square root.

Because the capture leads predominantly to the ground state of hydrogen [42] we use the value of $Q = -2.4$ a.u. In addition to that, the error in $k_{z,e2}(\Delta k_{z,e2})$ is of the order of $\Delta Q/v_P$, where the maximum error in Q is $\Delta Q = 0.5$ a.u. Thus, for the investigated projectile velocities this gives $0.07 < \Delta k_{z,e2} < 0.20$, which is in the range of the experimental resolution.

Concerning the sign in (20.3), we use the negative one for the calculation of $k_{z,e2}$, since only electrons with $k_{z,e2} > v_P$ correspond to the positive sign. The contribution of electrons with $k_{z,e2} > v_P$ is in the range of 1% to 3% of the total cross section [43], thus this approximation does not significantly affect the calculated electron distributions. The resolution of $k_{z,e2}$ due to calculation by (20.3) is $\Delta k_{z,e2} = \pm 0.2$ a.u. in the best case ($E_p = 0.15$ MeV), and $\Delta k_{z,e2} = \pm 0.3$ a.u. for the worst case ($E_p = 1.4$ MeV).

20.3   Experimental Results and Discussion of Observed Momentum Patterns

Our study of the TI process in p-He collisions was stimulated by the systematic work of Horsdal et al. [44] and Giese et al. [45] on TI processes for proton impact on He, who found a pronounced peak at about 6.5×10^{-4} rad in the H^0 scattering dependent ratio of TI to pure capture differential cross section. The peak maximum increased with projectile energy and reached about 25% at 1 MeV proton impact energy. Their observation contradicted all expectations, and was indeed very puzzling. Horsdal et al. explained their findings by a possible large contribution of eeTTI processes, whereas Olson et al. [46] as well as Gayet and Salin in their papers [43,47,48] showed (using CTMC and quantum-mechanical calculations in the independent-electron approximation) that multiple scattering might also produce such peak structures. Based on complete differential final-state momentum distributions, Mergel et al. [18] could clearly show that neither eeTTI nor multiple scattering is responsible for the observed peak in the cross section ratio, but could not present an explanation for this peak structure. As Mergel et al. [18] have shown, the main contributions to TI for the projectile velocities investigated here and thus also the puzzling structures observed by Horsdal et al. result from the highly correlated TuTI process.

The complete differential cross sections in momentum space of [18,39] show some even more puzzling features of the momentum patterns, namely:

1. Electron 2 is predominantly emitted into the backward and negative k_x direction, i.e. the emission of electron 2 with respect to the outgoing H^0 is completely asymmetric.
2. The He^{2+} ion momentum distribution and, therefore, also the electron 2 distribution peak in the H^0 scattering plane.
3. The ratio of TuTI to pure capture total cross sections increases with decreasing perturbation, i.e. increasing proton impact energy.
4. Electron 1, recoil He^{2+}, and electron 2 always share comparable momenta. In particular, none of these particles in the final state shows a momentum distribution peaking at zero momentum at the laboratory frame. According to theoretical predictions [40] and [41], the momentum of shake electron 2 should peak at zero and the recoil k_r momentum would be expected to peak near $k_r = (0, 0, -v/2)$.

As was shown in [18], these four features cannot be explained by noncorrelated particle dynamics of a proton interacting with a He nucleus and two uncorrelated s electrons. In particular, observation 2., i.e., the four-particle planar final-state motion, requires a strong four-particle correlation in angular momentum. This angular momentum must already be present in the initial He ground state, since the momentum transfer of the proton to the He is small and, thus, also the angular momentum transfer is small, i.e., $\ll 1$ a.u. This conclusion is supported by the observation that at high E_p electron 1 is nearly exclusively captured into the projectile 1s state and in the pure capture channel electron 2 is very rarely excited into any higher $He^+(n\ell)$ state. We will show below that the TuTI process for the proton-impact energies investigated here proceeds nearly exclusively via shake-off processes from correlated non-s^2 contributions to the He ground state. Thus, the fragmentation of the He^{1+} ion always occurs due to the angular momentum entanglement of the three He particles and not by an uncorrelated interaction of the proton with electron 2. From the final-state momentum pattern of H^0, He^{2+} and the electron 2 we can deduce and directly reveal the part of the initial momentum wave function that is dominated by non-s^2 contributions.

To compare our data with previous measurements we have to integrate our data over some degrees of freedom, since no fully differential cross sections for TI processes in momentum space have been reported previously. In Fig. 20.3, the total TI cross sections are plotted as a function of the proton-impact

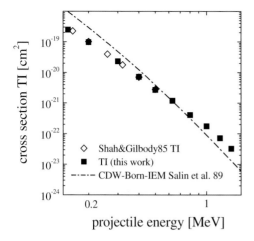

cross section TI [cm²]

projectile energy [MeV]

Fig. 20.3. Total TI cross sections as function of proton-impact energy

energy and compared with other published data [49]. The present data are
in very good agreement with the earlier results.

In Fig. 20.4, we show the single differential TI (right column) together
with the corresponding pure capture (left column) cross sections as a function of the H^0 transverse momentum k_{x,H^0} (i.e., the scattering angle θ_P) for
different proton impact energies from 150 to 1400 keV. Within the experimental uncertainty, these data agree with the results presented by Horsdal
et al. [50].

Plotting the ratio between TI and single capture cross sections as a function of the H^0 scattering angle (i.e., $k_{x,H^0}/k_0$) we find (see Fig. 20.5, open
squares) the same narrow peak structure at about 0.65 mrad as reported by
Giese and Horsdal [45] (solid circles). One observes in both experiments that
the peak ratio slightly increases with projectile energy.

In Fig. 20.6 the ratios of total cross sections between TI and the sum of
TI plus pure capture are plotted. Below $E_p = 600$ keV the ratio remains
constant at about 2.5% in good agreement with the earlier data of Shah et
al. [49]. Above $E_p = 600$ keV, however, the ratio increases linearly to about
4% for $E_p = 1.4$ MeV in agreement with recent data of [51]. Above 1400 keV
it starts to decrease again towards higher energies [51]. Such a proton energy
dependence for the total cross section ratio (smooth peak at about 1 MeV) is
indeed unexpected. If electron 2 has to be emitted by any uncorrelated two-
step process with the proton, we would expect a ratio similar to the result for
pure He double to single ionization (open triangles) [45,52]. In both papers
a ratio for the pure ionization channels far below 1% was found at 1 MeV,
which decreases with increasing proton energy, since the perturbation by the
proton decreases with increasing E_p. This comparison shows that the present
measured ratio is in clear contradiction to that reported for the uncorrelated

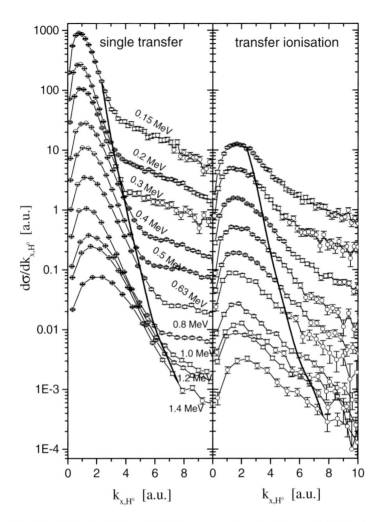

Fig. 20.4. Single differential TI cross sections $d\sigma/dk_{x,H^0}$ (*right column*) together with the corresponding pure capture cross sections (*left column*) as a function of the H^0 transverse momentum $k_{x,H}$ for different projectile energies

double-ionization processes. This supports our argument that the mechanism behind TuTI is mediated by strong electron correlation and not by two-step processes.

Considering the H^0 scattering-angle-dependent differential cross section for capture and TI (see Fig. 20.4), we see that both results show a large peak at very small scattering angles (below 0.6 mrad) and a smooth decrease of the cross sections above 1 mrad. This small-angle peak accounts for nearly all protons scattered by electrons of the He atom. This explanation has been

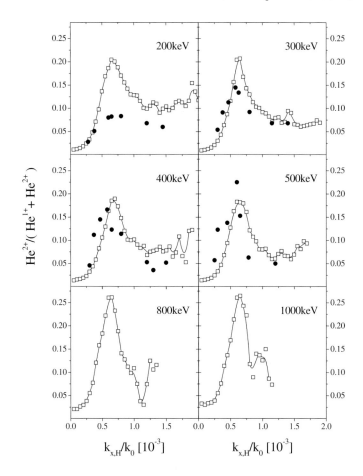

Fig. 20.5. H^0 scattering angle θ_P tatios of TI differential cross sections (this work, *open squares* and Giese and Horsdal, *solid circles*) to the sum of pure capture plus TI cross sections for different projectile energies. $\theta_P = k_{x,H}/k_0$, where $k_{x,H}$ is the H^0 transverse momentum and k_0 is the incoming projectile momentum

proven by calculations in which the nuclear-nuclear repulsion has been neglected [43,47,48]. As a result, the small scattering angle part of the capture cross section remains almost unchanged in the relative H^0 scattering angle dependence and the shape of each peak reflects the electron transverse velocity distribution for the given proton velocity v_p.

To present more evidence for the creation of the small-angle peak by proton scattering on the electrons, the data of DeHaven et al. [54] are presented in Fig. 20.7. The authors of [54] have investigated uncorrelated double-collision processes for the pure ionization channel of fast protons on He. In Fig. 20.7 the measured relation between the recoil transverse momentum $k_{x,\text{rec}}$ and the

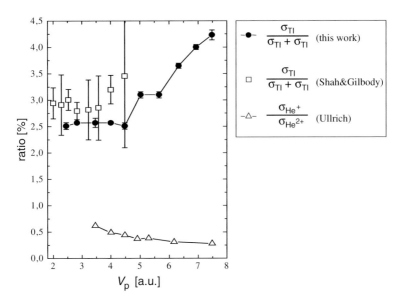

Fig. 20.6. Ratios of total TI cross section σ_{TI} to the sum of σ_{TI} and total single capture cross section σ_{SC}, *solid circles* and *open squares* (this work and [49]), ratio of total cross section of He^{2+} to He$^+$ production, *open triangles* (Kristensen et al. [53] and Ullrich et al. [52]), as a function of proton impact velocity v_p

projectile scattering angle θ_p is shown. For small angles they observe that the projectile transverse momentum $k_{x,p} = \theta_p m_p v_p$ results exclusively from close collisions with the electrons (vertical line at $k_{x,rec} = 0$). Due to the small ratio of electron-to-proton mass, the maximum angle of proton scattering by a free electron is 0.55 mrad. The nuclear momentum exchange (diagonal solid line) is for small θ_p much less probable but extends out to 180°. In the very small angle regime (the region of the peak) for most collisions the transverse nuclear-momentum exchange is below 0.2 a.u.. For Coulomb scattering at impact energies below 1 MeV this corresponds to impact parameters larger than the He K-shell radius. From the nuclear-transverse momentum exchange (relation between recoil and projectile) we thus obtain information on the nuclear-impact parameter range and indirectly also on the distance (close or distant) from the He nucleus where electron 1 is captured.

In the capture channel below 6×10^{-4} rad the H^0 transverse momentum is thus nearly exclusively determined by momentum transferred by the captured electron. For uncorrelated processes the probability that the proton shares comparable momentum with both the electron and the recoil ion is small (Fig. 20.7: the region between the two dashed-dotted lines). This behavior is well predicted by the small-angle multiple-scattering theory [55]. For the scattering of a proton from single atoms or thin solid targets even small

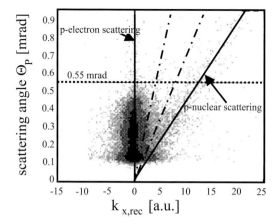

Fig. 20.7. Single ionization of He by 6 MeV proton impact He(p,p+e)He$^+$ measured by DeHaven et al. (adapted from [54]). Density plot of one component of the recoil transverse momentum $k_{x,\mathrm{rec}}$ vs. the projectile scattering angle θ_p. The *vertical solid line* represents H^0 scattering on the He electrons, where the recoil ion is a spectator and the diagonal solid line represents nuclear–nuclear transverse momentum exchange, when the electrons are spectators. The area between the *dashed-dotted lines* indicates the area where for the TuTI channel the measured recoil momenta are located

angle proton deflection is exclusively due to one close encounter with either an electron or a nucleus. The probability for the proton to be deflected by an angle θ_p on a single He atom by multiple scattering compared to single close-encounter scattering is always very small, since uncorrelated multiple scattering can go in any direction. Thus we can conclude that in uncorrelated processes equal momentum sharing of the proton with the emitted electron and the recoil is negligibly small.

For the discussion of the fully differential final-state momentum pattern of the TuTI, a coordinate system is defined where the z-axis is directed along the incoming projectile momentum $k_0 = m_\mathrm{p} v_\mathrm{p}$ and the H^0 projectile is always scattered into the positive x-direction. This coordinate system is obtained by rotating the laboratory frame around the z-axis so that the y-component of the projectile momentum $k_{y,\mathrm{H}}$ is always set to zero for each measured coincidence event. The solid lines in Figs. 20.8, 20.9 and 20.10 define the k_x and k_z zero positions.

The proton at large nuclear impact parameters would remain nearly non-deflected by the He nucleus, but the tunneling electron 1 carries some transverse momentum that has to be compensated by the outgoing H^0. When electron 1 approaches the proton with momentum $k_{x,\mathrm{e1}}$, the deflected H^0 must conserve this component, thus being deflected by $k_{x,\mathrm{H}} = k_{x,\mathrm{e1}}$. Since for kinematical capture the longitudinal momentum component of the initial-state electron velocity should match the projectile velocity (overlap with the

Compton profile of the 1s state of H^0), the initial-state momentum vector k_{el} of electron 1 can be approximately determined from the measured data by

$$k_{el} = \begin{pmatrix} k_{el,x} \\ k_{el,y} \\ k_{el,z} \end{pmatrix} = \begin{pmatrix} k_{x,H} \\ 0 \\ v_p \end{pmatrix}. \tag{20.5}$$

For the pure capture channel the maximum transverse momentum of H^0 due to scattering on the electron is therefore $k_{x,H} \approx m_e v_p$, in nearly perfect agreement with the data. For the TuTI process the proton is scattered from a correlated electron pair (a quasi-heavy boson) thus the peak regime of θ_P can extend to about 1 mrad, which is twice the angle of the maximum deflection by a single electron.

The recoil momentum in the beam direction $k_{z,rec}$ can be expected to be close to that for pure single capture. This momentum, which is given by energy and momentum conservation is

$$k_{z,He^+} = -m_e v_p/2 - Q/v_p, \tag{20.6}$$

where $Q = -2.9$ a.u. (see [19,39] for the kinematics). This k_{z,He^+} value is indicated by the dashed lines in Figs. 20.8 and 20.9. This equation can be interpreted in the following way: The He^+ recoil in the initial state has a $k_{z,rec}$ momentum of $-m_e v_p$ in order to balance the momentum of the forward directed electron 1. In respect to the laboratory frame, the electron 1 after being captured gains the kinetic energy $m_e v_p^2/2$. Furthermore, due to that process the electronic binding energy is changed by $-Q = E_H + E_{He^+} - E_{He}$. The energy $m_e v_p^2/2 + Q$ must be provided by the kinetic energy of the proton and therefore leads to an energy loss or gain of the projectile. In a tunneling process (virtual photon exchange) projectile and recoil must undergo in their center-of-mass a symmetric energy-gain or loss process to conserve energy and momentum. This yields for the z-component of the recoiling ion $k_{z,He^+} = -(m_e v_p^2/2 - Q)/v_p$ as given in formula 20.6 and for the projectile $k_{z,H^0} = +(m_e v_p^2/2 - Q)/v_p$. Thus, the total relative longitudinal momentum change between recoil and H^0 in the final state is $m_e v_p$. In a TTI process, however, the momentum exchange is a sequence of close binary Coulomb collisions and the recoil yields a completely different momentum pattern in the final state compared to TuTI processes.

After the capture of the first electron in the TuTI process the He^+ fragments into the nucleus and a free electron 2. The momentum projections presented in Figs. 20.8, 20.9, and 20.10 are obtained by a smoothing procedure of the measured statistical distributions. The shading represents the numbers of measured counts in a linear scale normalized to the maximum intensity. The absolute scale varies from figure to figure and can be obtained for each data set from Fig. 20.4, where absolute cross sections for each angle and impact energy E_p are given. Thus for small angles and for small E_p the count rate is highest.

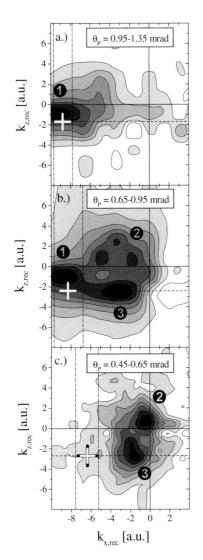

Fig. 20.8. Fully differential recoil ion TI cross sections projected on the H^0 scattering plane for selected projectile velocities and H^0 transverse momenta (**a**: 0.5 MeV, **b**: 0.8 MeV, **c**: 1 MeV). The *solid lines* represent $k_z = 0$, $k_x = 0$, the *dashed lines* the He^{1+} recoil momentum in the laboratory frame for the pure electron capture into the H^0 ground state, the areas between the *dashed-dotted lines* the window for the negative H^0 transverse momenta, the "+" cross the location calculated by the CTMC method for the channels neTTI and nTuTI. The locations of the observed peaks 1, 2, and 3 are discussed in the text

To prove the correctness of our momentum measurements, we first present data on those TI channels whose kinematics is well understood. In Fig. 20.8 data for large H^0 angles are shown, where transverse nuclear momentum exchange dominates. Only one recoil peak is seen and all recoil momenta are close to the location (+) predicted by CTMC calculations [21] for uncorrelated neTTI and n/eTuTI processes. The recoil momentum location agrees well with the expected location, but it is slightly shifted by about 1 a.u. in the forward direction k_z (longitudinal position). This small forward shift is expected, since the nuclear scattering is probably accompanied by an in-

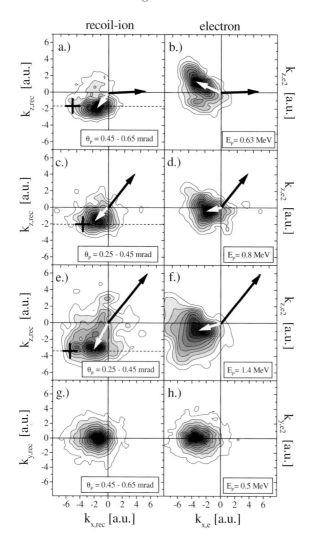

Fig. 20.9. Fully differential recoil ion (*left column*) and electron (*right column*) TI cross sections projected on the H^0 scattering plane for 630 keV impact energy at $\theta_P = 0.45$–0.65 mrad (**a,b**), in (**c,d**) for 800 keV $\theta_P = 0.25$–0.45 mrad, and in (**e,f**) for 1400 keV $\theta_P = 0.25$–0.45 mrad. The *solid lines* represent $k_z = k_x = 0$, the *dashed lines* the He^{1+} recoil momentum in the laboratory frame for the pure electron capture into the H^0 ground state. The *black vectors* indicate the mean electron 1 momentum vector in the initial state (20.7), the *white vectors* represent the mean measured recoil momenta, the "+" cross the expected location of neTTI and nTuTI channels. In (**g,h**) the differential recoil and electron cross sections perpendicular to the beam direction are shown

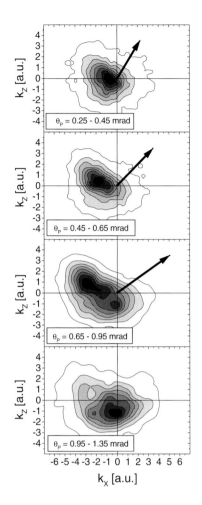

k_z [a.u.]

θ_P = 0.25 - 0.45 mrad

θ_P = 0.45 - 0.65 mrad

θ_P = 0.65 - 0.95 mrad

θ_P = 0.95 - 1.35 mrad

k_x [a.u.]

Fig. 20.10. Fully differential electron TI cross sections projected on the H^0 scattering plane for $300\,\text{keV}$ impact energy at different H^0 scattering angles (for explanation see Fig. 20.9)

dependent very small angle TuTI process, where the electron 2 is emitted slightly backward. The areas between the dashed-dotted lines represent the window corresponding to the negative H^0 transverse momenta. In Fig. 20.8 data for those H^0 angle are shown, where several TI channels can contribute. Indeed three peaks are seen, peak 1 represents the nTuTI and neTTI channel near the expected location (+), where again the H^0 transverse momentum results from nuclear scattering. Peak 2 represents the electron–electron Thomas channel (eeTTI), (see Mergel et al. [21]). Its measured kinematical location also agrees well with the expected values. For eeTTI the recoil ion is mainly a spectator and the recoil momentum location is expected at small positive k_z and small negative k_x, in agreement with our measurement. In Fig. 20.8c, the data for the small H^0 angular regime are shown, where the proton is mainly scattered by electron 1. Two peaks are seen, one at $k_x = -1\,\text{a.u.}$,

$k_z = +1$ a.u., which is the eeTTI channel [21]). The peak 3 in Fig. 20.8c, like the peak 3 in Fig. 20.8b, represents the TuTI channel. Their locations are contrary to the predictions for any known uncorrelated TI process, which should be located between the dashed-dotted lines.

Further results are shown in Fig. 20.9 where the corresponding fully differential recoil ion (left column) and electron (right column) cross sections projected on the H^0 scattering plane are presented. (In Fig. 20.9a and b for 630 keV impact energy at $\theta_p = 0.45$–0.65 mrad, in Fig. 20.9c and d for 800 keV at $\theta_p = 0.25$–$0,45$ mrad, and in Fig. 20.9e and f recoil and electron 2 momenta for 1400 keV at $\theta_p = 0.25$–$0,45$ mrad). Since for both the recoil ion and the electron one observes momentum distributions that are not rotationally symmetric with respect to k_{e1}, momentum exchange of the proton with the α nucleus and electron 2 must be correlated. It is obvious from Fig. 20.9 that the recoil and electron mean momenta have well-localized positions in the H^0 scattering plane. Furthermore both azimuthal momentum distributions peak always in the H^0 scattering plane. This is evident from Fig. 20.9g and h, where the corresponding azimuthal angular distributions of recoil and electron 2 momenta for a typical case of $E_p = 500$ keV are plotted.

In Fig. 20.10, the final-state momentum distributions of electron 2 are shown for 300 keV proton energy at different H^0 scattering angles. The black vectors in Figs. 20.9 and 20.10 indicate the momentum vector of electron 1 in its initial state calculated from (20.7), the white vectors indicate the corresponding recoil vectors (laboratory frame), respectively. From the discussions above we can conclude:

- The data presented here are reliable within the quoted error bars of about < 0.5 a.u.
- Besides the well-known TI channels we observe a TI channel, called TuTI, whose kinematics cannot be explained by any previously known TI mechanism.

20.4   Shake-Off Process From Non-s² Contributions

Before we discuss the measured fully differential momentum distributions, the shake-off ratios, i.e., ratios between TI and single-capture differential cross sections as function of the H^0 scattering angle, given in Fig. 20.5, will be compared with the theory presented in [40].

As in the standard shake-off theory [41], we estimate the probability of the cKTI process as a double overlap integral:

$$\langle k_1 k_2 | \Phi_0 \rangle = \sum_{nl} C^{00}_{lm,l-m} \langle k_1 | nlm \rangle \langle k_2 | nl - m \rangle , \tag{20.7}$$

where m denotes the projection of the orbital momentum ℓ. Here we make a multi-configuration Hartree–Fock expansion of the wave function of the

He atom ground state. Configuration interaction coefficients decrease rapidly with increasing n, l, the leading terms being $A_{1s} = 0.996$, $A_{2s} = -0.059$, $A_{2p} = 0.059$, $A_{3d} = -0.012$. The Clebsch–Gordan coefficients couple the two individual electron angular momenta to the zero angular momentum of the He atom. In the first overlap integral, we assume that the electron is picked up by the proton at a finite distance from the He nucleus:

$$\langle \boldsymbol{k_1} | nlm \rangle = C_{lm} \int_{b>0}^{\infty} \mathrm{d}x \, \mathrm{e}^{k_{1x}x} \int_{-\infty}^{\infty} \mathrm{d}z \, \mathrm{e}^{k_{1z}z} \, R_{nl}(r) \, \mathrm{e}^{im\phi}, \tag{20.8}$$

where b is the impact parameter. Here we also choose the angular momentum quantization axis in the y-direction and write the electron wave function in the scattering plane as

$$\Psi_{nlm}(r) = R_{nl}(r) \, Y_{lm}(\theta = \pi/2, \phi) = C_{lm} R_{nl}(r) \, \mathrm{e}^{im\phi}, \quad \tan \phi = x/z.$$

In the second overlap $\langle \boldsymbol{k_2} | nl - m \rangle$, the integration is expanded over the whole scattering plane and the final state $\langle \boldsymbol{k_2} |$ is treated as the Coulomb wave in the He^{2+} field.

In the standard shake-off theory, the x integration in (20.8) is expanded over the whole scattering plane and the integral becomes symmetric with respect to the sign reversal of m. In the TuTI theory there is a very large asymmetry between $\pm m$ components $\langle \boldsymbol{k_1} | nlm \rangle / \langle \boldsymbol{k_1} | nl - m \rangle \propto k_{1z} b \propto 1$. This asymmetry can be understood if one remembers that the departing electron carries away the classical angular momentum $k_{1z} b$ and the projection of this momentum on the quantization axis favors only one particular sign of m. The large angular momentum $k_{1z} b \gg 1$ has to be drawn from a ground-state orbital with a limited l, m. This makes the overlap integral exponentially small $\langle \boldsymbol{k_1} | nlm \rangle \propto \exp(-k_{1z} b)$. This smallness is offset by a growing power term $(\beta b)^1$ where β is the exponential fall-off parameter of the radial orbital $R_{nl}(r)$ (see [40] for more details). The power term compensates the small coefficients A_{nl} for $l > 0$. As a result, the strongest contribution to the amplitude equation (20.8) comes from the 2p$_{+1}$ and 3d$_{+2}$ terms but not the 1s one.

In Fig. 20.11 (right column) the experimental ratios for 500 and 1000 keV proton impact energy are shown as a function of the measured H^0 scattering angles ($\theta_p = k_{x,\text{H}^0}/k_0$, unit millirad). In the left column the theoretical predictions are presented (dashed line: only s^2 contributions, solid line: including non-s^2 contributions) as a function of the inverse impact parameter, which for pure nuclear scattering is proportional to the transverse momentum. The abcissa of both figures can only be qualitatively compared, since in the experimental data above 1.3 mrad the H^0 deflection is due to Rutherford scattering of both nuclei, (thus this regime corresponds to a small impact parameter ≈ 0.1 of the K-shell radius) and below 1 mrad the H^0 is scattered on the electrons only (thus the nuclear-impact parameter should be large (> 1 a.u.)). The striking difference in the calculations for pure s^2 and non-s^2 contributions proves that the puzzling peak first observed by Giese et al.

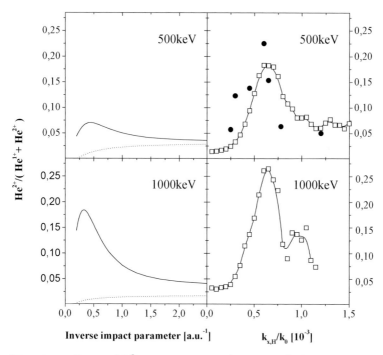

Fig. 20.11. Ratio of H^0 scattering angle ($\theta_p = k_{x,H}/k_0$) TI differential cross sections (this work, *open squares* and Giese and Horsdal, *solid circles*) to the sum of pure capture plus TI cross sections for different projectile energies. (*right column:* experiment, *left column:* theory (see text))

[45] can be related to capture and subsequent shake-off of paired non-s^2 electrons. The theory even gives nice tentative agreement in the absolute height. This indicates that the TuTI process indeed probes the non-s^2 contributions of the ground-state He momentum wave function. It is, however, not possible to reveal the details of the correlated non-s^2 wave function from the measured fully differential cross sections in comparison with theory. We find that the present calculations [40,41] can only partially describe the observed momentum pattern.

Based on a proton straight line trajectory the calculations predict four characteristic features:

1. Only $m = +1$ contributions can be captured into the fast-moving proton. Thus, electron 2 is very asymmetrically emitted (only opposite to the deflected proton towards negative k_x),
2. Momenta of recoil and electron 2 are coplanar in the H^0 scattering plane,
3. At large impact parameters for the impact energies investigated here the non-s^2 contributions to the cKTI process dominate, and
4. For both s^2 and non-s^2 components the emitted electron 2 momentum peaks near zero momentum in the laboratory frame.

Predictions 1., 2., and 3. are in agreement with the data:

- The measured final-state momentum distributions in the nuclear H^0 scattering plane (the nuclear angular momentum vector) are strongly asymmetric below 0.6 mrad, i.e., they show an orientation with respect to the deflected H^0.
- Electron 2 and recoil are coplanar in the H^0 scattering plane.
- The TuTI contribution yields more than 85% of the total TI cross section.

But

- the data are in clear contradiction with the theory, which predicts electrons with low shake-off energy (prediction 4.).

Experimentally we find that TuTI hardly yields electron momenta peaking near zero momentum. As seen in Figs. 20.9 and 20.10, the energy of electron 2 even increases with increasing impact energy and with increasing H^0 angle (below 0.6 mrad). The present theory can not explain why the shake-off electron 2 kinetic energy is of such a large magnitude and, in many cases, exceeding 200 eV.

We note that the non-s^2 angular momentum is not transferred from the proton to any electron, but is provided by the initial He ground state. This conclusion is supported by experimental and theoretical investigations of the pure electron capture process of fast protons on He [43,56]. These authors show that the internal electronic excitation, i.e., the excitation of electron 2 into the p-state of He and the capture of electron 1 into any excited H^0 state is negligibly small for the fast collision systems investigated here. Therefore, the required angular momentum transfer can only be provided from initial-state properties of the captured electron. If the electron is initially in an entangled p^2 or d^2 state the electron 1 can indeed provide the required angular momentum. Since the two electrons have to couple to an $^1S_0$ state, the angular momentum of electron 2 must be antiparallel to that of electron 1 at all times. In classical terms this p or d (or higher ℓ) electron angular momentum $\boldsymbol{l}_{e1}$ is pointing perpendicular to the initial plane of motion of the first electron and the motion of the second electron (and α-nucleus) is then confined to the same plane. Since this p or d electron is merged into the H^0 during the capture process, the H^0 must absorb $\boldsymbol{l}_{e1}$ and its deflection (scattering plane) must be perpendicular to $\boldsymbol{l}_{e1}$. Thus, classically, TuTI can only occur if the H^0 scattering plane and the He initial plane of motion are parallel, as observed in our experiments. A TuTI process proceeding via p^2 electrons (with negligibly small momentum and angular momentum exchange between proton and He) could thus indeed explain the observation of a 4-body (p+e1+e2+α-nucleus) coplanar fragmentation.

To further understand the physics behind this TuTI process, we semiempirically reduce the complex pattern to obtain a simpler scaling behavior. To do this, we calculate from the final-state H^0 momentum the initial-state electron 1 momentum. The vectors (black line) in Figs. 20.9 and 20.10 indicate the mean locations of the initial-state momentum vector (given by 20.2) of the captured electron 1, the vectors (white lines) indicate the mean location of the emitted electron 2 (right column) and the mean location of the emitted α-nucleus in the laboratory frame (left column), respectively.

It should be noted that the vector sum in the laboratory frame of the momenta of all three particles is not zero, since the proton also changes its k_z momentum component ($k_{z,p} = -m_e v_p/2$). The initial-state momentum relation between the three He particles derived from our data is presented in Fig. 20.12, where for nearly all investigated impact energies E_p at three H^0 scattering-angle regimes the measured momentum relation in the CM system is shown (for the smaller H^0 range the energies 150, 200, and 300 keV are not included, since the H^0 transverse momentum resolution was comparable with the measured deflection). It is striking to see that for the TuTI process one always yields more or less the same discrete momentum pattern between the two electrons and the recoil ion, whereas the mean momentum of electron 2 with respect to the projectile momentum in the final state varies strongly with projectile energy and H^0 deflection angle.

In Fig. 20.12, the initial-state vector k_{e1} is plotted. Its length for the different $E_p(m_e v_p)$ is set equal to one. For all systems investigated the relative angle θ_{e1-e2} between electron 1 and 2 appears constant with $\theta_{12} = 140° \pm 25°$ and the angle between electron 2 and the recoil ion is $\theta_{e2} - \theta_{recoil} = 70° \pm 25°$. Also the ratios of the momentum vector magnitudes are constant within the experimental uncertainty of about 30%.

20.5   Conclusions

We conclude that the puzzling structures observed by Mergel et al. [18] and Giese et al. [45] can be qualitatively explained by the TuTI process proceeding via selected shake-off processes from non-s^2 components in the asymptotic part of the He ground-state wave function. Several experimental observations can be qualitatively explained by the theory:

1. the puzzling peak in the angle-dependent ratio of TI to sum of TI+ capture
2. the observed asymmetry in electron 2 emission, and
3. the coplanar emission pattern of recoil electron 2 and scattered H^0.

However, the large electron 2 momenta and the striking general scaling of the momentum distributions shown in Fig. 20.12 are in clear contradiction to theory. It is interesting to note that in classical mechanics such a scaling was predicted for the He ground state by [57]. When the two He electrons move on two opposite (180°) elliptic orbits with the nucleus at rest, they can never fulfill simultaneously momentum and angular momentum conservation. They need a nucleus for compensation of momentum (strong phonon coupling), which is then more easily fulfilled, if the axes of the electron ellipses are not

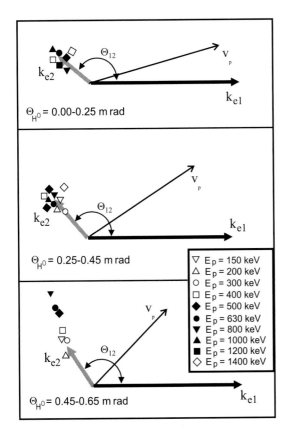

Fig. 20.12. The measured *initial state* momentum relation ($\Longrightarrow$ correlated asymptotic momentum wave function) of the two electrons is shown for three H^0 scattering-angle regimes for all the indicated proton energies. The momentum vector k_{e1} of the captured electron 1 is always set to $(1,0)$ for all proton impact energies E_p. Its determination is described in the text (see (20.2))

intersecting by $180°$ but, as suggested by Sommerfeld (1923) [57], by a smaller angle between $90°$ and $150°$.

Experimentally we find: electron 2 is always emitted into polar angles of about $140° \pm 25°$ with respect to electron 1 with a well-defined relative velocity, i.e., the entangled three-particle momentum wave function shows a semi-quantized structure. Furthermore, we have shown that the non-s^2 contributions in the He ground-state wave function are not purely mathematical constructs in the virtually exited space, but have measurable consequences. These off-diagonal non-s^2 components seem to hide interesting properties with respect to the secret world of correlation. These states have classically seen a huge amount of kinetic energy, thus they are called highly virtually excited continuum states. These very fast electrons at large distance from the He nucleus are those with the strongest dynamical e–e correlations.

Classically seen, here the negative Coulombic energy is more than a factor of 10 smaller than the positive kinetic energy of the fast electrons in a non-s^2 state. Since the kinematical capture observed here can only occur when the electron 1 velocity (absolute value) matches the projectile velocity, this TuTI process at a given v_p sets a narrow window on the captured electron velocity and therefore provides a powerful method for viewing selectively very high momentum components ($< 10^{-6}$ fractions of the global wave function) in the He ground state, which are not observable in even the most precise binding energy measurements. Both electrons occupy the non-s^2 state of motion together simultaneously with the nucleus, since the He ground state is a $^1S_0$ state.

Generally one would call such a two-electron system a pairing state (e.g., like a Cooper pair in a solid), however, this is misleading and overlooks the most important reason for that entanglement. It is the coupling of both electron momenta and angular momenta to the nuclear motion (the nucleus is never at rest). It is well known for superconductivity that phonon coupling to the solid (isotope effects) is very important. We see here for the He system that beside entanglement in momentum ($\Longrightarrow$ phonon coupling) the angular momentum entanglement is even more important. Therefore, also for superconductivity and the quantized Hall effect (in particularly, the fractional Hall effect) angular momentum entanglement might be crucial for the existence of such dynamically entangled systems.

Acknowledgements

We acknowledge the numerous fruitful disputes and discussions with our colleagues and friends: U. Ancarani, D. Belkić, J. Berakdar, S. Berry, J. Briggs, S. Brodzki, H. Cederquist, R. Dreizler, J. Feagin, B. Fricke, S. Fritsche, C. Greene, W. Greiner, J. McGuire, S. Hagmann, M. Horbatsch, O. Jagutzki, E. Horsdal, P. Hoyer, D. Ionescu, D. Madison, J. Macek, S. Manson, R. Moshammer, C.D. Lin, R. Olson, Yu. Popov, J. Reading, R. Rivarola, J.M. Rost, H. Schmidt, V. Schmidt, R. Schuch, J. Ullrich, and many other colleagues.

V.M. acknowledges the support by the Studienstiftung des Deutschen Volkes, and E.W. by the Humboldt-Stiftung. This work was supported by the DFG, the BMBF, GSI-Darmstadt, Graduiertenprogramm des Landes Hessen, DAAD and Roentdek GmbH.

References

1. T. Kinoshita: Phys. Rev. **115**, 366 (1959)
2. G. W. F. Drake: in *Long Range Casimir Forces: Theory and Recent Experiments on Atomic Systems*. Eds. F.S. Levin, D.A. Micha (Plenum, New York 1994) p. 107
3. S. D. Bergeson et al.: Phys. Rev. Lett. **80**, 3475 (1998)
4. A. T. Stelbovics: Invited paper, Proceedings of ICPEAC 1991, Eds. W. R. MacGillivray, I. E. McCarthy, M. C. Standage (IOP, Bristol) p. 21, SSBN 0-7503-0167-8
5. W. C. Fon, K. A. Berrington, P. G. Burke, A. E. Kingston: J. Phys. B **14**, 1041 (1981)
6. T. T. Scholz, H. R. J. Walters, P. G. Burke, M. P. Scott: J. Phys. B **24**, 2097 (1991)
7. G. Tanner, K. Richter, J. M. Rost: Rev. Mod. Phys. **72**, 497 (2000)
8. J. P. D. Cook, I. E. McCarthy, A. T. Stelbovics, E. Weigold: J. Phys. B **21**, 2415 (1984)
9. Y. F. Smirnov, A. V. Pavlitchenkov, V. G. Levin, V. G. Neudatschin: J. Phys. B **11**, 1587 (1978)
10. V. G. Levin, V. G. Neudatchin, A. V. Pavlitchankov, Y. F. Smirnov: J. Phys. B **17**, 1525 (1984)
11. J. S. Briggs, V. Schmidt: J. Phys. B **33**, R1 (2000)
12. A. Dorn, R. Moshammer, C. D. Schröter, T. J. M. Zouros, W. Schmitt, H. Kollmus, R. Mann, J. Ullrich: Phys. Rev. Lett. **82**, 2496 (1999)
13. A. Dorn, A. Kheifets, C.D. Schröter, B. Najjari, C. Höhr, R. Moshammer, J. Ullrich: Phys. Rev. Lett. **86**, 3755 (2001)
14. I. Taouil, A. Lahmam-Bennani, A. Duguet, L. Avaldi: Phys. Rev. Lett. **81**, 4600 (1998)
15. J. Macek: Z. Phys. D **3**, 31 (1986)
16. V.M. Efimov: Comments Nucl. Part. Phys. **19**, 271 (1990)
17. A. Lahmam-Bennani, C. C. Jia, A. Duguet, I. Avaldi: J. Phys. B **35** L215 (2002)
18. V. Mergel, R. Dörner, K. Khayyat, M. Achler, T. Weber, O. Jagutzki, H.J. Lüdde, C.L. Cocke, H. Schmidt-Böcking: Phys. Rev. Lett. **86**, 2257 (2001)
19. R. Dörner, V. Mergel, O. Jagutzki, L. Spielberger, J. Ullrich, R. Moshammer, H. Schmidt-Böcking: Phys. Rep. **330**, 96 (2000)
20. J. Ullrich, R. Moshammer, R. Dörner, O. Jagutzki, V. Mergel, H. Schmidt-Böcking and L. Spielberger: J. Phys. B: At. Mol. Opt. Phys. **30**, 2917 (1997)
21. V. Mergel, R. Dörner, M. Achler, K. Khayyat, S. Lencinas, J. Euler, O. Jagutzki, S. Nüttgens, M. Unverzagt, L. Spielberger, W. Wu, R. Ali, J. Ullrich, H. Cederquist, A. Salin, R.E. Olson, D. Belkic, C.L. Cocke, H. Schmidt-Böcking: Phys. Rev. Lett. **79**, 387 (1997)
22. J. Palinkas, R. Schuch, H. Cederquist, O. Gustafsson: Phys. Rev. Lett. **63**, 2464 (1989)
23. J. S. Briggs, K. Taulbjerg: J. Phys. B **12**, 2565 (1979)
24. T. Ishihara, J.H. McGuire: Phys. Rev. A **38**, 3310 (1988)
25. J. H. McGuire, J.C. Straton, W. J. Axmann, T. Ishihara, E. Horsdal: Phys. Rev. Lett. **62**, 2933 (1989)
26. J. H. McGuire, N. Berrah, R. J. Bartlett et al.: J. Phys. B **28**, 913 (1995)

27. R. Shakeshaft, L. Spruch: Rev. Mod. Phys. **51**, 369 (1979)
28. L. H. Thomas: Proc. Roy. Soc. A **114**, 561 (1927)
29. E. Horsdal-Pedersen, C. L. Cocke, M. Stöckli: Phys. Rev. Lett. **50**, 1910 (1983)
30. H. Vogt, R. Schuch, E. Justiniano, M. Schulz, W. Schwab: Phys. Rev. Lett. **57**, 2256 (1986)
31. J. C. Levin, D. W. Lindle, N. Keller, R. D. Miller, Y. Azuma, N. Berrah Mansour, H. G. Berry, I.A. Sellin: Phys. Rev. Lett. **67**, 968 (1991)
32. L. Spielberger, O. Jagutzki, R. Dörner, J. Ullrich, U. Meyer, V. Mergel, M. Unverzagt, M. Damrau, T. Vogt, I. Ali, K. Khayyat, D. Bahr, H. G. Schmidt, R. Frahm, H. Schmidt-Böcking: Phys. Rev. Lett. **74**, 4615 (1995)
33. L. Spielberger, O. Jagutzki, B. Krässig, U. Meyer, K. Khayyat, V. Mergel, T. Tschentscher, T. Buslaps, H. Bräuning, R. Dörner, T. Vogt, M. Achler, J. Ullrich, D.S. Gemmell, H. Schmidt-Böcking: Phys. Rev. Lett. **76**, 4685 (1996)
34. L. Spielberger, H. Bräuning, A. Muthig, J.Z. Tang, J. Wang, Y. Qui, R. Dörner, O. Jagutzki, T. Tschentscher, V. Honkimäki, V. Mergel, M. Achler, T. Weber, K. Khayyat, J. Burgdörfer, J. McGuire, H. Schmidt-Böcking: Phys. Rev. A **59**, 371 (1999)
35. J. C. Levin, G. B. Armen, I. A. Sellin: Phys. Rev. Lett. **76**, 1220 (1996)
36. F. W. Byron, C.J. Joachain: Phys. Rev. A **164**, 1 (1967)
37. L. R. Andersson, J. Burgdörfer: Phys. Rev. Lett. **71**, 50 (1993)
38. E. G. Drukarev, N.B. Avdonina, R.H. Pratt: J. Phys. B **34**, 1 (2001)
39. V. Mergel: PhD Thesis, Universität Frankfurt, Shaker Verlag, ISBN 3-8265-2067-X (1996)
40. A. S. Kheifets: private communication, 2003 (to be published)
41. T. Y. Shi, C. D. Lin: Phys. Rev. Lett. **89**, 163202 (2002)
42. R. Gayet, A. Salin: J. Phys. B **20**, L571 (1987)
43. R. Gayet, A. Salin: Nucl. Instrum. Methods B **56**, 82 (1991)
44. E. Horsdal, B. Jensen, K.O. Nielsen: Phys. Rev. Lett. **57**, 1414 (1986)
45. J. P. Giese, E. Horsdal: Phys. Rev. Lett. **60**, 2018 (1988)
46. R. E. Olson, J. Ullrich, R. Dörner, H. Schmidt-Böcking: Phys. Rev. A **40**, 2843 (1989)
47. R. Gayet: J. de Phys., **C1-53**, 70 (1989)
48. R. Gayet, A. Salin: in *High-Energy Ion-Atom-Collisions* Proc. of the 4th Workschop, Debrecen Sept. 1990, eds. D. Bereny and G. Hock (Lecture Notes in Physics; Vol. 376) (Springer, Berlin, Heidelberg, New York 1991) (1990)
49. M. B. Shah, H.B. Gilbody: J. Phys. B **18**, 899 (1985)
50. E. Horsdal, B. Jensen, K.O. Nielsen: Phys. Rev. Lett. **57**, 675 (1986)
51. H. T. Schmidt et al.: Phys. Rev. Lett. **89**, 163201 (2002)
52. J. Ullrich, R. Moshammer, H. Berg, R. Mann, H. Tawara, R. Dörner, J. Euler, H. Schmidt-Böcking, S. Hagmann, C.L. Cocke, M. Unverzagt, S. Lencinas, V. Mergel: Phys. Rev. Lett. **71**, 1697 (1993)
53. F. G. Kristensen and E. Horsdal-Pedersen: J. Phys. B **23**, 4129 (1990)
54. W.R. DeHaven, C. Dilley, A. Landers, E.Y. Kamber: Phys. Rev. A **57**, 292 (1998)
55. P. Sigmund, K.B. Winterbon: Nucl. Instrum. Methods **119**, 541 (1974)
56. V. Mergel: Diplomarbeit Universität Frankfurt (1994)
57. A. Sommerfeld: J. Opt. Soc. Am. **7**, 509 (1923)

21 Single and Multiple Ionization in Strong Ion-Induced Fields

J. Ullrich

21.1 Introduction

This Chapter summarizes a wealth of recent results, obtained within less than a decade, on correlated few-particle quantum dynamics explored in the fragmentation of atoms being exposed to strong (10^{15}–10^{21} W/cm^2), attosecond (10^{-18} s) electromagnetic half-cycle pulses that are generated in collisions with fast highly charged ions as illustrated in Fig. 21.1. With the development of "Reaction Microscopes" (see Chap. 2.1 and Fig. 2.8 in this book) in 1994 [1], a breakthrough in the experimental investigation of atomic fragmentation has been achieved, allowing for the first time access to the final-state vector-momenta of several electrons and of the recoiling target ion with high resolution and, to a large extent, independent of their relative energies and emission directions. In other words, these techniques enable simultaneous mapping of a major part, typically more than 80%, of the complete multi-dimensional (9D for single ionization) final-state momentum space, thus, visualizing the square of the many-particle final-state wave function, as will be illustrated below.

Hence, since then, basic questions on collision-induced single- and multiple-ionization dynamics, on the correlation between emitted electrons, on momentum balances between light and heavy fragments, i.e., between electrons and ions (target and projectile), on energy- and momentum-transfer mechanisms, on the connection to ionization by single photons as well as by strong laser fields have been explored in unprecedented detail and completeness. This Chapter exclusively reviews the present status on multiple differential experimental and theoretical results obtained with Reaction Microscopes in fast heavy-ion collisions. Readers interested in total cross sections, single or double-differential data measured with previous, less accurate versions of recoil-ion momentum spectrometers [2,3], in pure electron-emission cross sections [4], in more general articles on "recoil-ion momentum spectroscopy" (RIMS) [5,6] or cold-target RIMS (COLTRIMS) [7] have to be referred to the cited literature. The connection between atomic fragmentation dynamics in ion-induced attosecond pulses and in strong, femtosecond laser fields has recently been put forward by Ullrich and Voitkiv [8].

The basic interaction mechanisms between the field and the target atom will be described in some detail in Sect. 21.2, along with a selection of illustrative results for single ionization and the three-particle momentum balance. In Sect. 21.3 we present recent results for double ionization in strong fields followed by a discussion on multiple ionization processes in Sect. 21.4, addressing many-particle momentum balances and correlation between the emitted electrons. Some future developments are sketched in Sect. 21.5. Due to the short scope of the contribution and the explosive-like expansion of the field, only the main lines can be sketched and illustrative pictures are developed, sometimes at the expense of a rigorous theoretical treatment, for which the reader is referred to the literature.

21.2   Interaction of Ion-Generated Strong Fields with Atoms and Single Ionization

21.2.1   Ion-Generated Fields

It was recognized more than 20 years ago from satellites in high-resolution X-ray spectroscopy, that a target can be multiply ionized in a collision with a fast highly charged projectile. Subsequently, total cross sections for multiple target ionization have been explored in great detail for projectile (ionic) charge states Z_P up to 92 and for velocities v_P reaching the relativistic regime (1 GeV/u) where the typical collision time τ is on a sub-attosecond time scale. Target-ion charge states of up to fully stripped Ar^{18+} or Xe^{32+} have been observed to be produced with huge cross sections ($\sim 10^{-18}$ cm^2) in collisions with U^{75} at $v_P = 0.18c$ ($c = 137$ a.u.: speed of light) [9,10].

In order to obtain a physical picture as to what might happen in such a collision and come to some intuitive insight into the collision dynamics, we have calculated the transverse electric field (x-direction) in Fig. 21.1 for U^{92+} impinging on helium at a (relativistic) velocity of 120 a.u. and an impact parameter b of two atomic units, which is typical for double ionization. Here, at a Lorentz factor of $\gamma = (1 - \beta)^{-1/2} = 2$, with $\beta = v_P/c$, the intensity of the longitudinal field component (z-direction) is relativistically suppressed already by a factor of $\gamma^2 = 4$. In any case, it changes sign during the collision, resulting in a net force of zero and, therefore, is not important in most even non-relativistic situations [11]. Hence, as for a light pulse or for single photons, the direction of the electric field and, thus, the effective force during the collision mainly occurs transverse to the ion propagation. When the projectile approaches the target, the transverse field strongly rises to a peak value of about $\gamma Z_P/b^2 \sim 50$ a.u. and falls off again with a full width half maximum of $\tau = 0.2$ as, with $\tau \sim b/(\gamma v_P)$. The power density I, the atom is exposed to during this very short time, is close to 10^{20} W/cm^2.

Depending on the impact parameter b and the relativistic factor γ and the charge Z_P, it is easily seen that power densities between some 10^{13} to 10^{23} W/cm^2 can be realized in ion-atom collisions at typical collision times, i.e., full widths at half maxima (FWHM) of the electromagnetic pulses, between 1 and 10^{-4} a.u, or, in other words, between ten attoseconds and a

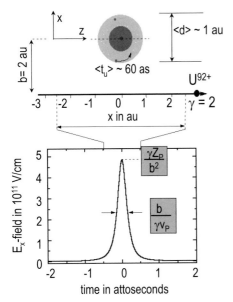

Fig. 21.1. Transverse (x-) component of the time-dependent electromagnetic field generated by a 1 GeV/u U^{92+} ion ($\gamma = 2$; $v_P = 0.87c = 120$ a.u.) at the position of a helium target atom, while passing it at an impact parameter of 2 a.u., outside the helium electron cloud with a diameter of about $< d > \sim 1$ a.u., $< t_u >$: classical average revolution time of the electrons in helium

few zeptoseconds in moderate relativistic encounters of $\gamma = 100$. As illustrated in Fig. 21.2a and detailed in Sect. 21.4, there is a momentum transfer $\boldsymbol{q} = \boldsymbol{P}_P^i - \boldsymbol{P}_P^f$, pointing essentially into the transverse x-direction since the longitudinal z-component, being equal to the minimum momentum transfer (see Chap. 1) at a given energy transfer $q_{\parallel} = q_{min} = \Delta E / v_P$, is negligibly small in most cases at large velocities v_P.

21.2.2   Single Ionization at Small Perturbations

Single ionization at small perturbations $Z_P / v_P < 1$ was explored in great detail, experimentally as well as theoretically over more than three decades for electron impact (for the first experiment see [12]) in kinematically complete investigations (for recent reviews see [13–15]). Typical power densities of the electromagnetic field in such a situation are of the order of some 10^{13} to 10^{15} W/cm^2 and the first Born approximation (FBA) has been successfully applied to theoretically describe the experimental data at large velocities with $Z_P / v_P \ll 1$. Measurements have nearly exclusively been performed in a "coplanar" geometry, where the ionized target electron is detected in the scattering plane of the fast electron. In the FBA, all cross sections scale with Z_P^2, and, hence, no differences are expected for positively charged ion impact,

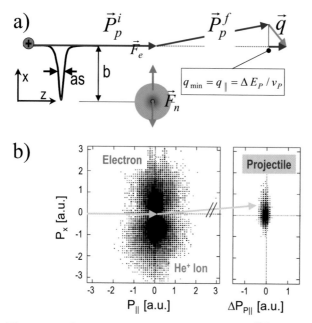

Fig. 21.2. Collision dynamics for $1\,\mathrm{GeV/u}$ U^{92+} ion impact on an atom. (**a**): Schematic illustration. $\boldsymbol{P}_{\mathrm{P}}^{i}$, $\boldsymbol{P}_{\mathrm{P}}^{f}$: incoming and scattered ion momentum, respectively, with momentum transfer $\boldsymbol{q} = \boldsymbol{P}_{\mathrm{P}}^{i} - \boldsymbol{P}_{\mathrm{P}}^{f}$ and minimum momentum transfer $q_{\mathrm{min}} = q_{\parallel} = \Delta E_{\mathrm{P}}/v_{\mathrm{P}}$ (ΔE_{P}: projectile energy loss). $\boldsymbol{F}_{\mathrm{n}}$, $\boldsymbol{F}_{\mathrm{e}}$: main forces acting on the nucleus and the electron cloud, respectively. (**b**): Experimental results. Two-dimensional final-state momentum distributions for the recoiling He^{+} target ion, the electron and momentum change of the projectile in singly ionizing $1\,\mathrm{GeV/u}$ U^{92+} on He collisions (logarithmic y-scale). The projectile propagates along the z-direction, its field mainly acts along the x-axis

where kinematically complete experiments only became feasible since 1994 [1] and where triply differential cross sections (TDCS) have not been reported before 2001 [16]. Therefore, in the present contribution we will restrict ourselves to a short intuitive illustration of the general features of the collision dynamics in the light of complete pictures in momentum space. In addition, a few novel aspects that have only been explored for ion impact up to now, will be mentioned, like the surprising results obtained in "out-of-plane" geometry, the relationship to photoionization, and relativistic effects observed at large velocities.

Comprehensive Pictures in Momentum Space and Surprising Deviations from First-Order Behavior.

In Fig. 21.2b, final-state momenta of the electron, the recoiling target ion, as well as the momentum change of the scattered projectile ($\Delta\boldsymbol{P}_{\mathrm{P}} = -\boldsymbol{q}$) are shown for helium single ionization in collisions with $1\,\mathrm{GeV/u}\ \mathrm{U}^{92+}$ projectiles [17], i.e., at a velocity of 120 a.u. in a situation that has been schematically illustrated in Figs. 21.1 and 21.2a. Even if the perturbation strength $Z_\mathrm{P}/v_\mathrm{P}$ is close to one in this case and, hence, higher-order contributions might be expected to emerge already, comparison of the experimental data with theoretical results obtained in a relativistic first-order treatment [18] demonstrated, however, that all essential features are reproduced [19]. This is mainly due to the fact that the field in the longitudinal direction is relativistically suppressed already by a factor of 2, decreasing the effective collision time – an important criterion for the applicability of a first-order theory – by a factor of two. As a result, the importance of the final-state interaction of the receding projectile with the ionized electron is decisively diminished [11] which is one important contribution of higher order, as will be shown below.

Exploiting azimuthal symmetry, all momenta are projected onto a plane defined by the incoming projectile momentum vector $\boldsymbol{P}_\mathrm{P}^\mathrm{i} = \boldsymbol{P}_{\mathrm{P}||}$ and the momentum vector $\boldsymbol{P}_\mathrm{R} = (-P_{\mathrm{R}x}, P_{\mathrm{R}||})$ of the recoiling ion. Following the intuitive scenario depicted in Fig. 21.2a, this should be the plane (with respect to the azimuth) containing the projectile-generated electric field. The momentum change of the projectile in the longitudinal direction, calculated from the collision dynamics $q_{||} = \Delta E_\mathrm{P}/v_\mathrm{P}$, is negligibly small ($\le 0.06$ a.u.) for typical electron energies $E_e < 200\,\mathrm{eV}$ (more than 90% of all events). The FWHM of the $\Delta P_{\mathrm{P}||}$-distribution in Fig. 21.2b is determined by the experimental resolution (mainly of the recoil ion) and is about 0.2 a.u. Thus, essentially "no" momentum is transferred to the target in the longitudinal direction and even the transverse momentum transfer is found to be small compared to the target fragment momenta. Scattering angles are typically less than 20 nanorad, never accessible by direct means.

Deeper insight into the collision dynamics might be obtained if fully differential cross sections (FDCS) are investigated [20] as illustrated in Fig. 21.3 for single ionization of helium in collisions with $100\,\mathrm{MeV/u}\ \mathrm{C}^{6+}$ projectiles at a significantly smaller perturbation of $Z_\mathrm{P}/v_\mathrm{P} = 0.1$. Here, the FBA is assumed to perfectly describe the cross sections. Shown are experimental (Fig. 21.3a) and theoretical (Fig. 21.3b) complete three-dimensional (3D) emission patterns for target electrons with well-defined energy ($E_e = 6.5\,\mathrm{eV}$) at a fixed momentum transfer $\boldsymbol{q}$ with $|\boldsymbol{q}| = 0.75$ a.u. as a function of the azimuthal (φ_e) and polar (θ_e) electron emission angles, respectively. The initial projectile momentum, directed along the z-direction in Figs. 21.1 and 21.2, now points upwards for better 3D illustration of the emission pattern.

As discussed in Fig. 21.2 for fast collisions, the momentum transfer essentially points in the transverse direction (x-axis in Figs. 21.2 and 21.3) and the electron emission in three dimensions exhibits a characteristic double-peak structure with two maxima. One is along the momentum-transfer direction and the second opposite, the well-known "binary" and "recoil" peaks, respectively (see, e.g., [21]). Binary electrons are primarily ejected along the

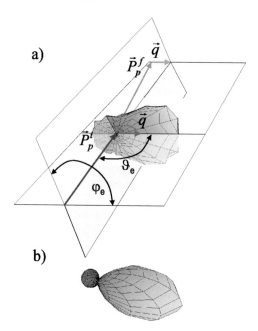

Fig. 21.3. Fully differential cross section (FDCS) in arbitrary units for target electrons with well-defined energy ($E_e = 6.5\,\mathrm{eV}$) at a fixed momentum transfer $\boldsymbol{q}$ with $|\boldsymbol{q}| = 0.75$ a.u. as a function of the azimuthal (φ_e) and polar (θ_e) electron emission angles, respectively. The initial projectile momentum direction $\boldsymbol{P}_P^i$, along the z-direction in Figs. 21.1 and 21.2, now points upwards for better 3D illustration of the emission pattern. (**a**) experiment, (**b**) theory (see text)

$\boldsymbol{q}$-axis as a result of a direct two-particle momentum exchange with the scattered projectile. The emission pattern (theory, Fig. 21.3b) is symmetrically broadened around the $\boldsymbol{q}$-axis due to the initial momentum distribution, the so-called "Compton profile" of the target electron. Whereas the target nucleus essentially remains passive in this process the "recoil peak" has been interpreted as a re-scattering [22], where the target electron is initially pushed along $\boldsymbol{q}$, but is back-scattered on its path out of the atom from the nucleus that, thus, picks up most of the momentum transfer.

The theoretical results, which include higher-order contributions in the interaction of the projectile with the target beyond the FBA (for details see [23]) are nearly cylindrically symmetric around the $\boldsymbol{q}$-axis, a feature characteristic for each first-order approach. The sharp 3D minimum at the origin indicates both the absence of higher-order processes as well as the dominance of dipole-like transitions, where the dipole part $\boldsymbol{q} \cdot \boldsymbol{r}$ from an expansion of the full transition operator $\exp(\mathrm{i}\boldsymbol{q} \cdot \boldsymbol{r})$ mainly contributes ($\boldsymbol{r}$: target electron coordinate). Whereas the experimental data, including the minima at $\theta_e = 0°$ and $180°$, respectively, are in close agreement with the predictions in the coplanar geometry, i.e., for a cut of the three-dimensional pattern along the plane defined by $\varphi_e = 0°$, dramatic deviations are observed out-of-plane. As is obvious from the figure, the experimental FDCSs do not shrink to zero close to the origin out-of-plane but exhibit a distinct structure with two maxima in their θ_e-distribution (counted within the plane tilted by φ_e) at $\theta_e = 90°$ and $270°$, respectively. In the plane perpendicular to the scattering plane and containing the initial momentum $\boldsymbol{P}_P^i$, i.e., for $\varphi_e = 90°$ the two maxima are of equal size, whereas theory predicts an angular-independent, constant behavior. Until now, the reasons for these discrepancies, which are discussed in detail by Madison et al. [23,24] are unclear: Presumably, they are due to surprisingly strong higher-order contributions in out-of-plane geometry at low perturbations, where the applicability of first-order theory was taken for granted up to now on the basis of electron-impact single-ionization studies that have been nearly exclusively investigated in the coplanar geometry. Thus, comprehensive pictures of the full final-state wave-function square reveal new insight into the collision dynamics.

In further experiments it should be clarified whether a similar behavior is observed for electron impact at identical perturbation strength, i.e. at considerably lower collision velocities and how this unexpected behavior develops with the perturbation for ion impact. Kinematically complete data have recently been obtained by Weber et al. (Figs. 30 and 31 in [7] and [25]) for a 0.5–MeV/u proton on helium collisions, i.e., at a perturbation of 0.22, but FDCSs have not been reported up to now.

Relativistic Effects and the Connection to Photoionization.

Relativistic effects in electron-impact ionization of light target atoms have never been explored up to now in kinematically complete (e, 2e) experiments, since traditional techniques applied rely on the direct measurement of the momentum transfer by the projectile by measuring its deflection angle and final energy. At large velocities, where modifications due to relativity are expected, the relative energy changes and scattering angles are so small that the experimental resolution is by far not sufficient to test theory. With the advent of Reaction Microscopes such studies have become feasible for the first time, since here the momenta of the recoiling target ion and of the emitted target electron are measured instead, obtaining the momentum transfer from momentum conservation $\boldsymbol{q} = \boldsymbol{P}_P^i - \boldsymbol{P}_P^f = -(\boldsymbol{P}_R + \boldsymbol{P}_e)$ which, hence, is accessible for any projectile velocity with the same resolution of presently about 0.1 a.u.

At the same time, the relationship to photoionization, that has been intensively discussed in the literature (see, e.g., [8,17,26–29] and references therein), can be explored in unprecedented detail. Similarities are expected to emerge at velocities v_P approaching the speed of light c and, simultaneously, in the limit of minimum momentum transfer with $q_{min} = q_{||} = \Delta E_P/v_P$ and $q_\perp = 0$ for $\boldsymbol{q} = (q_\perp, q_{||})$. Then, in the nonrelativistic dipole limit, the matrix element becomes identical to that for the absorption of a photon of $E_\gamma = \Delta E$ with the photon momentum $\boldsymbol{q} = E_\gamma/c$, except that the photon polarization vector $\boldsymbol{\epsilon}$ is replaced by $\boldsymbol{q}$. Accordingly, at $\gamma = 100$ and $q_\perp = 0.001$ a.u. (emitted electron energy $E_e = 27.2\,\mathrm{eV}$), which should be close to the "photon-limit", one finds for the FDCS (Fig. 21.4) in the nonrelativistic dipole limit a dipolar emission pattern with equally strong "binary" and "recoil" peaks (dashed line) oriented along $\boldsymbol{q}$ (dashed arrow), which is directed nearly along the beam propagation since $q_\perp \ll q_\parallel$. This has been shown recently to hold also for a fully relativistic treatment [30] leading to the "important conceptual result that collisions with minimum momentum transfer, which are often termed the optical limit and are regarded to be closest to photoabsorption [31], in fact can never be photon-like": Whereas the oscillator strength is the same, directions are just orthogonal with the dipole pattern oriented perpendicular to the photon propagation for photoabsorption and along the projectile direction for charged-particle impact.

It has been shown by Voitkiv and Ullrich [30] that the interaction with the charged-particle–induced field indeed becomes photon-like, meaning that the absorption of so-called "transverse" virtual photons dominates, for $q_{\min}/\gamma^2 \ll q_\perp \ll q_{\min}$. Then, the orientation of the emission pattern is no longer symmetric with respect to $\boldsymbol{q}$ but tends to be oriented more to the transverse direction along a new vector $\boldsymbol{G}$ [30], an effect that can be already observed for $\gamma = 2$ (full line in Fig. 21.4) and becomes striking for $\gamma = 100$ (dotted line). Here, the dipole-loop is oriented nearly perfectly along the $90°$-direction, transverse to $\boldsymbol{q} \approx q_\parallel$ and to the beam propagation. Nevertheless, "photon-like" transitions only dominate the total cross section for $\gamma \gg 100$.

Relativistic effects and deviations from the dipole approximation (non-relativistic photon-limit) have already been observed for $1\,\mathrm{GeV/u}\ \mathrm{U}^{92+}$ on helium collisions at $\gamma = 2$ as illustrated in Fig. 21.5. Here, doubly differential cross sections for electron emission are shown, obtained by integration of the two-dimensional data of Fig. 21.2b over certain transverse electron momenta as a function of the longitudinal electron momentum. Clear differences are observed between the dipole-approximation (dotted line), the non-relativistic FBA (dashed line) and the relativistic first Born results (full line) being in significantly better agreement with the experiment. Thus, using Reaction Microscopes, ionization of light targets by relativistic projectiles can be investigated for the first time in kinematically complete experiments.

21.2.3   Single Ionization for Strong Ion-Induced Fields at Large Perturbations

General Dynamical Features and Electron Spectra.

For strong ion-induced electromagnetic fields, i.e. at perturbations $Z_\mathrm{P}/v_\mathrm{P} \gg 1$, single ionization dynamics changes significantly as illustrated in Fig. 21.6 for highly charged fast ion impact (for a detailed discussion see [1,32–35]).

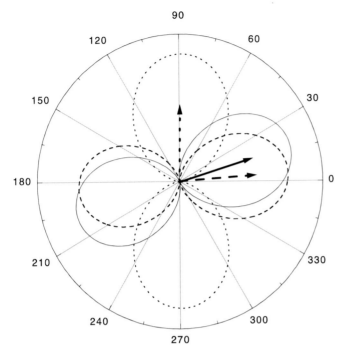

Fig. 21.4. FDCS (in arbitrary units) in the coplanar geometry for single ionization of H(1s) in collisions with protons as a function of the polar emission angle (θ_e) for an emitted electron energy $E_e = 27.2\,\mathrm{eV}$ and $q_\perp = 0.001\,\mathrm{a.u.}$ *Full and broken curves*: relativistic and nonrelativistic calculations, respectively, for a collision energy of $1\,\mathrm{GeV/u}$ ($\gamma = 2.07$). *Dotted curve*: result for $\gamma = 100$. Broken arrow: momentum transfer direction $\boldsymbol{q}$. *Full* and *dotted arrows*: polarization direction $\boldsymbol{G}$ of the "compound" virtual photon containing "transverse" as well as "longitudinal" contributions for $\gamma = 2.07$ and $\gamma = 100$, respectively (see text)

Again, this is a situation that is not accessible for electron impact where $|Z_\mathrm{P}| = 1$. While the general characteristics of the dynamics prevails, i.e., small momentum transfers by the projectile still dominate and a pronounced balancing of momenta between the recoil ion and emitted electron is observed, the final state, however, obviously becomes more and more "deformed" with increasing perturbation strength. This effect, which is very well described by continuum distorted wave (CDW) approaches (see, e.g., [36–39]; full lines in the right-hand panels of the figure [40]), has been interpreted as a "post-collision interaction" (PCI), an interaction of the slowly receding highly charged projectile ion with the emitted low-energy target electrons and ions in the final state: Electrons are dragged behind the projectile, yielding a pronounced forward shift of the otherwise forward-backward symmetric momentum spectra, whereas the recoil ion is pushed backwards with about the same force. In other words, positively and negatively charged target fragments seem to be

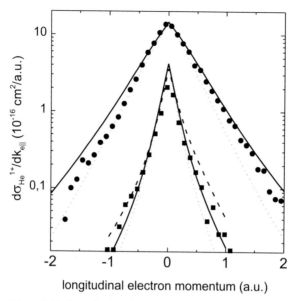

Fig. 21.5. Longitudinal momentum distributions for electrons emitted in singly ionizing 1 GeV/u U^{92+} on He collision. *Circles* and *squares*: experiment; *solid* and *dotted curves*: relativistic first Born and dipole approximation, respectively; *dashed curve*: nonrelativistic FBA ($c \to \infty$). *Upper part*: emitted electrons with transverse momenta restricted to $P_{e\perp} < 3.5$ a.u.; *lower part*: $P_{e\perp} < 0.25$ a.u.

dissociated in the field of the projectile ion without significant net-momentum transfer. Reversing the sign of the projectile charge should interchange the role of electrons and ions in the final state, an effect that has been hardly visible in the first experiments for single ionization of He by antiproton impact [41].

Even at strong perturbation, only very little momentum is transferred to the target system as a whole into the longitudinal direction $q_{||} = \Delta E_P / v_P$ for swift collisions. Thus, the width of the momentum distribution along the projectile propagation has been interpreted as a quantity, being inherent to the target itself, namely to the momentum distribution in the ground state, the so-called Compton profile. The visibility of bound-state properties is discussed in Fig. 21.7. Here, experimental longitudinal momentum distributions of electrons (DDCS) of electrons are shown for different cuts in $P_{e\perp}$ for single ionization of argon (symbols) by 3.6 MeV/u Au^{53+} impact along with theoretical CDW-EIS predictions (full line, [37,42]). Structures were identified near $v_{e||} = \pm 0.5$ a.u. in theory which are within the error bars of the experimental data. In this first calculation, the enhancements at $P_{e||} = \pm 0.5$ a.u. were interpreted to be related to the nodal structure of the $3p_0$ state. Whereas this direct signature of the ground-state momentum distribution (see inset in

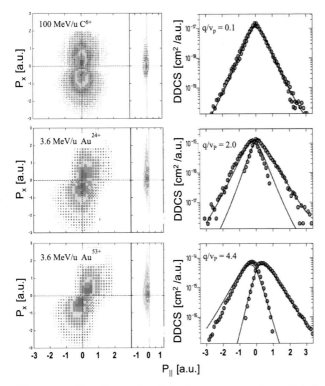

Fig. 21.6. Two-dimensional final-state momentum distributions for the recoiling He^+ target ion and the electron (*left column*) as well as singly differential cross sections for both particles as a function of the longitudinal momentum (*right column*) for different projectile charges and impact energies (perturbation strengths as indicated in the figure) in singly ionizing collisions (logarithmic z-scale). The projectile propagates along the $P_\parallel$ direction, its field mainly acts transverse along the x-axis

Fig. 21.7) could not be verified in more recent calculations [38] it was found, however, in agreement with [42], that different subshells lead to pronounced differences in the longitudinal electron-momentum distribution and that all substates have to be considered in order to reproduce the experimental spectrum.

In summary, strong evidence is provided that the longitudinal electron-momentum distributions reflect the properties of the respective bound-state wave functions, independent of the momentum transfer occurring mainly in the transverse direction. Moreover, since the transverse momentum transfer is small, the leading dynamic single-ionization mechanism for fast heavy-ion impact can be identified as a dipole interaction at large impact parameters dissociating the target atom in the strong field of the projectile as visualized in Fig. 21.2. Finally, precise quantum-mechanical CDW-EIS calculations beyond

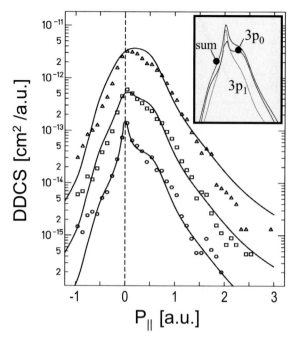

Fig. 21.7. Doubly differential cross sections DDCS $= \mathrm{d}^2\sigma/(\mathrm{d}P_{||}\mathrm{d}P_\perp 2\pi P_\perp)$ for electron emission in singly ionizing 3.6 MeV/u ($v_P = 12$ a.u.) Au^{53+} on Ar collisions as a function of the longitudinal momentum $P_{||}$ for fixed transverse momentum transfers $P_\perp$ as indicated in the figure. *Open circles*: experiment. *Full line*: CDW-EIS (see text). DDCS at different $P_\perp$ are multiplied by factors of 10, respectively. *Inset*: Theoretical results for different subshells and sum of all contributions

perturbation theory are at hand, which reliably predict the observed features in the electron spectra on an absolute scale, properly taking into account the influence of the slowly receding projectile in the final state.

Three-Body Dynamics and Fully Differential Cross Sections.

In general, even the simplest dynamical situation where the time evolution of three point-like particles that mutually interact via the nonrelativistic Coulomb-force is considered, i.e., the three-particle problem, remained one of the most fundamental and lively debated topics in atomic physics. Until recently, when a mathematically consistent, extremely time-consuming solution of the problem was presented [43–46], all theoretical approximate solutions had failed to describe absolute experimental data for electron impact at low velocities [47]. Along these lines, major difficulties have been found to arise when the full three-body dynamics of single ionization in strong ion-generated fields is addressed at large perturbations [48], where the experimental emission spectra of low-energy electrons have been demonstrated before to be

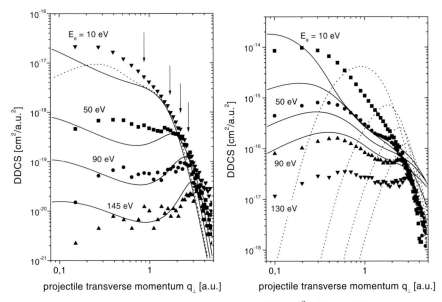

Fig. 21.8. Doubly differential cross sections DDCS $= \mathrm{d}^2\sigma/(\mathrm{d}q_\perp \mathrm{d}E_\mathrm{e})$ as a function of the projectile transverse-momentum transfer for specified and fixed electron energies for pure single ionization of He by 100 MeV/u C^{6+} impact with $Z_\mathrm{P}/v_\mathrm{P} = 0.1$ (*left figure*) and 3.6 MeV/u Au^{53+} impact with $Z_\mathrm{P}/v_\mathrm{P} = 4.4$ (*right figure*). *Left figure*: FBA (*solid line*); convolution of DDCS for $E_\mathrm{e} = 10\,\mathrm{eV}$ with the experimental resolution $\Delta q_\perp = 0.2\,\mathrm{a.u.}$ (*dotted line*). *Right figure*: CDW-EIS with screened nucleus-nucleus interaction (*solid line*); standard CDW-EIS (*dashed lines*)

accurately described on an absolute scale by non-perturbative theoretical approaches.

In Fig. 21.8, doubly differential cross sections for electrons of different energies are plotted as a function of the projectile transverse-momentum transfer $P_\perp = P^\mathrm{f}_{\mathrm{P}\perp} = \theta \times P^\mathrm{i}_\mathrm{P}$ or deflection angle θ for 100 MeV/u C^{6+} (Fig. 21.8a) and 3.6 MeV/u Au^{53+} (Fig. 21.8b) on helium collisions, i.e., for a perturbative as well as strongly non-perturbative situation, respectively. At moderate electron energies, $E_\mathrm{e} > 50\,\mathrm{eV}$, essentially two different dynamical contributions to the spectra can be identified: First, a broad, unstructured shoulder (the only contribution for $E_\mathrm{e} < 50\,\mathrm{eV}$) at momentum transfers by the projectile $q_\perp = -P_\perp$ smaller than the respective electron energy in a binary projectile-electron collision: $E_\mathrm{e} < P^2_\perp/2m_\mathrm{e}$ (m_e: electron mass). These are electrons that mainly balance their momenta with the recoiling target ions, emitted in a dipole- or "photon-like" interaction with the projectile field without significant momentum transfer. Second, a pronounced shoulder or even a broad peak is observed at transverse projectile momentum transfers that match the respective electron energies $E_\mathrm{e} = P^2_\perp/2m_\mathrm{e}$, the so-called "binary encounter electrons" (BEE). Whereas the experimental results at low perturbation for fast C^{6+} impact are reasonably well described by a FB calculation (full line in Fig. 21.8a), dramatic deviations between experiment and theory arise at large perturbations: A standard CDW calculation (dashed line in Fig. 21.8b) that does not include the interaction between the nuclei – from differential data at small perturbations commonly thought to be not important at all in the present situation, where only large impact parameters $b > 10$ a.u. contribute (see also the discussion in [6,7,49]) – fails completely. It only predicts binary-type electrons (which by itself is astonishing and not obvious at all). Various improvements of theoretical approaches [50], now implementing the internuclear interaction on different footings and, thus, trying to consider the full three-particle problem, tend to improve the situation (see, for example, the full line in Fig. 21.8b) but, in general, are in surprisingly poor agreement with the experiment. Classical calculations that are usually found to describe the many-body dynamics in strong ion-induced fields quite well do not find any binary electrons [50], in strong disagreement with experiment.

The reasons for these discrepancies are by no means clear at present. However, intuitively, one might convince oneself that we face a quite delicate situation: Since huge impact parameters dominantly contribute, typically fifteen times larger than the radius of the bound-state target-electron spatial probability distribution, the force on the target nucleus and the electrons is of similar magnitude but oppositely directed. Thus, target ionization might be viewed as effectively "ripping apart" the atom or dissociating it (see Fig. 21.2a) in the strong field of the projectile resulting in similar but oppositely directed final-state momenta of the electron and the recoil ion. Consequently, the net-force acting on the projectile is nearly zero and, thus, its deflection sensitively mirrors the details of a subtle balance of forces between all three particles, strongly depending on the exact, time-dependent target-electron density distribution in the incoming part of the trajectory as well as during the collision and in the final state. Hence, single ionization in swift collisions at large perturbations seems to represent an extremely challenging realization of the three-body Coulomb problem serving as a benchmark system for the development of non-perturbative theoretical approaches.

This has been further elucidated recently by Schultz et al. [51], who investigated fully differential cross sections for the same collision system, presented in Fig. 21.9 for a coplanar geometry, an electron emission energy of $E_e = 17.5$ eV (top) and 55 eV (bottom), and fixed momentum transfers of 0.65 a.u., 1.0 a.u., and 1.5 a.u.. At small q (left), the binary peak predicted by theory, along the q-direction for the FBA (dashed line) and shifted into the forward direction in CDW calculations (full line), is only barely visible in the experiment, whereas a strong new peak (recoil peak) emerges pointing exactly in the forward direction. Here, the absolute magnitude is underestimated by a factor of six by the standard CDW, where the interaction between the nuclei was not yet considered. Such calculations for FDCSs are not currently available. Increasing the momentum transfer (middle and right panels), the

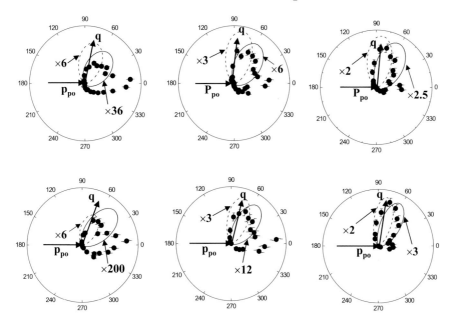

Fig. 21.9. FDCSs as a function of the polar angle for electrons of $E_e = 17.5\,\mathrm{eV}$ (*top*) and 55 eV (*bottom*) emitted in 3.6 MeV/u Au^{53+} on He singly ionising collisions in the coplanar geometry for momentum transfers (*left* to *right*) of 0.65, 1.0, and 1.5 a.u. *Dashed curve*: FBA; *solid curve*: CDW-EIS

binary peak dominates more and more, yielding increasingly better agreement between experiment and theory in the angular distribution as well as on the absolute scale.

In summary, the few here-mentioned topics addressing single ionization by strong ion-induced fields, namely comprehensive pictures in momentum space including out-of-plane geometries, relativistic effects at large velocities and strongly non-perturbative situations for highly charged, swift ion impact impressively demonstrate the power of new projection techniques, i.e., of Reaction Microscopes, to elucidate many-particle dynamics in situations that have not been accessible with traditional techniques. Surprisingly, even after 30 years of research, which, however, was strongly focused on electron-impact studies in the coplanar geometry, comparison with state-of-the-art theoretical approaches bring to light that even the simplest dynamical situation, the effective three-body Coulomb problem, still seems to be very poorly understood in a general sense.

21.3   Double Ionization in Ion-Generated Strong Fields

21.3.1   Basic Mechanisms of Double Ionization

Double ionization by ion, photon and electron impact has been excessively discussed in the past (for a recent review see, e.g., [29]), mainly on the basis of total and single differential cross sections. Only for photon-induced double ionization of helium, a substantial set of kinematically complete experimental data has been reported (for recent reviews see, e.g., [52,53]) over about one decade since the first pioneering experiment in 1993 [54]. Here, traditional techniques, the coincident detection of two target electrons, was successfully applied. For charged-particle-induced double ionization, however, kinematically complete studies enforce the detection of a third reaction product that, for a long time, has been considered to be impossible due to extremely small coincidence solid angles. Even if such experiments have recently been realized under strongly restricted geometrical conditions for electron impact in (e,3e) studies (see, e.g., [55]) a breakthrough for double ionization, the production of huge comprehensive data sets and, last, but not least, the accessibility of strong-laser [8] and ion-induced double-ionization reactions again is intimately related to the invention of Reaction Microscopes.

In a simplified, though often stressed illustrative picture based on a perturbation expansion, double ionization by charged-particle impact can either occur due to an independent interaction of the projectile field with both target electrons (often termed "two-step-2", TS-2) or, due to a single interaction of the field with the atom, where the second electron is emitted as a result of the electron-electron correlation. In collisions with charged particles as for photoionization the latter process is usually further subdivided in terms of many-body perturbation-theory diagrams: Two-step-1 (TS-1), a single interaction of the projectile with the target plus a second step, when the emerging first electron interacts with the second one, is distinguished from shake-off (SO) or ground-state (GS) correlation contributions (for details and the diagrams see [29]).

At small perturbations TS-1, SO, and GS contributions dominate, the interaction of the projectile field with the target can be treated in first order and, consequently, FDCSs should be identical for fast ion and electron impact. Fully differential cross sections for electron-impact double ionization, the comparison with first-order theories as well as the connection to photo double ionization, which again can be brought forward for small momentum transfers, are discussed in this book by Dorn in Chap. 19. Thus, even if higher-order contributions might show themselves in certain geometrical situations as has been demonstrated for single ionization out-of-plane data, this Chapter will be restricted to the discussion of non-perturbative, strong-field situations for fast highly charged ion impact. Experimental as well as theoretical results at small perturbations for $100\,\mathrm{MeV/u}$ C^{6+}, where dynamical mechanisms [56] as well as signatures of the correlated initial state [57,58] have been investigated by inspection of partially differential cross sections in the plane transverse to and along the pulse propagation direction, respectively, are not described in this Chapter. Recently, even fully differential cross sections for double ionization of helium by fast-proton impact have been reported [59].

21.3.2 Double Ionization at Strong Perturbations

General Dynamical Features and Electron Spectra.

At strong perturbations, double ionization is dominated by the TS-2 mechanism, a second-order interaction of the projectile with the target where the ion-induced field independently acts on both target electrons giving rise to their ejection. As of now, despite several attempts (see, e.g., [60,61]), no quantum-mechanical theoretical description of the full four-particle dynamics is at hand that consistently includes second-order contributions. Since the electron-emission spectra alone, where one integrates over all projectile scattering angles, have been found to be unaffected by the interaction between the nuclei [62] they can in principle be described within the independent particle model (IPM) if correlation effects between electrons are neglected. Following these lines, Moshammer et al. [17] were able to quantitatively describe single differential cross sections as a function of the energy of one "typical" electron (integrated over all energies of the second) for double ionization of helium by $1\,\mathrm{GeV/u}$ U^{92+} impact within the dipole approximation.

More recently, Kirchner et al. [63] have developed a much more sophisticated IPM-model, following basic concepts of Deb and Crothers [64], where the impact-parameter-dependent effective single-particle ionization probabilities $P_i(b)$ for certain subshells i are calculated within the CDW-EIS approach (continuum distorted wave – eikonal initial state). Since the $P_i(b)$ were found to exceed one at small impact parameters b for Ne and Ar targets in collisions with $3.6\,\mathrm{MeV/u}$ Au^{53+} (a consequence of the non-unitarity of the CDW-EIS approximation that shows the even limited validity of this non-perturbative method at small b where the perturbation is very large), two models were applied to preserve unitarity: In "model (a)" the $P_i(b)$ were simply set equal to one for $P_i(b) > 1$ whereas in "model (b)" a unitarity prescription proposed in [65] was used. As shown in Fig. 21.10a for neon double-ionization, good agreement is obtained between experimental and theoretical double differential cross sections (DDCS) for electron emission demonstrating that correlation effects between the electrons, which are completely neglected in the IPM are of minor importance if only a small fraction of all target electrons is emitted, i.e., for low final charge states of the target. Furthermore, models (a) and (b) are quite similar.

For the helium target instead, Fig. 21.10b, the IPM (full line) strongly overestimates the experimental DDCSs at all transverse electron energies, indicating that electron correlation cannot be neglected in this case. Inclusion

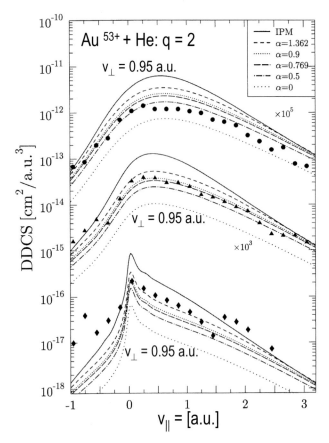

Fig. 21.10. Doubly differential cross sections DDCS = $\mathrm{d}^2\sigma/(\mathrm{d}v_{||}\mathrm{d}v_\perp 2\pi v_\perp)$ for electron emission in doubly ionizing collisions of 3.6 MeV/u Au^{53+} on neon (a) and helium (b) as a function of the longitudinal electron velocity (momentum) $v_{||} = P_{||}$ (in atomic units) for certain transverse-velocity ($v_\perp = P_\perp$) cuts. DDCSs for $v_\perp = 0.45$ and 0.95 are multiplied by the indicated factors. *Symbols*: experiment. (a) results for Ne; model (a) and (b): theoretical CDW results using different unitarization procedures (see text). (b) results for He; *curves*: IPM-based calculations with inclusion of the final-state correlation using different values of α in the correlation factor

of correlation effects via a simplified version [66] of the so-called BBK wave function [67] for different static average distances between the two electrons in the helium atom (various parameters α) yields better agreement (apart from the unrealistic value $\alpha = 0$). It was discussed in [63] that one might assume that α should be different for different transverse electron momenta corresponding to different impact parameters that mainly contribute to the respective DDCS. Thus, α might range from $\alpha = 1.362$, corresponding to an average distance between the two electrons in the unperturbed helium atom, being the appropriate choice at large impact parameters that contribute most to the DDCSs for $v_\perp = 0.05$ a.u., to $\alpha = 0.5$ (giving the correct ratio of total double to single ionization cross sections) for $v_\perp = 0.95$ a.u., where the electrons might, on average, be closer to each other due to the attraction by the projectile at smaller b. However, as of now, there are no general theoretical considerations for the optimal choice of α, giving this argumentation a somewhat arbitrary character.

Four-Body Dynamics

The full four-body dynamics has been investigated recently by Perumal et al. [68] for double ionization of helium by 3.6 MeV/u Au^{53+} projectiles. Considering the centre-of-mass motion of the two emitted electrons by adding their vector momenta the authors were able to show that the four-body problem can be reduced to a good approximation to an effective three-particle problem, essentially rediscovering all dynamical features that have been observed for single ionization.

This is shown in Fig. 21.11a and b, the longitudinal and transverse "three-particle" momentum balances, respectively. As for single ionization, a strong backward-forward asymmetry is found in the longitudinal direction where the electrons are pulled into the forward direction with their sum-momentum being essentially balanced by the recoiling He^{2+} ion that is emitted backwards. The projectile longitudinal momentum change (full histogram in Fig. 21.11) remains small. Since even for double ionization the emitted electron longitudinal velocities are typically small (less than 3 a.u., see Fig. 21.12a) compared to the projectile that recedes with 12 a.u., both electrons and the He^{2+} ion experience about the same force at a similar distance to the projectile in the final state. Consequently, the various target fragment fractions, negatively and positively charged, respectively, are accelerated to about the same final-state momenta as has been observed for single ionization.

Similarly, to a lower extent nevertheless, the transverse-momentum balance exhibits typical features of an effective three-particle situation. The projectile transverse-momentum distribution is nearly symmetric around zero with more than 75% of all ions being scattered to angles less than 1 μrad, i.e. $P_x < \pm 1$ a.u., indicating that neither electrons nor the target nucleus dominate the deflection in these cases. Since average impact parameters are smaller compared to single ionization, still, however, exceeding the Hartree–Fock helium radius by a factor of about ten [63], the momentum change of the projectile is not balanced as well as in the longitudinal direction to zero by the counteracting forces of positively and negatively charged target constituents, respectively. The present results might seem surprising on first glance, since the "two-step two" mechanism was commonly thought of as an independent, binary-like interaction of the projectile with both of the target electrons where the projectile transverse momentum would be largely balanced by the sum-momentum of the electrons. Instead, even for double

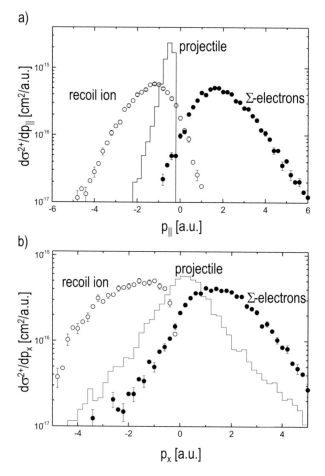

Fig. 21.11. Longitudinal (**a**) and transverse (**b**) momentum distributions for the recoil ions, the sum of both electrons along with the momentum change of the projectile for double ionization of He by 3.6 MeV/u Au^{53+}

ionization, typical impact parameters are so large that the target seems again to be dissociated in the strong projectile field, mainly transferring energy to the target in a dipole- or photon-like reaction with quite small net-momentum transfer.

Interestingly, even the transverse momenta of both individual electrons display a quite similar distribution, as shown in Fig. 21.12b, in contrast to the longitudinal ones (Fig. 21.12a), where fast and slow electrons are distinguished by the experiment. This indicates first, that longitudinal and transverse momentum components are independent to a large extent, and second, as has been suggested in [68], that a certain realization of the TS-2 mechanism might dominate where the momentum transfer to both target electrons and to the nucleus are not independent in each single collision but indeed are very similar in magnitude but into opposite directions. This then would describe a scenario where both electrons are effectively displaced from their nucleus during a time interval – the collision time – that is short compared to their classical orbital time in the atom. Thus, in a sense, one approaches a scenario suggested by Heisenberg [69] and recently picked up for neutron on helium collisions [70], which might be considered to be an ideal situation to investigate the short-time correlation of bound-state electrons. Indeed, as will be detailed in a forthcoming paper [71], distinct correlation effects can already be deduced from the comparison of Figs. 21.11b and 21.12b: Whereas both electrons (Fig. 21.12b) show a cusp-like feature at $P_x = 0$ (see also [34] for single ionization), which is less pronounced for "electron 1", the electron sum-momentum distribution (Fig. 21.11b) is smooth without any indication of an irregularity at the cusp. Or, in other words, folding a randomized distribution for a typical electron of Fig. 21.12b with itself would never yield the observed distribution in Fig. 21.11b indicating that the electrons must be strongly correlated.

Electron Correlation and the Correlation Function.

Along the lines discussed above, it is not surprising in retrospect that distinct correlation has been observed in recent experiments [72] for He and Ne double as well as Ne triple ionization, as shown in Fig. 21.13, where the correlated longitudinal momenta of two electrons are plotted in a two-dimensional density representation.

The observed structures were explained to some extent by "classical trajectory Monte Carlo" calculations (CTMC, see, e.g., [72]): Here, the target electrons move on classical Kepler orbits, with a microcanonical distribution each, bound with subsequent ionization potentials for many-electron atoms and Newton's equations are solved during the collision. In order to find similar patterns as in the experiment, electron correlation had to be included in the initial state by "dynamical screening", where the effective nuclear charge, seen by either one of the two electrons, dynamically varies as a function of the distance between "the other" electron and the nucleus. In addition, the final-state interaction between the two electrons has been "switched" on in the moment when both electrons are in the continuum, i.e., have positive energies during the collision.

Also shown in Fig. 21.13 (upper right frame) is the He ground-state probability distribution of the two electrons in the longitudinal momentum space, shifted by a longitudinal electron sum-momentum of 0.6 a.u. to account for the final-state post-collision attraction into the forward direction by the receding projectile. The latter is a reasonable approximation for collisions where the projectile emerges fast compared to the target fragment velocities, since all target fragments experience about the same electric field in the final state.

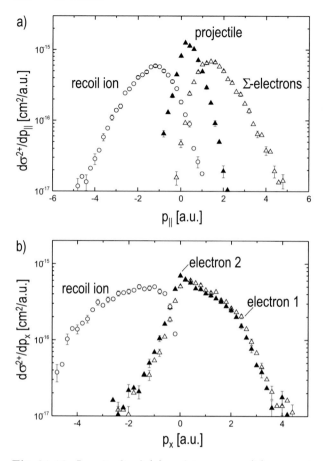

Fig. 21.12. Longitudinal (**a**) and transverse (**b**) momentum distributions for the recoil-ions, electron 1 and electron 2 for double ionization of He by 3.6 MeV/u Au^{53+}

If the initial-state correlation alone would determine the final-state observations as suggested in the Heisenberg scenario, where the nuclear potential is "suddenly" switched off, then, as a clearly not realistic benchmark, one would expect such a distribution.

Nevertheless, even in a more realistic perspective, there are good arguments that the final-state electron momenta observed in the experiment might indeed closely reflect properties of the initial-state correlated two-electron wave function, for the following reason. As has been demonstrated for single ionization in comparison with theory, the shape of the longitudinal electron-momentum distribution is determined by the momentum-density distribution of the one-electron bound-state wave function. Here, for double ionization, both helium electrons are independently ionizsed in a TS-2–like reaction dur-

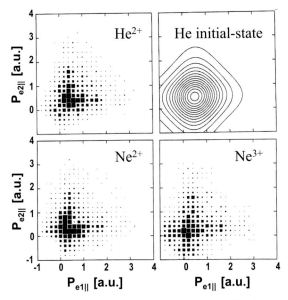

Fig. 21.13. Correlated two-electron longitudinal momentum distributions in a two-dimensional representation for He^{2+}, Ne^{2+}, and Ne^{3+} production in collisions with $3.6 \, \mathrm{MeV/u}$ Se^{28+} ($v_P = 12 \, \mathrm{a.u.}$). Experiment: different box sizes (15) represent doubly differential cross sections in $10^{-16} \, \mathrm{cm}^2$ (*largest box*) on a linear scale. *Upper right frame*: He ground-state distribution (see text)

ing a very short time interval of less than attoseconds. A negligibly small momentum is transferred and no significant momentum exchange between the electrons themselves or between each electron and the helium nucleus may take place during the collision since the pulse-time is short compared to the electron-revolution frequency in the bound state. Thus, the correlated longitudinal momentum spectrum of both emitted electrons might indeed mirror the short-time correlation of the bound electrons and the technique has been speculated to become an "attosecond-microscope" for the investigation of bound-state electron wave functions in general.

Recently, this idea has been further developed by Schulz et al. and others [74–78], by inspecting the so-called correlation function R between the emitted electrons. Here, the probability to find two electrons emitted in the same multiple ionization event with a certain momentum difference is compared to the corresponding probability for two independent electrons emitted in two different collisions. It was demonstrated that the correlation function is neither sensitive to the respective mechanism leading to double ionization (first-order or TS-2 interaction with the projectile) nor to the final-state post-collision interaction with the projectile, possibly making R an ideal tool to investigate ground-state properties of the correlated wave function. Recently, this was substantiated [66] by analyzing the differential correlation function

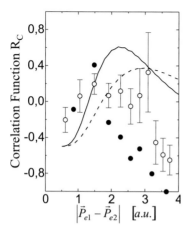

Fig. 21.14. Differential correlation function R_C for double ionization of He (*open circles*) and Ne (*full circles*) in collisions with 3.6 MeV/u Au^{53+} ($v_P = 12$ a.u.) as a function of the electron momentum difference for back-to-back emitted electrons. *Solid line*: correlated 16-term wave function. *Dashed line*: 3-term multiconfigurational Hartree–Fock wave function (see text)

R_C for back-to-back emission of electrons with equal energy. For this particular emission pattern it was found, in qualitative agreement with theory, that the maximum in R_C is not only sensitive on the mean initial-state separation between the two electrons but, moreover, its shape strongly depends on the correlated initial state used in the calculation (see Fig. 21.14).

21.4 Multiple Ionization in Ion-Generated Strong Fields

Up to now, only one kinematically complete pilot-experiment on multiple ionization has been performed for 3.6 MeV/u Au^{53+} impact on neon [79]. In Fig. 21.15 the momentum vectors of triply ionized Ne recoil ions are plotted along with the vector sum-momenta of all three emitted electrons. The collision plane is defined as in Fig. 21.2. No attempt has been made up to now to describe the complete five-particle dynamics in a strongly non-perturbative situation within a quantum-mechanical description, forcing us to compare to classical nCTMC results (n-body CTMC [73]) displayed in the right-hand panel of the figure.

Surprisingly, even for triple ionization we find the typical features observed for single- and double-electron emission: Little net-momentum is transferred at considerable energy transfer leading to a "dissociation" of the atom in the field. The classical calculations are found to be in remarkable agreement with the experimental results. Thus, the intuitive picture developed for single and double ionization, visualized in Fig. 21.2a, seems to hold even for triple-

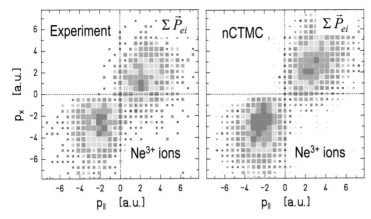

Fig. 21.15. Two-dimensional final-state momentum distributions for the Ne^{3+} recoil ion and the sum-momentum vector of all three emitted electrons for triple ionization of Ne by 3.6 MeV/u Au^{53+} impact. The collision plane is defined as in Fig. 21.2. *Left side*: Experiment. *Right side*: nCTMC (see text). Z-scale is logarithmic

electron emission. This is in accord with the work of Kirchner et al. [63] where the most likely contributing impact parameters for triple ionization in the present collision system were calculated to be still larger than 3 a.u., well outside the target electron cloud.

As before, at medium velocities and high Z_P, we find a strong effect by the PCI dragging each of the electrons behind but, at the same time pushing away the Ne^{3+} ions with nearly identical momenta. Thus, following the above ideas, the PCI can be seen as a dissociation of the target fragments in the field of the receding ion, again without any noticeable net-momentum transfer to the fragments. Implying that all the electrons are influenced by the PCI on the same footing after the collision (strongly supported by the fact that $\boldsymbol{P}_R \approx -\sum \boldsymbol{P}_{ei}$) independent on their momenta at the instant of ionization, one might separate the influence of the PCI from the relative motion of the three electrons by a transformation into the three-electron center-of-mass (CM) coordinate frame, where PCI is not present at all in the ideal case.

We have performed such a transformation and plotted the relative energies of the three electrons in the CM system $\epsilon_{ei} = E_i^{CM}$ (with E_i^{CM} being the CM energy of the i-th electron) in a modified Dalitz plot [80] in Fig. 21.16. This is an equilateral triangle where each triple-ionization event is represented by one point inside the triangle with its distance from each individual side being proportional to the relative energy of the corresponding electron as indicated in the figure. Only events in the inscribed circle are allowed due to momentum conservation of the three electrons in the CM frame $\sum \boldsymbol{P}_{ei}^{CM} = 0$. Numbering of the electrons is achieved by exploiting information on their emission angle: Electron 1 is the one with the smallest angle relative to the

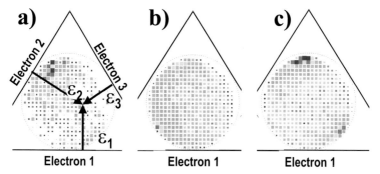

Fig. 21.16. Dalitz representation (see text) of the energy partitioning of three electrons emitted in triply ionizing 3.6 MeV/u Au^{53+} on Ne collisions in the electron center-of-mass (CM) coordinate system. ϵ_i: energy of the i-th electron in the CM system. Electrons are numbered according to their emission angle with respect to the projectile direction (see text). *Left*: Experiment; *middle*: nCTMC without electron-electron interaction; *right*: CTMC with fully correlated three-electron initial-state (see text)

projectile propagation direction in each triple-ionization event, electron 3 the one with the largest angle and electron 2 lies in between.

Obviously, the electron energies are not independent of each other and the many-electron continuum, explored for the first time experimentally, is found to be strongly correlated. There is an increased probability that electrons 1 and 3 have large energies compared to electron 2. Performing nCTMC calculations with the electron-electron interaction not included beyond an effective potential in the initial state these structures cannot be reproduced (Fig. 21.16b). This situation is similar to that described before for double ionization, where qualitative agreement between nCTMC and experiment was only achieved when the electron-electron interaction was explicitly implemented in the final state. Proceeding in the same way for triple ionization did lead to structures in the Dalitz plot but essentially with the role of electron 2 and 3 exchanged. Introducing in addition a completely correlated, three-electron classical initial state (P-electrons neglecting the spin), where the individual electrons move on Kepler ellipses but at equal distances relative to each other on the corners of an equilateral triangle in a plane, with the electron-electron interaction "switched on" during the entire collision, not only in the final state, brought the theoretical results surprisingly close to the experimental data (Fig. 21.16c).

Thus, in the light of the results for single ionization by the same projectiles, where the longitudinal electron-momentum spectra have been demonstrated to closely reflect features of the Hartree–Fock initial-state wave function (see Fig. 21.7) it does not seem to be too optimistic to expect that multi-electron momentum distributions might reveal direct information about the correlated many-electron bound-state wave function. Moreover, since the target disintegration occurs within attoseconds, i.e., on a time scale short compared to typical revolution times of ground-state electrons, one might even hope that such experiments will provide direct information on the short-time correlation between the electrons in the initial state.

21.5 A View Into the Future

21.5.1 Experiments in Storage Rings

Presently, work is in progress to perform such experiments at higher energies, i.e. at 500 MeV/u for projectile charge states between about 30+ and 92+ in the experimental storage ring ESR of GSI. Here, the conditions (Z_P/v_P) will be chosen such that the post-collision interaction will not notably influence the correlated two- (multi-) electron final state. To ultimately verify whether or not and to what extent the final-state electron momenta mirror the correlated initial state, i.e., whether an "attosecond-microscope" is realizable, experiments will be performed for ground-state as well as metastable excited helium targets. Due to the strongly increased luminosity in the storage ring as compared to single-pass experiments that have been exclusively performed up to now, we expect considerably (orders-of-magnitude) increased event rates, so that fully differential cross sections should become measurable not only for double but also for triple and quadruple ionization in attosecond fields.

21.5.2 Laser-Assisted Collisions

An interesting situation arises, and might be realized with the PHELIX laser at GSI [81], if both attosecond ion-induced fields and femtosecond strong laser fields act together. The ion-induced pulse efficiently brings a large number of electrons into the continuum, placing them "simultaneously" with little energy into the oscillating field of the laser, which then accelerates this bunch of electrons very effectively in a coherent way heating them tremendously. Thus, one might envisage that the most effective way to transfer energy to matter might be a concerted action between ion-induced and laser fields.

Unexpectedly strong coupling of an even weak ($F_0 = 0.005$ a.u.), low-frequency ($\omega_0 = 0.004$ a.u.) electromagnetic radiation field to matter has recently been predicted in laser-assisted collisions considering a nearly reversed situation, where target electrons are strongly accelerated in a direct collision with a fast ($v_P = 12$ a.u.) proton (binary encounter electrons: BEE) [82,83]. Whereas the laser field considered was not strong enough by far to noticeably disturb the hydrogen target-atom ground state alone, strong effects occur during the collision in the high-energy BEE emission, where thousands of laser photons were observed to couple to the system, strongly modifying the energy and angular distribution of the BEE.

In general, laser-assisted collisions, that have been theoretically explored for some time (see [84] for a recent review and other references therein on this topic), but were only accessible experimentally for elastic and resonant scattering up to now, may become feasible in the near future and are in preparation using Reaction Microscopes along with intense ns-pulsed YAG lasers.

Acknowledgements

I gratefully acknowledge support from the Bundesministerium für Forschung und Technologie BMFT, the Deutsche Forschungsgemeinschaft (DFG) within the Leibniz-Programm, from the Deutscher Akademischer Austauschdienst (DAAD), the Department of Energy (DOE), and from GSI.

References

1. R. Moshammer, J. Ullrich, M. Unverzagt, W. Schmidt, J. Jardin, R.E. Olson, R. Mann, R. Dörner, V. Mergel, U. Buck, H. Schmidt-Böcking: Phys. Rev. Lett. **73**, 3371 (1994)
2. C.L. Cocke C.L., R.E. Olson: Phys. Rep. **205**, 155 (1991)
3. J. Ullrich: Habilitationsschrift, Universität Frankfurt, GSI Report (1994)-08, ISSN 0171-4546
4. N. Stolterfoht, R.D. DuBois, R.D. Rivarola: *Electron Emission in Heavy Ion-Atom Collisions* (Springer, Berlin Heidelberg, New York 1994)
5. J. Ullrich, R. Dörner, V. Mergel, O. Jagutzki, L. Spielberger, H. Schmidt-Böcking: Comments At. Mol. Phys. **30**, 285 (1994)
6. J. Ullrich, R. Moshammer, R. Dörner, O. Jagutzki, V. Mergel, H. Schmidt-Böcking, L. Spielberger: J. Phys. B **30**, 2917, (1997), topical review
7. R. Dörner, V. Mergel, O. Jagutzki, L. Spielberger, J. Ullrich, R. Moshammer, H. Schmidt-Böcking: Phys. Rep. **330**, 95 (2000)
8. J. Ullrich, A. Voitkiv: in *Strong Field Laser Physics*, ed. by T. Brabec, H. Kapteyn (Springer, Berlin, Heidelberg, New York 2002) (in press)
9. C.L. Cocke: Phys. Rev. A **20**, 749 (1979)
10. S. Kelbch, J. Ullrich, R. Mann, P. Richard, H. Schmidt-Böcking: J. Phys. B **18**, 323 (1985); S. Kelbch, J. Ullrich, W. Rauch, H. Schmidt-Böcking, M. Horbatsch, R.M. Dreizler, S. Hagmann, R. Anholt, A.S. Schlachter, A. Müller, P. Richard, C. Stoller, C.L. Cocke, R. Mann, W.E. Meyerhof, J.D. Rasmussen: J. Phys. B **19**, L47 (1986)
11. C.J. Wood, R.E. Olson, W. Schmitt, R. Moshammer, J. Ullrich: Phys. Rev. A **56**, 3746 (1997)
12. H. Erhardt, M. Schulz, T. Tekaat, K. Willmann: Phys. Rev. Lett. **22**, 89 (1969)
13. I.E. McCarthy, E. Weigold: Rep. Prog. Phys. **54**, 789 (1991)
14. A. Lahmam-Bennani: J. Phys. B **24**, 2401 (1991)
15. M.A. Coplan, J.H. Moore, J.P. Doering: Rev. Mod. Phys. **66**, 985 (1994)
16. M. Schulz, R. Moshammer, D.H. Madison, R.E. Olson, P. Marchalant, C.T. Whelan, H.R.J. Walters, S. Jones, M. Foster, H. Kollmus, A. Cassimi, J. Ullrich: J. Phys. B **34**, L305 (2001)
17. R. Moshammer, W. Schmitt, J. Ullrich, H. Kollmus, A. Cassimi, R. Dörner, O. Jagutzki, R. Mann, R.E. Olson, H.T. Prinz, H. Schmidt-Böcking, L. Spielberger: Phys. Rev. Lett. **79**, 3621 (1997)

18. A.B. Voitkiv: J. Phys. B **29**, 5433 (1996); A.B. Voitkiv, N. Grün, W. Scheid: J. Phys. B **32**, 3929 (1999)
19. A.B. Voitkiv, B. Najjari, R. Moshammer, J. Ullrich: J. Phys. B **65**, 032707 (2002)
20. M. Schulz, R. Moshammer, D. Fischer, H. Kollmus, D.H. Madison, S. Jones, J. Ullrich: Nature **422**, 48 (2003)
21. G. Stefani, L. Avaldi, R. Camilloni: J. Phys. B **23**, L227 (1990)
22. C.T. Whelan, R.J. Allan, H.R.J. Walters, X. Zhang: (e, 2e), in: Whelan et al. (Eds.), (e,2e) related processes, (Kluwer, Dordrecht (1993) 1–32)
23. D. Madison, M. Schulz, S. Jones, M. Foster, R. Moshammer, J. Ullrich: J. Phys. B **35**, 3257 (2002)
24. D. Madison, D. Fischer, M. Foster, M. Schulz, R. Moshammer, S. Jones, and J. Ullrich: Phys. Rev. Lett. (2003), submitted
25. T. Weber, Diploma Thesis, Universität Frankfurt, 1998, unpublished
26. H. Bethe: Ann. Phys., (Lpz.), **5**, 325 (1930)
27. A. Dorn, A. Kheifets, C.D. Schröter, B. Najjari, C. Höhr, R. Moshammer, J. Ullrich: Phys. Rev. Lett. **86**, 3755 (2001)
28. J. Ullrich, B. Bapat, A. Dorn, S. Keller, H. Kollmus, R. Mann, R. Moshammer, R.E. Olson, W. Schmitt, M. Schulz: in: *X-Ray and Inner Shell Processes*, ed. by R.W. Dunford, D.S. Gambol, E.P. Kanter, L. Young (Melville, New York 2000) pp. 403–417
29. J. McGuire: *Electron Correlation Dynamics in Atomic Collisions*, (Cambridge Monographs on Atomic, Molecular and Chemical Physics) (Cambridge University Press, Cambridge 1997)
30. A.B. Voitkiv, J. Ullrich: J. Phys. B **34**, 4513 (2001)
31. M.E. Rudd, Y-K. Kim, D.H. Madison, T.J. Gay: Rev. Mod. Phys. **64**, 441 (1992)
32. J. Ullrich, R. Moshammer, M. Unverzagt, W. Schmitt, P. Jardin, R.E. Olson, R. Dörner, V. Mergel, H. Schmidt-Böcking: Nucl. Instrum. Meth. B **98**, 375 (1995)
33. R. Moshammer, J. Ullrich, H. Kollmus, W. Schmitt, M. Unverzagt, H. Schmidt-Böcking, C.J. Wood, R.E. Olson: Phys. Rev. A **56**, 1351 (1997)
34. W. Schmitt, R. Moshammer, F.S.C. O'Rourke, H. Kollmus, L. Sarkadi, R. Mann, S. Hagmann, R.E. Olson, J. Ullrich: Phys. Rev. Lett. **81**, 4337 (1998)
35. R.E. Olson, C.J. Wood, H. Schmidt-Böcking, R. Moshammer, J. Ullrich: Phys. Rev. A **58**, 270 (1998)
36. D.S.F. Crothers, J.F. McCann: J. Phys. B **16**, 3229 (1983)
37. P. Fainstein, L. Gulyás, F. Martín, A. Salin: Phys. Rev. A **53**, 3243 (1996)
38. L. Gulyás, T. Kirchner, T. Shirai, M. Horbatsch: Phys Rev. A **62**, 022702 (2000)
39. P.D. Fainstein, R. Moshammer, J. Ullrich: Phys. Rev. A **63**, 062720 (2001)
40. S.F.C. O'Rourke, R. Moshammer, J. Ullrich: J. Phys. B **30**, 5281 (1997)
41. K. Khayyat, T. Weber, R. Dörner, M. Achler, V. Mergel, L. Spielberger, O. Jagutzki, U. Meyer, J. Ullrich, R. Moshammer, W. Schmitt, H. Knudsen, U. Mikkelsen, P. Aggerholm, E. Uggerhoej, S.P. Moeller, V.D. Rodríguez, S.F.C. O'Rourke, R.E. Olson, P.D. Fainstein. J.H. McGuire, H. Schmidt-Böcking: J. Phys. B **32**, L73 (1999)
42. R. Moshammer, P.D. Fainstein, M. Schulz, W. Schmitt, H. Kollmus, R. Mann, S. Hagmann, J. Ullrich: Phys. Rev. Lett. **83**, 4721 (1999)

43. T.N. Rescigno, M. Baertschy, W.A. Isaacs, C.W. McCurdy: Science **286**, 2474 (1999)
44. G.D. Buffington, D.H. Madison, J.L. Peacher, D.R. Schultz: J. Phys. B **32**, 2991 (1999)
45. M.S. Pindzola, F. Robicheaux: Phys. Rev. A **54**, 2142 (1996)
46. J. Colgan, M.S. Pindzola, F. Robicheaux: J. Phys. B **34**, L457 (2001)
47. J. Röder, J. Rasch, K. Jung, C.T. Whelan, H. Ehrhardt, R. Allen, H. Walters: Phys. Rev. A **53**, 225 (1996)
48. R. Moshammer, A. Perumal, M. Schulz, V.D. Rodríguez, H. Kollmus, R. Mann, S. Hagmann, J. Ullrich: Phys. Rev. Lett. **87**, 223201-1 (2001)
49. W.R. DeHaven, C. Dilley, A. Landers, E.Y. Kamber, C.L. Cocke: Phys. Rev. A **57**, 292 (1998)
50. R.E. Olson, J. Fiol: J. Phys. B **34**, L625 (2001)
51. M. Schulz, R. Moshammer, A.N. Perumal, J. Ullrich: J. Phys. B **35**, L161 (2002)
52. V. Schmidt: *Electron Spectrometry of Atoms Using Synchrotron Radiation*, Cambridge Monographs on Atomic, Molecular and Chemical Physics, Vol. 6 (Cambridge University Press, Cambridge 1997)
53. J. Briggs, V. Schmidt: J. Phys. B **33**, R1 (2000)
54. O. Schwarzkopf, B. Krässig, J. Elminger, V. Schmidt: Phys. Rev. Lett. **70**, 3008 (1993)
55. I. Taouil, A. Lahmam-Bennani, A. Duguet, L. Avaldi, Phys. Rev. Lett. **81**, 4600 (1998)
56. B. Bapat, R. Moshammer, S. Keller, W. Schmitt, A. Cassimi, L. Adoui, H. Kollmus, R. Dörner, T. Weber, K. Khayyat, R. Mann, J.P. Grandin, J. Ullrich: J. Phys. B **32**, 1859 (1999)
57. B. Bapat, S. Keller, R. Moshammer, R. Mann, J. Ullrich: J. Phys. B **33**, 1437 (2000)
58. S. Keller, B. Bapat, R. Moshammer, J. Ullrich, R.M. Dreizler: J. Phys. B **33**, 1447 (2000)
59. D. Fischer, R. Moshammer, A. Dorn, J.R. Crespo López-Urrutia, B. Feuerstein, C. Höhr, C.D. Schröter, S. Hagmann, H. Kollmus, R. Mann, B. Dapal, and J. Ullrich: Phys. Rev. Lett. (2003) (accepted)
60. A. Dorn, A. Kheifets, C.D. Schröter, C. Höhr, G. Sakhelashvili, R. Moshammer, J. Lower, J. Ullrich: Phys. Rev. Lett. (2002) (submitted)
61. R.E. Mkhanter, C. Dal Cappello: J. Phys. B **31** 301 (1998); M. Grin, C. Dal Cappello, R.El. Mkhanter, J. Rasch: J. Phys. B **33**, 131 (2000)
62. P.D. Fainstein, V.H. Ponce, R.D. Rivarola: J. Phys. B **21**, 287 (1988)
63. T. Kirchner, L. Gulyás, M. Schulz, R. Moshammer, J. Ullrich: Phys. Rev. A **65**, 042727 (2002)
64. N.C. Deb, D.S.F. Crothers: J. Phys. B **23**, L799 (1990)
65. V.A. Sidorovich, V.S. Nikolaev: J. Phys. B **16**, 3243 (1983)
66. L.G. Gerchikov, S.A. Sheinermann: J. Phys. B **34**, 647 (2001)
67. M. Brauner, J.S. Briggs, H. Klar: J. Phys. B **22**, 2265 (1989)
68. A.N. Perumal, R. Moshammer, M. Schulz, J. Ullrich: J. Phys. B **35**, 2133 (2002)
69. W. Heisenberg: Z. Phys. **43**, 172 (1927)
70. J. Berakdar: J. Phys. B (2002) (submitted)
71. A.N. Perumal, R. Moshammer, M. Schulz, J. Ullrich: J. Phys. B (2002) (to be submitted)

72. R. Moshammer, J. Ullrich, H. Kollmus, W. Schmitt, M. Unverzagt, O. Jagutzki, V. Mergel, H. Schmidt-Böcking, R. Mann, C.J. Woods, R.E. Olson: Phys. Rev. Lett. **77**, 1242 (1996)
73. R.E. Olson, J. Ullrich, H. Schmidt-Böcking: Phys. Rev. A **39**, 5572 (1989)
74. M. Schulz, R. Moshammer, W. Schmitt, H. Kollmus, B. Feuerstein, R. Mann, S. Hagmann, J. Ullrich: Phys. Rev. Lett. **84**, 863 (2000)
75. B. Feuerstein, M. Schulz, R. Moshammer, J. Ullrich: Phys. Scr. **T92**, 447 (2001)
76. L.G. Gerchikov, S.A. Sheinermann: J. Phys. B **34**, 647 (2001)
77. M. Schulz, R. Moshammer, L.G. Gerchikov, S.A. Sheinermann, J. Ullrich: J. Phys. B **34**, L795 (2001)
78. L.G. Gerchikov, S.A. Sheinermann, M. Schulz, R. Moshammer, J. Ullrich: J. Phys. B **35**, 2783 (2002)
79. M. Schulz, R. Moshammer, W. Schmitt, H. Kollmus, R. Mann, S. Hagmann, R.E. Olson, J. Ullrich: Phys. Rev. A **61**, 022703 (2000)
80. R.H. Dalitz: Philos. Mag. **44**, 1068 (1953)
81. PHELIX proposal: http://www-aix.gsi.de/~phelix/
82. A.B. Voitkiv, J. Ullrich: J. Phys. B **34**, 1673 (2001)
83. A.B. Voitkiv, J. Ullrich: J. Phys. B **34**, 4383 (2001)
84. F. Ehlotsky, A. Jaron, J.Z. Kaminski: Phys. Rep. **297**, 63 (1998)

22 Coulomb-Explosion Imaging Studies of Molecular Relaxation and Rearrangement

R. Wester, D. Schwalm, A. Wolf, and D. Zajfman

22.1 Foil-induced Coulomb–Explosion Imaging

The concept of studying single molecules by rapidly destroying them in a Coulomb-explosion and detecting the momenta of all emerging fragments has appealed to scientists for many years. In principle, all the information about the spatial structure of a molecule is revealed in the fragments' momentum distribution. This is illustrated in Fig. 22.1 for a diatomic molecule, which can be conveniently described by a one-dimensional potential energy curve as a function of the internuclear distance. To initiate a Coulomb-explosion all binding electrons have to be removed from the molecule with high efficiency before the atomic fragments start to move apart. In this case, the probability distribution of molecular bond lengths, which for a single molecular eigenstate is given by the rovibrational wave function squared, is mapped through the repulsive Coulomb potential into a distribution of kinetic energies of the two charged atomic fragments.

Three important conditions are required to ensure that a clear picture of the molecular structure can be extracted from Coulomb-explosion experiments: (a) All binding electrons have to be removed on a timescale shorter than the rotational and vibrational times of the molecule. In this case, the Coulomb-explosion starts from a snapshot of the "frozen" spatial configuration of the atoms in the molecule (indicated by the vertical arrow in Fig. 22.1). (b) The dissociation has to follow the known Coulomb potential (or an otherwise known dissociation potential) to make a clear mapping of fragment momenta to intramolecular bond lengths and angles feasible. And (c): The efficiency for inducing the Coulomb-explosion should not depend on the specific spatial configuration to ensure that the observable distribution of fragment momenta is a true representation of the distribution of bond lengths and bond angles in the molecule.

To calculate the kinetic energy release distribution $P(E_{\mathrm{kin}})$ of two Coulomb-exploding fragments in a quantum mechanically exact way, the sudden approximation, condition (a) above, is employed. $P(E_{\mathrm{kin}})$ is then given by the overlap integral of the initial molecular wavefunction $\Psi_{v,J}$ with Coulomb wave functions $\Phi_{E_{\mathrm{kin}},J}$,

$$P_{v,J}(E_{\mathrm{kin}})\,\mathrm{d}E_{\mathrm{kin}} = \left| \int \Psi_{v,J}(R)\Phi_{E_{\mathrm{kin}},J}(R)\,\mathrm{d}R \right|^2 \mathrm{d}E_{\mathrm{kin}}. \tag{22.1}$$

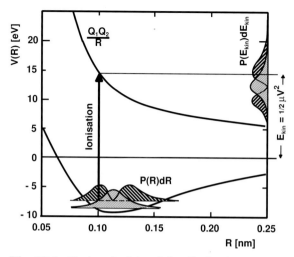

Fig. 22.1. Basic principle of the Coulomb-explosion imaging method, illustrated for a diatomic molecule. For prompt ionization, the kinetic energy release E_{kin}, and thus the relative velocity V between the fragments, is a direct measurement of the internuclear distance R of the two constituents at the moment of ionization. The corresponding kinetic energy release distribution $P(E_{kin})$ is an unambiguous reflection of the distance distribution $P(R)$

Often the classical approximation of (22.1) is used instead. The Coulomb wavefunctions then collapse to δ-functions located at the Coulomb potential and the kinetic energy release distribution is given by

$$P_{v,J}(E_{kin})\, \mathrm{d}E_{kin} = |\Psi_{v,J}(R)|^2 \, \frac{\mathrm{d}R}{\mathrm{d}E_{kin}}\mathrm{d}E_{kin} = P_{v,J}(R)\frac{\mathrm{d}R}{\mathrm{d}E_{kin}}\mathrm{d}E_{kin} \,, \quad (22.2)$$

where $E_{kin} = (Q_1 Q_2)/R$, Q_1, and Q_2 are the charge states of the two ionic fragments and R is the internuclear distance when the Coulomb-explosion is initiated. The classical approximation is valid when the de-Broglie wavelength associated to the Coulomb wave function is small compared to the de-Broglie wavelength of the molecular vibration, in other words when the relative velocity of the fragments after the explosion is large compared to the velocities of the vibrating atoms in the molecule. The classical approximation therefore neglects the initial velocities of the atoms, which is not the case in the quantum-mechanical description where these velocities are no independent dynamical variables. In practice the classical approximation is often good enough when applied to ground-state vibrational wave functions, compared to the experimental accuracy.

Several experimental approaches have been devised to study Coulomb-explosions of molecules, namely collisions with fast highly charged ions (H_2O, N_2 [1]), ionization in high-power laser pulses (I_2 [2]) and electron stripping of molecular beams by passage through thin target foils [3] (C_2H_2 [4], HD^+

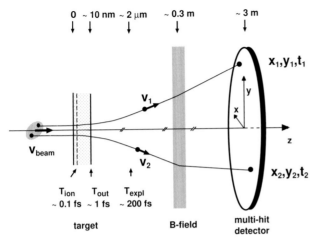

Fig. 22.2. Basic principle of a foil-induced Coulomb-explosion imaging measurement

[5], DCO$^+$ [6], etc. The latter technique of foil-induced Coulomb-explosion imaging (CEI) is covered in this article. The concept of this technique will be described in the following and the validity of the conditions (a)–(c) will be discussed. Important experimental and analysis details will be covered in Sect. 22.2 and a few applications will be presented in Sect. 22.3.

The basic principle of a CEI experiment is shown in Fig. 22.2. A fast molecule is passed through a very thin target foil, where its binding electrons are removed, thus dissociating the molecule into a set of charged atomic fragments. Mutual Coulomb repulsion between the fragments leads to a Coulomb-explosion. The momentum vector that each fragment acquires in the explosion is then measured using three-dimensional multiparticle imaging detectors. Before they hit one of the detectors the fragments usually pass through a magnetic deflection field, which allows for convenient mass and charge separation.

In order to swiftly remove the binding electrons of a molecule when it passes through matter, the velocity of the molecule has to be of the order of or larger than the typical Bohr velocity of the electrons. For hydrogen-like systems this velocity amounts to

$$v_{\mathrm{Bohr}} = \frac{Z\alpha c}{n}, \tag{22.3}$$

where Z is the nuclear charge, α the fine structure constant, c is the vacuum speed of light, and n the electron principal quantum number. The kinetic energy needed for an atom with mass m to strip all electrons from the n-th shell with high probability is thus estimated to be

$$E_{\mathrm{kin}} > 25 \,\mathrm{keV} \,\frac{Z^2}{n^2} m \,[\mathrm{amu}] . \tag{22.4}$$

Consequently, molecules to be studied by foil-induced Coulomb-explosion imaging have to be accelerated to similar kinetic energies, and thus only molecular ions, which can be efficiently accelerated in electric fields, are directly open for investigations. Neutral molecules can only be investigated by accelerating the corresponding negative ion, which is photodetached before the molecule impacts on the target foil.

The time it takes for a molecule to lose its binding electrons upon entering a solid target can be estimated by assuming a geometric scattering cross section $\sigma \approx 1\,\text{Å}^2$ and a target density $n \approx 10^{23}/\text{cm}^3$. With a beam velocity $v_{\text{beam}} = 4\alpha c$ a typical scattering time is

$$\tau = \frac{1}{\sigma n v_{\text{beam}}} = 10^{-16}\,\text{s} = 0.1\,\text{fs}, \tag{22.5}$$

which corresponds to a travel distance in the target of only 10 Å. Target foils with a thickness below 100 Å are therefore sufficiently thick to allow efficient electron stripping. At the same time the molecule spends only about 1 fs inside the target foil and almost all of the Coulomb-explosion dynamics of the atomic fragments, which starts when the binding electrons are removed, actually happens in the free space behind the target foil. 99% of the total Coulomb energy is released into kinetic energy within a few μm after the target.

While the electrons of the fast molecule are rapidly scattered, the much heavier atomic nuclei interact less with the atoms in the target foil. For a single atomic ion only two processes play a role inside the target, besides electron stripping. These are small angle-multiple scattering and energy loss. Multiple scattering leads to an angular spread of the incident beam of the order of one milliradian or less at FWHM. This is smaller than the scattering angle a fragment acquires in a Coulomb-explosion by typically a factor of 10, but leads to a noticeable broadening in the measured kinetic energy release distributions and needs to be accounted for when comparing CEI data to calculations (see Sect. 22.2). The relative energy loss in a CEI target foil, which occurs because the charged particle polarizes the target electrons in its wake and is thus decelerated, is typically of the order of 10^{-3}. The corresponding velocity changes of the fragments in a Coulomb-explosion are small compared to the velocities that the fragments acquire due to the explosion and can often be safely neglected. However, the wake field that a fragment creates inside the target has an additional effect in that it pulls a trailing second fragment into its wake and thus changes the orientation of the internuclear axis of the two fragments. While this wake field effect for diatomic Coulomb-explosions has been known for many years [7], the precise modeling of this effect in the Coulomb-explosion of a triatomic molecule (CH_2^+ [8]) has only recently been accomplished.

It remains to be shown that the foil-induced Coulomb-explosion technique is capable of extracting spatial structures of small molecules, based on the three conditions given above. Condition (a) is easily fulfilled, because the electron scattering time of 0.1 fs is much shorter than rotational or vibrational timescales of even small molecular ions. For example, the $\nu = 1$ vibrational level in the hydrogen molecular ion has an excitation energy of 290 meV, which corresponds to a vibrational time of 14 fs, two orders of magnitude faster than the stripping time. The rotational constant of H_2^+ is about 3.7 meV, leading to a rotational time of 1100 fs, another two orders of magnitude slower. It is an important advantage of the foil-induced Coulomb-explosion technique over femtosecond laser-induced CEI that up until recently femtosecond laser pulses were limited to pulse lengths of at best 4–5 fs and are thus not much faster than molecular vibrations. This reduces their ability to provide snapshots of molecular structure particularly for small, fast vibrating molecules. But the advent of attosecond laser pulses will very likely have an impact in this field.

The condition (b) is more difficult to fulfil in Coulomb-explosion imaging and this applies not only to foil-induced techniques, but also to laser and ion-induced Coulomb-explosion. The reason is that the dissociation potential is only Coulomb-like when all atomic fragments resemble point charges. This is easily achieved when the fragments are completely stripped of all their electrons; however, for a carbon-containing molecule such as CH_2^+ this requires e.g. a beam energy greater than 13 MeV, which is difficult to achieve using standard accelerator technology. When a carbon or oxygen fragment contains electrons on the $n = 1$ shell during the Coulomb-explosion the interaction between the fragments is still point-like to a good approximation, since the $n = 1$ wave functions are spherically symmetric and of much smaller dimension than molecular bond lengths. A more complicated situation occurs when electrons are excited to higher-bound levels, which leads to different non-Coulombic interaction potential curves. This effect is more pronounced in ion-impact ionization, where typically fewer electrons are removed from the molecule than in foil-induced electron stripping [1]. But also in foil-induced Coulomb-explosion this effect has been found to play a role [6,9] for the limited beam energies that have been available to experiments. An illustrative example is the Coulomb-explosion of HeH^+, where these effects have been investigated in more detail [9,10]. The consequence of this is that for heavier triatomic molecules only bond-angle distributions can be measured with the CEI technique, which are less sensitive to the steepness of the dissociation potential. This is a noteworthy feature, because rotational spectroscopy, which is the standard technique to determine the structure of small molecules, is less accurate for bond angles than for bond lengths and not directly sensitive to the distribution of bond lengths and angles. The bond-angle distribution of floppy molecules such as the quasilinear CH_2^+ and NH_2^+ molecules is therefore more accurately determined using the CEI technique than with standard spectroscopy (see Sect. 22.3.2).

The third condition needed to study molecular structure from Coulomb-explosion imaging, efficient dissociation independent of the internal molecular structure (noted above as condition (c)), is assumed to be fulfilled quite well in foil-induced Coulomb-explosion. This is essentially a consequence of the large scattering cross section for electrons in the foil, which does not depend on the distances to the comoving atoms in the molecule. This is e.g. not the case in single-photon photoionization, where the transition moment depends on the phase space of the ionized electrons and thus on the internuclear distance at which ionization occurred, as was shown for H_2 [11].

Based on the above discussion it can be seen that for small molecular ions and sufficiently high beam energies the foil-induced Coulomb-explosion imaging technique is capable of investigating molecular bond-length and bond-angle distributions. Such a probability distribution represents, for a single molecular energy level and within the Born–Oppenheimer approximation, the square of the molecule's rovibrational wave function. Assuming the molecules to be rather rigid rotors the dependence of the spatial structure on the rotational quantum number(s) is small and can often be neglected. The vibrational level, however, has a profound influence on the spatial spatial structure, because of the larger extension and the increased number of nodes of the vibrational wave function. This implies that in an experiment the vibrational excitation of the molecule under study has to be accurately known, which demands careful preparation of the molecular ions before they are passed to the Coulomb-explosion target foil. In the next section the heavy-ion storage ring technique will be discussed that achieves this preparation. The other implication of the CEI sensitivity to vibrational excitation is that the technique can be used as a diagnostic tool to determine the vibrational population of an ensemble of internally hot molecular ions. This procedure, which relies on known vibrational wave functions, has also been used successfully for a number of ions in a heavy ion storage ring (see Sect. 22.3).

22.2   Experimental Procedure

Experiments on foil-induced Coulomb-explosion imaging of fast molecular ions relies on an accelerator that delivers molecular ion beams at velocities of a few per cent of the speed of light (i.e. MeV energies). The fast molecular ions are then passed one by one through the CEI target foil and the Coulomb-explosion fragments are detected a few meters downstream with a dead-time-free three-dimensional imaging detector. This detector measures the velocity vectors of all fragments of an explosion event before the next molecule is passed through the target. A small magnetic field is used right after the target for mass and charge separation.

This basic CEI technique has been developed and employed at two experimental facilities: the Argonne National Laboratory, Chicago and the Weizmann Institute of Science, Rehovot [12]. The latter setup was used to study the CEI of neutral molecules, produced by photodetaching fast molecular anions. At both setups experiments have been hampered by the unknown

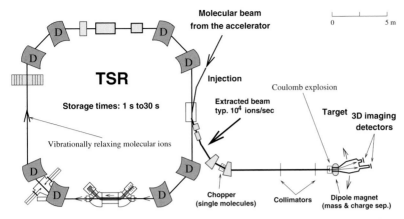

Fig. 22.3. Experimental setup at the Heidelberg Test Storage Ring TSR

vibrational excitation of the studied molecules. At our experimental facility at the Max-Planck-Institut für Kernphysik, Heidelberg, (see Fig. 22.3) the heavy-ion storage ring TSR is used to store the molecular ions for typically milliseconds to several seconds, before they are extracted from the storage ring and passed towards the CEI target foil and detector. During these storage times the molecular ions lose their vibrational excitation through spontaneous radiative emission, provided that molecular ions with dipole-allowed vibrational transitions are chosen. This not only allows preparation of vibrationally cold molecular ions, but it also offers the opportunity to study ions at different vibrational excitation during the first few hundred milliseconds of storage time. After several seconds the ions eventually reach a rovibrational level population in equilibrium with the 300 K black body radiation inside the vacuum chamber. Such a thermalization was shown in a photodissociation study of CH^+ [13,14].

The Coulomb-explosion imaging setup at the TSR, shown in Fig. 22.3 has been described in detail in [5]. Different molecular ions with masses between 2 and about 50 can be produced with either an electrostatic MP tandem van-de-Graaff accelerator or a radio-frequency quadrupole (RFQ) accelerator. The ions are injected into the TSR at their full energy within a few microseconds after their production. In the 55–m circumference magnetic storage volume of the TSR the ions then circulate for a variable storage time of between a few milliseconds and up to about one minute, limited only by the ion loss rate due to residual gas collision. A small fraction of the stored ions is extracted at a given time using a slow extraction scheme and is guided towards the CEI detector. Two collimators define the initial position and direction of the ion beam. An electrostatic high-voltage chopper, controlled by the imaging detector data acquisition system is used to pass the molecular ions to the target foil one by one.

The targets used so far in these experiments consist of thin Formvar [15] or diamond-like carbon (DLC) foils [16] on target holders covered by high transmittance (> 80%) meshes. Foil thicknesses down to 5 nm can be realized in this way, the DLC foils being superior to the Formvar foils of similar thickness as the percentage of pinholes of DLC foils can be kept well below a few per cent.

Behind the target foil, the different fragments from a Coulomb-explosion event are separated in a dipole magnetic field with respect to their charge-to-mass ratio. This allows separation of protons from heavier fragments and different charge states of, e.g., carbon, nitrogen or oxygen can be distinguished. After a drift length of 3 m each fragment hits one of the two imaging detectors (see Fig. 22.3). The position of the two detectors and the field strength of the dipole magnet are optimized to steer different fragments to separate areas on the two detectors. This allows the active detector area to be minimized compared to a design with a single detector.

The two imaging detectors are based on microchannel plates with phosphor screens and CCD cameras for position readout, as described in Chap. 3. The impact-time readout is performed using timing anodes underneath the phosphor screen that are digitized using constant-fraction discriminators and charge-ADCs. Each detector carries an additional CsI–coated aluminum foil in front of its MCP to produce an initial pulse of electrons for every impinging fragment. This increases the fragment-detection efficiency to near unity and leads to a stronger MCP signal, which enhances the impact-time resolution.

In a typical experiment a data sample of about 10^5 to 10^6 Coulomb-explosion events is acquired, where each event comprises the $3N$ impact position and time values for the N atoms in the parent molecule together with the storage time of the parent molecule in the storage ring. To simplify this high-dimensional event distribution relative velocity vectors are calculated and transformed into symmetry-adapted coordinates. The $3N$-dimensional space of the fragment velocities is often denoted V-space. It is contrasted by the $3N$-dimensional R-space in which each point describes a spatial structure of the parent molecule. In the case of a diatomic molecular ion the magnitude of the relative velocity between the two fragments is the decisive variable. Its probability distribution is related to the distribution of internuclear distances in the molecule in R-space. The other five available one-dimensional distributions reflect the overall distribution of the molecular center-of-mass and the orientation of the internuclear axis of the two fragments relative to the laboratory reference frame.

To compare experimental data to theoretical structure calculations a Monte Carlo model of the Coulomb-explosion is employed [17,18] which incorporates effects of the target foil and the finite resolutions of the imaging detectors. Such a forward simulation is used because a direct unambiguous calculation of the R-space distribution that corresponds to a measured V-space distribution is not feasible. The Monte Carlo simulation models the

Coulomb-explosion inside and behind the target foil using classical trajectory calculations and includes small-angle multiple scattering and a simple statistical description of charge exchange inside the target foil. Recently a model of the wake fields, which influence the relative orientation of the fragments inside the target foil, has been added successfully [8]. The essential parameters of the Monte Carlo model are the parent molecular ion velocity, which is measured with a Schottky pickup inside the TSR storage ring, the detector resolutions and the thickness of the target foil. The overall response of the detectors has been studied in detail using UV laser pulses and the position and time resolutions have been extracted from these data [19]. The target thickness is obtained by measuring the multiple scattering of atomic ion beams and fitting the scattering-angle distributions by the Monte Carlo model with the target thickness as the only free parameter.

The output of the Monte Carlo model is a $3N$-dimensional simulated data set of detector impact positions and times that is then analyzed in the same way as the experimental data. Direct comparisons of any of the V-space coordinates are therefore easily feasible.

22.3 Selected Results

A number of diatomic (H_2^+ [20], HD^+ [5,21,22], HeH^+, LiH^+ and CH^+ [9]) and triatomic (H_3^+ [23], D_2H^+, LiH_2^-, CH_2^+ [8,24], NH_2^+ [25], DCO^+ and DOC^+ [6]) molecular ions have been investigated so far with the combination of Coulomb-explosion imaging and heavy-ion storage ring technique. Results for two of these systems, HD^+ and CH_2^+, will be presented here, exemplifying the use of this technique to study relaxation processes as well as spatial structures of molecular ions.

22.3.1 Radiative Vibrational Relaxation of HD^+

HD^+ is the simplest molecular ion with a permanent dipole moment and therefore the simplest ion that exhibits radiative vibrational relaxation when stored in the TSR storage ring. The single electron is easily stripped when HD^+ is passed through the Coulomb-explosion target with a kinetic energy of 2 MeV, which allows for a precise investigation of its spatial structure, that is of its bond-length probability distribution in R-space, with the CEI technique. The goal of the experiment was to extract the storage-time dependence of the vibrational excitation of the stored HD^+ ions from the measured probability distributions in R-space. Such data, which can be determined using the well-known HD^+ vibrational wave functions, are needed to investigate electron-molecular ion collision processes, such as dissociative recombination, in a vibrational state-selective manner.

In this experiment, HD^+ molecular ions were produced in an electron-impact penning ion source on the terminal of a 2–MV single-ended van-de-Graaff accelerator and were then transferred to the heavy-ion storage ring

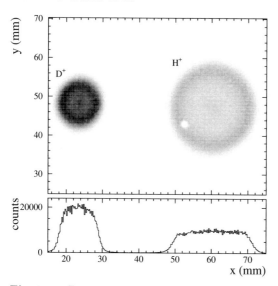

Fig. 22.4. Density plot of the transverse impact positions of H^+ (*right*) and D^+ fragments (*left*) on one of the Coulomb-explosion imaging detectors. The *lower panel* is a projection of the data on the x-direction (from [5])

TSR. The TSR was typically filled with 10^7 HD$^+$ ions, which are stored with a lifetime of about 10 s, limited by ion loss due to residual gas collisions. Ions with storage times between 2 ms and 1 s were continuously extracted from the storage ring, collimated by two circular apertures of 2 mm situated 3 m apart before hitting the CEI target, a 70–Åthick Formvar foil. The proton and deuteron fragments that explode after the binding electron is stripped are mass separated in a weak magnetic field of 7 mT, 150 mm downstream of the target, after which they fly straight onto one of the imaging detectors located 3 m behind the target foil. At this distance the explosion cones of the two Coulomb-explosion fragments have a diameter of 22 mm for protons and 11 mm for deuterons, horizontally separated by 37 mm due to the magnetic field. An example of the measured impact positions on the detector is displayed in Fig. 22.4. For about 10^6 coincident proton–deuteron events all impact positions and relative impact times have been measured in this experiment as a function of the storage time.

The circularly symmetric impact-position distributions in Fig. 22.4, which correspond to the projections of the two fragment velocity vectors onto the detector plane, already show qualitatively that the direction of the Coulomb-explosion and thus the molecular bond before the explosion is oriented isotropically in the laboratory system. This has been analyzed quantitatively using also the relative impact times of the two fragments [5].

The kinetic energy release in the HD$^+$ Coulomb-explosion is deduced for each event from the relative distance of the impact positions in the detector

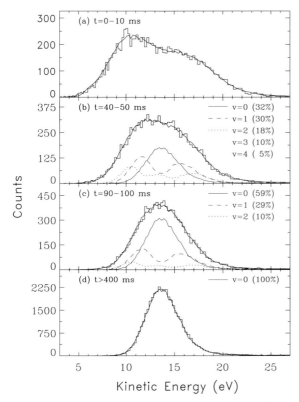

Fig. 22.5. Kinetic energy release spectra of HD^+ for different storage-time slices. The *thick solid lines* represent fits as described in the main text (from [22])

plane ΔD (after correcting for the deflection due to the magnetic field) and the impact time difference Δt using

$$E_{\mathrm{kin}} = \frac{E_{\mathrm{ion}}}{L^2} \frac{m_{\mathrm{H}} m_{\mathrm{D}}}{(m_{\mathrm{H}} + m_{\mathrm{D}})^2}((v_{\mathrm{ion}} \Delta t)^2 + \Delta D^2), \qquad (22.6)$$

where E_{ion} and v_{ion} are the HD^+ kinetic energy and velocity in the laboratory, L is the target–detector distance and $m_{\mathrm{H,D}}$ are the proton and deuteron masses. From the measured event sample a distribution $P(E_{\mathrm{kin}})$ is obtained that changes significantly when plotted for different storage-time intervals separately. This is shown in Fig. 22.5, where the measured distributions are plotted for three intervals of 10 ms each and for all storage times larger than 400 ms. After 400 ms, no further change in the kinetic energy release distribution was observed.

As described in Sect. 22.1, the kinetic energy release distribution reflects the distribution of bond lengths in the HD^+ molecule prior to the dissociation (see Fig. 22.1 and (22.1) and (22.2)). The changes in the measured energy distributions therefore reflect changes in the bond-length distributions, which

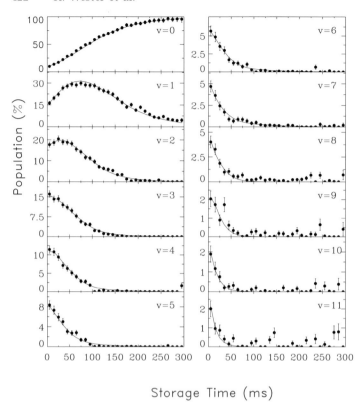

Storage Time (ms)

Fig. 22.6. Storage-time evolution of the vibrational level populations $p_v(t)$ of HD^+ for $0 \leq v \leq 11$, normalized to $\Sigma_v p_v(t) = 100\%$, as obtained by fitting the time-sliced kinetic energy release spectra. The *smooth solid lines* are the results from fitting $p_v(t)$ by a rate-equation model (from [22])

for HD^+ can only be caused by less and less strongly vibrating molecular ions in the storage ring; the possible influence of additional rotational excitation on the bond-length distributions is small compared to the differences between two distributions of adjacent vibrational levels, because for the rotational temperatures expected under the ion-source conditions used the estimated average J value amounts to only $< J > \sim 3\hbar$. After 400 ms of storage no further changes are observed, indicating that all HD^+ ions have cooled to the vibrational ground state (at 300 K ambient temperature the $v = 1$ state is not thermally populated). To test this the expected kinetic energy release distribution is calculated for the HD^+ ground-state vibrational wavefunction using the CEI Monte Carlo program. The good agreement of the calculation, plotted as a solid line in Fig. 22.5 (lowest panel), with the data shows that the HD^+ ions have indeed reached the vibrational ground state. It also indicates that the entire Coulomb-explosion processes for HD^+ are well reproduced with the Monte Carlo model.

In the next step, the vibrational excitation of the HD^+ ions during the first 400 ms of storage, before they reach $v = 0$, is extracted from the kinetic energy distributions. For this purpose, Monte Carlo simulations have been carried out for all HD^+ vibrational wave functions from $v = 0$ to $v = 11$ employing the full quantum-mechanical treatment of the Coulomb-explosion process (see (22.1)), and a linear superposition of the 12 corresponding kinetic energy distributions have been fitted to the measured distributions for each 10 ms storage-time interval. The only fit parameters were the relative populations $p_v(t)$ of the vibrational levels. For the three intervals shown in the top panels of Fig. 22.5 the resulting fits are shown as solid lines. In the second and third panel the contributions of the individual vibrational levels are plotted and their relative populations are given, the node structure of the wavefunctions being clearly reflected in the simulated kinetic energy distributions.

The fitted vibrational populations $p_v(t)$ for all storage times are shown in Fig. 22.6. Within 10 ms after their production in the ion source, the HD^+ ions still carry a lot of vibrational energy, but the highest vibrational levels decay quickly and already after 50 ms only the lowest five vibrational levels remain populated. Their decay leads to an increase of the lower vibrational states, best seen for $v < 3$, because the excited HD^+ ions are not lost from the storage ring, but decay by spontaneous radiative transitions into lower vibrational levels. Finally the vibrational ground state is populated to 100%.

The entire vibrational relaxation dynamics can be fitted with a rate-equation model that assumes only $\Delta v = 1$ transitions (solid lines in Fig. 22.6) [22]. This yields experimental vibrational lifetimes of the lowest 12 vibrational states, which are in good agreement with theoretical Einstein coefficients for these transitions. Furthermore, the vibrational diagnostics of the stored HD^+ ions, obtained with the Coulomb-explosion imaging technique, has been utilized to investigate vibrational-state-dependent rate coefficients for the dissociative recombination of HD^+ with free electrons [21,22] (see Chap. 26).

22.3.2   The Quasilinear Molecule CH_2^+

CH_2^+ is an example of a molecular ion with a very soft, "floppy" spatial structure in its bond angle between the two hydrogen atoms. As such it is very difficult to determine molecular-structure information by conventional spectroscopy, which makes it an interesting case for the Coulomb-explosion imaging technique. The electronic structure of CH_2^+ at the linear configuration is of $^2\Pi_u$ symmetry, the degeneracy of which is lifted due to vibronic coupling (Renner–Teller effect). As a consequence, the electronic ground state $\tilde{X}^2A_1$ has its minimum energy not at the linear configuration, but at a bond angle of about 140°, whereas the first electronically excited state $\tilde{A}^2B_1$ has its minimum at the linear configuration, only 130 meV higher in energy than the minimum of the $\tilde{X}^2A_1$ potential energy surface. Furthermore, the potential energy minimum of the $\tilde{X}^2A_1$ state is very shallow and anharmonic along the bond-angle coordinate, which leads to the "floppy" spatial structure [26].

In the experiments at the TSR [8,24], a beam of CH_2^+ ions is produced indirectly by starting with a beam of methoxy anions CH_3O^- from a cesium sputter source. The methoxy beam is then injected into a tandem accelerator, accelerated to 4.7 MeV and fragmented in a gas stripper on the high-voltage terminal of the tandem. The cationic fragments are then further accelerated towards the high-energy exit of the tandem, where the CH_2^+ fragments with a total kinetic energy of 6.7 MeV are selected in the dipole field of a deflection magnet. This CH_2^+ beam is then injected into the TSR storage ring, from which a small fraction of the ions is continuously extracted towards the Coulomb-explosion imaging experiment. CEI data were acquired for CH_2^+ storage times between 2 ms and 15 s using, in this case, diamond-like carbon (DLC) targets [16]. At a beam energy of 6.7 MeV, the maximum energy at which CH_2^+ ions can be stored in the TSR, not all the electrons are stripped inside the target foil. However, in the experiment the most likely charge state of the carbon fragment was C^{4+}, which leaves only two, most likely 1s, electrons with the carbon nucleus. With the magnetic field behind the target foil these C^{4+} fragments are steered towards one of the imaging detectors, while the two coincident protons are steered towards the other detector.

For each CEI event the bending geometry of the explosion is characterized in V-space by the angle ω_V between the two proton velocities relative to the carbon velocity, namely

$$\cos \omega_V = \frac{(\boldsymbol{V}_{H_a^+} - \boldsymbol{V}_{C^{4+}}) \cdot (\boldsymbol{V}_{H_b^+} - \boldsymbol{V}_{C^{4+}})}{|\boldsymbol{V}_{H_a^+} - \boldsymbol{V}_{C^{4+}}||\boldsymbol{V}_{H_b^+} - \boldsymbol{V}_{C^{4+}}|} \,, \tag{22.7}$$

where $\boldsymbol{V}_X$ is the laboratory velocity vector of fragment X after the Coulomb-explosion. These velocity vectors are directly obtained from the impact positions and times that are measured with the imaging detectors.

The measured distribution of the cosine of this bond angle $P(\cos \omega_V)$ is plotted in Fig. 22.7 for molecules with storage times larger than 2 s. This has been shown to be sufficient for all vibrational excitation to be removed by spontaneous radiative transitions [24]. To suppress possible disturbances of the V-space distributions due to imperfections in the description of the wake-field effects, only molecules are considered for which the normal $\boldsymbol{n}$ of the molecular plane before the explosion is oriented almost parallel to the beam axis $\boldsymbol{z}$ by requiring $|\cos(\boldsymbol{n}, \boldsymbol{z}) \geq 0.8|$. As depicted in Fig. 22.7, the angle ω_V in V-space is distributed over a large range of angles from $180°$ ($\cos \omega_V = -1$) to less than $90°$ ($\cos \omega_V > 0$), reflecting the broad distribution of bond angles in R-space.

When comparing this distribution to a Monte Carlo simulation based on a theoretical R-space distribution, such as that of the ground vibrational level, it is important that the target effects are accurately taken into account. In an earlier experiment at the TSR storage ring a discrepancy was found between CEI data [24] and a theoretically expected bond-angle distribution, which used the adiabatic approximation to calculate the nuclear wave function in

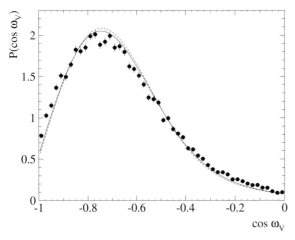

Fig. 22.7. Measured distribution of the explosion angle in V-space of CH_2^+ in comparison with a Monte Carlo simulation of the theoretical bond-angle distribution

the $\tilde{X}^2 A_1$ state [26]. It was recently resolved that this discrepancy was partly due to uncertainties in the strength of the multiple scattering in the target foil and, to a minor degree, due to the lack of a wake-field description when several ions traverse the target close to each other [8]. In the meantime, also a more accurate calculation of the bond-angle distribution in R-space was performed using the nonadiabatically coupled $\tilde{X}^2 A_1$ and $\tilde{A}^2 B_1$ states and a thermal average of the vibrational levels populated at $300\,\mathrm{K}$ [27]. For this calculation the corresponding angular distribution in V-space is calculated using the updated version of the Monte Carlo program and plotted in Fig. 22.7 as the solid line. The good agreement that is now found supports our improved understanding of the effects in the Coulomb-explosion target. The dashed line in Fig. 22.7 represents the V-space distribution if wake-field effects are neglected. For the selected orientations of the molecular plane, the influence of wake-field effects on the measured distribution is, as expected, rather small. However, for other orientations the inclusion of these effects is increasingly important in order to achieve a satisfactory agreement between experimental and calculated distributions [8].

22.4 Outlook

Coulomb-explosion imaging is a powerful tool to obtain an alternative, direct view of the distribution of spatial structures of small molecules and molecular ions. In recent years changes of molecular structure, e.g. due to vibrational relaxation, have become the focus of CEI experiments, making use of the unique combination of the CEI beamline with the heavy-ion storage ring TSR. Diagnosing the vibrational excitation of a molecular ion and measuring vibrational lifetimes will be an important aspect of CEI experiments in the future. Considering recent advances in the understanding of molecular ion-target interactions, also direct spatial structure measurements for more complex floppy molecules are an attractive perspective.

References

1. U. Werner, J. Becker, T. Farr, and H.O. Lutz: Nucl. Instrum. Methods B **124**, 298 (1997)
2. H. Stapelfeldt, E. Constant, H. Sakai, and P.B. Corkum: Phys. Rev. A **58**, 426 (1998)
3. Z. Vager, R. Naaman, and E.P. Kanter: Science **244**, 426 (1989)
4. J. Levin, H. Feldman, A. Baer, D. Ben-Hamu, O. Heber, D. Zajfman, and Z. Vager: Phys. Rev. Lett. **81**, 3347 (1998)
5. R. Wester, F. Albrecht, M. Grieser, L. Knoll, R. Repnow, D. Schwalm, A. Wolf, A. Baer, J. Levin, Z. Vager, and D. Zajfman: Nucl. Instrum. Methods A **413**, 379 (1998)
6. R. Wester, U. Hechtfischer, L. Knoll, M. Lange, J. Levin, M. Scheffel, D. Schwalm, A. Wolf, A. Baer, Z. Vager, D. Zajfman, M. Mladenović, and S. Schmatz: J. Chem. Phys. **116**, 7000 (2002)
7. Z. Vager and D.S. Gemmell: Phys. Rev. Lett. **37**,1352 (1976)
8. L. Lammich et al.: to be published
9. L. Knoll: Ph. D. thesis, Universität Heidelberg (2000)
10. L. Knoll et al.: to be published
11. R. Dörner, H. Bräuning, O. Jagutzki, V. Mergel, M. Achler, J.M. Feagin, R. Moshammer, T. Osipov, A. Bräuning-Demian, L. Spielberger, L.M.H. Prior, J.H. McGuire, N. Berrah, J.D. Bozek, C.L. Cocke, and H. Schmidt-Böcking: Phys. Rev. Lett. **81**, 5776 (1998)
12. D. Kella, M. Algranati, H. Feldman, O. Heber, H. Kovner, E. Miklazky, E. Malkin, R. Naaman, D. Zajfman, J. Zajfman, and Z. Vager: Nucl. Instrum. Methods A **329**, 440 (1993)
13. U. Hechtfischer, Z. Amitay, P. Forck, M. Lange, J. Linkemann, M. Schmitt, D. Schwalm, R. Wester, D. Zajfman, and A. Wolf: Phys. Rev. Lett. **80**, 2809 (1998)
14. U. Hechtfischer, C.J. Williams, M. Lange, J. Linkemann, D. Schwalm, R. Wester, and D. Zajfman: J. Chem. Phys. **117**, 8754 (2002)
15. G. Both, E.P. Kanter, Z. Vager, B.J. Zabransky, and D. Zajfman: Rev. Sci. Instrum. **58**, 424 (1987)
16. J. Levin, L. Knoll, M. Scheffel, D. Schwalm, R. Wester, A. Wolf, A. Baer, Z. Vager, D. Zajfman, and V.K. Liechtenstein: Nucl. Instrum. Methods B **168**, 268 (2000)
17. D. Zajfman, G. Both, E.P. Kanter, and Z. Vager: Phys. Rev. A **41**, 2482 (1990)
18. D. Zajfman, T. Graber, E.P. Kanter, and Z. Vager: Phys. Rev. A **46**, 194 (1992)
19. M. Scheffel: Master's thesis, Universität Heidelberg (1999)
20. S. Krohn, Z. Amitay, A. Baer, D. Zajfman, M. Lange, L. Knoll, J. Levin, D. Schwalm, R. Wester, and A. Wolf: Phys. Rev. A **62**, 032713 (2000)
21. Z. Amitay, A. Baer, M. Dahan, L. Knoll, M. Lange, J. Levin, I.F. Schneider, D. Schwalm, A. Suzor-Weiner, Z. Vager, R. Wester, A. Wolf, and D. Zajfman: Science **281**, 75 (1998)

22. Z. Amitay, A. Baer, M. Dahan, J. Levin, Z. Vager, D. Zajfman, L. Knoll, M. Lange, D. Schwalm, R. Wester, and A. Wolf: Phys. Rev. A **60**, 3769 (1999)
23. H. Kreckel et al.: to be published
24. A. Baer, M. Grieser, L. Knoll, J. Levin, R. Repnow, D. Schwalm, Z. Vager, R. Wester, A. Wolf, and D. Zajfman: Phys. Rev. A **59**, 1865 (1999)
25. A. Baer: Ph. D. thesis, Weizmann Institute of Science, Rehovot (1999)
26. W.P. Kraemer, P. Jensen, and P.R. Bunker: Can. J. Phys. **72**, 871 (1994)
27. G. Osmann, P.R. Bunker, W.P. Kraemer, and P. Jensen: Chem. Phys. Lett. **309**, 299 (1999)

23   Charged-Particle-Induced Molecular Fragmentation at Large Velocities

A. Cassimi, M. Tarisien, G. Laurent, P. Sobocinski, L. Adoui, J.Y. Chesnel,
F. Frémont, B. Gervais, and D. Hennecart

23.1   Introduction

A present major challenge is the understanding of molecular-fragmentation
dynamics. Numerous studies have already been devoted to molecular fragmentation induced either by electron impact (see, for example, [1] for a review), synchrotron radiation ([2–4] or [5] for a review), or femtosecond lasers
[6]. Some very first studies using fast multicharged ions have been reported
[7–10]. In the case of ion-induced fragmentation, one may address many different fundamental aspects. On the one hand, the study of the fragmentation
may be used as a probe of the electronic processes involved. In the field of
radiation damage, these processes dominantly contribute to energy loss of
swift projectiles in matter. Indeed, the energy is deposited on the electrons
[11] and is known to lead to the production of defects due to displacement
of atoms. One fundamental remaining question in this field is how electronic
energy is transformed into kinetic energy of the nuclei, which finally ends
as tracks in the bulk. Ion-impact-induced fragmentation of molecules is an
example of how this transfer may proceed. In the particular case of biological
tissue, water radiolysis is identified as an important effect leading to the observed damages (cell death or mutation) [12]. The study of the fragmentation
of H_2O is therefore important for that issue [13,14].

   On the other hand, ion impact is a way to produce molecular ions, in different excited states, which may dissociate. The chemical forces originating
from the remaining electrons govern this dissociation. Thus, a refined analysis
of the fragmentation patterns such as stability, branching ratios between the
different fragmentation pathways, and kinetic energy release (KER) distributions, plays a major role in assessing the validity of the computed theoretical
molecular potential energy surfaces [15].

   Multiple ionization of a molecule may be reached by different means:
single-photon photoionization (synchrotron radiation), multiphoton photoionization (high-power lasers), and fast charged-particle impact. Single photoionization may be seen as different from the two others. Let us recall only two
main characteristics of this process. First, the energy deposited on the molecule is known and equals the photon energy. Second, this process is resonant,
meaning that by choosing the photon energy, one may select the molecular
electron to be ionized. In the case of innershell photoionization it is even

possible to select the atomic site to which the electron first belongs [5]. For the two other types, the situation is more complicated in the sense that the amount of energy deposited on the molecule is not unique, depending either on the number of absorbed photons, in the case of lasers, or on the impact parameter, in the case of collisions. Furthermore, the resonant character of the excitation is also lost even if mainly valence electrons are concerned.

In the particular case of fast ion-impact ionization, almost no momentum is transferred to the target. Equivalence to photoionization has been discussed earlier in the case of atomic targets [16,17]. The main difference between ions and lasers remains in the interaction duration that is much shorter for ion impact. So one may consider fast ion impact as ultrashort laser pulses. While laser pulse-durations lie in the femtosecond range, ion-molecule interaction may be two orders of magnitude shorter. Such short time excitation means that nuclear motion of the target fragments is only due to the fragmentation process and is free from any interaction with the projectile ion. This is not true for laser excitation since the fragments start to move during the light pulse.

Comparing the collision time to the different molecular characteristic times (vibration, rotation) leads to the commonly accepted two-step picture for molecular fragmentation. In a first step, electron removal takes place on a fixed-in-space molecule and the transient molecular ion has the equilibrium internuclear distance of the neutral molecule. Then, in a second step, nuclear motion starts, driven by the fragmentation dynamics.

While dissociation of multi-ionized molecules by photons has been the subject of numerous studies [5], ion-impact ionization of molecules has really been investigated in detail only the few years. This has been the occasion to bring techniques that have been designed for ion-atom collision experiments to the field of photoionization [18,19]. As a consequence, the different communities have tighter connections, which will improve our understanding of the excited molecule-fragmentation dynamics.

The growing interest in the understanding of molecular fragmentation has strongly benefited from the recent evolution of the recoil ion momentum spectroscopy (RIMS), which now allows now multihit detection. Measurement of all fragment momenta in coincidence makes it possible to determine the complete kinematics of the process.

23.2   Molecular Fragmentation

23.2.1   Branching Ratios
and Multielectron Removal Cross Sections

An important characteristic of ion impact is the possibility to highly multi-ionize a molecule in a single collision. Thus, many of the earlier experiments were devoted to the measurement of ionization cross sections. While for an atomic target, a simple time-of-flight mass spectrometer is sufficient, molecular targets are more complicated to deal with since molecular ions mainly dissociate as soon as their charge state is higher than 2. Thus, coincident mass spectroscopy (similar to the photoion photoion coincidence (PIPICO) technique [20]) is needed in which both fragments (for a diatomic molecule) have to be identified. Consequently, both cross section and dissociation pathways are obtained in such experiments. The latter gives access to an important feature which is how the remaining electrons are shared among the fragments during the dissociation process, through the branching ratios between the different paths corresponding to the same transient molecular ion charge state. We performed such measurements in the case of $6.7\,\mathrm{MeV/u}$ Xe^{44+} on carbon monoxide (CO) molecules [21].

In the case of single or double ionization molecular ions are observed: CO^+ molecular ion is found to be the dominating final channel of the singly ionized molecule ($73.1\pm2.4\%$), while a few CO^{2+} molecular ions are also detected ($5.3\pm1.0\%$ of the total double ionization events). These latter molecular ions are still present since metastable states may be populated. Thus, their lifetime is longer than the $650\,\mathrm{ns}$ spent in the extraction zone. All molecular ions with charge states higher than 2 are dissociated in less than 300 ns. We present in Table 23.1 the weight of the different fragmentation channels. Charge-symmetric dissociations are found to be favored. Five channels involving a neutral fragment ($\mathrm{C/O^+}$, $\mathrm{C^+/O}$, $\mathrm{C^{2+}/O}$, $\mathrm{C/O^{2+}}$, $\mathrm{C^{3+}/O}$) are also observed. As reported by Becker et al. [22] on the q-fold ionization, the $\mathrm{C}^{q+}/\mathrm{O}$ channels are found to have larger branching ratios than the $\mathrm{C/O}^{q+}$ ones. This is usually attributed to the lower ionization potential of the carbon ion [9]. The comparison of these branching ratios with those obtained with $1\,\mathrm{MeV/u}$ F^{4+} [9] and $2.4\,\mathrm{MeV/u}$ Ar^{14+} [23] does not exhibit any strong difference. The relative cross sections for removing q electrons from the CO molecule are obtained after summing the contributions of all of the ion pairs detected coming from the same molecular-ion charge state. The results are presented in Fig. 23.1. Multiple ionization is found to represent about 40% of the ionizing events (about 50% more than in the $1\,\mathrm{MeV/u}$ F^{4+} projectile

Table 23.1. Branching ratios for each fragmentation channel for $6.7\,\mathrm{MeV/u}$ Xe^{44+}+CO collisions. The results are presented as % of total single ionization (dissociative or not) [21]

Channels	O	O^+	O^{2+}	O^{3+}	O^{4+}	O^{5+}
C		12.37 ±1.84	0.31 ±0.73			
C^+	14.49 ±1.24	14.29 ±1.08	2.27 ±0.23	0.16 ±0.05		
C^{2+}	1.96 ±1.21	4.83 ±0.67	3.14 ±0.29	1.35 ±0.28	0.32 ±0.08	0.14 ±0.04
C^{3+}	0.54 ±0.4	0.56 ±0.09	1.13 ±0.14	0.89 ±0.15	0.54 ±0.07	0.32 ±0.12
C^{4+}			0.14 ±0.03	0.34 ±0.06	0.32 ±0.06	0.34 ±0.05
C^{5+}					0.12±0.01	

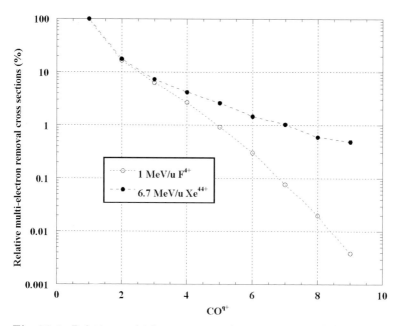

Fig. 23.1. Relative multielectron removal cross sections of CO molecule by two different projectiles [9,21]

case). Note that a qualitatively similar evolution of the multiple-ionization cross sections for atomic targets (Ar) by the same projectiles has already been reported [24,25]. It exhibits the high ionizing power of such heavy ions in the strong interaction regime, already pointed out in the atomic-target ionization case [26,27].

23.2.2   Orientation Effect

One degree of freedom that is not relevant in the case of atomic targets but appears for molecules is the orientation of the internuclear axis with respect to the projectile ion beam at the moment of the collision. This parameter, on which the primary process (i.e., the ionization probabilities) may depend, is specific to molecular targets. The cylindrical symmetry of the ion–atom collisions is lost and the impact parameter is no longer sufficient. Understanding the collision and fragmentation processes needs the knowledge of these angular distributions. Furthermore, a strong orientation effect may influence cross section measurements since many experiments rely on the strong assumption of an isotropic distribution [9,21].

On the basis of simple physical arguments, it was expected that a low degree of ionization of diatomic molecules may be enhanced when the molecular axis is transverse to the projectile beam: the geometric size of the target is larger in that case. Meanwhile, a molecular axis aligned with the projectile velocity direction may be preferred for multiple ionization: the projectile experiences a higher electron density when the impact parameter is small with respect to both target nuclei.

This alignment effect was observed in an early experiment with $1\,\mathrm{MeV/u}$ F^{9+} ions colliding with N_2 molecules [28]. A simple geometrical model was first proposed to predict this effect [29,30], which relies on two hypotheses. First, the molecule is considered as the association of two independent atoms and the resulting electron distribution is treated as the sum of the independent atomic ones. Secondly, the ionization cross sections are calculated within the independent electron approximation. This latter assumption and the use of binomial law allow the multiple ionization from one electron probabilities to be treated. In this way, a strong orientation effect has been predicted for swift low charge state projectiles [30]. However, single-electron ionization probabilities calculated by the classical trajectory Monte Carlo method have shown that fast highly charged projectile ions, such as Xe^{44+}, lead to orientation effects that may bearly be larger than a few per cent [21]. This prediction has been recently confirmed in the case of $5.9\,\mathrm{MeV/u}$ Xe^{43+} impact on N_2 in which the angular distribution starts to deviate from isotropy for a target charge state as high as 10 [31] (Fig. 23.2).

A more quantitative calculation based on the statistical-energy deposition (SED) model [32] compared favorably with a systematic experimental study for He collisions with N_2 in the 100–300 keV energy range [32,33]. This model has been recently extended in order not to be limited to small projectile charges and high velocity [34]. The main behaviors predicted by this calculation is that for a given degree of ionization, the orientation effect is vanishing with increasing projectile velocity and projectile charge (Fig. 23.2). This effect is expected to be weakly sensitive to the molecular properties. Today, it seems clear that this orientation dependence shows up if the impact parameter range at which the process occurs (i.e., multiple ionization) is of the order of the molecule size. Thus, the original idea [29,30] of a geometrical effect is confirmed. Using simple arguments, Kaliman et al. [34] were able to estimate qualitatively the lowest degree of ionization, $n_{\min}$, at which the orientation effect may be observed in the case of diatomic molecules. Two interaction strength ranges, characterized by the Sommerfeld parameter $\kappa = Z_P/v_P$, are separated:

$$\kappa \ll 1 \quad \text{(weak interaction)} \quad n_{\min} \simeq 3.3\,v_P^{1/3}\,, \tag{23.1}$$

$$\kappa \gg 1 \quad \text{(strong interaction)} \quad n_{\min} \simeq 5\kappa\,. \tag{23.2}$$

These scaling laws are in agreement with both extended SED calculation and experimental results [31,33]. If one assumes a continuous transition between these two scaling laws, one may obtain, for a given projectile charge state Z_P, the minimal number of electrons to remove from the molecule in order to observe an orientation effect and the most favorable collision veloc-

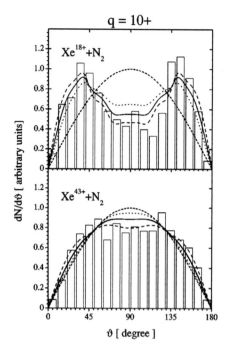

Fig. 23.2. Orientation dependence of the 10-fold ionization of N_2 molecules by 5.9 MeV/u Xe^{18+} and Xe^{43+} ions. The histograms correspond to the experimental results. *Short-dashed curve* shows a sine distribution that would be obtained for no angular dependence. The other *curves* are the results of the theoretical calculation within the extended statistical-energy-deposition model [31]. ϑ is the angle between the molecular axis and the projectile beam

ity v_P:

$$n_{min} \simeq 3.66 \sqrt[4]{Z_P} \tag{23.3}$$

$$\kappa = Z_P/v_P \simeq 0.73 \sqrt[4]{Z_P}. \tag{23.4}$$

An important parameter in these calculations is the molecule length, which, for a diatomic molecule, is close to 1 a.u.. Following the theoretical prediction, the orientation effect should be enhanced in the case of longer molecules. Further experiments are still needed to confirm this expectation.

23.3   Fragmentation Dynamics

Compared to atoms, molecules have an additional channel for relaxing the excess of internal energy that is the fragmentation process. This process is directly due to the interplay between electronic excitation and nuclear motion. The breakthrough came from the imaging techniques that recently have

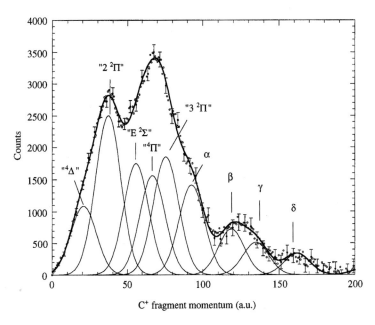

Fig. 23.3. C^+ momentum distribution corresponding to the $(CO^{+*} \to C^+ + O)$ fragmentation channel [39]

allowed access to the complete kinematics of the process through coincident measurement of all fragment momenta.

23.3.1   Non-Coulombic Fragmentation

The scenario of the explosion may be separated into two steps. First, a transient molecular ion is produced during the collision through the same processes as for atomic targets (ionization, capture, and excitation). Then, since most of these multiply charged molecular ions are not stable, fragmentation takes place and leads to atomic (or molecular) charged fragments. In the case of diatomic molecules, the transient molecular ion charge is preferentially equally shared between the two fragments [21,23,35]. The reason is that electron rearrangement is much faster than the dissociation process.

The origin of fragment-ion kinetic energy has already been assessed in experiments (e.g., [36,37]) but, because of the resolution, the mean kinetic energy release values could only be estimated. These results were in fair agreement with the predictions of the Coulomb-explosion model (CEM). This model assumes that, in the case of a diatomic molecule, the final kinetic energy of the fragments is equal to the initial Coulomb repulsion energy: $14.4q_1q_2/r$ (eV), where q_1 and q_2 are the asymptotic charges of the two fragments and r (Å) is the neutral molecule internuclear distance at equilibrium. This distance is justified by the fact that the electron removal proceeds much faster than the nuclear motion. More recently, although the resolution was again not sufficient to show any clear structure, it was observed that a unique KER value, as expected from the CEM, could no longer be invoked [8,10,21,35,38,39]. It should be pointed out that the experimental KER average values were found to be either higher [8,35,38,40] or lower [10,38,40] than the CEM predictions. This was interpreted by, respectively, the weaker screening of the nuclear charge by excited electrons or the formation of molecular ions in electronic excited states with non-Coulombic potential energy curves. An early experiment was able to answer this question [39].

In the excitation-ionization case (6.7MeV/u $Xe^{44+}+CO \rightarrow CO^{+*}$), only (27±3%) of the CO^{+*} ion dissociates. The main fragmentation pathway is the C^+/O channel, leading to the emission of just one charged fragment. For this one, we directly measured the KER distribution through the determination of the full momentum vector of the C^+ fragment by RIMS. The momentum resolution was around 8 a.u. for a C^+ fragment. The total C^+ momentum distribution corrected for collection efficiency is presented in Fig. 23.3 as well as the deconvolution performed to determine the contribution of all of the excited states. Note that the whole procedure has been detailed in [39]. An excellent agreement is found for the low-lying excited states and the calculation of Krishnamurthi et al. [41] – for the first five states – and those observed by Mathur and Eland [42] by photoionization and by Baltzer et al. [4] by high-resolution photoelectron spectroscopy – for the next two states referred to as α and β on Fig. 23.3.

The detection of both fragments in coincidence further improved the KER resolution, reaching a value of 250 meV, leading to clear structures on the spectra obtained with different projectiles impinging on CO molecules (Fig. 23.4). These structures have been successfully interpreted by the use of 'exact' molecular potential energies [43]. Such experiments have shown that, in conjunction with high-level computations of dication potential energy curves and application of time-dependent wavepacket dynamics make spectroscopy of molecular-ion energy levels feasible. Thus, fragmentation process is completely driven by dissociative excited states of the transient molecular ion produced during the collision.

23.3.2 Polyatomic Molecules

The dynamics of the fragmentation process, rather simple for diatomic molecules, becomes more interesting in the case of polyatomic molecules due to the coupling between the additional internal degrees of freedom (bending, stretching). In order to reveal the role of these couplings, we have chosen the linear triatomic molecule CO_2 as the target. In order to end with a complete dissociation of the molecule into atomic fragments, different scenarios may hold. According to Maul and Gericke [44] terminology, the different fragmentation mechanisms are distinguished as follows. Two independent two-body

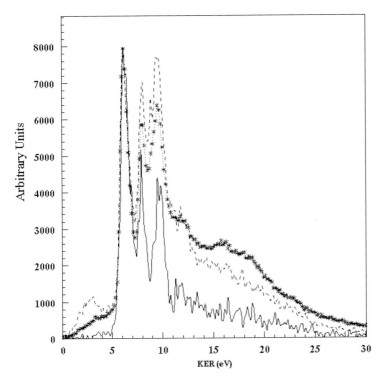

Fig. 23.4. Comparison between kinetic energy release (KER) distributions for the $C^+ + O^+$ fragmentation induced by $8\,MeV/u\ Ni^{24+}$ (*dashed curve with symbols*), $11.4\,MeV/u\ O^{7+}$ (*dashed curve*) and $4\,keV/u\ O^{7+}$ (*straight line*) projectiles. The spectra are normalized to the 6–eV peak intensity

dissociation step reactions will be referred to as sequential fragmentation. In that case, the first atomic fragment has no interaction with the other fragments when the second dissociation step of the intermediate diatomic fragment takes place. Beside this pure sequential behavior, all other mechanisms may be termed concerted and cover reactions in which the two bond breakages are correlated. Hsieh and Eland [45] make a further distinction between synchronous-concerted reactions for which the bond breakage are simultaneous as well as symmetric and asynchronous-concerted reactions. The latter category includes from instantaneous dissociation via asymmetric stretch to nonpure sequential dissociation, meaning a fast second step influenced by the primary fragment.

This is the starting point of our analysis of the fragmentation dynamics of CO_2 molecular ions induced by $8\,MeV/u\ Ni^{24+}$ projectile ions. One useful way to picture the three-body dynamics is Newton diagrams that show the magnitude and direction of the three momentum vectors. In Fig. 23.5 and

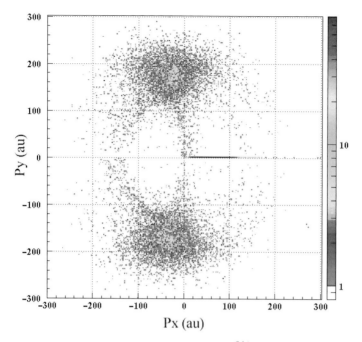

Fig. 23.5. Newton diagram for 8 MeV/u Ni^{24+}-induced CO$_2$ triple ionization and C$^+$+O$^+$+O$^+$ fragmentation channel. C$^+$ momentum is along the x-axis. First (second) detected O$^+$ is in the positive (negative) y-half-plane, respectively. Momenta are expressed in atomic units

Fig. 23.6, the C$^+$ momentum direction is chosen as the x-axis, while the second fragment momentum direction, O$^+$ in both figures, defines the positive y-half-plane.

The Newton diagram corresponding to the CO$_2$ triple ionization (C$^+$+O$^+$+O$^+$) dissociation channel, given in Fig. 23.5, shows a symmetric emission of the two oxygen ions in the molecular frame. This behavior is characteristic of a synchronous-concerted fragmentation. This is re-enforced by the fact that the carbon ion is nearly at rest and high momenta are measured for both oxygen ions.

The case of CO$_2^{2+}$ dissociation is more complex. From Fig. 23.6 we observe that the reaction is clearly not synchronous-concerted at all. Is the fragmentation sequential? It should be recalled that two types of sequential reactions have been observed earlier in the case of dissociative double ionization [46]: secondary decay (which would correspond to a CO$^+$+O$^+$ intermediate step) and deferred charge separation (a CO^{2+}+O intermediate step). Since Fig. 23.6 shows that the neutral oxygen fragment is emitted towards the C$^+$ momentum direction, the reaction is more likely to be of the secondary decay type. In order to clearly establish this statement one

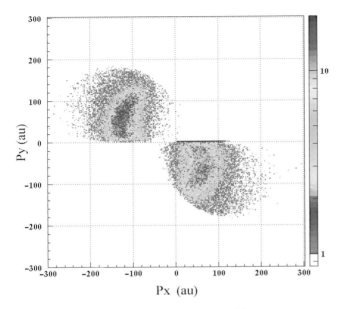

Fig. 23.6. Newton diagram for $8\,\mathrm{MeV/u}$ Ni^{24+}-induced CO_2 double ionization and $C^+ + O^+ + O$ fragmentation channel. C^+ momentum is along the x-axis. O^+ (O) fragment is in the positive (negative) y-half-plane, respectively. Momenta are expressed in atomic units

has to plot another diagram that quantitatively shows the momentum balance.

Figure 23.7 represents the momentum component of each fragment along the O^+ momentum direction, as a function of the kinetic energy release. This plot reflects the complete three-body momentum balance. Indeed, in the transverse direction, C^+ momentum completely compensates the O momentum. It appears that for the lowest KER value, O and C^+ momenta are equal (zero kinetic energy in their center-of-mass frame), which is clear evidence of a secondary decay reaction. Nevertheless, a refined analysis of the data shows that even if a sequential character is observed, the two steps of the dissociation are not completely independent and the fragmentation is of asynchronous-concerted type.

23.3.3   Projectile Momentum Transfer

As the projectile velocity is decreased, the collision time increases. The fragmentation is no longer free, as in the case of femtosecond lasers, but proceeds in the Coulomb field of the projectile ion. This additional interaction affects the molecular dissociation dynamics even in the simplest diatomic molecule case.

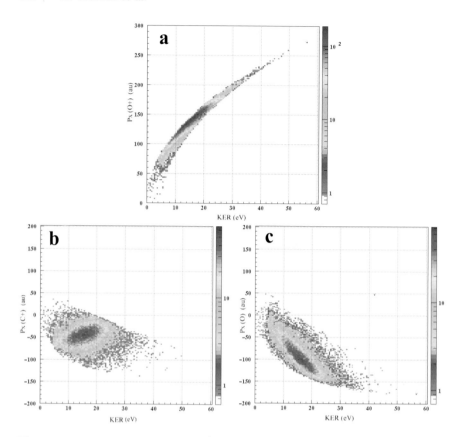

Fig. 23.7. Momentum along the O^+ fragment emission direction as a function of the KER for the double ionization $C^+ + O^+ + O$ fragmentation channel. These distributions are given for the O^+ (**a**), C^+ (**b**) and O (**c**) fragments. Momenta are expressed in atomic units and KER in eV

As shown for the system $Xe^{54+} + H_2$ using the classical trajectory Monte Carlo method [47,48], the collision has to be treated as an n-body problem, including the projectile, the fragments and the active electrons. It was observed [47] that the energy of the fragments H^+ depends strongly on the projectile velocity. The calculations revealed at least two sources for the fragment energy: the recoil energy E_r induced by the projectile on the center-of-mass of the ionized target, and the fragment energy E_f due to a pure Coulombic dissociation. At high projectile energies, the quantity E_r is negligible compared to E_f. Consequently, the energies of the protons ($\simeq 9.5\,\text{eV}$) originate only from the Coulomb-explosion of the molecule at the Franck–Condon limit. In contrast, at low impact energies, both slow and fast protons were predicted [47], resulting from the vector addition of the collisional momentum transfer

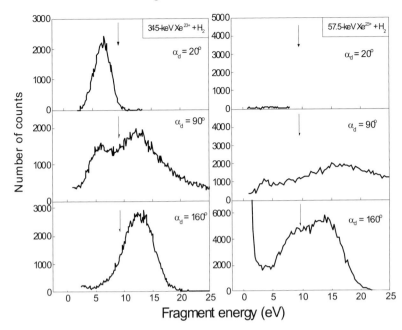

Fig. 23.8. Energy distribution of the H^+ fragments in 345 and 57.5 keV Xe^{23+}-induced H_2 dissociation

to the center-of-mass of the molecule and that due to the two-body Coulomb breakup of the dissociating molecular ion.

The influence of the projectile seems to be similar to the well-known postcollision interaction (PCI), which influences the energy spectra of Auger electrons emitted from the projectile in the field of the ionized target. Consequently, the fragment energy shift due to this effect is expected to depend on the projectile charge, the projectile velocity, and the orientation of the molecular target before the collision. Therefore, the fragmentation of H_2 has been studied for the two collision systems $Xe^{23+}+H_2$ and $O^{5+}+H_2$ in the 10–300 keV energy range. The energy of the fragments was analyzed at detection angles ranging from 20° to 160°. Figure 23.8 shows typical spectra for the system $Xe^{23+}+H_2$ at collision energies of 345 keV and 57.5 keV, and for detection angles of 20°, 90°, and 160°. First, it is seen that at both projectile energies, the mean energy of fragments differs significantly from that expected in the case of a vertical Franck–Condon transition (indicated by an arrow in Fig. 23.8). At an energy of 345 keV, the fragment energy increases for increasing detection angles (left side of Fig. 23.8). Moreover, two groups of peaks appear at detection angles ranging from 30° to 150°. Strong differences appear clearly when decreasing the projectile energy. A strong preferential backward emission of the fragments is evidenced. At a detection angle of 20°, no peak is observed in the case of the 57.5–keV projectile energy. This

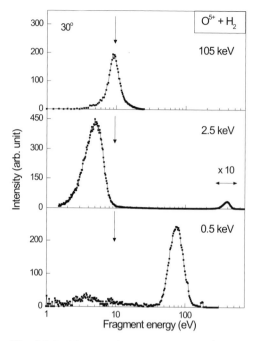

Fig. 23.9. Energy distributions of H^+ fragments following electron capture in $O^{5+}+H_2$ collisions at projectile energies of 105, 2.5, and 0.5 keV at an observation angle of 30°, with respect to the beam direction. The *arrows*, pointing at 9.5 eV, indicate the energy expected for a free fragmentation [52]

means that the projectile is sufficiently slow to repulse the fragments that are emitted in a forward direction. At 90° and 160°, the structures are similar to those found at the highest energy. However, the widths of the peaks increase considerably when decreasing the projectile energy. Another structure is found at 160°, due to the single-capture process. This observed structure is in accordance with the overbarrier model [49], which predicts a recoil of the H_2^{2+} ion following the single capture process at an angle of approximately 130° and with an energy of about 1 eV.

At these collision energies, the dynamics is fairly well reproduced by a two-step two-body model that gives the final H^+ momentum from the simple vector addition of the collisional momentum transfer to the molecular center-of-mass and the momentum of the H^+ fragment resulting from a free dissociation [50]. If the collision energy is further decreased, this simple picture fails and the collision follows a 3-body dynamics in which the projectile may interact strongly with one of the target atoms [51]. This is shown in Fig. 23.9 in which H^+ fragments with energies as high as 300 eV are evidenced. This peak is the signature of small impact parameter collisions between the projectile and one of the H atoms [52].

In order to better understand the specific role of the projectile on the fragmentation dynamics, experimental measurements at lower impact velocities and with higher energy resolution are desirable. Furthermore, a refined analysis of the momentum vectors of the collision partners using fragment-ion momentum spectroscopy may help to enlighten the different interactions experimentally [53].

23.4   Conclusion and Future Trends

Molecular fragmentation dynamics has strongly benefited from the imaging techniques during the last few years, either induced by ion impact or synchrotron radiation [18]. Refined details such as the origin of the kinetic energy released by the dissociation or the role of the molecular orientation on the ionization cross sections have been obtained and have improved our understanding of the process.

Polyatomic target fragmentation studies are just starting with triatomic molecules such as CO_2. Technical improvements will soon, if not already, allow many more particles to be handled opening the way to complex targets such as large molecules, clusters, and even surfaces.

Electron emission from ion–molecule collisions is still poorly investigated experimentally as well as theoretically. Interesting features such as interference effects are expected and motivate experiments and calculations in preparation in different groups. The understanding of the many-particle dynamics of this system will be one of the subjects of the next few years.

A growing interest is clearly visible in the field of very slow collisions ($\leq 100\,eV$) for which even the total cross sections are scarce. As we have seen, imaging techniques are certainly needed and promising. Their application to this kind of experiments is not obvious and constitutes a challenging race between different groups around the world.

References

1. T.D. Märk: *Electron Impact Ionization*, ed. by T.D. Märk, G.H. Dunn (Springer, New York 1985)
2. A.P. Hitchcock, P. Lablanquie, P. Morin, E. Lizon, A. Lugrin, M. Simon, P. Thiry, I. Nenner: Phys. Rev. A **37**, 2448 (1988)
3. P. Lablanquie, J. Delwiche, M.J. Hubin-Franskin, I. Nenner, P. Morin, K. Itoh, J.H.D. Eland, J.M. Robbe, G. Gandara, J. Fournier, P.G. Fournier: Phys. Rev. A **40**, 5673 (1989)
4. P. Baltzer, M. Lundvist, B. Wannberg, L. Karlsson, M. Larsson, M.A. Hayes, J.B. West, M.R.F. Siggel, A.C. Parr, J.L. Dehmer: J. Phys. B: At. Mol. Opt. Phys. **27**, 4915 (1994)
5. I. Nenner, P. Morin: "Electronic and Nuclear Relaxation of Core Excited Molecules". In: *VUV and Soft X-Ray Photoionization Studies in Atoms and Molecules*, ed. by U. Bexker, D.A. Shirley (Plenum, London 1995)

444   A. Cassimi et al.

6. S. Chelkowski, P.B. Corkum, A.D. Banbrauk: Phys. Rev. Lett. **82**, 3416 (1999)
7. H. Tawara: Phys. Rev. A **33**, 1385 (1986)
8. G. Sampoll, R.L. Watson, O. Heber, V. Horvat, K. Wohrer, M. Chabot: Phys. Rev. A **45**, 2903 (1992)
9. I. Ben-Itzhak, S.G. Ginther, K.D. Carnes: Phys. Rev. A **47**, 2827 (1993)
10. D. Mathur, E. Krishnakumar, K. Nagesha, V.R. Marathe, V. Krishnamurthi, F.A. Rajgara, U.T. Raheja: J. Phys. B: At. Mol. Opt. Phys. **26**, L141 (1993)
11. C. Lehmann: *Defects in Christalline Solids* **10**, ed. by S. Amelinckx, R. Gevers, J. Nihoul (North Holland, Amsterdam 1977)
12. J.E. Biaglow: *Radiation Chemistry: Principles and Applications*, ed. by Farhataziz, M.A.J. Rodgers (VCH, New York 1987) p. 527
13. U. Werner, K. Becord, J. Becker, H.O. Lutz: Phys. Rev. Lett. **74**, 1962 (1995)
14. G.H. Olivera, C. Caraby, P. Jardin, A. Cassimi, L. Adoui, B. Gervais: Phys. Med. Biol. **43**, 2347 (1998)
15. D. Mathur: Phys. Rep. **225**, 193 (1993)
16. R. Moshammer, W. Schmitt, J. Ullrich, H. Kollmus, A. Cassimi, R. Dörner, O. Jagutzki, R. Mann, R.E. Olson, H.T. Prinz, H. Schmidt-Böcking, L. Spielberger: Phys. Rev. Lett. **79**, 3621 (1997)
17. N. Stolterfoht, J.Y. Chesnel, M. Grether, B. Skogvall, F. Frémont, D. Lecler, D. Hennecart, X. Husson, J.P. Grandin, B. Sulik, L. Gulyas, J.A. Tanis: Phys. Rev. Lett. **80**, 4649 (1998)
18. T. Weber, O. Jagutzki, M. Hattass, A. Staudte, A. Nauert, L. Schmidt, M.H. Prior, A.L. Landers, A. Bräuning-Demian, H. Bräuning, C.L. Cocke, T. Osipov, I. Ali, R. Diez Muino, D. Rolles, F.J. Garcia de Abajo, C.S. Fadley, M.A. Van Hove, A. Cassimi, H. Schmidt-Böcking, R. Dörner: J. Phys. B: At. Mol. Opt. Phys. **34**, 3669 (2001)
19. A. Landers, T. Weber, I. Ali, A. Cassimi, M. Hattass, O. Jagutzki, A. Nauert, T. Osipov, A. Staudte, M.H. Prior, H. Schmidt-Böcking, C.L. Cocke, R. Dörner: Phys. Rev. Lett. **87**, 013002 (2001)
20. G. Dujardin, S. Leach, O. Dutuit, P.M. Guyon, M. Richard-Viard: Chem. Phys. **88**, 339 (1984)
21. L. Adoui, C. Caraby, A. Cassimi, D. Lelièvre, J.P. Grandin, A. Dubois: J. Phys. B: At. Mol. Opt. Phys. **32**, 631 (1999)
22. U. Becker, O. Hemmers, B. Langer, A. Menzel, R. Wehlitz, W.B. Peatman: Phys. Rev. A **45**, R1295 (1992)
23. K. Wohrer, G. Sampoll, R.L. Watson, M. Chabot, O. Heber, V. Horvat: Phys. Rev. A **46**, 3929 (1992)
24. O. Heber, G. Sampoll, B.B. Bandong, R.J. Maurer, R.L. Watson, I. Ben-Itzhak, J.M. Sanders, J.L. Shinpaugh, P. Richard: Phys. Rev. A **52**, 4578 (1995)
25. P. Jardin: PhD Thesis, Basse Normandie University, Caen (1995)
26. M. Unverzagt, R. Moshammer, W. Schmitt, R.E. Olson, P. Jardin, V. Mergel, J. Ullrich, H. Schmidt-Böcking: Phys. Rev. Lett. **76**, 1043 (1996)
27. P. Jardin, A. Cassimi, J.P. Grandin, H. Rothard, J.P. Lemoigne, D. Hennecart, X. Husson, A. Lepoutre: Nucl. Instrum. Methods B **98**, 363 (1996)
28. S.L. Varghese, C.L. Cocke, S. Cheng, E.Y. Kamber, V. Frohne: Nucl. Instrum. Methods B **40/41**, 266 (1989)
29. K. Wohrer, R.L. Watson: Phys. Rev. A **48**, 4784 (1993)
30. C. Caraby, A. Cassimi, L. Adoui, J.P. Grandin: Phys. Rev. A **55**, 2450 (1997)
31. B. Siegmann, U. Werner, R. Mann, Z. Kaliman, N.M. Kabachnik, H.O. Lutz: Phys. Rev. A **65**, 010704 (2001)

32. N.M. Kabachnik, V.N. Kondratyev, Z. Roller-Lutz, H.O. Lutz: Phys. Rev. A **57**, 990 (1998)
33. U. Werner, N.M. Kabachnik, V.N. Kondratyev, H.O. Lutz: Phys. Rev. Lett. **79**, 1662 (1997)
34. Z. Kaliman, N. Orlic, N.M. Kabachnik, H.O. Lutz: Phys. Rev. A **65**, 012708 (2001)
35. I. Ben-Itzhak, S.G. Ginter, V. Krishnamurthi, K.D. Carnes: Phys. Rev. A **51**, 391 (1995)
36. H. Tawara, T. Tonuma, T. Matsuo, M. Kase, H. Kumagai, I. Kohno: Nucl. Instrum. Methods A **262**, 95 (1987)
37. T. Matsuo, T. Tonuma, M. Kase, T. Kambara, H. Kumogai, H. Tawara: Chem. Phys. **121**, 93 (1988)
38. V. Krishnamurthi, I. Ben-Itzhak, K.D. Carnes: J. Phys. B: At. Mol. Opt. Phys. **29**, 287 (1996)
39. C. Caraby, L. Adoui, J.P. Grandin, A. Cassimi: Eur. Phys. J. D **2**, 53 (1998)
40. H.O. Folkerts, R. Hoekstra, R. Morgenstern: Phys. Rev. Lett. **77**, 3339 (1996)
41. V. Krishnamurthi, K. Nagesha, V.R. Marathe, D. Mathur: Phys. Rev. A **44**, 5460 (1991)
42. D. Mathur, J.H.D. Eland: Int. J. Mass Spectrom. Ion Process. **114**, 123 (1992)
43. M. Tarisien, L. Adoui, F. Frémont, D. Lelièvre, L. Guillaume, J.Y. Chesnel, H. Zhang, A. Dubois, D. Mathur, Sanjay Kumar, M. Krishnamurthi, A. Cassimi: J. Phys. B: At. Mol. Opt. Phys. **33**, L11 (2000)
44. C. Maul, K.H. Gericke: Int. Rev. Phys. Chem. **16**, 1 (1997)
45. S. Hsieh, J.H. Eland: J. Phys. B: At. Mol. Opt. Phys. **30**, 4515 (1997)
46. J.H. Eland: Mol. Phys. **61**, 725 (1987)
47. C.J. Wood, R.E. Olson: Phys. Rev. A **59**, 1317 (1999)
48. C.R. Feeler, R.E. Olson, R.D. Dubois, T. Schlathöter, O. Hadlar, R. Hoekstra, R. Morgenstern: Phys. Rev. A **60**, 2112 (1999)
49. A. Niehaus: J. Phys. B: At. Mol. Opt. Phys. **19**, 2925 (1986)
50. F. Frémont, C. Bedouet, M. Tarisien, L. Adoui, A. Cassimi, A. Dubois, J.Y. Chesnel, X. Husson: J. Phys. B: At. Mol. Opt. Phys. **33**, L249 (2000)
51. R.E. Olson, C.R. Feeler: J. Phys. B: At. Mol. Opt. Phys. **34**, 1163 (2001)
52. P. Sobocinski, J. Rangamma, G. Laurent, L. Adoui, A. Cassimi, J.Y. Chesnel, A. Dubois, D. Hennecart, F. Frémont: J. Phys. B: At. Mol. Opt. Phys. **35**, 1353 (2002)
53. I. Ali, R.D. Dubois, C.L. Cocke, S. Hagmann, C.R. Feeler, R.E. Olson: Phys. Rev. A **64**, 022712 (2001)

24 Electron–Interaction Effects in Ion-Induced Rearrangement and Ionization Dynamics: A Theoretical Perspective

T. Kirchner

24.1 Introduction

A considerable fraction of recent work in atomic- and molecular-collision physics was motivated by the quest for a better understanding of electron-interaction effects. Excitation, rearrangement, and fragmentation patterns mirror various facets of the mutual Coulomb repulsion and the indistinguishability of electrons and provide a wealth of information on the characteristics and peculiarities of interacting many-fermion systems.

Naturally, the many-electron quantum dynamics is most transparent and best understood in situations in which descriptions in terms of only a few coupled states or amplitudes of perturbation expansions are appropriate. In ion–atom collisions, which are considered in this chapter, these conditions are met either at low impact energies, where the dominant capture processes are caused by nonadiabatic couplings between quasimolecular states, or in fast collisions of weakly charged ions, in which rearrangement and ionization processes can be described in terms of first- and second-order Born amplitudes. Such situations and corresponding theoretical approaches are well documented in the literature [1,2], and will not be discussed here. Instead, the intermediate energy regime, in which the coupling of different reaction pathways calls for a nonperturbative solution of the Schrödinger equation is addressed.

This region poses high demands on the theoretical treatment even for one-electron scattering systems. Only recently has it become possible to implement approaches for the simultaneous calculation of accurate total cross sections for capture, ionization, and excitation processes, and for the calculation of differential electron emission patterns (see Sect. 12.3.2 and references cited therein). A few studies were also concerned with the fully correlated description of two-electron systems [3,4], in particular, with the ionization of helium atoms by protons (p) and antiprotons ($\bar{\text{p}}$) [5–8], which is regarded as a prime example for the manifestation of electron-correlation effects.

A simple estimate shows that such explicit many-electron calculations will be prohibitive for systems with more than two active electrons for some time to come [9]. The outstanding interpretation of the rich experimental data accumulated in recent years [10], therefore, calls for simplified approaches to the many-body problem of atomic collision physics. A few such models, namely, the *forced impulse method* [5], the *frozen correlation approximation* [11], and the *independent time approximation* [9] include some aspects of the electron–electron interaction explicitly without being computationally as costly as fully correlated calculations. Up to now, however, no applications to systems with more than two active electrons have been reported, which indicates the still demanding nature of these methods.

One may thus conclude that the only practicable approach to treat true many-electron systems in atomic collisions is some kind of effective single-particle description, i.e., the solution of single-particle equations for all active electrons. The usefulness of such independent particle models (IPMs) has been demonstrated in several studies, but also their limited validity has been emphasized [2]: By definition they are not suited to discuss electron-correlation effects.

It is explained in Chap. 12 of this book how density functional theory (DFT) puts these issues into perspective. The central theorems of DFT ensure the existence of an *exact* mapping of the true many-body problem to an effective single-particle description, thus providing a sound basis for the application of IPM-type approaches. Admittedly, the promise to solve the many-body problem exactly is neglected in practice, since the exact expressions for many important quantities are not known and have to be approximated, but DFT proved to be a powerful tool for the description and understanding of dynamic [12] and, of course, stationary [13] many-body quantum systems.

It is the purpose of this chapter to demonstrate the usefulness of DFT for ion–atom collisions at intermediate impact energies. The key concepts of the time-dependent version of DFT, which provide the basis of this discussion are summarized in Chap. 12 and will not be explained in detail here. Only a few remarks at the beginning of Sect. 24.2 are in order to keep the discussion self-contained for a reader who is familiar with the basic ideas of DFT. After that, some specific approximations are introduced, which are both necessary to make the numerical solution of the problem at hand feasible, and helpful to classify the role of electron-interaction effects in the collision dynamics. Some results, which are selected from a series of recent papers [14–21] are discussed in Sect. 24.3. Emphasis is given to the analysis of different facets of the electron–electron interaction, which can be identified from the comparison of theoretical results with experimental data. Some concluding remarks are provided in Sect. 24.4.

24.2   Classification of Electron–Interaction Effects: A Density Functional Approach

Ion–atom collisions at not too low impact energies can be discussed within the impact parameter model [22]: the motion of the nuclei is described in terms of a classical straight-line trajectory $\boldsymbol{R}(t)$, which gives rise to a time-dependent external potential in the Hamiltonian (12.5) of the many-electron system (atomic units with $\hbar = m_e = e = 1$ are used)

$$\hat{V}(t) = \sum_{j=1}^{N} \left(-\frac{Z_T}{r_j} - \frac{Z_P}{|\boldsymbol{r}_j - \boldsymbol{R}(t)|} \right) . \tag{24.1}$$

Here, Z_T and Z_P denote the nuclear charges of the target and projectile nuclei, respectively, and $\boldsymbol{r}_j$ is the coordinate of the j-th electron. By virtue of the Runge–Gross theorem [23] the time-dependent many-electron problem can be mapped to an effective single-particle description, in which the *exact* density $n(\boldsymbol{r}, t)$ of the system is obtained from the solutions of single-particle equations (12.9). Moreover, the many-electron wave function as well as practically all observables of the system are uniquely determined by $n(\boldsymbol{r}, t)$. Hence, the so-called Kohn–Sham (KS) scheme is formally equivalent to solving the N-electron Schrödinger equation (12.1).

As discussed in Chap. 12 the downside of the structural simplicity of the KS scheme is the lack of knowledge of the true density dependences of the effective KS potential in the single-particle equations and of many important observables, such as multiple-ionization yields. Therefore, approximations are necessary on two fronts in practical calculations provided that the time-dependent KS equations can be solved accurately for a given potential. Such approximations and their physical implications are sketched in the next two subsections.

24.2.1 Effects Associated with the Kohn–Sham Potential

Some general properties of the KS potential as well as the time-dependent *local-density approximation* (LDA) and the time-dependent *optimized-potential method* (OPM) are discussed in Sect. 12.3.1. These two approaches were used for the calculation of charge transfer in Ar^{8+}–Ar collisions [24], but results for ionization were not reported. Ionization has been considered very recently in $\bar{p}$–He collision [25].

The work to be discussed in this chapter relies on a decomposition of the KS potential into the external Coulomb potential of the projectile and target nuclei and two different contributions to the remaining effective electron–electron interaction v_{ee}: A stationary part v_{ee}^0 that accounts for the electron–electron interaction in the undisturbed ground state of the target atom, and a response potential δv_{ee} that depends on the time-dependent density and reflects the change of the electronic interaction in the presence of the projectile:

$$v_{ee}(\boldsymbol{r}, t) = v_{ee}^0(\boldsymbol{r}) + \delta v_{ee}(\boldsymbol{r}, t) . \tag{24.2}$$

In the *no-response* approximation, δv_{ee} is neglected completely. This is well justified for fast collisions, in which the electronic density does not change considerably during the short interaction time, and, more generally for one-electron transitions over a broad range of impact energies. One is then left with the question, which level of accuracy is needed for the ground-state potential v_{ee}^0 in order to obtain reliable results in these situations.

To investigate this issue, the effective potential v_{ee}^0 has been split into the classical Hartree and the exchange-correlation potentials, and two different approximations have been considered for the latter [14,15], which were tested extensively in atomic-structure calculations [13,26]:

- the LDA [27], which similarly to the Hartree–Fock–Slater (HFS) approximation [28] rests on the density dependence of the homogeneous electron gas. As a consequence, the exchange part decreases exponentially in the asymptotic region rather than exhibiting the correct $-1/r$ tail. This deficiency is cured a posteriori by the Latter correction [30] or by more sophisticated self-interaction correction schemes [31], but exchange potentials of this type are, of course, only approximate. The smaller correlation contribution is given in terms of accurate analytic interpolation formulae [29].
- the OPM, in which exchange effects are treated nearly exactly [26,32]. It was also demonstrated that correlation can be included in the OPM in a systematic fashion [33], but test calculations with a semiempirical correlation potential indicated that the static correlation contribution is too small to influence total scattering cross sections significantly [15].

¿From this short discussion it can be expected that the comparison of results obtained with the LDA and the OPM sheds light on the role of static exchange effects in collision processes. This is indeed the case and will be demonstrated in Sect. 24.3.1.

In the next step, response effects have been included in the potential v_{ee} in a global fashion [18]. An important property of response is the increased attraction of the total target potential as soon as capture and ionization processes set in during the collision. This feature can be modeled by approximating the target potential as a linear combination of ionic ground-state potentials with time-dependent q-fold electron removal probabilities $P_q(t)$ as weighting factors. When the ionic potentials are related to the stationary potential v_{ee}^0 by simply adjusting the asymptotic tail of the latter one arrives at the target response potential [18]

$$\delta v_{ee}(r,t) = \frac{-1}{N-1} \sum_{q=1}^{N} (q-1) P_q(t) v_{ee}^0(r) . \qquad (24.3)$$

The effects of target response on collision cross sections are discussed in Sect. 24.3.2. A similar model was devised to study projectile response effects that are associated with the unscreening of the projectile ion due to capture processes, but for the cases investigated so far projectile response turned out to be of minor importance [20].

A somewhat different response model was introduced recently to investigate $\bar{\mathrm{p}}$-He collisions [21]. It is motivated by the fact that the no-response approximation predicts an instable quasimolecule ($\bar{\mathrm{p}}$He) at internuclear distances $R \leq 0.5$ a.u., whereas the true adiabatic ground-state energy level lies below the threshold for single ionization for all R [34,35]. As a consequence, single ionization at low impact energies is too high when calculated in the no-response model.

In order to remedy this flaw an *adiabatic* response potential was constructed in [21] such that the binding energy of the lowest single-particle level reproduces the true ionization potential of the quasimolecule at small R. The potential was taken to be of the form

$$\delta v_{\mathrm{ee}}(\boldsymbol{r}, R) = -a(R)\mathrm{e}^{-r} - \frac{p(R)r\cos\theta}{[d(R)]^3 + r^3}, \qquad (24.4)$$

with parameters $p(R)$, $d(R)$, and $a(R)$ adjusted to fulfil the condition mentioned above.

Such modeling of particular features of the electron–electron interaction poses the question of how to construct a response potential that comprises the desired properties automatically in a systematic fashion. The most promising approach is certainly the time-dependent OPM (Sect. 12.3.1). In the exchange-only limit the scheme has been worked out theoretically and has been applied to various physical problems, but a full implementation that allows the description of ionization and capture processes in ion–atom collisions has not been reported yet. A time-dependent correlation potential beyond the *adiabatic* approximation (Sect. 12.3.1) has only been proposed on the level of the LDA so far [36].

24.2.2   Effects Associated with the Density Dependence of Observables

In Sect. 12.4.1 it is explained how particle numbers for net ionization and net capture can be calculated exactly from the density at asymptotic times after the collision. Corresponding cross sections are well suited to study the effects associated with the KS potential, since they are not contaminated with approximations on the second front: the extraction of observables from the solutions of the KS equations.

The density dependence of more detailed observables, such as probabilities for finding q out of N electrons in the continuum is only known approximately. These probabilities can be expressed in terms of q-particle densities, which are given as $q \times q$ determinants of the one-particle density matrix in the exchange-only limit (see Sect. 12.4.2). The determinantal structure reflects the antisymmetry of the N-electron wave function, i.e., the Pauli principle is still taken into account on the level of the density-matrix analysis.

Very often, the determinantal structure is neglected in practice, usually without analyzing this additional approximation. All probabilities of interest

can then be calculated by multinomial expressions of single-particle probabilities, which are sometimes modified to circumvent certain problems, such
as nonzero probabilities for unphysical multiple-capture events [17]. It will
be demonstrated in Sect. 24.3.3 for one obvious and one subtle example that
the effects of the Pauli principle in the final states become apparent when
data obtained from multinomial and density-matrix analysis are compared.

In the case of a two-electron spin-singlet system evolving from its initial
ground state the determinantal structure of the q-particle densities breaks
down and the single- and double-ionization probabilities are reduced to the
simple binomial formulae (12.36). That this exchange-only analysis is not
compatible with double-ionization probabilities obtained from more elaborate
methods is pointed out in Sect. 24.3.4, in which some of the limitations of the
present approach are sketched.

24.3   Identification of Electron–Interaction Effects: Comparison with Experiment

As mentioned before, the accurate solution of the time-dependent KS equations for ion–atom collision systems is a delicate numerical problem even in
the simplest case of the no-response approximation. Therefore, the developement of an efficient propagation method was a prerequisite for a meaningful investigation of electron interaction effects. The *basis-generator method*
(BGM) [37] proved to be such a method in a number of succesful studies
[14–21,38]. The idea of the BGM is the representation of the single-particle
orbitals in a dynamically adapted model space, i.e., in terms of a basis that
spans that part of the Hilbert space that is relevant for the specific problem
under investigation.

Typically, the BGM basis sets used consist of undisturbed target eigenstates of the 1s through 4f states and a set of approximately 100 pseudostates
generated by the repeated application of a regularized Coulomb potential at
the projectile center on the undisturbed functions. The observables discussed
below are extracted from the numerical solutions by the methods explained
in Chap. 12.

24.3.1   Static Exchange Effects

The influence of static exchange effects on the ion–atom collision dynamics
was studied by solving the time-dependent KS equations on the level of the
no-response approximation with target ground-state potentials obtained from
the OPM and the Latter-corrected LDA, respectively [14,15]. Let us focus on
net ionization and capture cross sections at first in order to disentangle effects
associated with the form of the KS potential from effects, which correspond
to the approximate extraction of less global observables from the solutions of
the KS equations.

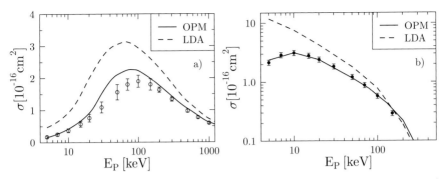

Fig. 24.1. (a) Net ionization and (b) net capture cross sections as functions of impact energy for p–Ne collisions. *Lines*: BGM calculations with different atomic potentials [48]. Experiment: *open circles* [49], *closed circles* [50]

In Fig. 24.1 results for the p–Ne collision system are presented. Clearly, the LDA leads to capture and ionization cross sections that are too large over the entire range of impact energies, whereas the agreement with the experimental data is convincing when the OPM potential is used. The failure of the LDA solutions can be traced back in part to the inaccurate prediction of the first ionization potential, i.e., to an underestimation of the binding energy of the outermost 2p electrons. However, it is not sufficient to correct this binding energy by simply adjusting the overall strength of the LDA exchange potential. This can be inferred from results for total [14] and differential [15,39] electron removal cross sections obtained with a HFS target potential whose exchange contribution differs from the LDA expression by the factor $\alpha = 3/2$. In particular, a thorough analysis of heavy-ion-induced electron-emission patterns demonstrated that the sudden switching from the exponentially decaying exchange potential to the asymptotic $-1/r$ tail introduced by the Latter correction is an additional source of errors and produces artificial structures in the doubly-differential cross section [39].

Two conclusions can be drawn at this point:

- Inaccurate treatments of the static-exchange potential of the target atom lead to wrong electron-removal cross sections. This demonstrates that static exchange effects are indeed mirrored in such data.
- Response effects are of minor importance for proton-induced transitions, at least as long as net-electron processes are considered: Otherwise, the results obtained with the OPM description of the target atom would not be in good agreement with the experimental data. Only at intermediate impact energies do we observe a theoretical net ionization cross section that lies above the experimental result (Fig. 24.1a). From the discussion in the next subsection it will become clear that this discrepancy is a signature of target-response effects.

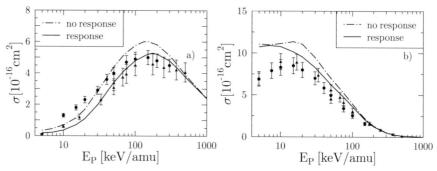

Fig. 24.2. (a) Net ionization and (b) net capture cross sections as functions of impact energy for He^{2+}–Ne collisions. *Full* and *dash-dotted lines:* BGM calculations with and without target response, respectively [18]. Experiment: *closed triangles* [40], *closed circles* [41]

24.3.2 Response Effects

Multiply charged projectile ions induce multiple-electron removal processes with higher probabilities than singly charged ions. They produce higher average recoil-ion charge states and should thus be more sensitive to response effects. Therefore, let us consider the He^{2+}–Ne collision system as a first application for the target response model (24.3) based on the static OPM potential.

As expected, the results for net ionization and capture (Fig. 24.2) obtained from the no-response approximation deviate more strongly from the experimental data than in the case of proton impact. Both cross sections are reduced at low and intermediate impact energies E_P when target response is included, while they are insensitive to response effects at higher E_P. This behavior confirms the earlier assumption that response is not effective when the projectile moves considerably faster than the outershell target electrons. In the case of ionization (Fig. 24.2a) the reduction of the cross section below $E_P = 500\,\mathrm{keV/amu}$ leads to an almost perfect agreement of the target response data with the experimental results of [40]. These measurements are believed to be more accurate than the earlier data of [41] that are also included in the figure.

For the capture channel (Fig. 24.2b), however, the agreement holds only down to $E_P = 20\,\mathrm{keV/amu}$, while the theoretical cross section lies above the experimental one at slower collisions. In this region, the influence of the response potential (24.3) on the capture cross section is rather small. This is partly due to a compensation of the behavior of electrons of different subshells: while capture of the 2s electrons is reduced by response effects the 2p electrons are transferred to the projectile with higher probability when δv_{ee} is included in the description. This is a consequence of the changing energy differences and coupling strengths between the relevant channels when response

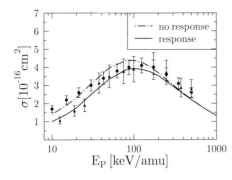

Fig. 24.3. Net ionization cross section as a function of impact energy for He$^+$-Ne collisions. *Full* and *dash-dotted lines*: BGM calculations with and without target response, respectively [19]. Experiment: *closed triangles* [42], *closed circles* [43]

is or is not taken into account. This suggests that the electron dynamics in slow collisions are rather sensitive to the specific form of the response potential. First attempts to include a spherical model for time-dependent projectile screening did not improve considerably upon the data of Fig. 24.2b. At present, it is not clear whether a *microscopic* exchange-only response model, e.g., based on the time-dependent OPM would resolve the discrepancy with the experimental data or whether correlation plays a crucial role for electron capture in this energy region.

As a further example, the net ionization cross section for *singly* charged He-ion impact on Ne atoms is shown in Fig. 24.3. In this case, free electrons are produced by target and by projectile ionization, so that the time evolution of the active projectile electron driven by the target potential also has to be taken into account. This was done on the level of the no-response approximation by an analogous BGM expansion (details are discussed in [19]). It was found that 20–30% of the cross section displayed in Fig. 24.3 stems from such projectile-ionization events. The contribution due to target ionization is again reduced by target response, and the resulting net ionization cross section is in good agreement with the experimental data of [42]. Similar to the case of He^{2+}-impact these measurements are in conflict with earlier data obtained from a different experimental setup [43], and are believed to be more accurate, although no clear explanation for possible errors in the earlier measurements was provided [42].

Finally, results of the adiabatic response model for the ionization in $\bar{p}$– He collisions are discussed. In Fig. 24.4 the single-ionization cross section obtained with and without adiabatic response is compared with experimental data and results of correlated two-electron calculations. Note that with the consideration of single instead of net ionization we are faced with the second source of errors besides the approximate form of the KS potential: we have to calculate the probability to remove *exactly* one electron from the target, which is only known in the exchange-only approximation (12.36). However, the effects of the exchange-only analysis should be of minor importance in the present case, in which multiple ionization events occur only with small probabilities and single ionization P_1 does not differ considerably from net

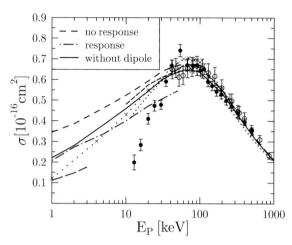

Fig. 24.4. Single-ionization cross section as a function of impact energy for p̄–He collisions. Theory: *lines*: BGM calculations with full adiabatic response, with adiabatic response without dipole contribution, and without response [21]; *short-dashed curve* [7], *long-dashed curve* [35], *short-dotted curve*, multi-cut FIM [5], *long-dotted curve with crosses* [6]. Experiment: *closed circles* [51], *open circles* [52]

ionization $P_{net} = P_1 + 2P_2$. Evidently, such an argument does not apply when one considers double ionization (see Sect. 24.3.4).

Returning to the discussion of potential effects we first observe in Fig. 24.4 that there is a problem with the full adiabatic response potential of (24.4): The results lie considerably below the experimental and the other theoretical cross sections in the region above $E_P \approx 40$ keV, in which adiabatic effects should not play any role, as the electron cloud does not have enough time to adjust to the two-center potential of the nuclei. When the calculations are repeated with the dipole contribution in (24.4) turned off the results do merge with the no-response cross section at intermediate impact energies. At lower E_P, however, they lie significantly below the no-response data, and are in close agreement with the cross section obtained with the full adiabatic response potential (24.4). Hence, the dipole part of (24.4) should be omitted entirely as it is relatively unimportant at low E_P and reduces the cross section wrongfully at higher E_P.

Remarkably, the results obtained with the dipole correction turned off are in good agreement with the much more elaborate two-electron calculations of [5–7] down to $E_P = 10$ keV. This demonstrates that the reduction of the no-response cross section in the 10 keV $\leq E_P \leq$ 50 keV range is not caused by explicit correlation effects, such as the deviation of the true two-electron wave function from a simple product form, but is mainly due to a global response of the electron cloud in the presence of the antiproton. Note that this conclusion cannot be deduced from the results of the two-electron models, since it is difficult to trace different aspects of the electron–electron interaction in these 'complete' calculations.

At even lower impact energies the situation is less clear, since the coupled-channel calculation of [6] and the hidden-crossing calculation of [35] predict a significantly smaller cross section than the adiabatic response model, but are in conflict with each other. The picture becomes even more confusing when the experimental results are also taken into consideration as they lie below all theoretical data at $E_P \leq 30$–40 keV and do not seem to approach the results of the hidden-crossing calculation below 10 keV. New measurements are planned for the near future and may help to clarify this situation.

24.3.3   Pauli Blocking

In this section, effects associated with the antisymmetry of the final many-electron state are discussed. As mentioned in Sect. 24.2.2, the antisymmetry and the Pauli principle are maintained when the analysis is based on q-particle densities in the exchange-only approximation (see Sect. 12.4.2). By contrast, widely used multinomial formulae for q-particle probabilities rest on the assumption that the many-electron wave function can be expressed as a simple product state.

To demonstrate the role of the Pauli principle results for target electron capture and projectile-electron loss in He^+-Ne collisions obtained from both types of analyses are presented in Fig. 24.5. More specifically, the multinomial results were calculated with the analysis in terms of products of binomials proposed in [17]. Standard multinomial statistics for capture suffers from the problem that one obtains nonzero probabilities for unphysical multiple capture events that correspond to the formation of negatively charged projectile ions. This problem is avoided in the products-of-binomials analysis. The unphysical channels are eliminated, and the net capture probability is distributed statistically over the physically allowed ones. All cross sections of Fig. 24.5 are extracted from the same set of single-particle calculations, which are based on the target-response model defined by (24.3) and the no-response approximation for the propagation of the active projectile electron [19].

The results for electron capture corresponding to the neutralization of the projectile ion (Fig. 24.5a) are easily explained. In the 'products' analysis, both spin-up and spin-down electrons contribute to the cross section in the same fashion, since they are not hindered by the fact that the dominant final $He(1s^2)$ state is a spin-singlet. As a consequence, the experimental cross section is overestimated by almost a factor of two. The density-matrix analysis, termed 'Pauli' in Fig. 24.5, takes this aspect into account and leads to nearly perfect agreement with experiment. Hence, the neutralization cross section reflects Pauli blocking very directly.

The situation is more subtle for the case of projectile-electron removal, for which the density matrix analysis reduces the cross section again by almost a factor of two from low to intermediate impact energies, and leads to very

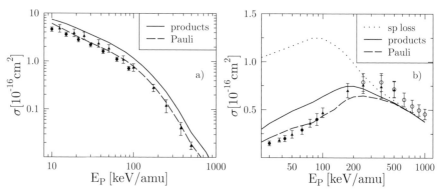

Fig. 24.5. Total cross sections for (**a**) neutralization of and (**b**) electron loss from the projectile as functions of impact energy for He$^+$–Ne collisions. *Lines*: BGM calculations with target response and final-state analysis in terms of products of binomials (*full curve*) and density-matrix analysis (*broken curve*); The *dotted curve* in the *right panel* corresponds to a single-particle calculation for the active He$^+$ electron [19]. Experiment: *closed triangles* [42], *closed circles* [43], open circles [53]

good agreement with experiment in this region (Fig. 24.5b). At first sight, one might be surprised that the two analyses yield different results at all, since many configurations contribute to this process, and 'blocking' of final states should not be important. In fact, the observed reduction of the cross section in the density matrix analysis can only be understood with a formal argument based on probability conservation [19]: By incorporating the antisymmetry into the state vector the transferred electron density is redistributed such that the correct balance for He^0, He$^+$, and He^{2+} formation is obtained.

The redistribution is ineffective at high impact energies, where He^0 formation is very unlikely (see Fig. 24.5a). In this region, the data obtained from both analyses merge, and are also in agreement with the result of a simple calculation with one active He$^+$ electron and frozen target electrons. All theoretical data lie somewhat below experiment in this region, which probably indicates the lack of a correlation contribution in the KS potential (see Sect. 24.3.4). The single-electron model overestimates the cross section badly at lower impact energies, demonstrating that simultaneous electron-transfer processes from the target to the projectile affect the final projectile charge state decisively and cannot be neglected.

Summarizing the discussion of this section it can be stated that a careful analysis of the final wave function is crucial for the understanding of cross sections, for which the final charge state of the projectile (and of the target) are well defined. The exchange-only analysis appears to be sufficient to explain quantitatively the formation of He^0 and He^{2+} in He$^+$–Ne collisions.

24.3.4   What Lies Beyond: Correlation Effects

All results presented in this chapter have been obtained on the level of the exchange-only approximation. Correlation effects in the KS potential and in the analysis of the final wave function have been ignored. This subsection has the purpose to exemplify the limitations of this treatment.

At first, we consider a case for which correlation effects associated with the KS potential can be isolated. A critical look at the net ionization cross section in He^+–Ne collisions (Fig. 24.3) indicates that the theoretical results slightly underestimate the experimental data at impact energies above 200–300 keV/amu. Based on a perturbative description of the scattering process it was argued that ionization of the projectile electron, which contributes to net ionization can be induced either by an interaction with the (screened) target nucleus or by an interaction with one of the target electrons [2]. In particular, the second process, sometimes termed *antiscreening* has been the subject of many studies in recent years (see, e.g., [44,45] and references therein). When calculated in the plane-wave Born approximation and added to the theoretical net ionization cross section displayed in Fig. 24.3 it leads to an improved agreement with experiment [19].

¿From the DFT viewpoint antiscreening must be contained in the *exact* time-dependent KS potential, since the net ionization cross section is calculated from the density without approximation in the analysis. Furthermore, since the calculations displayed in Fig. 24.3 are based on the exchange-only OPM target potential and since response effects are unimportant at impact energies above 300 keV/amu we have strong reason to believe that it is the time-dependent correlation potential that gives rise to the desired enhancement of the net ionization cross section. Its inclusion should also remove the more apparent discrepancies between the theoretical results and experiment in projectile ionization at high impact energies (Fig. 24.5b), for which the role of the antiscreening contribution is normally discussed rather than for net ionization [2,44]. Conceptually, however, this case is somewhat less clear, since the projectile electron-loss cross section is not calculated exactly from the electron density, but in the exchange-only approximation, which might also influence the results.

The failure of the exchange-only *analysis* becomes apparent for the case of a two-electron system, for which it reduces to the simple binomial formulae (12.36) that predict a fixed relation between the one- and two-electron probabilities (12.37). This appears to be in conflict with the fact that the total single-ionization cross section is very similar in p– and p̄–He collisions at high energies $E_P \geq 1$ MeV, while double ionization by antiprotons is significantly more efficient than by protons [46]. In principle, such a pattern is not prohibited by the exchange-only analysis as the impact-parameter dependence of p– and p̄–He ionization events could be such that an equal single-ionization and a different double-ionization cross section is obtained. However, this contradicts the expectation of perturbation theory, and, in fact, results of the correlated forced impulse method at $E_P = 2.31\,\text{MeV}$ [47] showed that the single-ionization probability is almost indistinguishable, while the double ionization probabilities exhibit pronounced deviations for p and p̄ impacts. The relation (12.37) is clearly violated.

This suggests that the important part of electron correlation in the regime of high impact energies is hidden in the unknown true density dependence of the double-ionization yield, and not in the KS potential. At lower E_P, dynamical effects should become more important, and both aspects of correlation might contribute to the fact that pronounced deviations from experiment and from correlated calculations occur when double ionization is calculated with the adiabatic response potential (24.4) and the binomial formulae (12.36) [21].

24.4   Concluding Remarks

It was the purpose of this chapter to show that various facets of the many-electron dynamics in ion–atom collisions can be explained from the viewpoint of time-dependent density functional theory. The effects of exchange and response contributions in the Kohn–Sham potential, and the role of Pauli blocking in the final states were analyzed by comparing calculations performed on different levels of approximation with experiment. This has only become possible with the development of the basis generator method for the accurate solution of the KS equations.

Given the reliability of the BGM and the accuracy of the experimental data remaining discrepancies can be attributed to *microscopic* response and correlation in the KS potential and correlation in the density dependence of the observables considered. In some cases, both aspects can be separated, which might be helpful for the developement of functionals that include correlations. Some ideas were reported for the case of the time-dependent KS potential in the literature, but – to the author's knowledge – no approaches for the calculation of observables, such as multiple-ionization yields beyond the exchange-only approximation have been developed so far. This would certainly be a major breakthrough for the application of DFT to ion–atom collisions, and, more generally, for the understanding of the time-dependent many-electron problem.

Acknowledgements

I thank Hans Jürgen Lüdde and Marko Horbatsch for our fruitful and enjoyable collaboration.

References

1. R. Suzuki, A. Watanabe, H. Sato, J.P. Gu, G. Hirsch, R.J. Buenker, M. Kimura, and P.C. Stancil: Phys. Rev. A **63**, 042717 (2001)
2. J.H. McGuire: *Electron Correlation Dynamics in Atomic Collisions.* (Cambridge University Press, Cambridge 1997)
3. C. Pfeiffer, N. Grün, and W. Scheid: J. Phys. B **32**, 53 (1999)
4. I.F. Barna: Ionization of helium in relativistic heavy-ion collisions. PhD Thesis, Universität Gießen, Germany (2002)
5. T. Bronk, J.F. Reading, and A.L. Ford: J. Phys. B **31**, 2477 (1998)
6. T.G. Lee, H.C. Tseng, and C.D. Lin: Phys. Rev. A **61**, 062713 (2000)
7. A. Igarashi, A. Ohsaki, and S. Nakazaki: Phys. Rev. A **62**, 052722 (2000); **64**, 042717 (2001)
8. C. Díaz, F. Martín, and A. Salin: J. Phys. B **35**, 2555 (2002)
9. A.L. Godunov and J.H. McGuire: J. Phys. B **34**, L223 (2001)
10. J. Ullrich, R. Moshammer, R. Dörner, O. Jagutzki, V. Mergel, H. Schmidt-Böcking, and L. Spielberger: J. Phys. B **30**, 2917 (1997); R. Dörner, V. Mergel, O. Jagutzki, L. Spielberger, J. Ullrich, R. Moshammer, and H. Schmidt-Böcking: Phys. Rep. **330**, 95 (2000)
11. C. Díaz, F. Martín, and A. Salin: J. Phys. B **33**, 4373 (2000)
12. N.T. Maitra, K. Burke, H. Appel, E.K.U. Gross, and R. van Leeuwen: Ten Topical Questions in Time-Dependent Density Functional Theory. In: *Review in Modern Quantum Chemistry: A Celebration of the Contributions of R.G. Parr* ed. by K.D. Sen (World Scientific, Singapore 2001)
13. R.M. Dreizler and E.K.U. Gross: *Density Functional Theory* (Springer, Berlin, Heidelberg, New York 1990)
14. T. Kirchner, L. Gulyás, H.J. Lüdde, A. Henne, E. Engel, and R.M. Dreizler: Phys. Rev. Lett. **79**, 1658 (1997)
15. T. Kirchner, L. Gulyás, H.J. Lüdde, E. Engel, and R.M. Dreizler: Phys. Rev. A **58**, 2063 (1998)
16. T. Kirchner, H.J. Lüdde, and R.M. Dreizler: Phys. Rev. A **61**, 012705 (2000)
17. T. Kirchner, H.J. Lüdde, M. Horbatsch, and R.M. Dreizler: Phys. Rev. A **61**, 052710 (2000)
18. T. Kirchner, M. Horbatsch, H.J. Lüdde, and R.M. Dreizler: Phys. Rev. A **62**, 042704 (2000)
19. T. Kirchner and M. Horbatsch: Phys. Rev. A **63**, 062718 (2001)
20. T. Kirchner, M. Horbatsch, and H.J. Lüdde: Phys. Rev. A **64**, 012711 (2001)
21. T. Kirchner, M. Horbatsch, E. Wagner, and H.J. Lüdde: J. Phys. B **35**, 925 (2002)
22. B.H. Bransden and M.R.C. McDowell: *Charge Exchange and the Theory of Ion-Atom Collisions* (Clarendon Press, Oxford 1992)
23. E. Runge and E.K.U. Gross: Phys. Rev. Lett. **52**, 997 (1984)
24. R. Nagano, K. Yabana, T. Tazawa, and Y. Abe: J. Phys. B **32**, L65 (1999); Phys. Rev. A **62**, 062721 (2000)
25. X.M. Tong, T. Watanabe, D. Kato, and S. Ohtani: Phys. Rev. A **66**, 032709 (2002)
26. E. Engel and R.M. Dreizler: J. Comp. Chem. **20**, 31 (1999)
27. W. Kohn and L.J. Sham: Phys. Rev. **140**, A1133 (1965)
28. J.C. Slater: Phys. Rev. **81**, 385 (1951)

T. Kirchner

29. S.H. Vosko, L. Wilk, and M. Nusair: Can. J. Phys. **58**, 1200 (1980); J.P. Perdew and Y. Wang: Phys. Rev. B **45**, 13244 (1992)
30. R. Latter: Phys. Rev. **99**, 510 (1955)
31. J.P. Perdew and A. Zunger: Phys. Rev. B **23**, 5048 (1981)
32. J.D. Talman and W.F. Shadwick: Phys. Rev. A **14**, 36 (1976)
33. A. Facco Bonetti, E. Engel, R.N. Schmid, and R.M. Dreizler: Phys. Rev. Lett. **86**, 2241 (2001)
34. R. Ahlrichs, O. Dumbrajs, H. Pilkuhn, and H.G. Schlaile: Z. Phys. A **306**, 297 (1982)
35. G. Bent, P.S. Krstić, and D.R. Schultz: J. Chem. Phys. **108**, 1459 (1998)
36. J.F. Dobson, M.J. Bünner, and E.K.U. Gross: Phys. Rev. Lett. **79**, 1905 (1997)
37. H.J. Lüdde, A. Henne, T. Kirchner, and R.M. Dreizler: J. Phys. B **29**, 4423 (1996); O.J. Kroneisen, H.J. Lüdde, T. Kirchner, and R.M. Dreizler: J. Phys. A **32**, 2141 (1999)
38. H.J. Lüdde, T. Kirchner, and M. Horbatsch: Quantum Mechanical Treatment of Ion Collisions with Many-Electron Atoms. In: *Photonic, Electronic, and Atomic Collisions*. eds. J. Burgdörfer, J. Cohen, S. Datz, and C.R. Vane (Rinton Press, Princeton 2002) p. 708
39. L. Gulyás, T. Kirchner, T. Shirai, and M. Horbatsch: Phys. Rev. A **62**, 022702 (2000)
40. R.D. DuBois: Phys. Rev. A **36**, 2585 (1987)
41. M.E. Rudd, T.V. Goffe, and A. Itoh: Phys. Rev. A **32**, 2128 (1985)
42. R.D. DuBois: Phys. Rev. A **39**, 4440 (1989)
43. M.E. Rudd, T.V. Goffe, A. Itoh, and R.D. DuBois: Phys. Rev. A **32**, 829 (1985)
44. E.C. Montenegro, A.C.F. Santos, W.S. Melo, M.M. Sant'Anna, and G.M. Sigaud: Phys. Rev. Lett. **88**, 013201 (2002)
45. H. Kollmus, R. Moshammer, R.E. Olson, S. Hagmann, M. Schulz, and J. Ullrich: Phys. Rev. Lett. **88**, 103202 (2002)
46. H. Knudsen and J.F. Reading: Phys. Rep. **212**, 107 (1992)
47. A.L. Ford and J.F. Reading: J. Phys. B **23**, 2567 (1990)
48. T. Kirchner: Quantum-theoretical description of many-electron processes in ion–atom collisions. PhD Thesis, Universität Frankfurt a.M., Germany (1999)
49. M.E. Rudd, Y.K. Kim, D.H. Madison, and J.W. Gallagher: Rev. Mod. Phys. **57**, 965 (1985)
50. M.E. Rudd, R.D. DuBois, L.H. Toburen, C.A. Ratcliffe, and T.V. Goffe: Phys. Rev. A **28**, 3244 (1983)
51. P. Hvelplund, H. Knudsen, U. Mikkelsen, E. Morenzoni, S.P. Møller, E. Uggerhøj, and T. Worm: J. Phys. B **27**, 925 (1994)
52. L.H. Andersen, P. Hvelplund, H. Knudsen, S.P. Møller, J.O.P. Pederson, S. Tang-Pederson, E. Uggerhøj, K. Elsener, and E. Morenzoni: Phys. Rev. A **41**, 6536 (1990)
53. M.M. Sant'Anna, W.S. Melo, A.C.F. Santos, G.M. Sigaud, and E.C. Montenegro: Nucl. Instrum. Methods Phys. Res. B **99**, 46 (1995)

25 Ionization Dynamics in Atomic Collisions

S.Y. Ovchinnikov, J.H. Macek, Y.S. Gordeev, and G.N. Ogurtsov

25.1 Introduction

In this chapter we consider inelastic processes, in particular, ionization, that are accompanied by large energy transfer. Such processes play an important role in a broad variety of natural phenomena, e.g., they govern parameters of gaseous media and plasmas. Their understanding is of interest not only from the fundamental point of view but also as a source of information needed for many applications, such as plasma physics, laser techniques, aeronomy, spectroscopy, etc.

At high collision velocities, $v \gg 1$ (atomic units are used unless otherwise stated), ionization and other inelastic processes are described very well by the Born approximation, so that in many cases Born calculations are used as standards for calibration of experimental data. The Born approximation is not reviewed in this chapter, rather we concentrate on the low-energy region where alternative theoretical frameworks are needed.

At low and intermediate velocities, $v \leq 1$, the situation is not so fortunate. In spite of much work on the theory of slow atomic collisions (e.g. see [1] and references therein) the mechanisms for large energy transfer from particles whose velocity is much less than the mean velocity of initially bound electrons were not known for many years. Surprisingly, this related mostly to the systems with a single or one "active" electron, such as H^+–H or A^{z+}–H, where A^{z+} is a bare nucleus or highly charged ion. In these cases, ionization through decay of autoionization states does not occur and the only channel for production of free electrons is by direct coupling of the initial discrete level with continuum, which was claimed to give a negligible contribution, in contradiction with available experimental data.

In recent years, much progress has been achieved in calculations of static characteristics of atoms and molecules, e.g., energies of atomic and molecular states and probabilities of electron transitions in isolated atomic particles. However, an adequate theoretical description of the dynamics of colliding atomic systems remained unavailable for a long time. The largest conceptual difficulty of theoretical consideration is connected with the strong time dependence of the parameters characterizing the collision process, where the initial conditions are set up at $t \to -\infty$ while the solutions should be obtained at $t \to +\infty$. As a result, solving of the time-dependent Schrödinger equations for the simplest one-electron systems, using straightforward numerical methods was not possible even for up-to-date computers. An alternative way to solve such problems is expansion of total wave functions of the systems over a basis set. However, in this case considerable difficulties also arise.

Atomic and molecular basis sets are commonly used. Atomic bases accurately describe inelastic processes occurring at large internuclear distances R. In these bases, it is easy to set up initial conditions and the asymptotic translational motion of the electrons can be accurately described. However, these approaches fail at small R, since atomic bases do not suitably describe topology of electron motion in quasimolecules. Many attempts, some very successful, were made to improve the situation [2–6], but the validity ranges of the approximations used were limited and it was impossible to obtain exact solutions of the time-dependent Schrödinger equation in a broad range of collision parameters unless some *ad hoc* translation factors were introduced.

The molecular basis provides a correct description of the topology of electron motion. However, the use of this basis makes it very difficult to provide Galilean invariance of eigenfunctions and to take into account translational motion of electrons at $t \to \pm\infty$. As a result, the nonadiabatic coupling matrix elements do not vanish as $R \to \infty$ that leads to unphysical transitions between the adiabatic states. A rigorous method for construction of Galilean invariant basis functions was developed by Solov'ev and Vinitsky [7] who used a time-dependent scaling of the electron coordinates, $q = r/R(t)$, and an additional transformation of the wave function to preserve the form of the Schrödinger equation. However, practical calculations in this representation were very difficult due to the divergence of some matrix elements at $R = 0$.

Ovchinnikov and Macek [8] developed a method for adequate theoretical description of inelastic processes in atomic collisions based on the Solov'ev–Vinitsky scale transformation combined with an integral representation and Sturmian expansion of the total wave function. This makes it possible to perform *ab initio* calculations of probabilities and cross sections for various inelastic processes in a broad range of collision velocities and internuclear distances. The main ionization mechanisms and the available calculations will be discussed below.

25.2 Theory of Hidden Crossings

The ionization problem has both theoretical and experimental premises. The study of direct ionization was largely stimulated by the work of Woerlee et al. [9] who measured energy spectra of electrons ejected in collisions of ions and atoms of noble gases in the keV ion energy range. It was found that the high-energy parts of the doubly differential cross sections were well described by a rather simple empirical formula:

$$\frac{d^2\sigma}{dE_e d\Omega_e} = \frac{R_0^2}{4E_e} \exp\left[-\frac{\alpha_0\,(E - E_0)}{v}\right], \tag{25.1}$$

where E_e is the ejected electron energy, v is the incident ion velocity, Ω_e is the ejection solid angle, R_0, E_0, and α_0 are constants. The right side of (25.1) resembles very much the generic result of the adiabatic approximation since the exponent has the sense of the Massey parameter. It was tempting to interpret (25.1) within the framework of adiabatic approximations, to find the physical meaning of the constants and to develop adequate quantitative descriptions of direct ionization processes at low energies. Such approaches are based on continuing of the adiabatic potential curves (terms) of the collision system to the complex R plane [10]. This method was used earlier (e.g., see [4,10,11]) for studying transitions between two neighboring levels having crossing points close to the real R axis. In the case of ionization the situation is complicated since the crossings often occur far from the real R axis, so that they have little effect on the behavior of energy levels at real internuclear distances ("hidden crossings"). Moreover, the initial diabatic term must cross an infinite number of Rydberg levels. For this reason, the conventional adiabatic theory predicts negligible probability for ionization, in contradiction with the experimental findings. The solution of this problem was suggested by Demkov [12] who put forward the idea that quasimolecular eigenenergies $\varepsilon(R)$ could be considered as analytical functions defined on a multisheeted Riemann surface. The analytical features of $\varepsilon(R)$ in the complex plane R are closely related to the probabilities for transitions between an infinite succession of the quasimolecular levels [13].

25.2.1   General Formalism

The Schrödinger equation for the two-center Coulomb problem, $Z_1 e Z_2$, e being the electron charge, in the Born–Oppenheimer approximation is written as

$$\left(-\frac{1}{2}\nabla_r^2 - \frac{Z_1}{\left|r - \frac{R}{2}\right|} - \frac{Z_2}{\left|r + \frac{R}{2}\right|}\right) \Phi(r, R) = E(R)\Phi(r, R) , \qquad (25.2)$$

where r and R are the electron coordinate and internuclear distance, respectively, $\Phi(r, R)$ is the molecular wave function and $E(R)$ is the molecular energy. Equation (25.2) admits separation of variables in the prolate spheroidal coordinates ($r_1 = |r - R/2|$, $r_2 = |r + R/2|$, $r = \{x, y, z\}$, the z-axis being directed along the internuclear axis):

$$\xi = \frac{r_1 + r_2}{R} , \quad \eta = \frac{r_1 - r_2}{R} , \quad \phi = \arctan\frac{y}{x} ,$$
$$1 \leq \xi \leq \infty , \quad -1 \leq \eta \leq 1 , \quad 0 \leq \phi \leq 2\pi . \qquad (25.3)$$

When taking the wave function in the form

$$\Phi(r, R) = \left[(\xi^2 - 1)(1 - \eta^2)\right]^{-1/2} U(\xi) V(\eta) \, e^{im\phi} \qquad (25.4)$$

and substituting it into (25.2), one obtains the system of equations for the functions $U(\xi)$ and $V(\eta)$:

$$\left[\frac{d^2}{d\xi^2} - p^2 + \frac{a'\xi - \lambda}{\xi^2 - 1} + \frac{1 - m^2}{(\xi^2 - 1)^2}\right] U(\xi) = 0,$$

$$\left[\frac{d^2}{d\eta^2} - p^2 + \frac{b'\eta + \lambda}{1 - \eta^2} + \frac{1 - m^2}{(1 - \eta^2)^2}\right] V(\eta) = 0, \tag{25.5}$$

where $p = (-2E)^{1/2}R/2$, $a' = (Z_1 + Z_2)R$, $b' = (Z_2 - Z_1)R$, and λ is the separation constant.

The eigenenergies (potential curves) $E_n(R)$ with the same m values are connected at the branch points R_C in the complex plane in the vicinity of which the following relation holds [13]:

$$E(R) = E(R_C) + \text{const}\sqrt{R - R_C}. \tag{25.6}$$

Positions and analytical behaviors of the branch points reflect the particular mechanisms for the inelastic atomic collisions.

25.2.2   S-Ionization and Superpromotion

Solov'ev [14] discovered branch points $R_C = R_{nlm}^{(n+1)lm}$ connecting the terms $E_{nlm}(R)$ and $E_{(n+1)lm}(R)$, successively for all $n \geq l + 1$. The set of branch points $R_{nlm}^{(n+1)lm}$ with different n values but fixed set $\{lm\}$ forms an infinite series of points localized in a small region Ω of the complex R plane and has a limit point:

$$R_{lm} = \lim_{n \to \infty} R_{nlm}^{(n+1)lm}. \tag{25.7}$$

The branch points connect all terms of the given series $\{lm\}$ to form a unique analytical function $E_{lm}(R)$. In the domain Ω, the energy surface $E_{lm}(R)$ resembles a corkscrew, so that a single turn around the branch point R_{lm} promotes the given state $E_{nlm}(R)$ to the neighboring state $E_{(n+1)lm}(R)$ (Fig. 25.1). The series $\{lm\}$ was designated $S_{(l+1)lm}$, and the mechanism of electron production due to promotion of a diabatic term to the continuum through a succession of these points was called S-ionization (from the word "superpromotion"). In the work of Janev and Krstic [15] higher-order series S_{lm}^{κ} were revealed.

Analytical expressions for the coordinates of the limit branch points were obtained in [16], namely,

$$R_{lm} = \frac{1}{Z}\left\{\left(l + \frac{1}{2}\right)^2 - \frac{(m+1)^2}{2}\right.$$

$$\left. \pm i(m+1)\sqrt{2\left(l + \frac{1}{2}\right)^2 - \frac{(m+1)^2}{4}}\right\}, \tag{25.8}$$

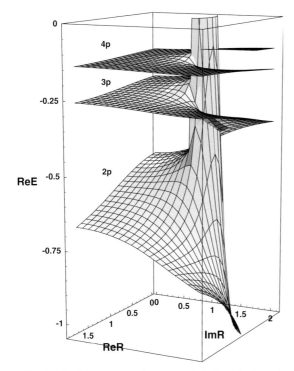

Fig. 25.1. Riemann surface associated with S-ionization in H^+–H collisions

where Z is the united atom charge. In addition to the branch points $R_{(l+1)lm}$ connecting neighboring levels, branch points $R_{(\tilde{l}+1)\tilde{l}\tilde{m}}$ connecting the lowest state of the series with quasistationary and virtual states $\{\tilde{l}\tilde{m}\}$ were revealed in [17] corresponding to coupling of the initial term with the continuum. Because the unstable state decays to produce free electrons, this discovery made it possible to compute ionization within the framework of the adiabatic approximation.

Classical interpretation of S-ionization is based on the topology of the electron motion. At close approach of the colliding particles, a united atom centrifugal barrier appears. This barrier keeps the electron out of the region between nuclei so that the trajectory along the line between nuclei becomes unstable. Oscillation of the electron along this unstable trajectory transfers energy from nuclear to electron motion until the electron acquires enough energy to escape. In terms of the quantum theory S-promotion corresponds to rearrangement of quasimolecular wave functions to united atom wave functions.

The electron energy distribution is given by the probability $P(E)$ [18]:

$$P(E, \hat{\boldsymbol{k}}) = \frac{1}{2\pi v} \left| \frac{\mathrm{d}R(E)}{\mathrm{d}E} C^2 \left(E, \hat{\boldsymbol{k}} \right) \exp \left\{ \frac{2\mathrm{i}}{v} \int^E R(E)\mathrm{d}E \right\} \right|, \qquad (25.9)$$

where $R(E)$ is the function reciprocal of $E(R)$ and $C(E, \hat{\boldsymbol{k}})$ is the normalization coefficient of the adiabatic wave function. The angular distributions will be discussed later, here we consider cross sections integrated over electron direction $\hat{\boldsymbol{k}}$. We denote $C^2(E) = \int |C(E, \hat{\boldsymbol{k}})|^2 \mathrm{d}\hat{\boldsymbol{k}}$. At high enough electron energies the following approximations can be used:

$$C^2(E) = 4\pi \mathrm{Im}R(E) \left| \frac{\mathrm{d}E}{\mathrm{d}R} \right|, \qquad (25.10)$$

$$\mathrm{Im}R(E, b) = \mathrm{Im}R(E) \left(1 + \frac{b^2}{2|R(E)|^2} \right). \qquad (25.11)$$

Integration of (25.9) over the impact parameter b gives the following expression for the differential cross section for electron ejection:

$$\frac{\mathrm{d}\sigma}{\mathrm{d}E} = 2\pi \int_0^\infty P(E, b)\, b \,\mathrm{d}b = A(E)\exp \left[-\frac{\alpha(E)}{v} \right], \qquad (25.12)$$

where

$$A(E) = \frac{4\pi |R(E)|^2 \mathrm{Im}R(E)}{\alpha(E)}, \quad \alpha(E) = 2 \int_{E_0}^E \mathrm{Im}R(E')\,\mathrm{d}E'. \qquad (25.13)$$

In some cases the function $\mathrm{Im}R(E)$ can be approximated by the constant R_{lm}, then (25.13) can be simplified to the form

$$A(E) = \frac{2\pi |R(E)|^2}{(E - E_0)}, \quad \alpha(E) = 2\,\mathrm{Im}R_{lm}(E - E_0). \qquad (25.14)$$

By comparing (25.14) with the empirical formula (25.1) we find the physical meanings of the constants: R_0 corresponds to the absolute value of the R_{lm}, $\alpha_0 = 2\,\mathrm{Im}R_{lm}$ to the stability exponent of the unstable trajectory [19], and E_0 to the energy of the ionized quasimolecular level.

25.2.3 T-Ionization and Saddle-Point Electrons

The T-type branch points were discovered by Ovchinnikov and Solov'ev [20] and used to compute transitions between bound states. It was later recognized that these transitions were closely related to the ionization mechanism now called top-of-barrier promotion or T-promotion. These mechanisms are effective at large internuclear distances as the colliding particles separate.

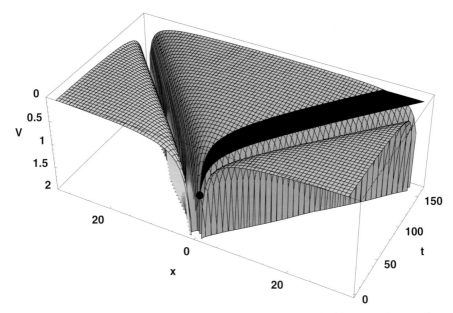

Fig. 25.2. Time evolution of an electron acceleration via the T-promotion mechanism. V is the potential energy, x is the internuclear separation, and t is the time

At such distances the Coulomb potential barrier between the nuclei becomes comparable to the electron energy. When the particles approach, the Coulomb potential barrier decreases, so that the atomic electrons begin to move in the field of both centers. When the particles recede from each other, the barrier increases, and the electrons can be captured on the top of the barrier (the saddle point of potential) and, finally, promoted into the continuum. The motion of the electrons on the top of the barrier is unstable and therefore energy is transferred from nuclear to electron motions. Time evolution of the motion of an electron accelerated via the T-promotion mechanism is illustrated in Fig. 25.2.

The corresponding branch points $R_{n_1 n_2 m}^T$ (n_1 and n_2 are the parabolic quantum numbers) connect states with the same n_1 but different n_2. Electrons ejected due to T-ionization (from the words "*top of barrier*") are often called "saddle-point electrons". In terms of quantum theory T-promotion corresponds to the rearrangement of the quasimolecular wave functions to the separated-atom wave functions. The Riemann surface calculated for T-ionization in H^+–H collisions [21] is shown in Fig. 25.3.

The eigenenergies $\varepsilon(R)$ in the vicinity of the saddle point behave like harmonic oscillators, while on the real axis they represent the Rydberg states. In the case of equal charges, $Z_1 = Z_2$, the velocity of an electron ejected via T-ionization equals one half of the incident ion velocity.

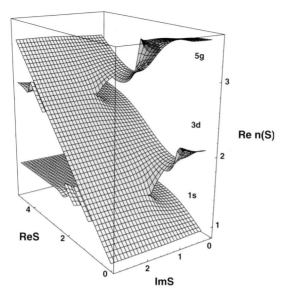

Fig. 25.3. Riemann surface associated with T-ionization in H^+–H collisions. The function $n(R) = \varepsilon^{-1/2}(R)$ vs. $S = R^{1/2}$

T-ionization occurs at finite internuclear distances R_∞, which can be estimated using the expressions for the total energy:

$$E_{\mathrm{Tot}} = V_{\mathrm{pot}} + U_{\mathrm{kin}} ,\tag{25.15}$$

$$V_{\mathrm{pot}} = -\frac{C_0}{R} , \qquad U_{\mathrm{kin}} = \frac{v_{\mathrm{T}}^2}{2} ,\tag{25.16}$$

$$C_0 = \left(\sqrt{Z_1} + \sqrt{Z_2}\right)^2 - Z_1 Z_2 , \qquad v_{\mathrm{T}} = v_0 \frac{\sqrt{Z_1}}{\sqrt{Z_1} + \sqrt{Z_2}} ,\tag{25.17}$$

where v_0 is the velocity of the incident particle Z_2. Using the condition $E_{\mathrm{Tot}} = 0$ for ionization one obtains

$$R_\infty = \frac{2C_0}{v_{\mathrm{T}}^2} .\tag{25.18}$$

Rydberg states with mean radius greater than R_∞ are considered to be ionized with unit probability. The corresponding separated atom principal quantum number n_∞ can be estimated by substituting $R_{\mathrm{T}}(Z_2) \approx 2n_\infty^2/Z_2$ ($Z_1 = 1$, $Z_2 > Z_1$) into (25.17) and (25.18) to obtain:

$$n_\infty \geq \frac{\left(1 + \sqrt{Z_2}\right)\left(1 + 2\sqrt{Z_2}\right)^{1/2}}{v_0} .\tag{25.19}$$

For $Z_1 = Z_2$, the substitution of $R \approx \left(\frac{\pi^2}{2} \right) n_\infty^2$ yields:

$$n_\infty \geq \frac{4\sqrt{3}}{\pi v_0} \, . \tag{25.20}$$

The probability of T-ionization from the initial $1s\sigma$ orbital can be written as

$$P_\sigma^{(T)} = (1 - P_{000}) \, P_{00}^{(T)} \, , \tag{25.21}$$

where

$$P_{0m}^{(T)} \equiv \prod_{i=0}^{n_\infty} P_{0im} = \mathrm{e}^{-\frac{2}{v_0} \Delta_{0m}^T} \, , \quad \Delta_{0m}^T = \sum_{i=0}^{n_\infty} \Delta_{0im} \, ,$$

P_{0im} is the probability of transition between the terms E_{0im} and $E_{0(i+1)m}$ connected by the first branch point of the series T_{0im}, Δ_{0im} is the relevant Massey parameter. After integration over the impact parameter the cross section is obtained:

$$\sigma_\sigma^{(T)} \cong \frac{\pi v_0 R_0^2}{\Delta_{000}} P_\sigma^{(T)} \, , \tag{25.22}$$

where R_0 and Δ_{000} are the coordinate of the branch point and the Massey parameter for the transition $1s\sigma$–$3d\sigma$, respectively, $\Delta_{00}^{(T)}$ is the sum of the Massey parameters for the superpromotion T_{00}^s. For another succession of transitions, via the rotational coupling $2p\sigma$–$2p\pi$ followed by T_{01}^s promotion, one gets:

$$P_\pi^{(T)} = P_{2p\sigma-2p\pi} P_{01}^{(T)} \, , \tag{25.23}$$

$$\sigma_\pi^{(T)} = \sigma_{2p\sigma-2p\pi} P_{01}^{(T)} \, , \tag{25.24}$$

$$\sigma_{2p\sigma-2p\pi} = 2\pi \int_0^\infty P_{2p\sigma-2p\pi}(b) \, b \, \mathrm{d}b \, . \tag{25.25}$$

The total cross section for ionization due to S- and T-promotion is

$$\sigma = \sigma^{(S)} + \sigma_\sigma^{(T)} + \sigma_\pi^{(T)} \, . \tag{25.26}$$

The contributions of S- and T-ionization in H$^+$–H collisions are shown in Fig. 25.4. Because impact parameters are integrated over, there is no interference between components with different symmetry. It is seen that T-ionization gives the major contribution to the low-energy part of the energy spectrum.

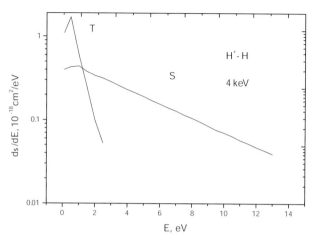

Fig. 25.4. Differential cross sections for ionization in H^+–H collisions: contributions of S- and T-ionization

25.2.4   D-Ionization and Radial Decoupling

D-ionization (from the word "decoupling") is a purely quantum mechanism discovered by Ovchinnikov and Macek [22] and studied using the zero-range model potentials [23]. Vanishing of coupling between the electron and nuclear motion at very small internuclear distances causes this mechanism.

A given classical orbit of an electron moving in the combined Coulomb field of two nuclei scales with R. So, as $R \to 0$, the classical orbit contracts accordingly. However, in quantum mechanics an electronic wave function cannot be reduced in size without limit because at $R = 0$ it has to coincide with the united atom wave function, which has a fixed spatial extension. Therefore, as the internuclear distance decreases, at some (very small) distance R_0 (which is well within the united atom limit) the electron will effectively move in the Coulomb field of only one (united) nucleus, and the quasimolecular wave function will become independent of R.

The contribution of D-ionization increases considerably on decreasing incident proton energy and becomes essential for ionization in H^+–H collisions at $E < 2\,\mathrm{keV/amu}$ [23]. The calculation results for the sum of contributions of all mechanisms agree well with the experimental data on cross sections for ionization [24] that may indicate that all the mechanisms essential for ionization in the energy range under study are taken into account. It should be noted that this is D-ionization that causes the v^2 dependence of cross sections at low collision energies.

25.3   Sturmian Theory

In spite of an evident progress in calculations of total cross sections for ionization in slow atomic collisions, application of the adiabatic approximation met principal difficulties in calculations of energy spectra of ejected electrons. Namely, the adiabatic approximation could not provide a correct description of the low-energy parts of the electron energy spectra, $k < v$, which give major contributions to the total energy spectra. Thus, it has become necessary to develop a theory of ionization applicable both in a wide range of collision velocities and in a wide range of ejected electron energies.

In this part of the chapter we will discuss the development of a method that provides a solution of the two-center problem in a broad velocity range and includes all possible mechanisms of inelastic atomic collisions. The method is based on three principal transformations: (i) scale transformation of Solov'ev–Vinitsky, (ii) expansion of the total wave function over eigenfunctions of the Sturm–Liouville problem, and (iii) presentation of the wave function in the form of the Fourier integral.

25.3.1   Scale Transformation of Solov'ev–Vinitsky

This transformation [7] allows one to take into account translational and rotational effects. It includes the change of variables:

$$q = \frac{r}{R(t)}, \tag{25.27}$$

$$\tau = \int_{-\infty}^{t} \frac{dt'}{R^2(t')}, \tag{25.28}$$

and the transformation of the wave function:

$$\Psi(r,t) = \frac{1}{R^{3/2}(\tau)} \exp\left\{ \frac{i\dot{R}(\tau)}{2R(\tau)} q^2 \right\} \Phi(q,\tau), \tag{25.29}$$

then in a reference frame rotating around the vector $n = v \times b$ with the angular velocity Ω we obtain a new Schrödinger equation:

$$\left[-\frac{1}{2}\nabla_q^2 + \frac{1}{2}\Omega^2 q^2 + \Omega \hat{L}_y + R(\tau)V(q,R) \right] \Phi(q,\tau) = i\frac{\partial \Phi(q,\tau)}{\partial \tau}. \tag{25.30}$$

The most important features of the transformation are as follows:

1. In the original physical coordinates $\{r,t\}$ the nuclei move along the trajectory $R(t)$, therefore the interaction between the electron and the nuclei depends on the direction of the vector $R(t)$. In the scaled space $\{q,\tau\}$ the nuclei are fixed, and the potential in the transformed coordinates does not depend on the direction of $R(t)$. The dynamical effects are described by the scalar function $R(\tau)$ and two additional terms in the new Hamiltonian, the simple harmonic oscillator and the angular momentum operator.

2. If $|vt| \gg b$, then the functions $\Phi(q, \tau)$ are Galilean invariant under translations in the plane of v and R, therefore there is no need for translational factors.

3. For straight-line trajectories, the angular velocity $\Omega = vb$ is constant in the scaled representation, therefore it is convenient to introduce the angle of rotation $\theta = \Omega \tau$, $0 \leq \theta \leq \pi$, as a new "time" in the Schrödinger equation, so that $R(\theta) = b \sin \theta$.

25.3.2   Sturmian Basis

In most cases, representation of bases of wave functions involves eigenenergies of a certain model potential. The representation of the Sturmian basis is different in that the Sturmian eigenfunctions do not correspond to the eigenenergies. In this representation, the parameter $\omega = E(R)R^2$ is introduced instead of the energy $E(R)$. A Sturm–Liouville problem is furnished by the system of differential equations:

$$[H_0(q) + \rho_\nu V(q) - \omega] S_\nu(\omega; q) = 0, \tag{25.31}$$

with proper boundary conditions. The values of internuclear distance at fixed potential energy are taken as new eigenvalues $\rho_\nu(\omega)$, solutions of the equation

$$\varepsilon(\rho) \rho^2 = \omega = \text{const}. \tag{25.32}$$

The corresponding Sturmian eigenfunctions $S_\nu(\omega, q)$ are defined for all values of ω, including negative, positive, and even complex values. In contrast to the adiabatic functions, the Sturmian functions do not depend, even parametrically, upon the internuclear distance R. The Sturmian bases for particular cases of zero-range potentials and two Coulomb centers have been defined in [25–27].

25.3.3   Wave Functions and Transition Amplitudes
in Fourier Space

The total time-dependent wave function expanded along the Sturmian basis is written as [8]:

$$\Phi(q, \tau) = \frac{1}{\sqrt{-2\pi i v}} \sum_\nu \int_{-\infty}^{\infty} \exp(-i\omega\tau) B_\nu(\omega) S_\nu(\omega; q) \, d\omega. \tag{25.33}$$

The expansion coefficients $B_\nu(\omega)$ are solutions of the coupled equations (25.30) depending on initial conditions. Integration over ω from $-\infty$ to $+\infty$ ensures that the total wave function and each term of the expansion contain the whole energy spectrum, including both discrete states and the continuum.

The amplitude for electron transition to a continuum state $\boldsymbol{k}$ is obtained by projecting our time-dependent wave function onto plane waves:

$$A(\boldsymbol{k}) = \lim_{t\to+\infty} \int \varphi_{\boldsymbol{k}}^*(\boldsymbol{r},t)\psi(\boldsymbol{r},t)\,\mathrm{d}^3 r\,, \tag{25.34}$$

where

$$\varphi_{\boldsymbol{k}}(\boldsymbol{r},t) = \frac{1}{(2\pi)^{3/2}}\mathrm{e}^{\mathrm{i}\boldsymbol{k}\cdot\boldsymbol{r}-\mathrm{i}\frac{k^2}{2}t}\,. \tag{25.35}$$

Performing the scale transformation and taking the limit we obtain

$$A(\boldsymbol{k}) = \left(\frac{\mathrm{i}}{v}\right)^{\frac{3}{2}} \varphi(\boldsymbol{q},\theta)|_{\theta=\pi,\boldsymbol{q}=\boldsymbol{k}/v} = \left(\frac{\mathrm{i}}{v}\right)^{\frac{3}{2}} \varphi(\boldsymbol{k}/v,\pi)\,. \tag{25.36}$$

The above consideration reveals evident advantages of the Sturmian basis sets, which can be summed up as follows:

1. The Sturmian functions form a complete discrete set for all values of ω, both negative and positive, i.e., they describe in a unique way both discrete and continuum states of the colliding system.
2. The Sturmian functions do not depend on the internuclear distance, therefore matrix elements for dynamical interaction do not contain singularities at $R \to 0$.
3. In the Sturmian basis, positions of nuclei are fixed, therefore the solutions obtained are Galilean invariant and there is no need for additional translational factors.
4. Integrals in the Fourier presentation of wave functions converge very rapidly, therefore a finite (and small) number of Sturmian functions is usually sufficient to describe particular collision processes.

So, the use of the Sturmian bases combined with the scaled Solov'ev–Vinitsky transformation and Fourier presentation of wave functions provides exact solution of the problem of ionization in atomic collisions.

25.4 Results of Calculations

Now we proceed to the problems connected with the use of Sturmian bases in practically important particular cases of zero-range potentials and two Coulomb centers. The first one is realized in electron detachment in negative ion–atom collisions and D-ionization. The second one involves a great variety of collision processes in which one "active" electron can be discerned.

25.4.1   Differential Cross Sections

Electron energy and angular distributions are widely used as probes of atomic dynamics. At high energy, where Born approximation is applicable, two features dominate the energy spectrum, namely, the binary encounter peak and the continuum capture cusp [27]. At low energy, S, T, and D mechanisms have been revealed. So it is important to check if the Sturmian theory could incorporate all these features. This problem has been successively studied for the cases of two zero-range potentials with $b = 0$ [29] and $b \neq 0$ [27], and two Coulomb centers [25].

The electron distributions are given in the Galilean invariant form:

$$\frac{\mathrm{d}^3 P}{\mathrm{d}k^3} = |A(\boldsymbol{k})|^2 . \tag{25.37}$$

It has been shown [29] that the main features of the ejected-electron spectra can be obtained using only one Sturmian function. This single function connects high-energy and low-energy regions. Calculation results for the system H$^-$–H are presented in Figs. 25.5 and 25.6. At high velocity, $v = 10$ a.u. (Fig. 25.5), there are two peaks, one centered at the target ($\boldsymbol{k} = 0$) and another one at the projectile ($\boldsymbol{k} = 1$), separated from an oval-shaped binary encounter ridge at $|\boldsymbol{k} - \boldsymbol{v}| = v$. These features reflect those observed in high-energy ion–atom collisions.

At low velocity, $v = 0.1$ a.u. (Fig. 25.6), there is only one peak for g states centered at $\boldsymbol{k} = \boldsymbol{v}/2$, and a small ridge at $k \approx 7$ a.u. In comparison with the high-velocity distributions one notes that the binary encounter peak completely disappears, and separate target and projectile peaks merge into one peak centered between target and projectile. The situation is similar for u states, where there is a node at the midpoint owing to symmetry requirements. In this case, there is also no binary encounter peak, and the separate target and projectile peaks have merged into a peak with a node exactly

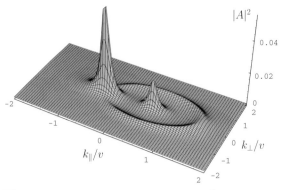

Fig. 25.5. Electron distributions $|A(\boldsymbol{k})|^2$ vs. k/v for H$^-$–H collisions at $v = 10$ a.u. in the center-of-mass frame

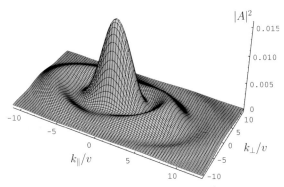

Fig. 25.6. Electron distributions $|A(\boldsymbol{k})|^2$ vs. k/v for $\mathrm{H^-\!-\!H}$ collisions at $v = 0.1$ a.u. in the center-of-mass frame

midway between target and projectile. The distribution with "atomic" initial conditions exhibits the peak at the center-of-mass velocity and a series of interference features.

In the case of two Coulomb centers, (25.31) takes the form [25]:

$$\left[-\frac{1}{2}\nabla_{\boldsymbol{q}}^2 + \rho_\nu \left(\frac{Z_1}{q_1} - \frac{Z_2}{q_2}\right)\right] S_\nu(\omega; \boldsymbol{q}) = \omega S_\nu(\omega; \boldsymbol{q}), \tag{25.38}$$

where the Sturmian eigenvalues are solutions of (25.32). In the course of computations [8] two different types of Sturmian functions at $\omega > 0$ have been revealed: T- and S-type Sturmian functions. Together, these two types of Sturmian functions form a complete set, but they otherwise have rather different properties. The T-type Sturmian functions are analytic continuations of the negative-energy Sturmian functions and therefore exist at all ω. Alternatively, the S-type Sturmian functions are defined only for $\omega > 0$ and $\rho_\nu^S(0) \neq 0$. The spectra of ejected electrons associated with two different types of Sturmians are displayed in Figs. 25.7 and 25.8. The spectrum related to the S promotion for $v = 0.4$ a.u. (Fig. 25.7) has two peaks at $k_\perp = 0$ and $k_\parallel = v/2$ in the center-of-mass frame. The energy distribution of the fast electrons is exponential. The spectrum related to the T promotion of $2p\pi$ orbital (Fig. 25.8) has two peaks at zero center-of-mass velocity, in accordance with the π symmetry of the T_{01} transition.

The Sturmian theory proved to be successful in explaining an interesting effect observed in experiments performed using the cold-target recoil-ion mass-spectrometry (COLTRIMS) [30]; namely, rapid oscillations of electron angular distributions in slow ion–atom collisions. It has been shown [31] that these oscillations are associated with T-ionization and, in particular, with the behavior of the real part of the potential energy $\varepsilon(R)$ in the harmonic-oscillation region.

Using the Solov'ev–Vinitsky transformation, expanding the Fourier transform of the wave function in terms of Sturmian eigenfunctions, and expanding

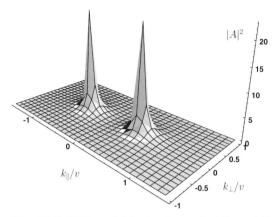

Fig. 25.7. Differential ionization probabilities $|A(\boldsymbol{k})|^2$ at $v = 0.4$ a.u. and $b = 0$ for S-promotion

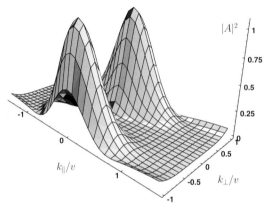

Fig. 25.8. Differential ionization probabilities $|A(\boldsymbol{k})|^2$ at $v = 0.4$ a.u. and $b = 0$ for T-promotion

the potential in the vicinity of the saddle point, one arrives at the following expression for the differential cross section:

$$\frac{\mathrm{d}^3\sigma}{\mathrm{d}k^3} \propto |A_\sigma(\boldsymbol{k}/v) + a\exp[\mathrm{i}\phi/v]A_\pi(\boldsymbol{k}/v)|^2 \,, \tag{25.39}$$

where

$$\phi = -\int\limits_{R_0}^{\infty} [\varepsilon_\sigma(R) - \varepsilon_\pi(R)]\mathrm{d}R \,, \tag{25.40}$$

$\boldsymbol{k}$ is the electron wave vector, A_σ and A_π are the amplitudes for σ_g and π_u ionization, a is the expansion coefficient. The amplitude A_σ is nodeless, but

$H^+ + H \rightarrow H^+ + e^- + H^+$ ($b = 1.8$)

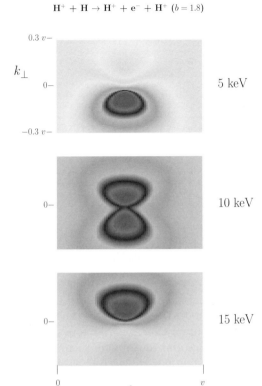

Fig. 25.9. Density plot of the electron distributions, $\boldsymbol{k}/v$, for proton impact on atomic hydrogen at fixed impact parameter $b = 1.2$ a.u. and ion energies of 5, 10 and 15 keV

A_π has a node at $k_\perp = 0$. For this reason, the angular distribution is not symmetric about the z-axis and changes rapidly with ion velocity.

The electron distributions for 5, 10 and 15 keV H^+–H collisions at an impact parameter $b = 1.2$ a.u. are shown as density plots in Fig. 25.9. In these plots, the z-axis is taken along the ion velocity so that $k_\parallel + k_z$ and the z-axis lie in the scattering plane along the direction of the impact parameter vector. The phase ϕ changes by nearly 2π over the 5–15 keV energy range, which gives rise to drastic change in the electron distributions, in accordance with the experimental findings.

25.4.2 Total Cross Sections

The total ionization cross sections in a broad energy range have been calculated for the systems H^+–H and H^0–He [32,33]. The first system is clearly of fundamental importance, since it is the simplest ion–atom system that can be

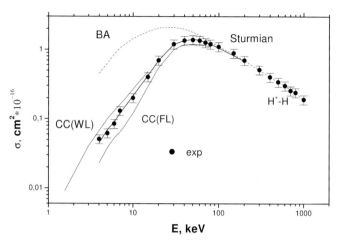

Fig. 25.10. Total cross sections for ionization in H$^+$–H collisions. BA – Born approximation, CC(WL) and CC(FL) – close-coupling calculations [34,35], *points* – recommended experimental data [36], the *bold solid curve* – Sturmian theory

used as a probe for testing theoretical methods. Development of the crossing beam techniques has made it possible to obtain experimental data on cross sections for ionization in a wide energy range and to compare them directly with the calculation results. The second system has an "active" electron initially belonged to the hydrogen atom and promoted along the orbital $2p\,\sigma$. Moreover, the Coulomb barrier is absent in this case, so that the main mechanism for electron production is S-ionization and the approach described in Sect. 25.2.2 can be used. The system H^0–He can be considered as a probe for testing the applicability of the developed theory to the systems with non-Coulomb interaction.

The total cross sections for ionization in H$^+$–H collisions are shown in Fig. 25.10, where the calculation results are presented together with those using different methods [34,35] and with the recommended experimental data [36]. As seen from the figure, the Sturmian calculations provide the best agreement with the experimental data in a wide energy range, from the adiabatic region up to the high-energy region, where the Born approximation is applicable.

Figure 25.11 shows experimental data and calculation results on the cross sections for stripping in H^0–He collisions. The good agreement between theoretical and recommended experimental data [37] is seen in this case as well. This offers evidence that the developed theoretical approach can give satisfactory results for non-Coulomb systems with one "active" electron.

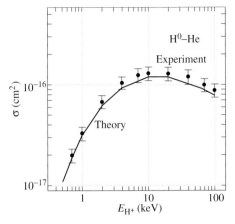

Fig. 25.11. Cross sections for stripping in H^0–He collisions. *Solid curve* – calculations [32], *points* – recommended experimental data [37]

25.5 Conclusions

The discussion presented above enables us to draw the following conclusions about the present state of the problem of ionization in atomic collisions. First, it has proved possible to develop a theoretical approach ensuring Galilean invariance of solutions of the time-dependent Schrödinger equation and describing in a unique way inelastic collision processes, including ionization, in a wide energy range. Secondly, the developed theory allows one to perform exact *ab initio* calculations of ionization in the $Z_1 e Z_2$ system and has a perspective to be extended to more complicated systems with an "active" electron. Thirdly, the theory provides obtaining reliable data on both gross parameters, such as the total cross sections, and on fine details characterizing ionization dynamics, such as cusps, ridges, and interference structures.

The results obtained during the past decade have even more general fundamental significance, bearing in mind the situation with dynamical problems that existed in quantum mechanics for a long time. If, in the solution of stationary problems (e.g., calculation of energies and widths of atomic and molecular states) the results could be obtained within the given accuracy, then in the case of dynamical problems (e.g., calculation of cross sections and transition probabilities) the results obtained, as usual, had an accuracy of tens and even hundreds of per cent. This resulted from the essentially many-body character of the dynamical problems. The theory reviewed in this chapter can be considered as a definite step towards solution of these problems.

Acknowledgements

This research is sponsored by the Division of Chemical Sciences, Office of Basic Energy Sciences, U.S. Department of Energy, under Contract No. DE-AC05-00OR22725 with UT-Battell, LLC. One of the authors (GNO) is supported by INTAS, grant No 2001-0155.

References

1. J.-T. Hwang and P. Pechukas: J. Chem. Phys. **67**, 4640 (1977)
2. D.R. Bates and R. McCarroll: Proc. Roy. Soc. London A **245**, 175 (1958)
3. W. Thorson and J.B. Delos: Phys. Rev. A **18**, 117 (1978)
4. J. Vaaben and K. Taulbjerg: J. Phys. B **14**, 1815 (1981)
5. R. McCarroll and D.S.F. Crothers: Adv. At. Mol. Opt. Phys. **32**, 253 (1994)
6. J. Grosser, T. Menzel and A.K. Belyaev: Phys. Rev. A **59**, 1309 (1999)
7. E.A. Solov'ev and S.I. Vinitsky: J.Phys. B **18**, L557 (1985)
8. S. Ovchinnikov and J. Macek: Phys. Rev. Lett. **75**, 2474 (1995)
9. P.H. Woerlee, Yu.S. Gordeev, H. deWaard and F. Saris: J. Phys. B **14**, 527 (1981)
10. E.C.G. Stueckelberg: Helv. Phys. Acta **5**, 369 (1932)
11. D.R. Bates and D.A. Williams: Proc. Phys. Soc. **83**, 425 (1964)
12. Yu.N. Demkov: *Inv. Papers of V Intern. Conf. on the Physics of Electronic and Atomic Collisions, Leningrad 1967*, ed. L. Branscomb (University of Colorado, Boulder 1968) p. 186
13. L.D. Landau and E.M. Lifshitz: *Quantum Mechanics: Non-relativistic Theory*, 3d edn. (Pergamon, Oxford 1965) p. 194
14. E.A. Solov'ev: Zh. Eksp. Teor. Fiz. **81**, 1681 (1981) (Sov. Phys. – JETP **54**, 893, 1981)
15. R.K. Janev and P.S. Krstic: Phys. Rev. A **44**, R 1435 (1991)
16. E.A. Solov'ev: Zh. Eksp. Teor. Fiz. **90**, 1165 (1986) (Sov. Phys. JETP **63**, 678, 1986)
17. S.Y. Ovchinnikov and E.A. Solov'ev: Zh. Eksp. Teor. Fiz. **91**, 477 (1986) (Sov. Phys. JETP **64**, 280, 1986)
18. E.A. Solov'ev: Usp. Fiz. Nauk, **157**, 437 (1989) (Sov. Phys. – Uspekhi **32**, 228, 1989)
19. S.Y. Ovchinnikov: Phys. Rev. A **42**, 3873 (1990)
20. S.Y. Ovchinnikov and E.A. Solov'ev: Zh. Eksp. Teor. Fiz. **90**, 921 (1986) (Sov. Phys. JETP **63**, 538, 1986)
21. M. Pieksma and S.Y. Ovchinnikov: J. Phys. B **27**, 4573 (1994)
22. S.Y. Ovchinnikov and J.H. Macek: *XVIII Int. Conf. on the Physics of Electronic and Atomic Collisions (Aarhus 1993)* eds. T. Andersen, B. Fastrup, F. Folkmann and H. Knudsen (Aarhus: IFA Print) Abstracts of Contributed Papers, p. 676
23. M. Pieksma, S.Y. Ovchinnikov and J.H. Macek: J. Phys. B **31**, 1267 (1998)
24. M. Pieksma, J. VanEck, W.B. Westerveld, A. Niehaus and S.Y.Ovchinnikov: Phys. Rev. Lett. **73**, 46 (1994)
25. S.Y. Ovchinnikov and J.H. Macek: Phys. Rev. A **55**, 3605 (1997)

26. S.Y. Ovchinnikov, J.H. Macek and D.B. Khrebtukov: Phys. Rev. A **56**, 2872 (1997)
27. S.Y. Ovchinnikov, D.B. Khrebtukov and J.H. Macek: Phys. Rev. A **65**, 032722 (2002)
28. J. Macek and K. Taulbjerg: J. Phys. B **26**, 1353 (1993)
29. J.H. Macek, S.Y. Ovchinnikov and E.A. Solov'ev: Phys. Rev. A **60**, 1140 (1999)
30. R. Dörner, H. Klemliche, M.H. Prior, C.L. Cocke, J.A. Gary, R.E. Olson, V. Mergel, J. Ulrich and H. Schmidt-Böcking: Phys. Rev. Lett. **77**, 4520 (1996)
31. J.H. Macek and S.Y. Ovchinnikov: Phys. Rev. Lett. **80**, 2298 (1998)
32. Y.S. Gordeev, G.N. Ogurtsov and S.Y. Ovchinnikov: Direct Ionization in Atomic Collisions, in *Ioffe Institute Prize Winners' 97*, St.-Petersburg 1998, p. 12
33. G.N. Ogurtsov, A.G. Kroupyshev, M.G. Sargsyan, Y.S. Gordeev and S.Y. Ovchinnikov: Phys. Rev. A **53**, 2391 (1996)
34. W. Fritch and C.D. Lin: Phys. Rev. A **27**, 3361 (1983)
35. T. Winter and C.D. Lin: Phys. Rev. A **29**, 3071 (1984)
36. S.V. Avakyan, R.N. Il'in, V.M. Lavrov and G.N. Ogurtsov: *Collision Processes and Excitation of UV Emission from Planetary Atmospheric Gases*, (Gordon and Breach, New York 1998) p. 97
37. S.V. Avakyan, R.N. Il'in, V.M. Lavrov and G.N. Ogurtsov: *Cross Sections for the Processes of Ionization and Excitation in Collisions of Electrons, Ions and Photons with Atoms and Molecules of Atmospheric Gases*, State Opt. Inst. Publ. 2000, p. 136 (in Russian)

26 Fragment–Imaging Studies of Dissociative Recombination

A. Wolf, D. Schwalm, and D. Zajfman

26.1 Dissociative Recombination

In binary collisions with an electron, molecular ions (of positive charge) can recombine much more efficiently than atomic ions. This is caused by the presence of the molecular degrees of freedom. The excess energy that must be released in the neutralization of the positive and negative charges can, in particular, be transformed into kinetic energy of neutral fragments, opening up the possibility of *dissociative recombination* [1,2]. Thus, molecular ions can be neutralized through a mere rearrangement of the electrons and nuclei participating in the reaction, while the binding of an electron to an atomic ion in free space requires the spontaneous emission of a photon (radiative recombination), which occurs much less frequently.

By its large cross section, dissociative recombination (DR) is an important process in low-density ionized media, controlling, for example, the abundance of molecular species in interstellar gas and dense interstellar clouds [3,4] and also in planetary ionospheres [5]. Most DR reactions are highly exothermic; kinetic or internal excitation energy given to the fragments can have significant consequences, such as the escape of atoms from planetary atmospheres [6] or prominent radiative-emission features [7].

As an elementary example, consider the molecular ion H_2^+ colliding with electrons of initial kinetic energy E,

$$H_2^+(v, J) + e(E_e) \rightarrow H_2^{**} \rightarrow H(n) + H(n') + E_k, \qquad (26.1)$$

yielding the atomic fragments $H(n)$ and $H(n')$ in final states n and n' together with a final kinetic energy release E_k in the center-of-mass (cm) frame. As indicated, the reaction process is governed by a highly excited neutral molecular complex H_2^{**} formed by the collision partners, whose nature strongly depends on the initial electron energy and also on the initial rovibrational excitation of the molecule (quantum numbers v and J). At least for some final states n, n', the reaction (26.1) is exothermic and hence can occur even for asymptotically zero incident electron energies E_e. Because of the attractive Coulomb potential between an electron and a positive ion, the reaction cross section at low electron energy can become very large (asymptotic scaling $\propto E_e^{-1}$) and reach 10^{-16} cm^2 at about $E_e \sim 0.1$ eV for H_2^+ in the rovibrational ground state.

In order to explore the detailed mechanism driving DR reactions, it is of interest to bring together electrons and molecular ions of a given species at well-defined kinematics in the single-collision regime. The quantities of interest to be extracted in such experiments range from the energy dependence of the cross section to the final states of the fragments and the detailed fragmentation dynamics. We will here discuss in particular those properties of DR reactions that have been investigated by multiparticle fragment imaging in fast merged-beams experiments.

26.2   Experimental Method

The strong dependence of the DR cross section on the vibrational excitation of the molecular ions calls for the application of internally relaxed fast ion beams and hence for the use of either sophisticated special sources for cold molecular ions or trapping and storage techniques, giving the ions time to relax internally. Furthermore, experiments with velocity-matched electron beams are largely facilitated by the use of ion beams with laboratory energies in the MeV range (corresponding to electron-beam energies above a few hundred eV). Ion storage rings [8] with magnetic deflection and focusing elements open up the possibility of recirculating such fast molecular ion beams over times of many seconds (up to a few minutes), which is long enough to enable vibrational relaxation by infrared emission for heteronuclear diatomic molecules or for the infrared-active modes of polyatomics (typical times for complete vibrational relaxation are often below $0.5\,\mathrm{s}$). As an example, the test storage ring (TSR) facility, coupled to MeV ion accelerators, at the Max-Planck-Institute for Nuclear Physics in Heidelberg, Germany, is shown in Fig. 26.1. Other ion storage rings that have been used for DR experiments are the ASTRID ring at the University of Aarhus, CRYRING at the University of Stockholm, and TARN at the Insitute of Nuclear Study, Tokyo. The high velocity of the copropagating beams, in fact, helps to obtain a very low spread in collision velocities, mainly determined by the electron temperature T_e in a reference frame moving with the electron beam. These temperatures reach down to $kT_e \sim 1\,\mathrm{meV}$, corresponding to $T_e \sim 10\,\mathrm{K}$, while the electron densities are typically 10^6–$10^7\,\mathrm{cm}^{-3}$. Collision energies can be varied over a wide range, down to a few meV and up to $> 100\,\mathrm{eV}$, by varying the electron acceleration voltage, while keeping the ion-beam velocity fixed.

The recombination products (neutral atoms, and possibly molecules for the DR of polyatomic molecular ions) are observed in the forward direction behind the next bending magnet of the storage ring, as shown in Fig. 26.1. The procedure and main results of molecular-recombination measurements at ion storage rings have been reviewed earlier [9]. A silicon surface-barrier detector serves to identify and count the DR events and is used to obtain absolute recombination cross sections and their dependence on the electron–ion collision energy E_e. With the surface-barrier detector removed, fragmentation

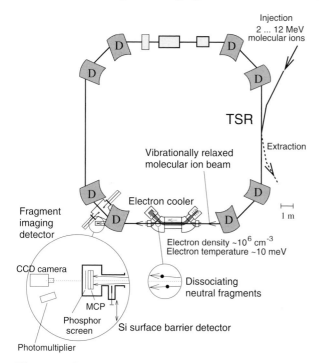

Fig. 26.1. Scheme of the fragment detection for dissociative recombination between electrons and molecular ions at ion storage rings, shown for the setup at the TSR facility in Heidelberg. The circular beam tube, the electron beam device (electron cooler) and the detectors are operated at $\sim 3 \times 10^{-11}$ mbar to ensure sufficient storage time and low collision rates with the background gas

patterns of the DR reactions can be analyzed on multichannel-plate (MCP) imaging detectors using the methods described in more detail in Chap. 3.

The internal excitation of the ions used in these experiments can be verified to some extent by analyzing the DR process itself, considering either the DR rates as a function of the electron energy [9] or the fragment imaging patterns discussed below. At the TSR facility, in addition, the method of Coulomb-explosion imaging (see Chap. 22) is available for monitoring the vibrational populations of the stored molecular ions [10]. All these diagnostic methods indicate a reliable radiative cooling to the vibrational ground state for heteronuclear diatomics and, in general, also for the polyatomic species studied so far. Information of the rotational excitation of the stored ions is more difficult to obtain; however, laser photodissociation could be used to demonstrate radiative relaxation also for the rotations of specific diatomic ions (CH^+ [11]), reaching the ambient (300 K) temperature of the storage ring. Mostly, only small contributions to the kinetic energy release due to rotational excitation were found in fragment-imaging spectra; only recent ex-

periments [12–14] on the DR of H_3^+ have revealed a considerable rotational
excitation of these ions, even after long storage times, which probably reflects
a strong excitation of high-lying, radiatively stable rotational levels in the ion
source.

26.3   Diatomic Molecules

Fragment imaging for DR reactions was established first on light diatomic
molecules. For such systems, the kinematic observables in the cm frame are
restricted to the size of the fragmentation momentum (or the kinetic energy release E_k, see (26.1)) and its direction. The fragmentation direction
in the cm frame can be assumed to be close to the internuclear axis of the
incident molecular ion, as the electronic interaction and the subsequent dissociation usually occur fast in comparison with the rotation of the molecule.
The angular distribution of the fragmentation momentum vectors then represents essentially the dependence of the DR cross section on the relative
orientation between the molecular axis and the incident electron direction. A
well-defined direction of the ensemble of incident electrons is realized if the
average (longitudinal) relative velocity between the merged beams is larger
than the irregular relative motion, dominated by the (transverse) electron
velocity spread.

The kinetic energy release carries valuable information about the DR reaction. In particular, for a well-defined initial state of the molecular ion and
a well-known electron-impact energy it reveals the internal excitation energy
carried away by the fragments; the higher their internal excitation, the lower
the relative momentum in the fragmentation. From the measured E_k and
from the knowledge of the energy levels in the fragment species it is in most
cases possible to characterize their final quantum states and to determine
the branching ratios for different product excitations. Together with the relaxation in the storage ring, which prepares a well-defined initial state (at
least with respect to vibrational excitation, i.e., $v = 0$), fragment-imaging
experiments at such facilities therefore have the capability of following DR
reactions in a state-to-state manner. As shown below, final-state branching
ratios are being studied to obtain a basic understanding of DR reaction mechanisms in simple molecules, but also to determine the product excitation for
DR processes important in astrophysics or atmospheric physics.

26.3.1   Branching Ratios for the Hydrogen Molecular Ion

Potential curves relevant for the DR of the hydrogen molecular ion H_2^+, as
given by (26.1), and its isotopomers HD^+, D_2^+, etc., are shown in Fig. 26.2.
For collisions of $v = 0$ $H_2^+(1s\sigma_g)$ ions with electrons of a given energy E_e
the initial state (indicated by the dashed line in Fig. 26.2) lies in the electronic continuum of the H_2 system, far above its ground state. Considering

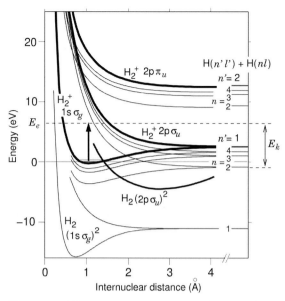

Fig. 26.2. Electronic energy levels of the hydrogen molecular ion and the neutral hydrogen molecule as functions of the internuclear distance. Rydberg states $H_2(1s\sigma_g\, nl\lambda)$ below the H_2^+ ground state and the corresponding states below the $2p\sigma_u$ and $2p\pi_u$ states of H_2^+ are shown schematically

the Franck–Condon (FC) overlap, which depends on E_e, doubly excited neutral states below the first excited ionic configurations $2p\sigma_u$ and $2p\pi_u$ can be formed more or less efficiently by electronic coupling [15]. At low E_e, a special role is played by the $(2p\sigma_u)^2$ doubly excited curve that has the largest FC overlap. After formation of the doubly excited neutral state, the dissociation along the new electronic potential curve enters into competition with the autoionization of the system back to the electronic continuum. In a simplified picture it is tempting to assume that the system will become electronically stabilized when it reaches the internuclear distance where the electronic energy of the doubly excited state lies below that of the ionic ground state; however, detailed experimental results on the energy dependence of the cross section show clear deviations from this simple two-step picture. During the dissociation phase, the $(2p\sigma_u)^2$ state is strongly coupled electronically to the singly excited Rydberg states $1s\sigma_g\, nl\lambda$, which in the diabatic picture leads to a series of avoided level crossings. Hence, the fragmentation process can populate several of the molecular Rydberg states by adiabatic transitions at these multiple level crossings, and at sufficient energy E_e all final channels $H(1s)+H(nl)$ can be reached. Only a weak coupling is predicted between the $(2p\sigma_u)^2$ state and the two lowest neutral potential curves that correlate to two atomic fragments in the $1s$ ground state. The potential curves in Fig. 26.2 also show that at higher energies many additional dissociative

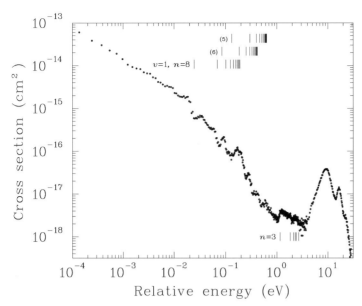

Fig. 26.3. Electron-energy dependence of the dissociative recombination cross section for vibrationally relaxed HD$^+$ ions ($v = 0$; rotational temperature ~ 500 K) measured [19] at the TSR facility. The *lower vertical lines* show the final-state thresholds for $n' = 1$ and $n \geq 3$; the *upper vertical lines* indicate HD Rydberg resonances with a vibrationally excited core HD(v; $1s\sigma_g$ $nl\lambda$) for $v = 1 \ldots 3$, which lead to the prominent window resonances in the cross section (the lowest n of the Rydberg series for each v are indicated)

potential curves [16] come into play. Starting from ~ 7 eV, higher doubly excited Rydberg states $2p\sigma_u$ $nl\lambda$ (commonly denoted as the Q_1 states) enter into the FC overlap with the initial continuum state, opening up new paths towards H($1s$)+H(nl). Even above, starting from ~ 13 eV, also connections towards the H($n' = 2$)+H(nl) final channels open via the $2p\pi_u$ $nl\lambda$ doubly excited Rydberg states (Q_2 states).

To obtain relaxation of the hydrogen molecular ions to $v = 0$ by infrared emission, most of the storage-ring experiments on this DR reaction were performed on HD$^+$. The energetics in HD/HD$^+$ are essentially the same as in H$_2$/H$_2^+$ except for shifts in the rovibrational energies. Recent additional studies on H$_2^+$ and D$_2^+$ ions in specific vibrational states [17,18] have revealed a number of quantitative differences between the three light molecular hydrogen isotopomers with respect to the DR process; nevertheless, the basic dynamics can be assumed to be similar for all three systems. Note that fragment channels such as H($1s$)+D(nl) and D($1s$)+H(nl) (for the same nl) cannot be separated in the present experiments. The observed energy dependence of the DR cross section for HD$^+$ (Fig. 26.3) is dominated at low energies E_e by the electron capture process forming the $(2p\sigma_u)^2$ doubly excited neutral state.

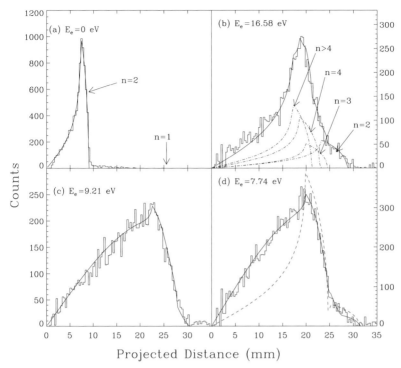

Fig. 26.4. Distributions of the projected fragment distance obtained [20] for the DR of vibrationally relaxed HD^+ ions at the indicated electron impact energies E_e. The electron temperature in this experiment was $kT_e \sim 100\,$meV. Fitted distributions, obtained by linear combination of the theoretical shapes for known values of the energy release E_k for the relevant final channels, are shown together with some contributions for single channels; in panel (**d**) the *dashed line* indicates the expected shape for an isotropic angular distribution (see text)

Resonances caused by vibrationally excited Rydberg states $1s\sigma_g\,nl\lambda$ lead to some structure [2,15] in the cross section below $\sim 1\,$eV, which will not be discussed in more detail here. For increasing E_e the DR via the HD $(2p\sigma_u)^2$ doubly excited state fades out although, considering the final-state energies, an increasing number of excited product channels can be reached. On the other hand, an increasing number of autoionization channels yielding vibrationally excited HD^+ ions also open up, which tends to reduce the cross section. Fragment-imaging results from this interesting energy region (~ 1–$2.6\,$eV) will be discussed below. Above $\sim 5\,$eV, the cross section clearly shows the opening of the new pathways via the Q_1 ($2p\sigma_u\,nl\lambda$) and Q_2 ($2p\pi_u\,nl\lambda$) doubly excited Rydberg manifolds of HD. At these high-energy peaks of the cross section already an infinite number of atomic Rydberg states are energetically accessible as product channels.

Branching ratios for the DR of vibrationally cold HD^+ ions were determined at various energies E_e using fragment imaging. In Fig. 26.4, measured distributions of the transverse projected distance between the H and D fragments are shown at $E_e \sim 0$ and in the region of the high-energy peaks of the cross section [20]. The measurement labeled $E_e \sim 0$ corresponds to matched beam velocities with effective impact energies of $kT_e \sim 100\,\mathrm{meV}$; only the final channels with $n' = 1$ and $n = 1$ or 2 are energetically open. As expected for a dominance of the $(2p\sigma_u)^2$ curve in the capture process (see Fig. 26.2) the branching ratio into the $n = 2$ final channel is practically 1, i.e., one of the products is always produced in an excited state (see Fig. 26.4a). The observed distribution is very well fitted by the calculated function (see Chap. 3, Fig. 3.4) for the expected fixed-energy release of 0.73 eV; the shape also agrees with high accuracy with that expected for an isotropic distribution of fragmentation angles. The shapes of the fragment-distance distributions change considerably in going to high energies, where many channels are energetically open and the electron capture is dominated by the Q_1 and Q_2 manifolds of doubly excited neutral states. At $E_e = 16.6\,\mathrm{eV}$, where final states with $n' = 2$ and $n \geq 2$ can be reached via the Q_2 states, the branching ratios are quite evenly distributed; high Rydberg states with $n > 4$ take $> 70\%$, but also the $H(n' = 2)+D(n = 2)$ [or $H(n' = 2)+D(n = 2)$] channel is quite frequent (see Fig. 26.4b). On the other hand, for the Q_1 states dominating at $E_e = 9.2$ and $7.7\,\mathrm{eV}$ (see Fig. 26.4c and d), the overwhelming branching fraction ($> 70\%$) is found for $n = 3$, while the $n = 2$ and the high-Rydberg contributions are quite small. In the energy region where the Q_1 manifold dominates, the $n = 3$ contribution, moreover, shows a pronounced anisotropy of $\cos^2 \theta$ character (see Fig. 3.4). Symmetry considerations suggest [20] that electron capture in this case leads into a state of Σ_u^+ symmetry. As many symmetries occur within the Q_1 manifold, the large influence of a single symmetry on the fragment distribution may indicate that the reaction is dominated by a single state.

At E_e between ~ 1 and 2.6 eV, the final Rydberg channels $H(1s)+D(nl)$ (or $H(1s)+D(nl)$) with $n \geq 3$ open one by one for the fragmentation following electron capture into the HD $(2p\sigma_u)^2$ state. This should reflect the transition probabilities by diabatic transitions at the multiple-level crossings between the $(2p\sigma_u)^2$ state and the $1s\sigma_g\, nl\lambda$ Rydberg states, as mentioned above. Observation of fragment-distance distributions at these energies is complicated by small countrates, due to the small value of the total cross section (see Fig. 26.3). At the resulting low counting statistics, contributions with different energy release E_k (and hence different final states) can only be extracted if they are clearly separated in the data. Determination of branching ratios in this case has become possible [21] by recording the total (3D) distance between the fragments. This 3D imaging then, for a given E_k, yields an isolated peak broadened only by the finite length of the interaction region and the timing resolution (see Chap. 3). A series of distributions was recorded for energies between 1.2 and 2.4 eV and the corresponding branching ratios

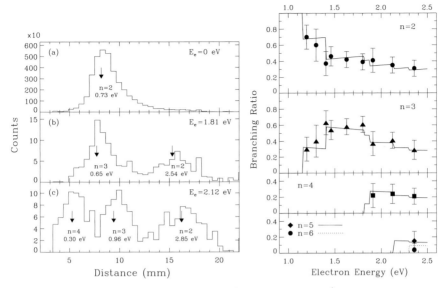

Fig. 26.5. 3D fragment-distance spectra [21] for the DR of HD$^+$ at various electron-impact energies E_e (*left*) and the derived final-channel branching ratios (*right*) in the energy region where the $^1\Sigma_g$ electron-capture channel dominates, as caused by population transfer in the Rydberg curve-crossing region. The branching ratios are compared to model calculations (*full curves*; see text)

extracted (Fig. 26.5). Remarkable agreement can be obtained in a comparison with a multistate curve-crossing model (full curves in the right panel of Fig. 26.5).

The electron-energy region characterized by the successive opening of new high-n dissociation channels is particularly challenging theoretically, as high vibrational and high electronic excitations must be considered at the same time. Advanced theoretical calculations by the multichannel quantum defect (MQDT) method [22] that very well reproduce the measured HD$^+$ DR cross section at most other energies presently cannot describe the experimental results in the region of the high-n dissociation thresholds. Thus, in the experimental cross section (Fig. 26.3) the opening of the Rydberg final channels (especially $n = 3$) is seen to be connected with threshold structures [19] that do not appear in the MQDT calculation. They are also in contradiction with the simple two-step picture described above, where the dissociating system first stabilizes electronically and then distributes over the energetically open neutral channels. However, in a still simplified but more appropriate picture, both the relative changes of the DR cross section at the $n \geq 3$ thresholds and the branching ratios can be reproduced by a multipass, multistate curve-crossing model [19], which also considers oscillations of the internuclear distance R in the highly excited molecule formed by the electron capture. Here, the molecule is first assumed to propagate out to large distances R, distributing over several electronic channels. The "probability flux" in channels closed to dissociation will then oscillate back and forth, repeatedly "probing" the re-emission of the captured electron at small R, which reduces the total DR cross section below the corresponding threshold. Hence, both the cross-section and fragment-imaging data in this electron energy region still seem to reflect relatively simple time-dependent dynamics of the molecular system.

26.3.2 Noncrossing Mode Recombination (HeH$^+$)

For a number of small molecular ions, the potential-curve patterns analogous to Fig. 26.2 do not feature the "crossing" of a doubly excited potential curve through the ionic ground state near its minimum – the condition to obtain FC overlap for the electron capture at $E_e \sim 0$. Recombination can then proceed only by nonadiabatic coupling between the continuum state and vibrationally unbound, neutral molecular Rydberg states just below the ionic ground state. DR processes driven by such mechanisms are termed "noncrossing" DR [2,23]. The final-state branching ratios for such DR reactions can indicate, in the case of several energetically open channels correlated to various neutral molecular states, to which of these states the most efficient nonadiabatic coupling occurs. A rather selective behavior with respect to the fragmentation channels has been observed for HeH$^+$, which represents one of the prototype "noncrossing"-DR species. At $E_e \sim 0$, only the final channels He(1s^2)+H(1s) and He(1s)+H($n = 2$) are energetically open. The experimental result [24] at this energy (upper panel of Fig. 26.6) indicates the almost exclusive production of H($n = 2$). At $E_e = 0.33\,\mathrm{eV}$, the H($n = 3$) final channel opens, and already slightly above this energy (but still below the next-higher, H($n = 4$) threshold) nearly all DR reactions lead again to the highest available final state. The investigations of this behavior for "noncrossing" DR have been pushed to much higher precision in recent experiments [18]. The onset of the $n = 3$ contribution at the corresponding threshold is consistent with a step-like behavior and the branching ratio into this channel above the limiting energy was found to be $\sim 70\%$. The energy dependence of the HeH$^+$ DR cross section in the threshold region, together with the branching-ratio measurement, indicates that DR events leading into $n = 3$ constitute an additional contribution to the total cross section with a step-like onset, while the trend of $n = 2$ contribution as a function of E_e continues across the $n = 3$ threshold.

26.3.3 Metastable States in CH$^+$

As the smallest carbon-containing molecular ion, the species CH$^+$ plays an important role in the chemistry of interstellar clouds and its creation and destruction processes (the latter including DR) are of great interest for the modeling of these environments. The molecule was one of the first species to be studied by DR fragment imaging [25] with the main goal of identifying the

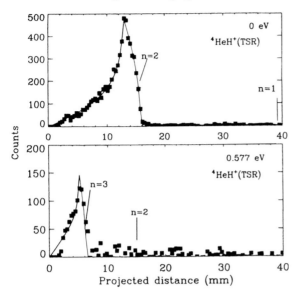

Fig. 26.6. Fragment-distance spectra [24] for DR via nonadiabatic coupling to neutral molecular Rydberg states, as found for HeH$^+$ at near-zero electron-impact energy (*top*) and slightly above the threshold to the He(1s^2)+H($n=3$) final channel (*bottom*)

excitation states of the fragments. At near-zero (i.e., thermal, $kT_e \sim 20$ meV) electron energy and for CH$^+$ ions in the electronic and vibrational ground state ($X^1\Sigma^+$, $v=0$), the final channels C(2p$^2\,^1D$)+H(1s) and C(^1S)+H(1s) could be identified by 2D imaging, deriving a branching ratio of 0.79±0.1 for C(2p$^2\,^1D$) and 0.21±0.1 for C(2p$^2\,^1S$), while the channel with ground-state carbon atoms, C(2p$^2\,^3P$), was absent in the data. Remarkably, the DR fragment spectra of the stored ion beam (Fig. 26.7, left panel) showed a time-dependent contribution with much smaller energy release. It has been attributed to a large fraction of CH$^+$ ions in the electronically excited state $a^3\Pi$, arriving from the accelerator with an initial abundance of $\sim 50\%$. The CH$^+$($a^3\Pi$, $v=0$) energy level lies ~ 1.2 above the $X^1\Sigma^+$, $v=0$ ground state. By analyzing the fragment spectra, it was concluded that the DR of $a^3\Pi$ leads predominantly (namely with $85^{+15}_{-25}\%$) to the channel C(2s2p^3 13D°)+H(1s), which is not energetically open for the ground-state CH$^+$ molecules. It was possible to extract two important spectroscopic properties of CH$^+$ from these data. First, from the maximum of the observed 2D fragment distance distribution the excitation energy from $X^1\Sigma^+$, $v=0$ to $a^3\Pi$, $v=0$ was derived to be (1.21±0.05) eV, in good agreement with (but much more precise than) available theoretical predictions [25]. Secondly, from the time variation of the relative peak areas in the fragment spectra, the excited $a^3\Pi$, $v=0$ ions were found to have a mean lifetime of (7.0±1) s metastable ions, which is in

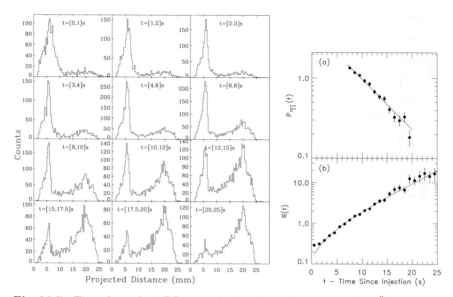

Fig. 26.7. Time-dependent DR contribution from the metastable $a^3\Pi$ state of CH$^+$ in 2D fragment-imaging spectra [25]. *Left:* Fragment spectra observed in various time intervals t after injection into the storage ring; *right:* decay curve of the metastable population $P_{3\Pi}(t)$ (given in arbitrary units) as derived from the peak-area ratio $R(t)$ in the fragment spectra, as shown in the *lower graph*. The fit to $P_{3\Pi}(t)$ and $R(t)$ as indicated yields an $a^3\Pi$ decay lifetime of $(7.0\pm1)\,\mathrm{s}$.

the range of rough estimates on the radiative decay induced by spin-orbit coupling.

26.3.4   O$_2^+$ and Similar, Atomspherically Relevant Species

Ionization of the upper earth atmosphere by ultraviolet solar radiation creates the O$_2^+$, NO$^+$ and also N$_2^+$ molecular ions that are efficiently removed by DR. It was suggested that the DR of O$_2^+$ ions might be responsible for the prominent green emission (corresponding to the transition O$(2p^4\,^1S)$ $\rightarrow$O$(2p^4\,^1D)$) seen in phenomena such as airglow and aurorae in the night sky. However, rather low quantum yields (< 0.002) were predicted theoretically for O$_2^+$ ions in low v-states, so that speculations about other sources for the O(^1S) excitation or a strong vibrational excitation of atmospheric O$_2^+$ ions were put forward. A DR fragment-imaging measurement on O$_2^+$ [26] could provide a laboratory confirmation of DR as a source of the green airglow emission. As the common, symmetric ^{16}O$_2^+$ ions do not cool by infrared emission, this experiment was performed with the isotopomer ^{18}O^{16}O$^+$, for which radiative cooling over 10 s before starting the measurement allowed partial vibrational relaxation to take place. The final channel O$(2p^4\,^1S)$+O$(2p^4\,^1D)$ did show up in the 2D fragment imaging spectra and a quantum yield of 0.05 ± 0.02, much

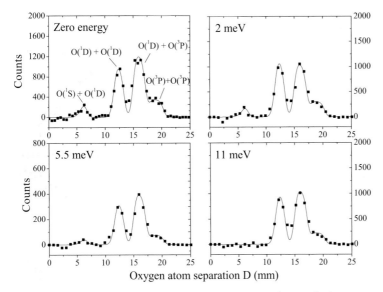

Fig. 26.8. Fragment-imaging spectra for DR of O_2^+ ions [27] for various adjusted collision energies E_e as indicated, showing contributions ascribed to O-atom final states as labeled. Hit-time differences were restricted to $< 800\,\mathrm{ps}$ to select fragmentation directions nearly parallel to the detector surface. *Lines* show fitted curves assuming $v = 0$ ions, from which branching ratios were deduced

larger than the theoretical predictions, was derived. A more recent experiment [27], for which typical fragment distance spectra are shown in Fig. 26.8, could detect a sharp dependence of the $O(^1S)$ yield on the electron-impact energy, as controlled by the longitudinal relative velocity between the merged beams (at a typical electron temperature close to 1 meV in this experiment). While the $O(^1S)$ and also the $O(^1D)$ and $O(^3P)$ yields at near-zero energy were consistent within the given errors with those of the earlier experiment [26], the ^1S contribution is, in fact, undetectable at 11 meV; it is found to traverse a broad minimum between ~ 5 and 30 meV. This dependence suggests [27] that emission-line ratios of airglow observations might be used as a diagnostic of electron temperatures in the upper atmosphere. It should be pointed out that the this experiment was performed on symmetric $^{16}O_2^+$ ions, where vibrational cooling by infrared emission is not only slow (as in [26]) but completely absent. The authors here used a special ion source that for this species allowed the fractions of ions in excited vibrational states to be minimized, as rather well confirmed by the fit shown in Fig. 26.8.

Further DR fragment-imaging experiments on species relevant for terrestrial and for planetary atmospheres were performed on NO^+ [28], N_2^+ [29], CO^+ [30] and OH^+ [31].

26.4   Small Polyatomic Molecules

The DR of small polyatomic molecules, in comparison to diatomic systems, offers a number of interesting additional aspects. First, the highly excited complex formed by binding an incident electron has an increasing number of alternative paths for its fragmentation. Full disintegration into atomic fragments may occur, or the formation of smaller molecular fragments, which themselves of course may be rovibrationally excited. Depending on the capture mechanism, the molecule can find itself either on a well-defined, electronically doubly excited potential surface (similar to the crossing-mode recombination discussed above), or nonadiabatic coupling mechanisms may already create vibrationally unbound, electronically singly excited molecular states in the capture step. Due to non-adiabatic coupling, the molecular system may evolve on a superposition of adiabatic electronic potential surfaces, as often found also in photoinduced, femtosecond-timescale dynamics of polyatomic molecules [32]. Small polyatomic molecular ions play important roles in the chemical networks of low-density ionized media and their DR rate coefficients, but also the chemical nature of the fragments and the kinetic energies given to them, are important properties for the understanding of these media.

Event-by-event DR studies on molecular species in well-defined initial states can be performed by means of ion storage rings with similar advantages for polyatomic as for diatomic systems. Focusing on product branching ratios, using nonimaging methods [33], a number of systems have already been studied, including triatomic hydrogen H_3^+, as the simplest polyatomic molecule, but also oxygen-, carbon- and nitrogen-containing species such as H_2O^+, H_3O^+ [34], $HCNH^+$ [35], and hydrocarbons up to CH_5^+ [36]. In these product-branching-ratio measurements, a remarkable propensity for the fragmentation into three separate atoms, not theoretically explained until now, was found, in particular for triatomic dihydrides (e.g., CH_2^+, NH_2^+, H_2O^+) and also H_3^+. The issue of branching ratios in polyatomic DR has been discussed at some length elsewhere [37,38]. Here, we will focus on the first genuine fragment-imaging experiments on the DR of polyatomic molecular ions that have been performed only recently and studied the species H_3^+ (and isotopomers) and H_2O^+.

26.4.1   General Experimental Aspects

At present, the fragment-imaging detectors used for DR studies only indicate the presence of a fragment at a certain transverse position and (possibly) arrival time; the possibilities to obtain independent information about intrinsic properties of the individual fragments (such as their impact energy on the detector, which would reveal their mass) are very limited. In specific cases (see Sect. 26.4.3) the energy loss in thin foils was used to make certain areas of an imaging detector inaccessible for lighter fragments, thus constraining the possible fragment masses for events observed in these regions. However, none of the detector schemes applied so far allowed an additional energy (or mass) signal to be obtained for each fragment impact in a multihit-imaging mode.

Facing these restrictions, mostly kinematical arguments (momentum conservation in connection with knowledge of the cm motion of the fragments) must be used to identify the fragment masses in the observed hit patterns. The use of such arguments is, moreover, complicated by the fact that the imaging detectors have large, but finite detection efficiencies, typically $\sim 60\%$ for individual fragments. Hence, a sample of two-hit images measured in the fragmentation of triatomic molecules will not only contain events with two-body final states (atom plus molecule), but also a considerable number of incompletely detected three-body decay events. In this context, the high quality of a phase-space cooled ion beam in a storage ring, with beam diameters down to $\sim 1\,mm$, facilitates the kinematical analysis by fixing the cm position of fragmentation events in the transverse detector plane to a narrow range.

Other restrictions for fragment-imaging studies on polyatomic ions come from the high requirements on the time resolution in the energy range relevant for merged-beam DR experiments, which makes the implementation of full 3D imaging very difficult (see Chap. 3). Although the development of suitable 3D imaging systems is being pursued, the few polyatomic DR imaging studies performed so far are essentially limited to the analysis of 2D patterns representing projections of the 3D fragmentation geometry onto the plane transverse to the motion of the center-of-mass.

26.4.2 Triatomic Hydrogen (H_3^+)

The triatomic hydrogen ion H_3^+ is abundant in cold hydrogen plasmas and highly reactive in collisions with neutral species to which it readily donates a proton, thus contributing to the formation of more complex molecules within networks of ion chemistry [39]. Knowledge of the DR rate of H_3^+ in collisions with cold ($\sim 10\,K$ or $\sim 1\,meV$) electrons is of great interest for predicting its abundance in cold low-density environments such as the interstellar medium. In contrast to H_3^+, the neutral system H_3 has no stable bound states, as the lowest potential surface of this system, comprising two sheets as indicated in Fig. 26.9, is dissociating. Higher-lying, dynamically bound Rydberg states of H_3, forming below the ground-state surface of H_3^+, all decay by predissociation, autoionization, or radiative transitions to the dissociating lower surface. Rydberg states of H_3 and their fragmentation via the ground-state surfaces of the neutral system have been studied in considerable detail by analyzing UV spectra emitted by a neutralized H_3^+ ion beam [40] and recently by detailed fragment-imaging experiments following laser excitation of individual H_3 Rydberg states [41]. It is interesting to compare the results of these studies (see Chap. 17) with those from DR fragment-imaging experiments.

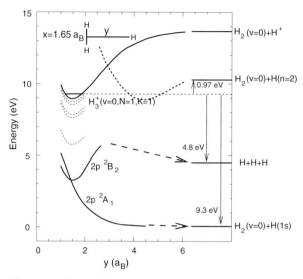

Fig. 26.9. Cut of the ground-state potential surfaces of H_3^+ and H_3 in C_{2v} (isosceles triangle) geometry. Rydberg states of H_3 and the doubly excited $^2A_1$ state of H_3 relevant for DR at higher collision energies are shown schematically (*dashed lines*). The energies of the lowest DR channels with respect to the H_3^+ ground state are indicated

The DR of H_3^+ ions with low-energy electrons currently attracts strong experimental and theoretical attention. From the disposition of the potential surfaces near the H_3^+ ground state, the Franck–Condon overlap for a DR process via purely electronic capture is very small for incident electrons of $< 1\,\text{eV}$ energy with ions in the vibrational ground state. The DR of H_3^+ in a cold environment should, therefore, be dominated by nonadiabatic coupling. The complexity of the theoretical description of such processes on the one hand, and the difficulties of ensuring a low rovibrational excitation of the H_3^+ ions in DR rate measurements on the other side, still cause much uncertainty in the study of this process. In fact, large discrepancies exist not only between different theoretical predictions, but also between different experimental results for the DR rate coefficient [42]. Fragment-imaging studies of this reaction were performed recently to provide a more detailed view of the process; they yield new information on the dissociation dynamics and also offer a possibility to verify the internal excitation of the H_3^+ ions used in a particular experiment.

Assuming small internal excitation of the H_3^+ ion, the DR reaction at low electron energy E can lead into a three-body channel α and a two-body channel β,

$$H_3^+ + e(E_e) \rightarrow \begin{cases} H(1s) + H(1s) + H(1s) & (\alpha) \\ H_2(X^1\Sigma_g^+, vJ) + H(1s) & (\beta). \end{cases} \qquad (26.2)$$

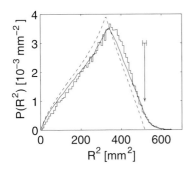

Fig. 26.10. Distribution of the sum R^2 of squared fragment distances for the three-body events observed after the DR of H_3^+ with low-energy electrons [13]. The data (shown as *histogram*) are compared to a simulation for internally cold H_3^+ ions (*dashed line*) and for H_3^+ ions with a mean internal energy of 0.3 eV (*full line*). The *arrow marks* the predicted endpoint for cold H_3^+ ions

The next electronically excited final channel, $H_2(X^1\Sigma_g^+, vJ)+H(n = 2)$, opens when the initial energy (electron energy E_e plus internal excitation of H_3^+) exceeds ~ 1 eV; more channels open at even higher initial energy. Using the experimental procedure described in Sect. 26.2, measurements of the DR rate coefficient and the branching ratios into the channels α and β were performed at ion storage rings [42]. In particular, the branching ratios for the two- and three-body channels β and α were measured at CRYRING to be 25% and 75%, respectively [33].

In recent 2D-imaging experiments [12,13] at the TSR facility, two-body fragmentation events following the DR of H_3^+ could be identified (and distinguished from incompletely detected three-body fragmentations, see Sect. 26.4.1) by selecting events for which the cm position lay within a narrow (~ 1 mm diameter) region of the detector plane, defined by the initial motion of the phase-space cooled ions. Following relaxation to the vibrational ground state and using matched electron and ion-beam velocities (i.e., $E_e < 5$ meV), large ensembles of two- and three-body fragmentation events could be collected.

The analysis of the three-body events splits into two aspects. The first one is the total (projected) kinetic energy of the three fragments, as reflected by the sum R^2 of the squared fragment distances R_i^2 from the cm in the detector plane. The transverse distance R_i of a fragment is determined by the size of its momentum and, in addition, by its direction in the cm frame and by its flight distance from the fragmentation point to the detector (see (3.16) in Chap. 3). Hence, similar to the transverse fragment-distance spectra of diatomic molecules discussed above, the R^2 distributions for triatomic decay assume a characteristic shape (see Fig. 26.10) following from the spread of fragmentation directions and flight distances; however, the endpoints of such spectra always correspond to the largest available fragmentation energy in connection with the largest flight distance to the detector and a transversely oriented fragmentation plane. For the case of low-energy DR of H_3^+, the three-body channel α has well-defined quantum states as given in (26.2) and hence the expected endpoint of the R^2 distribution for $E_e = 0$ and vanishing rovibrational excitation of the colliding H_3^+ ion can be safely predicted.

The observed R^2 distribution (Fig. 26.10) has a tail beyond this predicted endpoint and this result was interpreted [13], in connection with other observations [14], as a sign of substantial rotational excitation of the H_3^+ ions even though they are prepared in an ion storage ring. From the extent of the tail in Fig. 26.10 an average rotational energy of $\sim 0.3\,\mathrm{eV}$ is derived. The analysis of the R^2 distribution hence strongly contributes to the understanding of the experimental conditions of DR experiments on triatomic molecular ions, which is of particular interest considering the discrepencies found between different experimental results on the H_3^+ DR rate coefficient [42]. For other systems, in which more than three-body final channels are accessible, the R^2 distributions also can be used to determine the branching ratios between them; an example will be discussed in Sect. 26.4.3. It should be possible to use such projected distributions in a similar way also for higher fragment multiplicities originating from more complex polyatomic molecules, as long as the decay takes place in a planar geometry.

The second aspect in analyzing the three-body fragmentation events is the fragment momentum geometry. Observed momentum geometry distributions reflect the character of the forces acting between the fragments during the dissociation process and the propagation of the molecular system on the dissociative potential surface. The predominance of specific interparticle forces during the dissociation will lead to *correlations* that manifest themselves in a particularly frequent occurrence of certain fragmentation geometries. Conversely, the forces felt during the dissociation by each of the fragments may also be so unspecific that their net effect becomes independent of the other fragments' momenta, which is equivalent to an *uncorrelated* or random distribution of fragmentation geometries. For three fragments the momentum geometry can always be specified as a triangular shape and hence by two coordinates (understanding "shape" irrespective of the size of the momentum triangle, which is expressed by the total energy release, and the overall orientation). An ensemble of uncorrelated fragment momentum geometries yields a two-dimensional scatter plot of uniform density if the two coordinates representing the triangular shape are chosen as the Dalitz coordinates η_1, η_2, defined in (3.13). The three-body fragmentation events observed for the low-energy DR of H_3^+, using 2D imaging, were processed as described in Chap. 3 to obtain an experimental scatter plot of the momentum geometries. All possible shapes (apart from overall rotations of the momentum triangle and exchanges of the three identical H fragments) are contained in the region of the Dalitz plot shown in Fig. 26.11. It can be seen that nearly linear dissociation geometries (the region of the Dalitz plot with $\eta_2 \sim -0.3$ and $\eta_1 \sim 0$ occur considerably more frequently than expected for an uncorrelated distribution. A momentum correlation of this type can arise if during the dissociation two of the fragments effectively decouple from the third one, the force between the two becoming markedly repulsive. As detailed calculations on the fragmentation after DR are not yet available, simplified arguments

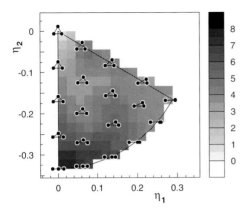

Fig. 26.11. Density plot of the momentum geometry distribution for the DR of H_3^+ with low-energy electrons obtained from the deconvolution of 2D–imaging data [13]

may help to explain such a configuration of forces. Thus, the H_3^+ ion has saturated electron spins so that, after electron capture, the dissociating H_3 complex will find itself in a three-electron doublet state. Considering now H_2 pairs within the dissociating complex, a triplet electron spin combination (that should be associated with a repulsive force, while a singlet would rather be attractive) would only be possible between one of the three pairs. A dominance of equilibrated repulsive forces between all three H_2 subsystems would rather be characteristic of an H_3 complex in a three-electron quartet state, which cannot be formed by the DR on an H_3^+ ion in its electronic ground state.

The Dalitz histograms observed for the predissociation of laser-excited H_3 Rydberg states (see Chap. 17) correspond to dissociation processes on the ground-state surface of H_3 that, regarding their initial molecular configurations, should be rather similar to those formed by the capture of a low-energy electron on the H_3^+ ion, as considered here. In the comparison, it must be kept in mind that the analysis of the energy release in the three-body fragmentation discussed above indicates a substantial internal excitation of the stored H_3^+ ions, corresponding to the population of rotational levels up to $J \gtrsim 10$. Moreover, even disregarding the rotational excitation, the initial electronic states formed in the electron capture on H_3^+ may involve several symmetries and much wider resonant states than those addressed by the laser excitation of H_3 Rydberg states. Within these limitations, it may be interesting to observe that a tendency to avoid the equilateral shape, as found in the fragmentation pattern for the H_3^+ DR, is also seen in the predissociation patterns of Fig. 17.3; the predominance of a linear dissociation is particularly clear in the fragmentation data of the $3d^2 E''$ Rydberg state of H_3, for which rotational coupling is invoked [43] as a predissociation mechanism.

For the two-body fragmentation events observed in the H_3^+ DR experiment, the distribution $P(D)$ of transverse fragment distances D is shown in Fig. 26.12a. Recombination events forming a molecular fragment $H_2(X^1 \Sigma_g^+, vJ)$ together with a ground-state H atom form the dominant part of this

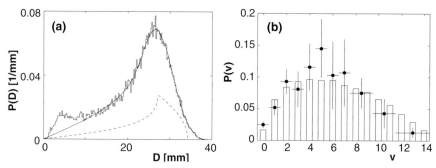

Fig. 26.12. (a) Distribution of projected distances D observed for the two-body fragmentation events following DR of H_3^+ ions with low-energy electrons [13]. The data (shown as *histogram*) are compared to a sum of theoretical contributions for single v states of the H_2 product (*full line*); the contribution for $v = 5$ is shown as an example (dashed line). (**b**) Vibrational-level populations of the H_2 product obtained from a fit to the projected-distance distribution [12] compared to results of a simple model calculation [38]

distribution ($D \gtrsim 10\,\mathrm{mm}$); the shape at $D < 10\,\mathrm{mm}$ is discussed in [13]. The energy release in the various recombination pathways significantly depends on the vibrational excitation v of the H_2 fragment; by fitting a superposition of components to $P(D)$ for different v levels, the vibrational population fractions of the H_2 product as formed by the low-energy DR of H_3^+ can be determined (Fig. 26.12b). The distribution is wide and follows remarkably well the shape estimated for a statistical distribution of fragmentation energies in the dissociation process within a simpified model [38]. Also, these data can, within the limitations discussed above, be compared to corresponding results from the predissociation of laser-excited H_3 Rydberg states (Chap. 17). A rotational resolution with respect to the final H_2 states, as obtained for the predissociation data, cannot be obtained in the DR data at the present state of the measurements. The v distributions for the dissociation products of the investigated H_3 Rydberg states also indicate a wide range of final states, but the v dependence of the relative populations appears to be more strongly modulated than in the DR results, depending on the Rydberg state symmetry. The peak of the v distribution for the $3\mathrm{d}^2E''$ Rydberg state (Fig. 17.6) appears in a similar range as the average v found for the H_3^+ DR.

In addition to H_3^+ the isotopomers D_3^+ and D_2H^+ were also recently studied by DR-fragment imaging [44]. In particular, the pronounced tail found in the R^2 distribution, as shown for H_3^+ in Fig. 26.10, was reproduced for D_3^+ but was absent for D_2H^+, which can probably be explained by the different radiative relaxation behavior of the heteronuclear as opposed to the homonuclear species. The analysis of the observed fragmentation geometry distributions is in progress.

26.4.3   The Water Ion

Another triatomic system for which DR-fragment imaging studies have been performed and analyzed in some detail is H_2O^+. These imaging studies were carried out at near-zero incident electron energy ($E_e \lesssim 0.002\,eV$) and restricted to three-body fragmentation, for which the branching ratio was determined earlier [34] to be close to 0.7. In these processes, the oxygen atom is produced (together with two ground-state H atoms) in the states $O(2p^4\,^3P)$ or $O(2p^4\,^1D)$, and the kinetic energy released to all three fragments then amounts to 3.04 or 1.07 eV, respectively. The imaging experiments [45,46] used essentially a 2D detector (although a time information of limited resolution, insufficient to construct 3D images, was used in some steps of the evaluation). The identification of the oxygen atom was possible by covering the central (5 mm diameter) region of the 77-mm diameter imaging detector by a thin foil that, given the ion-beam velocity, stops the hydrogen atomic fragments but lets the oxygen fragments pass through. By selecting images with a fragment hit in the central region plus two other fragments observed in the remaining area of the detector, three-body fragmentation events including a 'tag' on the O fragment could be collected. The method works efficiently for hydrides such as H_2O^+ because in most events the hydrogen fragments obtain much higher transverse velocities than the heavy atom, whose motion is restricted to a narrow range around the ion-beam direction.

The imaging data for H_2O^+ could be analyzed to reveal some basic features of the fragmentation process [45]. First, histograms equivalent to the R^2 distributions discussed above for H_3^+ clearly show two components with different kinetic energy release, from which the branching ratio $O(^3P):O(^1D)$ was determined to be 3.5(0.5):1. Secondly, the energy sharing between the two H atoms is essentially random, which indicates that the three-body dissociation does not proceed via an unstable intermediate OH^* or H_2^* product. Thirdly, distributions of the opening angles spanned by the two H fragments with respect to the O fragment were extracted; by selecting events from appropriate regions of the R^2 distributions, the angular distributions for the $O(^3P)$ and $O(^1D)$ channels can be separated to a certain extent (within the limitations of the 2D imaging, which may affect the purity of what is identified as the $O(^1D)$ component). For the $O(^3P)$ channel, the H–O–H angles are mostly found to differ strongly from the H_2O^+ bond angle of $\sim 110°$; small-angle and linear geometries appear to be favored [46]. On the other hand, the angular data assigned to the $O(^1D)$ channel [45] show a different trend with a wide angular distribution peaking near 90°, not too distant from the ground-state bond angles in H_2O^+ or H_2O. Regarding the angular distribution, only the $O(^3P)$ channel was considered in a recent more detailed analysis of the H_2O^+ data [46], which also presents a comparison with quasiclassical trajectory calculations on the dissociation process. A rather large number of excited potential surfaces of H_2O is considered in these calculations, and those surfaces that are found to produce the $O(^3P)$ three-body channel are also found to lead the fragments to geometries very different from that of the initial molecule. However, the calculations still seem to be far from reproducing the finer details of the fragment-imaging data.

26.5   Conclusions and Outlook

Imaging of individual fragmentation events following the dissociative recombination of molecular ions with slow electrons has been realized in several experiments using fast, internally relaxed ion beams in ion storage rings. The fragment-momentum distributions obtained from such experiments reveal detailed information about the final product states – essential data considering the importance of the DR process in gas-phase ion chemistry – and about the internal molecular mechanisms by which this reaction takes place. They also reflect the internal excitation of the molecular ions taking part in the DR; therefore, in connection with the long observation times for a given ion ensemble in a storage ring, fragment-imaging experiments also offer a sensitive method to follow the slow decay dynamics for various types of excitation in molecular ions. In particular for light molecules, including the small polyatomic species addressed by recent work, the fragment-imaging experiments provide detailed information about the quantum dynamics underlying the dissociation of highly excited molecular systems, offering sensitive checks of theoretical calculations. A better definition of the initial electron energies, more precise control of the initial molecular excitation, and full 3D imaging, yielding improved resolution in the measurement of individual fragment momenta and a higher sensitivity on angular distributions, are on the horizon for future experiments.

References

1. D.R. Bates: Phys. Rev. **78**, 492 (1950)
2. D.R. Bates: Adv. At. Mol. Opt. Phys. **34**, 427 (1994)
3. A. Dalgarno: in *Dissociative Recombination. Theory, Experiment and Applications IV*, M. Larsson, J.B.A. Mitchell, I.F. Schneider, eds. (World Scientific, Singapore 2000) p. 1
4. T.J. Millar, D.J. DeFrees, A.D. McLean, E. Herbst: Astron. Astrophys. **194**, 250 (1999); E. Herbst, in: *Dissociative Recombination: Theory, Experiment and Applications*, J.B.A. Mitchell, S.L. Guberman, eds. (World Scientific, Singapore 1989) p. 303
5. S.J. Bauer: *Physics of Planetary Ionospheres* (Springer, Berlin, Heidelberg, New York 1973); S.K. Atreya: *Atmospheres and Ionospheres of the Outer Planets and their Satellites* (Springer, Berlin, Heidelberg, New York 1986)
6. J.L. Fox, A. Hac: J. Geophys. Res. **104**, 24729 (1999)
7. S.L. Guberman: Science **278**, 1276 (1997), and references therein
8. A. Wolf: in *Atomic Physics with Heavy Ions*, H.F. Beyer, V.P. Shevelko, eds. (Springer, Berlin, Heidelberg, New York 1999) p. 1

9. M. Larsson: Annu. Rev. Phys. Chem. **48**, 151 (1997)
10. Z. Amitay, A. Baer, M. Dahan, J. Levin, Z. Vager, D. Zajfman, L. Knoll, M. Lange, D. Schwalm, R. Wester, A. Wolf, I.F. Schneider, A. Suzor-Weiner: Phys. Rev. A **60**, 3769 (1999)
11. U. Hechtfischer, Z. Amitay, P. Forck, M. Lange,J. Linkemann, M. Schmitt, U. Schramm, D. Schwalm, R. Wester, D. Zajfman, A. Wolf: Phys. Rev. Lett. **80**, 2809 (1998)
12. D. Strasser, L. Lammich, S. Krohn, M. Lange, H. Kreckel, J. Levin, D. Schwalm, Z. Vager, R. Wester, A. Wolf, D. Zajfman: Phys. Rev. Lett. **86**, 779 (2001)
13. D. Strasser, L. Lammich, H. Kreckel, S. Krohn, M. Lange, A. Naaman, D. Schwalm, A. Wolf, D. Zajfman: Phys. Rev. A **66**, 032719 (2002)
14. H. Kreckel, S. Krohn, L. Lammich, M. Lange, J. Levin, M. Scheffel, D. Schwalm, J. Tennyson, Z. Vager, R. Wester, A. Wolf, D. Zajfman: Phys. Rev. A (2002) (in print)
15. J.N. Bardsley: J. Phys. B **1**, 365 (1968)
16. S.L. Guberman: J. Chem. Phys. **78**, 1404 (1983)
17. S. Krohn, H. Kreckel, L. Lammich, M. Lange, J. Levin, D. Schwalm, D. Strasser, R. Wester, A. Wolf, D. Zajfman: in *Proceedings of the Symposium on Dissociative Recombination of Molecular Ions*, Chicago, August 27–30, 2001 (to be published)
18. S. Krohn, Ph. D. thesis, University of Heidelberg, 2001
19. M. Lange, J. Levin, U. Hechtfischer, L. Knoll, D. Schwalm, R. Wester, A. Wolf, X. Urbain, D. Zajfman: Phys. Rev. Lett. **83**, 4979 (1999)
20. D. Zajfman, Z. Amitay, C. Broude, P. Forck, B. Seidel, M. Grieser, D, Habs, D. Schwalm, A. Wolf: Phys. Rev. Lett. **75**, 814 (1995)
21. D. Zajfman, Z. Amitay, M. Lange, U. Hechtfischer, L. Knoll, D. Schwalm, R. Wester, A. Wolf, X. Urbain: Phys. Rev. Lett. **79**, 1829 (1997)
22. I.F. Schneider, C. Strömholm, L. Carata, X. Urbain, M. Larsson, A. Suzor-Weiner: J. Phys. B **30**, 2687 (1997)
23. B.K. Sarpal, J. Tennyson, L.A. Morgan: J. Phys. B **27**, 5943 (1994); S.L. Guberman, Phys. Rev. A **49**, R4277 (1994)
24. J. Semaniak, S. Rosén, G. Sundström, C. Strömholm, S. Datz, H. Danared, M. af Ugglas, M. Larsson, W.J. van der Zande, Z. Amitay, U. Hechtfischer, M. Grieser, R. Repnow, M. Schmidt, D. Schwalm, R. Wester, A. Wolf, D. Zajfman: Phys. Rev. A **54**, R4617 (1996)
25. Z. Amitay, D. Zajfman, P. Forck, U. Hechtfischer, B. Seidel, M. Grieser, D. Habs, D. Schwalm, A. Wolf: Phys. Rev. A **54**, 4032 (1996)
26. D. Kella, L. Vejby-Christensen, P.J. Johnson, H.B. Pedersen, L.H. Andersen: Science **276**, 1530 (1997); **277**, 167(E) (1997)
27. R. Peverall, S. Rosén, J.R. Peterson, M. Larsson, A. Al-Khalili, L. Vikor, J. Semaniak, R. Bobbenkamp, A. Le Padellec, A.N. Maurellis, W.J. van der Zande: J. Chem. Phys. **114**, 6679 (2001)
28. L. Vejby-Christensen, D. Kella, H.B. Pedersen, L.H. Andersen: Phys. Rev. A **57**, 3627 (1998)
29. J.R. Peterson, A. Le Padellec, H. Danared, G.H. Dunn, M. Larsson, A. Larson, R. Peverall, C. Strömholm, S. Rosén, M. af Ugglas, W.J. van der Zande: J. Chem. Phys. **108**, 1978 (1998)

508   A. Wolf et al.

30. S. Rosén, R. Peverall, M. Larsson, A. Le Padellec, J. Semaniak, A. Larson, C. Strömholm, W.J. van der Zande, H. Danared, G.H. Dunn: Phys. Rev. A **57**, 4462 (1998)
31. C. Strömholm, H. Danared, A. Larson, M. Larsson, C. Marian, S. Rosén, B. Schimmelpfennig, I.F. Schneider, J. Semaniak, A. Suzor-Weiner, U. Wahlgren, W.J. van der Zande: J. Phys. B **30**, 4919 (1997)
32. H. Köppel, M. Döscher, S. Mahapatra: Int. J. Quantum Chem. **80**, 942 (2000); D.R. Yarkony: Rev. Mod. Phys. **68**, 985; W. Domcke, G. Stock: Adv. Chem. Phys. **100**, 1 (1997)
33. S. Datz et al.: Phys. Rev. Lett. **74**, 896 (1995); Phys. Rev. A **52**, 2901 (1995)
34. L. Vejby-Christensen, L.H. Andersen, O. Heber, D. Kella, H.B. Pedersen, H.T. Schmidt, D. Zajfman: Astrophys. J. **483**, 531 (1997)
35. J. Semaniak, B.F. Minaev, A.M. Derkatch, F. Hellberg, A. Neau, S. Rosen, R. Thomas, M. Larsson, H. Danared, A. Paal, M. af Ugglas: Astrophys. J. Suppl. Ser. **135** 275 (2001)
36. J. Semaniak, Å. Larsson, A. Le Padellec, C. Strömholm, M. Larsson, S. Rosén, R. Peverall, H. Danared, N. Djuric, G.H. Dunn, S. Datz: Astrophys. J. **498**, 886 (1998)
37. S. Rosén, A. Derkatch, J. Semaniak, A. Neau, A. Al-Khalili, A. Le Padellec, L. Vikor, R. Thomas, H. Danared, M. af Ugglas, M. Larsson: Faraday Discuss. **115**, 295 (2000)
38. D. Strasser, J. Levin, H.B. Pedersen, O. Heber, A. Wolf, D. Schwalm, A. Wolf, D. Zajfman: Phys. Rev. A **65**, 010702(R) (2001)
39. A. Dalgarno: Adv. At. Mol. Opt. Phys. **32**, 57 (1994)
40. R. Bruckmeier, C. Wunderlich, H. Figger: Phys. Rev. Lett. **72**, 2550 (1994); D. Azinovic, R. Bruckmeier, C. Wunderlich, H. Figger, G. Theodorakopoulos, I.D. Petsalakis: Phys. Rev. A **58**, 1115 (1998)
41. U. Müller, T. Eckert, M. Braun, H. Helm: Phys. Rev. Lett. **83**, 2718 (1999)
42. M. Larsson: Philos. Trans. Roy. Soc. A **358**, 2433 (2000), and references therein
43. U. Müller, P.C. Cosby: Phys. Rev. A **59**, 3632 (1999)
44. For a preliminary account see L. Lammich, H. Kreckel, S. Krohn, M. Lange, D. Schwalm, D. Strasser, A. Wolf, D. Zajfman: in *Proceedings of the 2nd Conference on Elementary Processes in Atomic Systems*, Gdansk, September 2–6, 2002, to appear as special issue of Radiation Physics and Chemistry (RPCh)
45. S. Datz, R. Thomas, S. Rosen, M. Larsson, A.M. Derkatch, F. Hellberg, W.J. van der Zande: Phys. Rev. Lett. **85**, 5555 (2000)
46. R. Thomas, S. Rosen, F. Hellberg, A. Derkatch, M. Larsson, S. Datz, R. Dixon, W.J. van der Zande: Phys. Rev. A **66**, 032715 (2002)

Index

Springer Series on
ATOMIC, OPTICAL, AND PLASMA PHYSICS

Printing: Saladruck Berlin
Binding: Stürtz AG, Würzburg